Recent Advances in Anionic Polymerization

Recent Advances in Anionic Polymerization

Proceedings of the International Symposium on Recent Advances in Anionic Polymerization, held April 13–18, 1986 at the American Chemical Society Meeting in New York, New York, U.S.A.

Editors:

Thieo E. Hogen-Esch, Ph.D.
Department of Chemistry
University of Florida
Gainesville, Florida, U.S.A.

Johannes Smid, Ph.D.
Chemistry Department
State University of New York
College of Environmental Science and Forestry
Syracuse, New York, U.S.A.

Elsevier
New York • Amsterdam • London

Elsevier Science Publishing Co., Inc.
52 Vanderbilt Avenue, New York, New York 10017

Distributors outside the United States and Canada:

Elsevier Applied Science Publishers Ltd.
Crown House, Linton Road, Barking, Essex IG11 8JU, England

Softcover reprint of the hardcover 1st edition 1987

Library of Congress Cataloging-in-Publication Data

Recent advances in anionic polymerization, carbanions, mechanism and synthesis.

Sponsored by the American Chemical Society Polymer Division.
Includes indexes.
1. Addition polymerization – Congresses. 2. Carbanions – Congresses. I. Hogen-Esch, Thieo E. II. Smid, Johannes. III. American Chemical Society. Division of Polymer Chemistry.

QD281.P6R398 1987 541.3'93 87-9119

ISBN-13: 978-94-010-7924-2 **e-ISBN-13: 978-94-009-3175-6**
DOI: 10.1007/978-94-009-3175-6

Current printing (last digit)
10 9 8 7 6 5 4 3 2 1

CONTENTS

PREFACE

Over the years the field of anionic polymerization has attracted numerous outstanding scientists, and today it still is being pursued by many researchers all over the world. The exciting discovery of termination-less polymerization processes and living polymers culminating in the development of narrow molecular weight polymers, star polymers, and tailor-made block and graft copolymers, contributed immensely to the rapid expansion of polymer science. Areas of active research in anionic polymerization presently include the structure of ion pairs and their role in regulating polymer structure, ring opening polymerization of heterocyclic monomers, synthesis of well-defined block and graft copolymers including the application of macromers in such systems, telechelic polymers with functional end groups, and other topics. New developments in the organic chemistry of carbanions such as dipolar carbanions impinge on the field of anionic polymerization. More sophisticated characterization techniques have been instrumental in obtaining better correlations between the structure of polymers and that of intermediates leading to their formation.

This book contains the proceedings of the international symposium on "Recent Advances in Anionic Polymerization and Related Processes" which was held at the 1986 spring meeting of the American Chemical Society. It was the first Polymer Division-sponsored meeting exclusively devoted to anionic polymerization since the Houston ACS meeting in the spring of 1980. The proceedings of that meeting were published in the book "Anionic Polymerization", ACS Symposium Series No. 166, edited by Dr. J.E. McGrath. The chief emphasis of the recent New York symposium was directed towards synthetic and mechanistic aspects of anionic polymerization, and included a session on carbanion chemistry and one on group transfer polymerization. The latter was a joint session with the symposium on Unconventional Mechanisms and Methods of Polymerization. However, these lectures are not included in this book. We trust that the critical reviews and the many topics of current interest presented in the ensuing pages will be helpful to interested students and scientists.

Two researchers who made important contributions to the field of anionic polymerization and who had been invited to participate in this symposium are not listed among the authors. Dr. Ralph Milkovich, a former student of two of the most prolific researchers in anionic polymerization, professors Michael Szwarc and Maurice Morton, passed away as the symposium was being organized. Milkovich discovered and developed the thermoplastic styrene-diene-styrene and other triblock copolymers which started a new era in the synthesis of tailor-made block copolymers. He also did pioneering work in developing macromonomers useful in the synthesis of graft copolymers of controlled structure. Both topics were covered by Milkovich in the book edited by McGrath. For his outstanding and original research Ralph Milkovich received the American Chemical Society Award for Creative Invention at the Miami ACS meeting in the spring of 1985. Dr. David H. Richards from England, known to many of us as Rick, died in the spring of

1986. He was well recognized for his very interesting research on specially designed block copolymers which were synthesized by changing the mode of propagation from an active anionic center to a cationic or radical center. Both these individuals will be greatly missed by their colleagues and friends.

We like to thank all those who contributed to this book and gratefully acknowledge the secretarial assistance of Ragan Feidt and Lorraine Williams. A grant from the Petroleum Research Fund administered by the American Chemical Society was extremely helpful in attracting to the symposium participants from outside the United States.

Chemistry Department
University of Florida
Gainesville, Florida

Thieo E. Hogen-Esch

Chemistry Department
Polymer Research Institute
State University of New York
Syracuse, New York

Johannes Smid

CARBANIONS

CATION EFFECTS IN ORGANOALKALI METAL CHEMISTRY

ERLING GROVENSTEIN, JR.
School of Chemistry
Georgia Institute of Technology
Atlanta, Georgia 30332

I. INTRODUCTION

Organoalkali metal compounds can be regarded as substances in which the organic anion is joined to the counter alkali metal cation primarily by electrostatic forces between cations and anions. This generalization [1] is believed to hold even for organolithium compounds with localized negative charge, although here some small carbon-lithium covalency may enhance bonding [2]. From the simplified structure R^-M^+, it is tempting to think that the nature of M^+ is unimportant, in other words, that one is concerned only with the chemistry of "carbanions" in discussing the reaction of R^-M^+.

As the Arrhenius theory of ionization emphasizes, the chemistry of salts in dilute aqueous solution can be discussed in terms of kinetically "free" cations and anions; however, organoalkali metal compounds have inappreciable lifetimes in protic solvents and ordinarily are studied in ethereal or hydrocarbon solvents. In such media organoalkali metal compounds exist as ion pairs and aggregates of ion pairs as Michael Szwarc [3] and others have emphasized (see **Scheme 1**). It should be understood that, in ethereal

Scheme 1

$$(R{:}^-M^+)_n \rightleftharpoons R{:}^-M^+ \rightleftharpoons R{:}^- \mid M^+$$

ionic aggregate — contact ion pair — solvent-separated ion pair

or

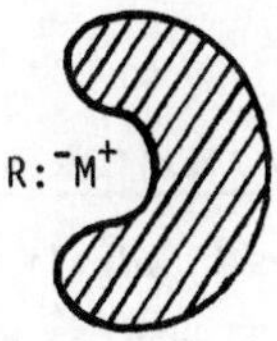

peripherally solvated
contact ion pair

solvents, all of these ionic species are solvated such that the coordination shell of the alkali metal cation is normally filled. Thus the contact ion pair of lithium would be expected to contain three ethereal oxygens in its periphery.

Published 1987 by Elsevier Science Publishing Co., Inc.
Recent Advances in Anionic Polymerization
T.E. Hogen-Esch and J. Smid, Editors

Some of the most informative studies concerning cation-anion interactions have come from careful measurements upon the physical properties of organoalkali metal compounds in ethereal solvents. For example, in the pioneering work of Hogen-Esch and Smid [4] upon the absorption spectra of 9-fluorenyl salts in tetrahydrofuran (THF) at 25°, the absorption maximum shifted from 374 to 349 nm in going from the free anion to the contact ion pair with lithium cation. This spectral shift most likely results from greater stabilization of the ground state than the excited state by the counter cation. A spectral shift from 374 to 349 nm is an energy change of 5.4 kcal/mole brought about by the lithium cation. This is a significant energy change which if operating in an equilibrium or rate process can effect a 10^4 fold change. Hence cation effects and the related factor of solvation or complexation of cations with ligands should be of great importance in the chemistry of organoalkali metal compounds.

The following is a brief summary of examples of cation effects in organoalkali metal chemistry. For convenience many of the examples will be taken from work in the author's laboratory. The first area for consideration will be chemical studies upon equilibria.

II. CHEMICAL STUDIES UPON EQUILIBRIA

(A) Reaction of Alkali Metals with Carbon

Lithium and sodium when heated with carbon form the acetylides Li_2C_2 and Na_2C_2. The heavier alkali metals including cesium react with carbon but give non-stoichiometric interstitial compounds where the metal atoms enter between the planes of carbon atoms in the lamellar graphite structure [5]. Thermodynamic calculations suggest that Cs_2C_2 is unstable relative to its elements and attempts to make it have not met with success [6], contrary to early reports in the literature.

(B) Metalation of Hydrocarbons by Alkali Metal Alkoxides

The metalation of hydrocarbons by alkali metal alkoxides [7,8] in tetrahydrofuran (eq. 1) is an equilibrium process whose equilibrium constant

$$Ph_3CH + MOC(CH_3)_3 \rightleftharpoons Ph_3C{:}^-M^+ + HOC(CH_3)_3 \qquad (1)$$

increases with the alkali metal M along the series Li < Na < K < Cs. While exact values of the equilibrium constant are not known, the differences are quite large between lithium *t*-butoxide and cesium *t*-butoxide.

(C) Metalation of Hydrocarbons by Alkali Metal Amides

From studies upon equilibria in the reactions of diphenylmethane with alkali metal amides in liquid ammonia or diethyl ether (eq. 2), the alkali

$$Ph_2CH_2 + MNH_2 \rightleftharpoons Ph_2CH{:}^-M^+ + NH_3 \qquad (2)$$

amides increase in metalating power along the series: Li < Na < K. With respect to kinetic basicity, cesium cyclohexylamide is up to 35,000 times more reactive than lithium cyclohexylamide in cyclohexylamine as solvent [9].

(D) Metalation of Hydrocarbons by Alkali Metal Acetylides

The equilibrium of equation (3) is shifted toward the right [10] by

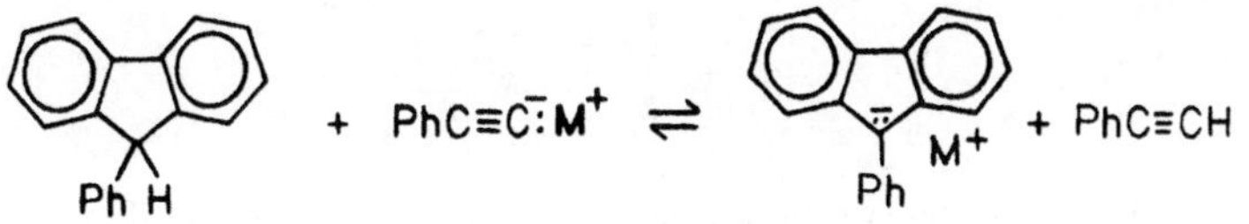

(3)

increasing the size of the alkali metal cation and by increasing the solvating ability of the solvent [Et_2O < $MeOCH_2CH_2OMe$ < $MeO(CH_2CH_2O)_2Me$ < $(Me_2N)_3PO$].

(E) Metal-Metal Exchange of Organolithium Compounds with Alkali Metal Alkoxide, Amides, and Halides

Organosodium and potassium compounds may be formed by reaction [11] of sodium and potassium alkoxides with organolithium compounds in hydrocarbon solvents (eq. 4). The conversion of organolithium compounds to organosodium,

$$RCH_2{:}^-Li^+ + R'O{:}^-M^+ \rightleftharpoons RCH_2{:}^-M^+ + R'O{:}^-Li^+ \qquad (4)$$

-potassium, and -cesium compounds by sodium, potassium, and cesium t-butoxide respectively in THF is evidently the basis for catalysis of the rearrangement of organolithium compounds by alkali metal alkoxides [12-14].

The cation effects on the equilibrium of eq. 4 and also that on the reactions of eq. 1, 2, and 3 may be summarized by a simple generalization.

For the chemical equilibrium,

$$R:^{-}M_1^{+} + R'Z:^{-}M_2^{+} \rightleftharpoons R:^{-}M_2^{+} + R'Z:^{-}M_1^{+} \qquad (5)$$

the larger cation prefers to be associated with the larger anion (frequently an anion with delocalized charge) and the smaller cation with the smaller anion (frequently an anion with localized charge). This generalization also rationalizes the reactions of alkali metals with carbon given above.

An explanation for this generalization can be found in electrostatic attraction between cations and anions. If for simplicity we assume that cations and anions behave electrostatically as point charges centered on the atoms which bear the charges in their usual structural formulas, then we can calculate the change in electrostatic energy (in atomic units) for equilibrium (5) from the expected interionic distances of the ion pairs as given in equation (6), where R_p is the distance (bohr units) between centers of

$$-\Delta E = \sum_p \frac{1}{R_p} - \sum_r \frac{1}{R_r} \qquad (6)$$

positive and negative charge in the product p and R_r is the similar distance in the reactant r, the summations being over all ion-pair products and reactants. We take 1.65 Å for the bond radius of carbon in an anion [15], 1.26 Å for the radius of oxygen in an alkoxide [16], and Pauling's crystal radii of the alkali metal cations [17]. If we assume that these radii are constant in all compounds, then $-\Delta E$ for equilibrium (4) may be calculated to be 8.4, 14.2, 15.8, and 17.8 kcal/mole for M equal to Na, K, Rb, and Cs respectively. These values appear adequate to account for the large values of the equilibrium constants of these synthetically useful reactions.

For the related reaction of eq. (7), if we take the anionic radius of

$$RCH_2:^{-}Li^{+} + R'_2N:^{-}M^{+} \rightleftharpoons RCH_2:^{-}\,M^{+} + R'_2N:^{-}Li^{+} \qquad (7)$$

nitrogen in an amide [18] to be 1.40 Å and the other values as previously specified, the calculated values of $-\Delta E$ are 4.8, 8.3, 9.2, and 10.4 kcal/mole for M equal sodium, potassium, rubidium, and cesium respectively. These numbers are about 58% of the corresponding values calculated for equilibrium (4) and help show how sensitive the calculation is to the separation of charge in the ion pairs. The data cited for equilibrium (2) is in qualitative agreement with this calculation but involves the diphenylmethyl anion which has a delocalized charge. Such charge delocalization effectively increases the radius of the carbanion and drives equilibrium (7) further to the right.

Another metal-metal exchange reaction is the conversion of organopotassium [19] or organocesium [20] compounds to organolithium compounds by reaction with lithium bromide (eq. 8) in ethereal solvents. If we take

$$-\overset{|}{\underset{|}{C}}{:}^{-}M^{+} + Li^{+}Br^{-} \longrightarrow -\overset{|}{\underset{|}{C}}{:}^{-}Li^{+} + M^{+}Br^{-} \qquad (8)$$

Pauling's crystal radius of bromide (1.95 Å) and the other values as previously specified, the calculated values of $-\Delta E$ from eq. 6 for equilibrium (8) are 4.1, 7.2, 8.0, and 9.2 kcal/mole for M equal Na, K, Rb, and Cs respectively. Again the calculated values appear large enough to agree with the synthetic utility of eq. (8).

It must be emphasized that, in addition to the approximations already cited, these calculations assume that the ion pairs are in the vapor phase at infinite separation from one another. In practice the ion pairs are located in a solvent whose dielectric constant is larger than that of free space; this factor will somewhat decrease the exothermicities which have been calculated from equation (6). Also the solvent, in principle, may preferentially solvate either the products or the reactants. A more important problem is that of aggregation. Since aggregation is but further Coulombic interaction between ions, the ions which form tighter ion pairs would be expected to form tighter and larger aggregates. Aggregation, therefore, would be expected to increase further the exothermicities which have been calculated by equation (6). In extreme cases the aggregates may become so large that insoluble products precipitate from solution and drive the equilibria further to completion.

(F) Ring-Chain Tautomerism of Unsaturated Organoalkali Metal Compounds

4,4-Diphenyl-3-butenylalkali compounds undergo reversible cyclization [19] as shown in eq. (9). For M = Li in diethyl ether as solvent only the

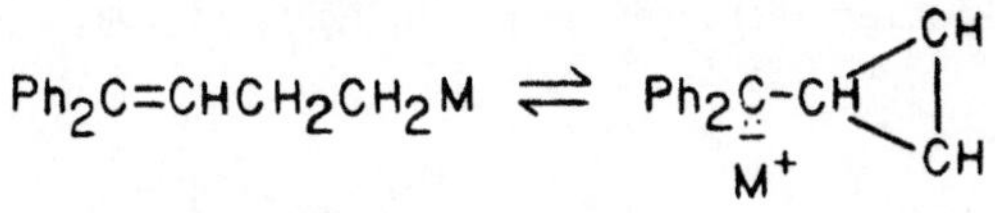

(9)

open form of the organometallic is observable at equilibrium (near room temperature) while with M = Na or K only the cyclic (cyclopropyl) compound is detectable. In THF only the cyclic compound is detectable with any of these cations. Equilibria of this type are not only cation and solvent dependent but also temperature and structure dependent. For example reaction (10) is found [21] to give the cyclic compound with lithium in diethyl ether at

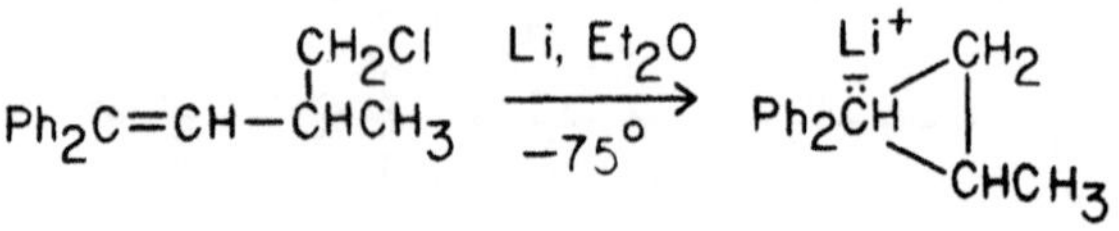

(10)

-75°C. Also in the system [20,22] of eq. (11) in THF at -75°C, with M = Li

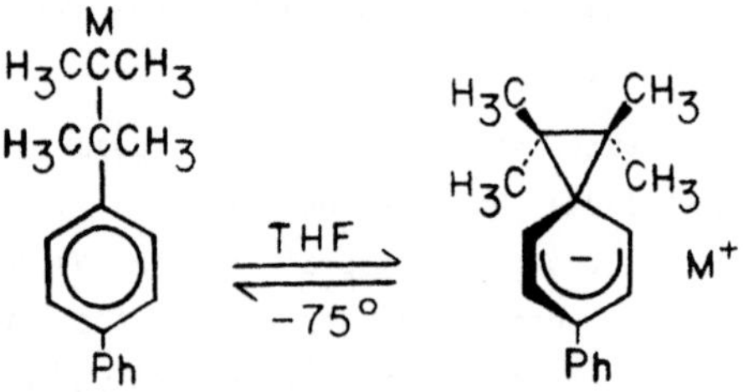

(11)

only the open organoalkali compound is observed while with M = Cs about a 30 to 70 ratio of open to cyclopropyl compound is found. These equilibria exemplify well the generalization given above: the larger cation prefers to be associated with the larger anion, i.e. the anion with delocalized charge, and the smaller cation with the smaller anion with localized charge. Also alkali metal cations can be made effectively larger by solvation which is enhanced by lowering the temperature. These equilibria are also affected by substitution, methyl groups on the cyclopropyl ring (especially gem-dimethyl groups) help stabilize the cyclic structure and cycloaddition to a double bond is more favorable than to an aromatic ring.

(G) cis, trans-Equilibria of Allylic Anions

The equilibrium Z/E isomeric composition of various 2-alkenylmetallic compounds is dependent upon the metal. Thus for the 2-butenyl system [23] of eq. (12) the % (Z)-isomer increases as M = H, MgBr, Li, Na, K, and Cs, the %

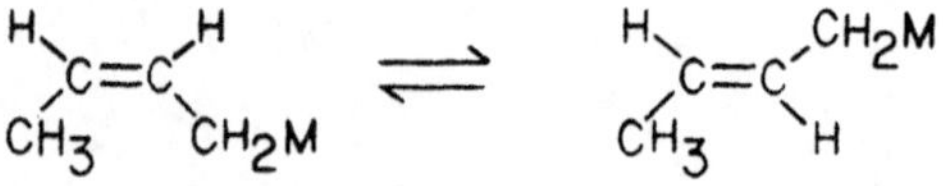

(12)

(Z)-isomer being 23, 54, 67, 93, 96, and 99.9 respectively in hexane (except for lithium in diethyl ether). The larger cation favors the (Z)-isomer. In THF Li gives 85% and K 99.2% of (Z)-isomer; again increased solvation makes the cations effectively larger. The larger cations give decreased electrostatic stabilization of the anion which accordingly must rely more on its internal structure for stabilization. The (_Z_)-isomer is likely stabilized by charge-induced dipole interaction of the anionic center with the _cis_-methyl group [24]. It should be noted that according to _ab initio_ calculations [25] monomeric allyllithium in the vapor phase has structure (**1**) whereas crystalline allyllithium as the tmeda complex [26] has

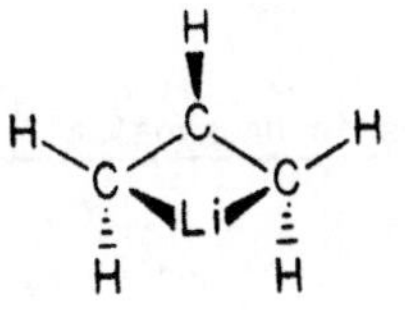

(**1**)

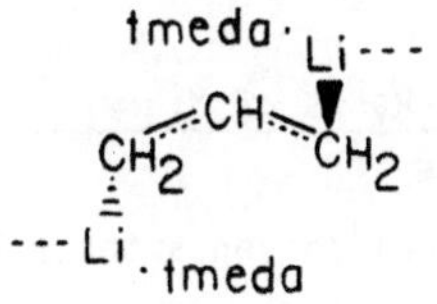

(**2**)

the polymeric structure (**2**). If the butenyllithium in the equilibration studies under discussion here has a structure similar to (**1**), then for the (_Z_)-isomer the methyl group would NOT be in the plane of the allylic carbon atoms and would be on the side of the allyl group away from the negative charge which is located largely near lithium. In this conformation charge-induced dipole interaction with the methyl group would be minimal.

III. CHEMICAL STUDIES UPON RELATIVE REACTION RATES

(A) Isotopic Exchange between Deuterium and Cyclohexylamine

Deuterium exchange in cyclohexylamine, eq. (13), is catalyzed by

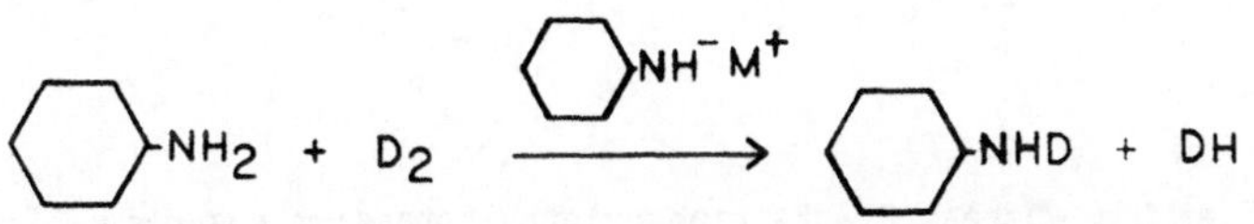

(13)

cyclohexylamide alkali metal salts (MCHA) with the observed order being LiCHA << NaCHA < KCHA $\leq$ CsCHA. The cesium salt is at least 10^3 times more reactive than the lithium salt [27]. The exchange likely proceeds via a transition state such as (**3**) in which the metal to nitrogen bond is stretched and the metal to deuterium bond is also extended. Hence more electrostatic energy is required to stretch the tight bonds of the smaller alkali metals to attain the transition state structure than the loose bonds of the larger alkali metals and little of this energy difference is offset by forming new metal to deuterium bonds in the transition state.

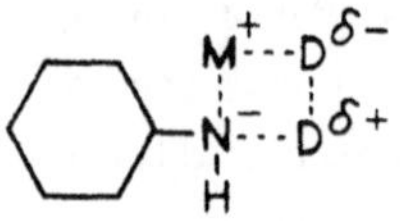

(3)

(B) <u>Relative Rates of Migration of Aryl Groups in Organoalkali Metal Compounds</u>

In ethereal solvents the relative rates of [1,2] and [1,4] migration of aryl groups, eq. (14), versus the rate of protonation of the organoalkali

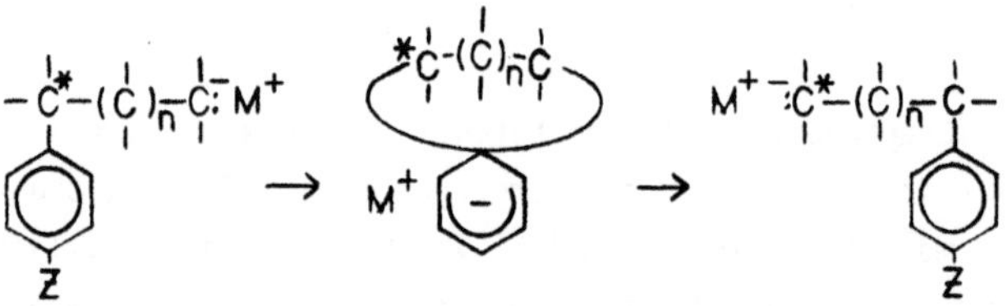

(14)

compound by solvent increases [12,28] along the series: Li << Na << K or Cs. This order seems likely to be the same as the relative rate of rearrangement of the organoalkali compounds themselves. Thus 2,2,2-triphenylethyllithium rearranges [29], eq. (15), with a half-life of about 40

$$Ph_3CCH_2Li \xrightarrow[0^{\circ}]{THF} Ph_2\ddot{C}^{-}CH_2Ph \quad Li^+ \qquad (15)$$

min. in THF at 0°; whereas 2,2,2-triphenylethylpotassium rearranges [30] as rapidly as it is formed, eq. (16), even at -75°. The qualitative explanation

$$Ph_3CCH_2Li \xrightarrow[THF,-75^{\circ}]{KO\text{-}\underline{t}\text{-}Bu} Ph_3CCH_2K \xrightarrow[-75^{\circ}]{<10\ min} Ph_2\ddot{C}^{-}CH_2Ph \quad K^+ \qquad (16)$$

of these results is that given in the prior example (A). Of course the chemistry of organoalkali compounds is complicated by aggregation into dimers, trimers, tetramers, etc.; however the most tightly bound aggregates are expected to be formed from the tightest ion pairs. The aggregates are expected to be less reactive than the monomeric ion pairs.

(C) Variation of Migratory Aptitudes of Groups in Organoalkali Rearrangements with Alkali Metal Cation and Solvent

1. Competitive Rearrangement of Phenyl versus Benzyl

Benzyl migration in 2,2,3-triphenylpropyllithium has been shown [31] to proceed by cleavage of benzyllithium followed by its readdition to 1,1-diphenylethene, whereas phenyl migration proceeds via an intramolecular process as shown in **Scheme 2.** The organolithium compound undergoes exclusive

Scheme 2

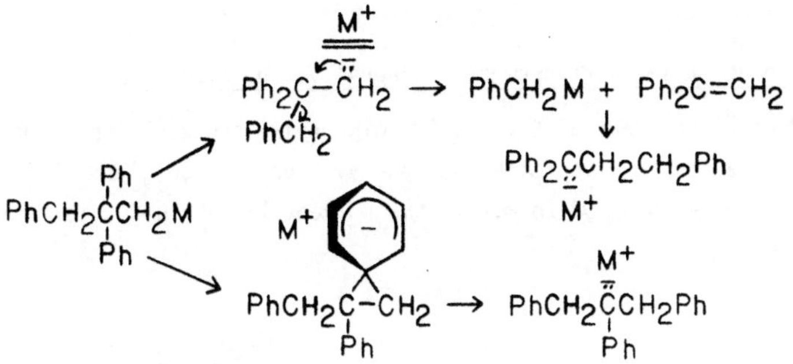

benzyl migration in THF at 0°, as does the sodium compound in THF at -75° or the potassium and cesium compounds which are complexed with 18-crown-6. In contrast the lithium compound rearranges in diethyl ether at 35° with exclusive phenyl migration and the cesium compound in THF at -75° with predominant (72%) phenyl migration. These results [13] can be understood on the basis that benzyl cleavage occurs in a loose or solvent-separated ion pair whereas phenyl migration proceeds in a tight ion pair. Thus conditions which favor loose ion pairs (small alkali metal cations, good solvents or ligands, low temperatures) favor benzyl migration whereas conditions which favor tight ion pairs (large alkali metal cations, poor solvents, high temperatures) favor phenyl migration.

2. Competitive Rearrangement of Phenyl versus Alkynyl

5,5-Dimethyl-2,2-diphenyl-3-hexynyllithium undergoes exclusive migration [32] of the alkynyl group by addition to the triple bond as shown in **Scheme 3.** In contrast the corresponding cesium compound undergoes phenyl migration [32] by addition to the phenyl group with no detectable (< 2%) migration of the alkynyl group (**Scheme 3**).

Scheme 3

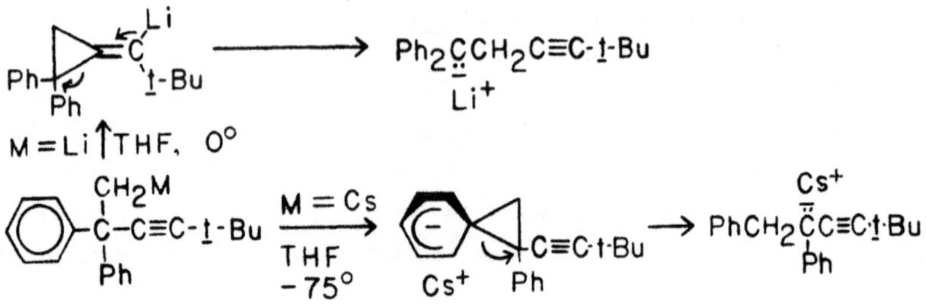

3. Competitive Rearrangement of Phenyl versus Alkenyl

In diethyl ether at 35°C trans-2,2-diphenyl-3-pentenyllithium undergoes propenyl migration to give (**5**) versus phenyl migration to give (**7**) (see **Scheme 4**) in a ratio of 92 to 8. In the solvent THF at 10°C this ratio

Scheme 4

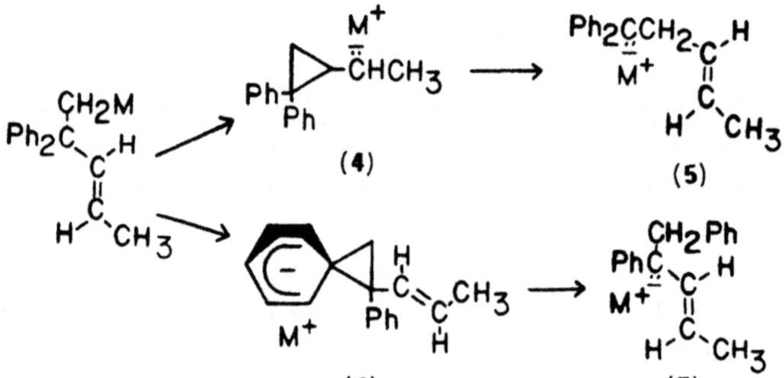

becomes 54 to 46. The corresponding potassium compound rearranges in THF at -75°C to give the product of phenyl migration (**7**) with no detectable propenyl migration [33].

According to **Scheme 4** propenyl migration occurs via an intermediate (**4**) with essentially localized negative charge while phenyl migration occurs via intemediate (**6**) with delocalized charge. Similar intermediates are suggested in **Scheme 3** for alkynyl versus phenyl migration. These results suggest a simple generalization. Rearrangements of organoalkali metal compounds tend to occur in the sense that compounds with large alkali metal cations preferentially rearrange via transition states (or reactive intermediates) with delocalized negative charge whereas compounds with small alkali metal cations rearrange via transition states (or intermediates) with localized negative charge. This generalization is similar to that suggested earlier for cation effects upon equilibria and has a similar rationale. With small cations and carbanions of localized negative charge, the transition state or intermediate is stabilized by strong electrostatic interaction between cation and anion. With large cations, cation anion interaction is greatly reduced and hence the anion must rely on its own structure for stabilization; such stabilization is generally greater in a delocalized anion. However, the effective size of an alkali metal cation is dependent upon its state of solvation. Thus the effect upon trans-2,2-diphenyl-3-pentenyllithium of rearrangement in diethyl ether at 35° versus rearrangement in THF at 10° is that under the later set of conditions increased solvation is expected and hence more phenyl migration should and does occur. Increased solvation implies a "looser" ion pair, not necessarily a solvent-separated ion pair.

4. Competitive Rearrangement of Phenyl versus Allyl

In the solvent THF at -70 to -75°C 2,2-diphenyl-4-pentenylpotassium and cesium compounds undergo migration either of the allyl group or the phenyl group [14]. With potassium complexed with 18-crown-6 a 92 to 8 ratio of allyl to phenyl migration is observed. Without 18-crown-6 this ratio is 32 to 68 for potassium or 16 to 84 for cesium. These arrangements are believed to occur by way of the intermediates or transition states shown in **Scheme 5.**

Scheme 5

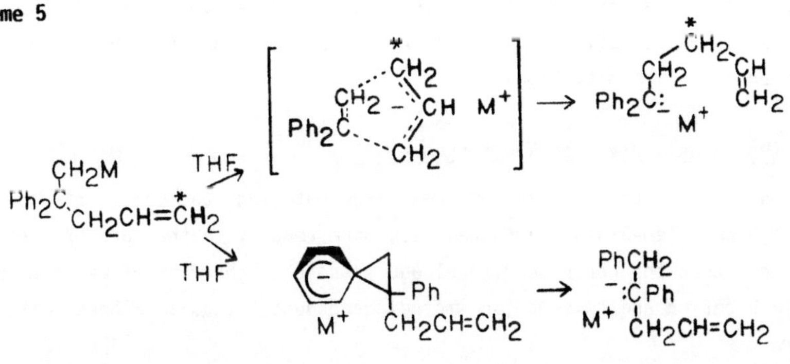

It appears that allyl migration occurs in a highly solvated transition state, likely in a solvent-separated ion pair, while phenyl migration occurs in a more weakly solvated or tight ion pair. With lithium as counter cation, rearrangement does not occur at an appreciable rate at -75°C but upon warming to near 0°C gives 96% allyl migration and only 4% phenyl migration. At this temperature the allyl group breaks free from the diphenylethylene moiety, evidently as an allyl anion, such that a carbon-14 label on the terminus of the allyl group becomes 67% scrambled and only 33% inverted (unlike the low temperature [2,3] sigmatropic process of **Scheme 5** which gives only inversion of the allyl group as shown). Yet the non-radioactive lithium compound when allowed to rearrange in the presence of a six fold excess of allyl-^{14}C-lithium showed only 14% incorporation of radioactive allyl group in the product of allyl migration. Hence the cleavage-readdition of allyl anion evidently occurs very rapidly largely within a solvent cage. These results show how cation and solvent (or ligand) effects can be used to control the nature of chemical reactions.

Since both allyl and phenyl migration in **Scheme 5** occur via transition states or intermediates with delocalized negative charge, prediction of the ratio of allyl to phenyl migration is somewhat difficult. For the low-temperature [2,3] sigmatropic pathway, the negative charge evidently is involved in binding the allyl group to the diphenylethylene group at both termini in an aromatic (6 π-electron) transition state in which the negative charge is concentrated in the region between the two groups and hence is shielded (by carbon and hydrogen) from close proximity to the cation. Hence strong solvation is required to attain this transition state which necessarily has the properties of a loose or solvent-separated ion pair. For the 0° elimination-readdition mechanism, the state of solvation is less clear. Perhaps now lithium ion binds the allyl anion at both termini as shown in structure (**1**) in a tight ion pair. Since the allyl anion is smaller than the cyclohexadienyl anion and has more concentrated negative charge, lithium prefers binding to allyl rather than cyclohexadienyl and thus brings about preferential allyl migration.

(D) The Anionic Oxy-Cope Rearrangement

In studies upon the oxy-Cope rearrangement Evans and Golob [34] found that 1,5-hexadien-3-ols rearranged much more readily in the form of their alkoxides than the corresponding alcohols and that the rate of rearrangement was both cation and solvent (or ligand) dependent. Thus in **Scheme 6** the rate

Scheme 6

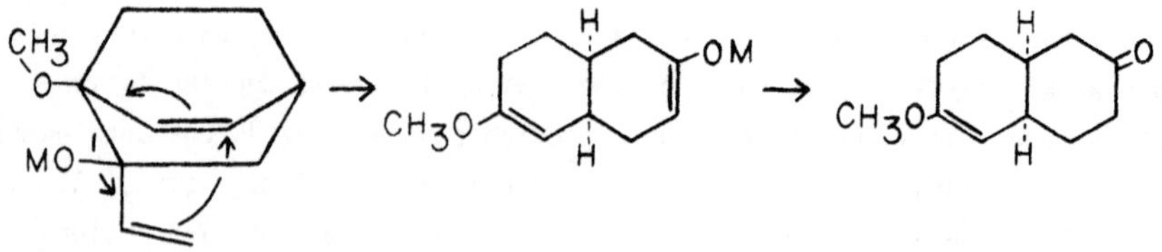

of rearrangement increased along the series $M^+ = H^+ << Li^+ << Na^+ < K^+ << K^+$ -

18-crown-6 with the acceleration for potassium complexed with 18-crown-6 10^{12} times that of the alcohol ($M^+ = H^+$). For a related system (**Scheme 6**, CH_3O replaced by H) the corresponding acceleration was 10^{17} fold. Similar anionic rate enhancements are observed in [3,3]-Claisen rearrangements of the enolates of α-allyloxy ketones [35], in [1,3] sigmatropic rearrangements [36], and in vinylcyclopropane rearrangements [37]. These rate enhancements are reminiscent of those discussed earlier for rearrangements of organoalkali metal compounds and likely have a similar explanation. Namely the transition states for the rearrangements have more charge delocalization than the reactants. Charge delocalization is better accommodated by large cations than by small since less energy is required to pull negative charge away from large than small cations (the proton can be viewed as the smallest cation).

A related explanation has been given by Steigerwald, Goddard, and Evans [38] who by an ab initio method have calculated C-H bond energies (see eq. (17) for CH_3OH, CH_3ONa, and CH_3O^- of 90.7, 70.6, 79.0 and 74.2 kcal/mole respectively. The authors note that the metal ketyl product of cleavage of

$$CH_3OM \longrightarrow \cdot CH_2OM + H\cdot \qquad (17)$$

CH_3OM is a hybrid of the two structures **8** and **9** and hence is charge-

$$\underset{\textbf{8}}{\dot{C}H_2\text{-}\ddot{\underset{..}{O}}\text{:}^-\ M^+} \longleftrightarrow \underset{\textbf{9}}{\overset{=}{C}H_2\text{-}\dot{\underset{..}{O}}\text{:}\ M^+}$$

delocalized whereas the reactant alkali metal alkoxide has its charge essentially localized on oxygen. Hence large alkali metal cations offer less electrostatic resistance to bond cleavage of the alkoxide than do small cations. A similar explanation is proposed to account for cation effects upon the anionic oxy-Cope rearrangement.

(E) NMR Rotational Barriers in Benzylic and Allylic Alkali Metal Compounds

Rotational barriers about the partial double bonds of benzylic and allylic anions are dependent on the counter ion. Some of the available data [39-41] for methylene rotation are summarized in **Table I.** In all cases cited the barrier to rotation increases along the series Li < Na < K < Cs. Somewhat similar barriers are found in 9-(1-naphthyl)methyleneoctatrienyl-alkali metal compounds [42].

Why does the rotational barrier increase or the rate of rotation decrease with increasing size of the alkali metal? In the other cation

Table I. NMR Rotational Barriers in THF

M		 $\Delta G^{\ddagger}$ kcal/mole	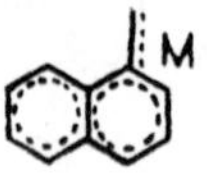
Li	17.0[a]	10.7[b]	12.9[c]
Na	> 17.8	—	14.8
K	> 20.1	16.7	(17.8)
Cs	—	18.0	—

[a]Ref. 39. [b]Ref. 40. [c]Ref. 41.

effects [see Sections (A), (B), and (D)], the opposite order is observed. Schleyer and coworkers [43] have made calculations of the reaction barrier for allyl compounds based on the mechanism shown in **Scheme 7.** These

Scheme 7

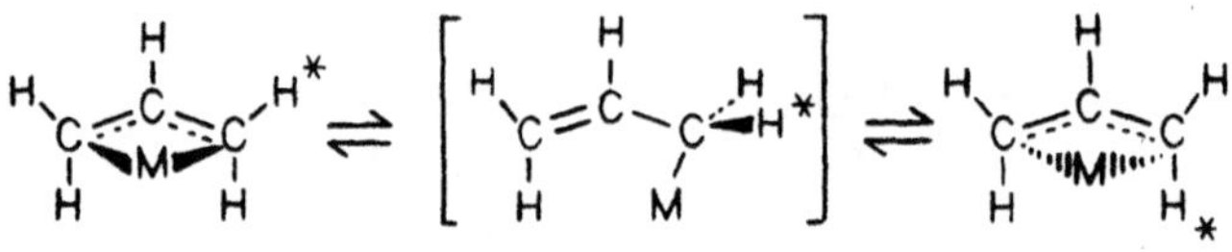

calculations pertain to monomeric species in the gas phase. The barrier for the free anion is estimated to be 22.2 kcal/mole, a value somewhat higher than that reported in Table I for allylcesium in THF. The value for allyllithium is calculated to be 17.7 kcal/mole, a barrier which is significantly higher than the observed barrier for allyllithium in THF. The value for allylsodium is 11.9 kcal/mole, not far from that which might be interpolated for the missing experimental value for allylsodium in Table I. Why is the experimental value for allyllithium much smaller than the calculated barrier? The effect under discussion is so large that the calculated barrier for lithium versus sodium is opposite the general trend of Table I. Schleyer and co-workers suggest [43b] that aggregation of allyllithium in THF may be responsible for the observed barrier being much less than that calculated. The situation may be similar to that in the inversion of alkyllithium for which aggregation has been calculated to lower the barrier [45]. The examples cited in this section should serve as a warning that the aggregation of organoalkali metal compounds is a factor which must be considered sometimes even in simple qualitative discussions.

The observed order of increasing rotational barriers in allylalkali metal compounds can be understood on the basis that the highest barrier is expected for the solvent-separated ion pair whose barrier should be nearly the same as that calculated for the free anion. Cation-anion interactions are expected to lower the barrier because of greater stabilization of the transition state (which has localized negative charge) than the reactant (which has delocalized negative charge) - this is opposite the general situation described for rearrangements in Sections (A), (B), and (D). Two (or more) cations should have a greater effect than one in this differential stabilization; hence aggregation serves further to lower the barrier. Where aggregation stabilizes the reactant more than the transition state as in the rearrangements of Sections (A), (B), and (D), it does not have to be invoked in simple qualitative arguments since aggregates are likely to be so unreactive as to play little role in reaction. The effect of aggregation should be greatest for lithium and least for cesium. Thus the observed order of rotational barriers is commonly Li < Na < K < Cs. This is not an invariant order. The barrier to rotation of the phenyl group in trans, trans-1,3-diphenylallyl anion in liquid ammonia containing 15% of methyl t-butyl ether is essentially independent of the counter-ion (Li^+, Na^+, K^+), doubtlessly because these compounds exist as solvent-separated ion pairs or possibly free anions [46]. Also methylene rotation in 2-methyl-1,3-diphenylallyl anion in THF occurs more readily in the potassium salt than in the lithium salt, likely because the potassium salt exists as a contact ion pair while the lithium salt occurs as a solvent-separated ion pair [47]. Thus both solvation and aggregation play important roles in determining rotational barriers in organoalkali metal compounds.

IV. OTHER EXAMPLES OF CATION EFFECTS

Many other instances of cation effects in organoalkali metal chemistry are known and could have been cited in this brief review. For example, the disproportionation of tetraphenylethylene (TPE) radical anion into dianion, eq. (18), has a formal equilibrium constant which increases along the series

$$2TPE^-,M^+ \rightleftharpoons TPE + TPE^{2-},2M^+ \qquad (18)$$

Li < Na < K < Cs in THF and varies with solvent in the order 1,2-

dimethoxyethane < THF < diethyl ether [48]. For the disproportionation of stilbene radical anion into dianion, the order of cation effects is reversed while the solvent effects are similar [49]. Related cation and solvent effects have been observed in disproportionation of many other radical anions [50].

In another example, alkali metal cations have been found to catalyze the reaction of ethylene oxide with fluoradenyl anion, the reaction rate increasing at least 10^5 fold from cesium to lithium. The catalysis has been attributed to the behavior of alkali metal cations as Lewis acids in coordination with the oxygen of ethylene oxide [51].

This review has not included cation effects which are combined with the differing reducing power of the alkali metals in various solvents. This area is rich in possibilities [52].

Finally interesting cation effects [53] are observed in anionic polymerization but these effects are too well known to polymer chemists for discussion here. It is hoped that the examples of cation effects cited have been representative of those possible in organoalkali metal chemistry and that these examples will encourage chemists, including polymer chemists, to use cation effects as a convenient tool to control chemical reactions and divert them in directions useful to mankind.

V. CONCLUSIONS

The factors involved in "cation effects" in organoalkali metal chemistry are diverse and interrelated but include (a) electrostatic attraction between cation and anion, (b) the size (distance of nearest approach) of cation and anion, (c) the charge distribution and geometry of the anion, (d) solvation of cation and anion and the related problem of whether contact or solvent-separated ion pairs are involved (note that contact ion pairs are frequently solvated), (f) specific interactions between cations and anions, including mutual polarization and covalent interactions. It frequently is not possible *a priori* to weigh all the factors and predict the outcome but general patterns are beginning to emerge. Cations frequently have *large* effects upon reaction rates and equilibrium constants of anions, sometimes so large as to alter the nature of a chemical reaction. Cations suitably coordinated to solvent or ligand can be used by chemists to exercise considerable control over chemical reactions of anions. More systematic studies on the effects of cations upon the behavior of carbanions would likely prove profitable.

ACKNOWLEDGMENTS

This work was supported in part by the National Science Foundation Grant No. CHE-8009810. Acknowledgment is also made to the Donors of the Petroleum Research Fund, administered by the American Chemical Society for support of the latter part of this research under PRF Grant No. 17174-AC1.

REFERENCES

1. (a) Streitwieser, A., Jr. Acc. Chem. Res. **1984**, 17, 353. (b) Streitwieser, A. Jr.; Williams, J. E., Jr.; Alexandratos, S.; McKelvey, J. M. J. Am. Chem. Soc. **1976**, 98, 4778. (c) Bushby, R. M.; Tytko, M. P. J. Organomet. Chem. **1984**, 270, 265.
2. (a) Bachrach, S. M.; Streitwieser, A., Jr. J. Am. Chem. Soc. **1984**, 106, 5818. (b) Clark, T.; Rohde, C.; Schleyer, P. v. R. Organometallics **1983**, 2, 1344. (c) Schleyer, P. v. R., Royal Society of Chemistry, International Symposium on Chemistry of Carbanions, Univ. of Durham, July 20, 1984, Plenary Lecture, "Searching for the True Nature of 'Carbanions'."
3. Thus see "Ions and Ion Pairs in Organic Reactions," M. Szwarc, Ed., Wiley-Interscience; New York, Vol. 1, 1972; Vol. 2, 1974.
4. Hogen-Esch, T. E.; Smid, J. J. Am. Chem. Soc. **1965**, 87, 699; **1966**, 88, 307, 318.
5. Cotton, F. A., and Wilkinson, G. "Advanced Inorganic Chemistry," 3rd ed., Wiley-Interscience, New York, 1972, p. 190.
6. Ader, M.; Hubbard, W. N. J. Chem. Thermodynamics **1973**, 5, 607.
7. Steiner, E. C.; Gilbert, J. M. J. Am. Chem. Soc. **1963**, 85, 3054.
8. Schlosser, M. "Struktur und Reaktivitat polarer Organometalle," Springer-Verlag, Berlin, 1973, p. 102.
9. (a) Streitwieser, A., Jr.; Hammons, J. H. Prog. in Phys. Org. Chem. **1965**, 3, 41. (b) Streitwieser, A., Jr.; Caldwell, R. A. J. Am. Chem. Soc. **1965**, 87, 5394.
10. Terekhova, M. I.; Kruglyak, L. I.; Petrov, E. S.; Shatenshtein, A. I. Chem. Abst. **1976**, 85, 93613.
11. (a) Lochmann, L.; Pospisil, J.; Lim, D. Tetrahedron Lett. **1966**, 257. (b) Lochmann, L.; Lim, D. J. Organometal. Chem. **1971**, 28, 153.
12. Grovenstein, E., Jr.; Rhee, J.-U. J. Am. Chem. Soc. **1975**, 97, 769.
13. Grovenstein, E., Jr.; Williamson, R. E. J. Am. Chem. Soc. **1975**, 97, 646.
14. Grovenstein, E., Jr.; Cottingham, A. B. J. Am. Chem. Soc. **1977**, 99, 1881.
15. This value is from the average value (2.25 Å) of the C-Li bond distance in crystalline $(CH_3Li)_4(tmeda)_2$ (Koster, H.; Thoennes, D.; Weiss, E. J. Organomet. Chem. **1978**, 160, 1) or the value (2.61 Å) of the C-Na bond distance in $(CH_3Na)_4$ cluster in crystalline methylsodium (Weiss, E.;Sauermann, G.; Thirase, G. Chem. Ber. **1983**, 116, 74) less Pauling's crystal radius of lithium or sodium respectively.
16. This value is from the Li-O bond distance (1.86 Å) in 1,3-dilithiobenzylketone $(tmeda)_2$ (Dietrich, H.; Mahdi, W.; Wilhelm, D.; Clark, T.; Schleyer, P. v. R. Angew. Chem. Int. Ed. Engl. **1984**, 23, 621) or the Li-O bond distance (1.87 Å) in benzophenonedilithium•tmeda•THF (Bogdanovic, B.; Kruger, C.; Wermeches, B. Angew. Chem. Int. Ed. Engl. **1980**, 19, 817) less Pauling's crystal radius of lithium.
17. Pauling, L. The Nature of the Chemical Bond, 3rd ed.; Cornell Univ.: Ithaca, N.Y., 1960, p 514.
18. This value is from the average value of 2.00 Å for Li-N in (2,2,6,6-tetramethylpiperidinato)lithium and $LiN(SiMe_3)_2$ (Lappert, M. F.; Slade, M. J.; Singh, A.; Atwood, J. L.; Rogers, R. D.; Shakier, R. J. Am. Chem.

Soc. **1983**, 105, 302) or in $(PhCH_2)_2NLi\ OEt_2$ and $(PhCH_2)_2NLi \cdot HMPA$ (Barr, D.; Clegg, W.; Mulvey, R. E.; Snaith, R. J. Chem. Soc. Chem. Commun. **1984**, 285).
19. Maercker, A.; Roberts, J. D. J. Am. Chem. Soc. **1966**, 88, 1742.
20. Grovenstein, E., Jr.; Lu, P.-C. J. Org. Chem. **1982**, 47, 2928.
21. Unpublished experiments with Dr. S. C. Goel.
22. Grovenstein, E., Jr.; Lu, P.-C. J. Am. Chem. Soc. **1982**, 104, 6681.
23. (a) Schlosser, M.; Hartmann, J. J. Am. Chem. Soc. **1976**, 98, 4674. (b) Stahle, M.; Hartmann, J.; Schlosser, M. Helv. Chim. Acta **1977**, 60, 1730.
24. Cf. the o-methylbenzylanion: Grovenstein, E., Jr., Bhatti, A. M.; Quest, D. E.; Sengupta, D.; VanDerveer, D. J. Am. Chem. Soc. **1983**, 105, 6290.
25. Clark, T.; Rohde, C.; Schleyer, P. v. R. Organometallics **1983**, 2, 1344.
26. Koster, H.; Weiss, E. Chem. Ber. **1982**, 115, 3422.
27. Symons, E. A.; Powell, M. F.; Schnittker, J. B.; Clermont, M. J. J. Am. Chem. Soc. **1979**, 101, 6704.
28. (a) Grovenstein, E., Jr. Adv. Organomet. Chem. **1977**, 16, 167. (b) Grovenstein, E., Jr. Angew Chem. Int. Ed. Eng. **1978**, 17, 313.
29. Grovenstein, E., Jr.; Williams, L. P., Jr. J. Am. Chem. Soc. **1961**, 83, 2537.
30. Unpublished experiments with Dr. R. E. Williamson.
31. Grovenstein, E., Jr.; Wentworth, G., J. Am. Chem. Soc. **1967**, 89, 1852; **ibid. 1967**, 89, 2348.
32. Grovenstein, E., Jr.; Chiu, K.-W.; Patil, B. B. J. Am. Chem. Soc. **1980**, 102, 5848.
33. Unpublished research with Dr. Subash C. Goel, Randall L. Hughes, John H. Northrop, and Douglas L. Streeter.
34. Evans, D. A.; Golob, A. M. J. Am. Chem. Soc. **1975**, 97, 4765. See also Evans, D. A.; Baillargeon, D. J.; Belson, J. V. J. Am. Chem. Soc. **1978**, 100, 2242.
35. Koreeda, M.; Luengo, J. I. J. Am. Chem. Soc. **1985**, 107, 5572.
36. (a) Thies, R. W.; Seitz, E. P. J. Org. Chem. **1978**, 43, 1050. (b) Wilson, S. R.; Mao, D. T. J. Chem. Soc., Chem. Commun. **1978**, 479. (c) Franzus, B.; Scheinbaum, M. L.; Waters, D. L.; Bowlin, H. B. J. Am. Chem. Soc. **1976**, 98, 1241.
37. Danheiser, R. L.; Martinez-Davila, C.; Morin, J. M., Jr. J. Org. Chem. **1980**, 45, 1341.
38. Steigerwald, M. L.; Goddard, W. A.; Evans, D. A. J. American Chem. Soc. **1979**, 101, 1994. See also Evans, D. A.; Baillargeon, D. J. Tetrahedron Lett. **1978**, 3319.
39. Sandel, V. R.; McKinley, S. V.; Freedman, H. M. J. Am. Chem. Soc. **1968**, 90, 495.
40. Thompson, T. B.; Ford, W. T. J. Am. Chem. Soc. **1979**, 101, 5459.
41. Kronzer, F. J.; Sandel, V. R. J. Am. Chem. Soc. **1972**, 94, 5750.
42. Staley, S. W.; Dustman, C. K. J. Am. Chem. Soc. **1981**, 103, 4297.
43. (a) Clark, T.; Jemmis, E. D.; Schleyer, P. v. R.; Binkley, J. S.; Pople, J. A. J. Organomet. Chem. **1978**, 150, 1. (b) Clark, T.; Rohde, C.; Schleyer, P. v. R. Organomettallics **1983**, 2, 1344.
44. Chandrasekhar, J.; Andrade, J. G.; Schleyer, P. v. R. J. Am. Chem. Soc. **1981**, 103, 5609.
45. Clark, T.; Schleyer, P. v. R.; Pople, J. A. J. Chem. Soc. Chem. Comm. **1978**, 137.
46. Bushby, R. J.; Ferber, G. J. J. Chem. Soc. Perkin II **1976**, 1688. See also (a) Boche, G., Schneider, D. R. Tetrahedron Lett. **1976**, 3657. (b) Boche, G.; Martens, D.; Wagner, H.-V. J. Am. Chem. Soc. **1976**, 98, 2668.
47. Burley, J. W.; Young, R. N. J. Chem. Soc. Perkin II **1972**, 1843.
48. (a) Garst, J. F.; Zabolotny, E. R. J. Am. Chem. Soc. **1965**, 87, 495. (b) Garst, J. F.; Zabolotny, E. R.; Cole, R. S. ibid. **1964**, 86, 2257. (c) Lundgren, B., Levin, G.; Claesson, S.; Szwarc, M. ibid. **1975**, 97, 262 and references therein cited.

49. (a) Zabolotny, E. R.; Garst, J. F. J. Am. Chem. Soc. **1984(b)** Jachimovicz, F.; Levin, G.; Szwarc, M. ibid., **1978**, 100, 5426. (c) Wang, H. C.; Levin, G.; Szwarc, M., ibid., **1978**, 100, 6137.
50. (a) Rainis, A.; Szwarc, M. J. Am. Chem. Soc. **1974**, 96, 3008. (b) Levin, G.; Holloway, B. E.; Szwarc, M. ibid., **1976**, 98, 5706. (c) Levin, G.; Szwarc, M. ibid., **1976**, 98, 4211. (d) Pola, J.; Levin, G.; Szwarc,M. J. Phy. Chem. **1976**, 80, 1690. (e) Hirota, N.; Weissman, S. I. J. Am. Chem. Soc. **1964**, 86, 2538. (f) Garst, J. F.; Walmsley, D.; Hewitt, C.; Richards, W. R.; Zabolotny, E. R. J. Am. Chem. Soc., **1964**, 86, 412.
51. Chang, C. J.; Kiesel, R. F.; Hogen-Esch, T. E. J. Am. Chem. Soc. **1973**, 95, 8446.
52. For example see: (a) Grovenstein, E., Jr.; Longfield, T. H.; Quest, D. E. J. Am. Chem. Soc. **1976**, 99, 2800. (b) Grovenstein, E., Jr.; Bhatti, A. M.; Quest, D. E.; Sengupta, D.; VanDerveer, D. ibid., **1983**, 105, 6290.
53. For example see: (a) Bhattacharyya, D. N.; Smid, J.; Szwarc, M. J. Phy. Chem. **1965**, 69, 624. (b) Dyball, C. J.; Worsfold, D. J.; Bywater, S. Macromolecules **1979**, 12, 819.

STRUCTURE AND DYNAMIC BEHAVIOR OF ORGANOLITHIUM COMPOUNDS

GIDEON FRAENKEL
Department of Chemistry, Ohio State University, 120 W. 18th Avenue,
Columbus, Ohio 43210 USA

INTRODUCTION

The study of carbanions necessitates investigating organometallic compounds. Except for the highly conjugated ones, most carbanions are not free but exist as the partially negative part of some organometallic compound. Ideally, in order to learn about carbanions one should study organocesium or organorubidium compounds since these materials are most likely to contain carbanions. Unfortunately, such compounds do not form stable solutions; they either react violently with potential solvents or are inert and insoluble, as with alkanes.

Organolithium compounds behave like carbanions chemically [1], so one would expect their physical properties to reflect useful information about carbanions. Some early indications as to the validity of this proposal come from studies of phenyllithium and related species [2]. Pyridine is isoelectronic with phenyl anion. Phenyllithium and pyridine exhibit very similar ^{13}C and ^{1}H NMR parameters, down to the smaller 1,4(p) proton-proton coupling constants, and quite different from parameters for covalently substituted benzenes such as silylbenzenes. The UV spectrum of phenyllithium tails into the blue in similar fashion to the n -> π^* absorption seen for pyridine. Evidently there is sufficient ionic character associated with the carbon lithium bond in phenyllithium to permit an observable σ_{C-M} -> π^* transition. The more ionic arylmetal compounds should exhibit pseudo n -> π^* absorption at increasingly higher wavelengths. Consistent with this is the lower wavelength tailing observed for phenylmagnesium bromide with its higher carbon metal covalency compared to phenyllithium [2].

Just exactly how ionic is the carbon lithium bond has been the subject of extensive controversy with the theoreticians coming out at this writing in favor of much ionic character [3]. Certainly electron density difference mapping calculations argue for ionic bonding. On the other hand, some difference plots (RLi cluster versus an assembly of free atoms) show lithium ions to occupy "holes in a sea of electrons," [4] which is not the traditional view of ionic character as in salts.

Organolithium compounds form the largest single group of synthetically useful organometallic compounds with ca. two new ones being announced every day [1]. Clearly to understand the rich and colorful chemistry of these versatile reagents it is necessary to know something about their structures, that is how these species hold together.

Lithium compounds exhibit an astonishing variety of structures from different kinds of ion-pairs to covalent clusters such as cubic tetramers [5], **1**, and octahedral hexamers [6], **2**.

The unusual bonding arrangements in organolithium compounds have attract-ed the attention of theoreticians who have, via calculations, commended the virtues of such unusual species as planar dilithiomethane,

Recent Advances in Anionic Polymerization
T.E. Hogen-Esch and J. Smid, Editors

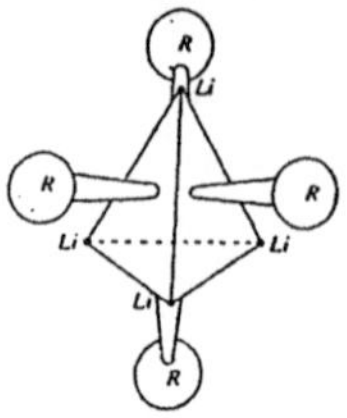

1

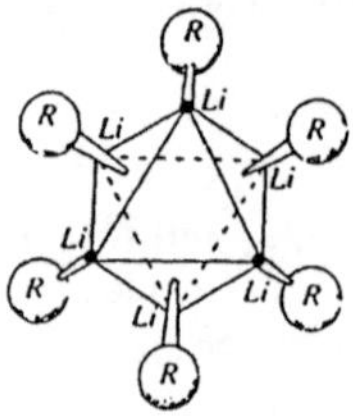

2

3, bridged dilithium acetylide, **4**, and twisted 1,1-dilithioethene, **5**, albeit monomeric, unsolvated and as yet not synthesized [7-9].

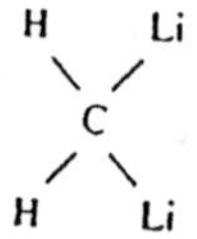

3

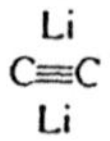

4

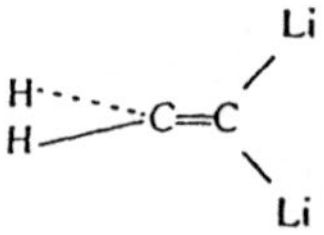

5

Interesting as species **3-5** might be, this article is addressed to what we know physically about the simple workhorse "RLi" species of everyday chemistry.

First we describe, with examples, the NMR methods used, then summarize results obtained for unsolvated alkyllithium compounds and finally present some insights into the structure and C,Li bond exchange behavior of solvated organolithium species, especially as revealed by studies of a new model solvated system, lithium t-butylacetylide.

THE METHOD

The method we have developed [10], which has been extensively used in other laboratories [11] is to prepare the organolithium compound using ^{6}Li (96% ^{6}Li, 4% ^{7}Li), readily available from the U.S. Department of Energy, and investigate ^{13}C and ^{6}Li NMR as a function of temperature, concentration of "RLi" species and of possible added potential ligands [10].

It has been known for some time that certain organolithium species exhibit scalar coupling between ^{13}C and directly bonded ^{7}Li [12]. More often than not ^{7}Li nuclear electric quadrupole induced relaxation, especially at low temperatures, is fast enough to partly average this coupling [10]. Consequently the ^{13}C resonance, at low temperatures, for carbon bonded to ^{7}Li in many "R^{7}Li" species consists of broad featureless absorption.

The less abundant stable isotope of lithium is ^{6}Li (4%); its quadrupole moment is ca. one hundredth that of ^{7}Li [13]. In the expressions for T_1 (quad.) of ^{6}Li or ^{7}Li, at the extreme narrowing limit, all terms are the same except for the respective nuclear electric quadrupole moments. Hence we argued that under conditions when ^{13}C, ^{7}Li coupling is just obscured by ^{7}Li quadrupole relaxation ^{13}C, ^{6}Li coupling in the same system should be readily observable; there are now several examples of "$R^{6}Li$" species where this is indeed the case [14], see below.

The ^{13}C, NMR (22.03 MHz) of propyllithium-1-^{7}Li, 0.6 M in cyclopentane solution at 173°K consists of two broad overlapping featureless resonances [10]. The ratios of their respective areas varies with increasing temperature and at the same time the resonance narrows due to increasing rates of interaggregate carbon-lithium bond exchange, see Figure 1. In contrast, ^{13}C, NMR of the isotopomer, propyllithium-1-^{13}C-1-

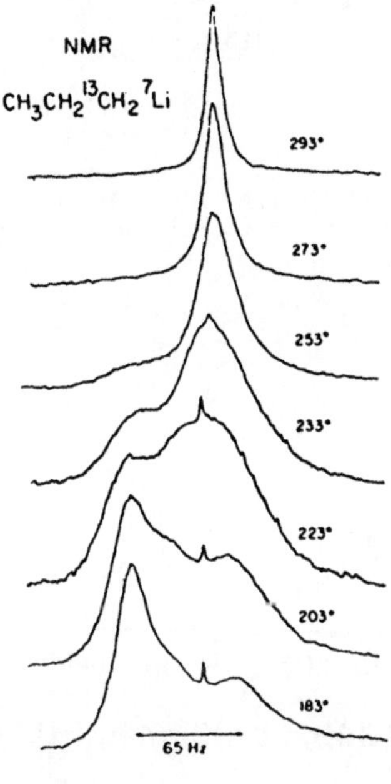

FIG. 1. ^{13}C NMR, 22.03 MHz, proton decoupled, of C_1 of 1-propyllithium-1-^{13}C-^{7}Li, 0.6M in cyclopentane, at different temperatures, K°, Reproduced by permission of the American Chemical Society.

^{6}Li, under the same chemical conditions as above, exhibits three resonances, one with well defined splitting due to ^{13}C, ^{6}Li coupling. NMR at a somewhat higher field, 75 MHz reveals the existence of **five** species, all of whose ^{13}C resonances are split or broadened by ^{13}C, ^{6}Li coupling [15]. Resolution enhancement (sine bell) of these data show all these ^{13}C resonances to be split, Figure 2. That this fine structure comes from ^{13}C, ^{6}Li coupling is amply confirmed by the results of ^{6}Li decoupling experiments.

NMR data for "$R^{6}Li$" species is best obtained at high field in order to resolve the existence of several species. At higher fields much beyond 7.049 Tesla relaxation of ^{6}Li via anisotropy of the chemical shift [16] becomes fast enough to partly average the ^{13}C, ^{6}Li coupling. This effect is readily observed among ^{13}C NMR data for 1-lithium$_6$-3,3-dimethyl-1-butyne in THF solution, 173°K, obtained at three magnetic fields, Figure

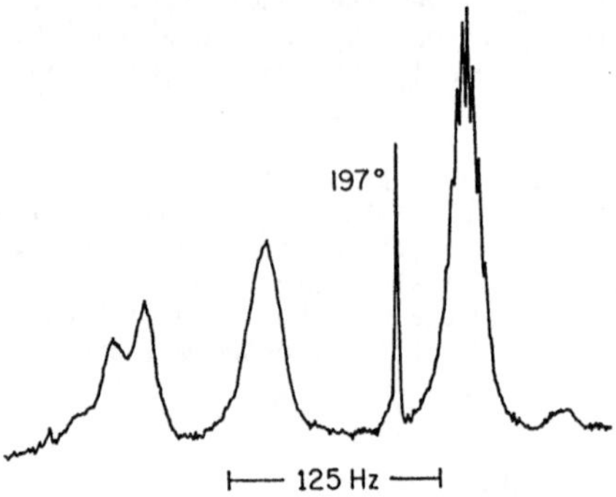

FIG. 2. ^{13}C NMR, 22.03 MHz of C_1 of 2-propyllithium-2-^6Li, 0.8M in cyclopentane, 197°K.

3. The two heptets around 114 ppm come from C_1 of differently solvated tetramers. By 10.57 Tesla the ^{13}C, ^6Li coupling is almost obscured and

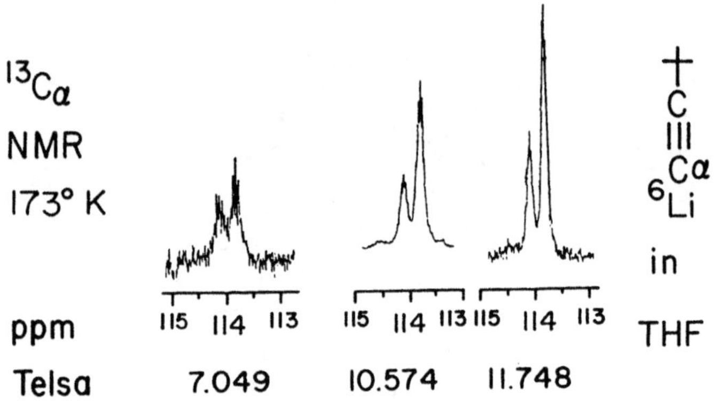

FIG. 3. ^{13}C NMR, different magnetic fields, C_1 of 1-lithio$_6$-3,3-dimethyl-1-butyne, 0.7M in THF at 183°K.

the ^{13}C resonance distinctly narrowed compared to the data obtained at lower fields. Since all other ^{13}C lines in these spectra are quite narrow even at 11.74 Tesla the averaging of the ^{13}C, ^6Li coupling must be due to ^6Li relaxation via anisotropy of the chemical shift. In the event that ^{13}C undergoes fast relaxation, the lines in the spectrum would broaden but the ^{13}C, ^6Li coupling would not average.

By use of the relaxation operator for chemical shift anisotropy [16,17] one can calculate the ^{13}C NMR line-shape for a ^{13}C bonded to one ^6Li. At slow ^6Li relaxation this should consist of a 1:1:1 triplet.

Figure 4 shows the coupled density matrix equations in matrix form required to draw such a ^{13}C NMR line-shape. The absorption comes from

$$\begin{bmatrix} i(\Delta\omega - J) - 4j(\pm) & 4j(\pm) & 0 \\ 4j(\pm) & i\Delta\omega - 8j(\pm) & 4j(\pm) \\ 0 & 4j(\pm) & i(\Delta\omega + J) - 4j(\pm) \end{bmatrix} \begin{bmatrix} \rho_{\alpha+,\beta+} \\ \rho_{\alpha o,\beta o} \\ \rho_{\alpha-,\beta-} \end{bmatrix} = \frac{iC}{6kT} \begin{bmatrix} \frac{1}{2} \\ \frac{1}{2} \\ \frac{1}{2} \end{bmatrix}$$

FIG. 4. Density Matrix equation for calculating ^{13}C absorption for one ^{13}C coupled to one ^{6}Li, the latter subject to relaxation via anisotropy of the chemical shift.

plotting the imaginary part of the sum of the transition probabilities $\rho_{i,j}$ as a function of frequency, $\Delta\omega$, (radians/sec.). In the matrix equation J stands for J(^{13}C, ^{6}Li), $\Delta\omega$ is the frequency point and $j(\pm)$ the spectral density, a first order rate constant for relaxation is given by

$$j(\pm) = \frac{1}{24}(\omega_{ol} - \omega_{ot})^2\tau_c \qquad (1)$$

(at the extreme narrowing approximation) where the term in brackets is a constant, as is also the imaginary term on the right hand side of the equation in Figure 4. Finally τ_c is the molecular correlation time. Reference to the central diagonal element shows that at slow but observable ^{6}Li relaxation rates the central resonance will be twice as wide as the two outer ones. Also one can see how the spectral densities in the off-diagonal elements of the coefficient matrix bring about averaging of the ^{13}C transition frequencies.

Under conditions when ^{6}Li relaxation in "R^{6}Li" species is slow , one can determine how many ^{6}Li's are coupled to each $^{13}C_\alpha$ from the multiplicity of the $^{13}C_\alpha$ resonance, hence learn about the structures of the aggregates. For example, in bridged dimers, often solvated, two ^{6}Li's coupled equivalently to the same $^{13}C_\alpha$, generate a ^{13}C NMR multiplet of relative intensities 1:2:3:2:1 whereas in the case of the well known cubic tetramer where each $^{13}C_\alpha$ has three ^{6}Li nearest neighbors these produce a $^{13}C_\alpha$ heptet, 1:3:6:7:6:3:1, see Figure 5.

Now, consider for a moment the utility of ^{6}Li NMR of $(RLi)_n$ samples enriched in ^{6}Li. Despite a quadrupole moment that is close to that of deuterium, the ^{6}Li NMR at low temperature of solutions of $(R^{6}Li)_n$ species,

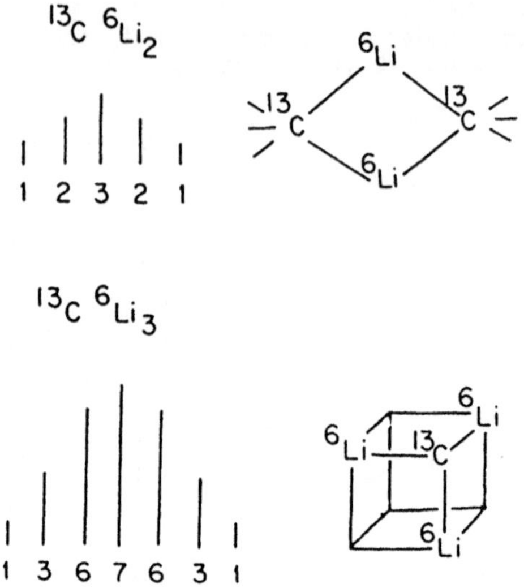

FIG. 5. ^{13}C NMR multiplet due to coupling to two or three ^{6}Li atoms.

(with ^{13}C in natural abundance and under conditions of proton decoupling) consists of several sharp single lines [13], 0.1 to 0.5 Hz wide, spanning a range of not more than 0.5 ppm. There is often a direct correspondence between the ^{6}Li and ^{13}C NMR for the same sample implying that each $(RLi)_n$ species gives rise to single ^{13}C and ^{6}Li, resonances, respectively [15]. Since 99% of ^{6}Li in these samples is bonded to ^{12}C the ^{6}Li spectra show no absorption for ^{6}Li bonded to ^{13}C. In addition, up to the time of this writing there has been no evidence for scalar coupling between ^{6}Li's or between ^{6}Li and ^{7}Li.

Above 183°K in the ^{13}C NMR of $(R^{6}Li)_n$ species there is observed signal averaging indicative of the operation of fast exchange processes. In carbon-13 NMR averaging of the ^{13}C, ^{6}Li coupling is seen, the result of fast intra-aggregate and especially inter-aggregate carbon lithium bond exchange [15]. Another dynamic phenomenon which can influence the proton NMR line-shapes is inversion at carbon bonded to lithium [18].

The effects of interaggregate C, Li bond exchange are also observed in ^{6}Li NMR spectra of $(R^{6}Li)_n$ with ^{13}C in natural abundance, that is above 173°K there is signal averaging among the resonances which come from different $(RLi)_n$ species. So using methods of NMR line-shape analysis, described by us previously [19] we can investigate the kinetics of these exchange and inversion processes and obtain information about their respective mechanisms.

These organolithium systems discussed here are always in a state of equilibriuim so the results of our line-shape studies throw light on mechanisms of dynamic processes at equilibrium, a hitherto inaccessible goal.

The following sections describe examples of $(RLi)_n$ systems which have been studied using the general procedures described above. Consider first the qualitative behavior of unsolvated alkyllithium compounds.

UNSOLVATED ALKYLLITHIUM COMPOUNDS

The $^{13}C_2$ NMR (natural abundance ^{13}C) of t-butyllithium-^6Li in cyclopentane at 173°K consists of a heptet with relative intensities 1:3:6:7:6:3:1, $J(^{13}C,^6Li)$=5.3, Hz, implying the existence of a tetramer in which $^{13}C_2$ is scalar coupled to the three nearest neighbor ^6Li's but not to the fourth [20], Figure 6. Above 240°K these C_2 resonances coalesce

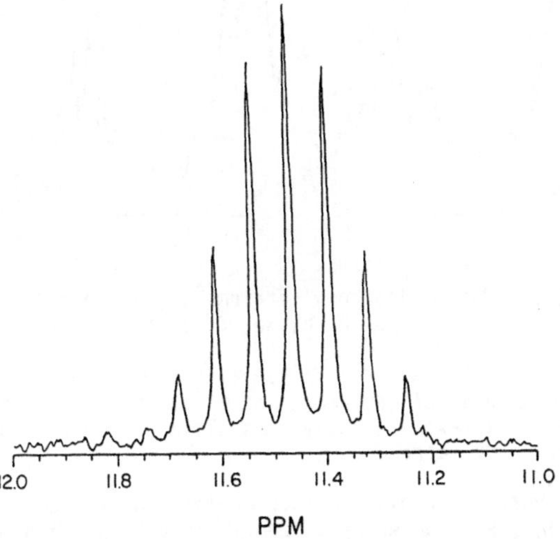

FIG. 6. ^{13}C NMR, 75.47 MHz of C_2 of tertiary butyllithium-^6Li, 0.7M in cyclopentane at 173°K, Kashi Ram, Ohio State University.

but by 295°K there appears a nine line spectrum, closely 1:4:10:16:19:16:10:4:1 indicating all four lithiums to be equivalently coupled to each $^{13}C_2$. Since the splitting in the nine line spectrum is just 3/4 that observed in the heptet at 173°K, the higher temperature data (nine lines) must represent a tetramer undergoing fast intra-aggregate carbon-lithium bond exchange. Line shape analysis of the $^{13}C_2$ resonance shows that $\Delta H^{\ddagger}$ = 25 kcal and $\Delta S^{\ddagger}$ = 44 eu. Above 310°K the $^{13}C,^6Li$ coupling begins to average as a result of interaggregate carbon-lithium bond exchange, fast relative to the NMR time scale.

Isopropyllithium in cyclopentane exhibits two main resonances for carbon bonded to lithium one at 6.4 ppm. clearly 1:3:6:7:6:3:1 $J(^{13}C, ^6Li)$ = 6.1 Hz diagnostic of a non-fluxional tetramer and the other at 10.8 ppm. with $J(^{13}C,^6Li)$ = 3.05 Hz closely similar to that assigned to $^{13}C_1$ of the unsolvated fluxional hexamer of propyllithium, Figure 7 [10,15]. With

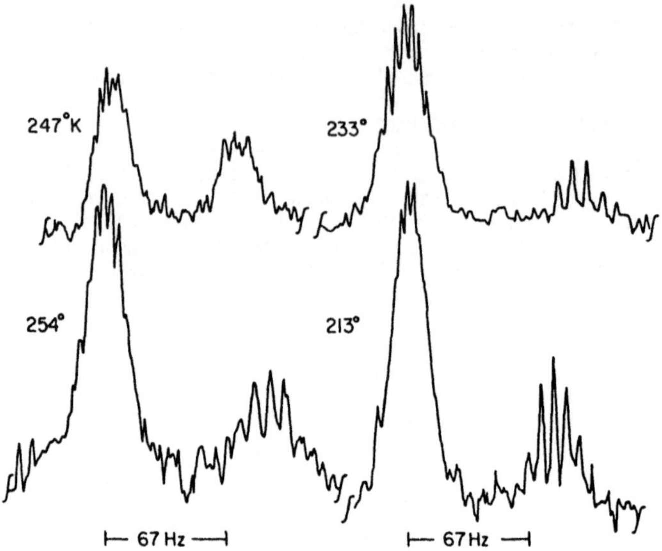

FIG. 7. ^{13}C NMR, 22.03 MHz of isopropyllithium-^6Li, 2M in cyclopentane, different temperatures, °K, data of B. M. Su, Ohio State University.

decreasing temperature, below 300°K, isopropyllithium hexamer increases in concentration at the expense of tetramer, the effect being reversible with temperature. Below 200°K hexamer slowly crystallizes from the solution increasing the fraction of tetramer.

An equilibrium between unsolvated isopropyllithium tetramer and hexamer was proposed by Russian workers from IR data [21]. A similar conclusion ensued from the results of vapor pressure mesurements of pentane solutions of isopropyllithium at 298°K [22]. These experiments yielded association numbers in the range 4.9 to 5.5. The data in Figure 7 do not yield consistent equilibrium constants. Further investigations at higher fields (^{13}C NMR at 75 MHz) revealed that the resonance previously assigned to "hexamer" is actually two different hexamers, Figure 8. The ^{13}C, NMR correlate directly with the ^6Li NMR i.e. each aggregate species gives rise to a single resonance in each kind of spectrum. Just exactly what form these hexamers assume is still a mystery whose resolution must await X-ray crystallographic investigation.

Above 245°K the ^{13}C NMR spectra of these unsolvated isopropyllithium samples exhibit line-broadening and averaging of the $^{13}C,^6Li$ coupling constants the result of interaggregate C, Li bond exchange at rates fast compared to the $^{13}C,^6Li$ coupling constants. This exchange rate is still too slow by 330°K to average the C_2 chemical shifts of the three species.

Results reported by us for unsolvated 2-butyllithium-2-^6Li [23] are similar to those for isopropyllithium - an equilibrium between hexamer and

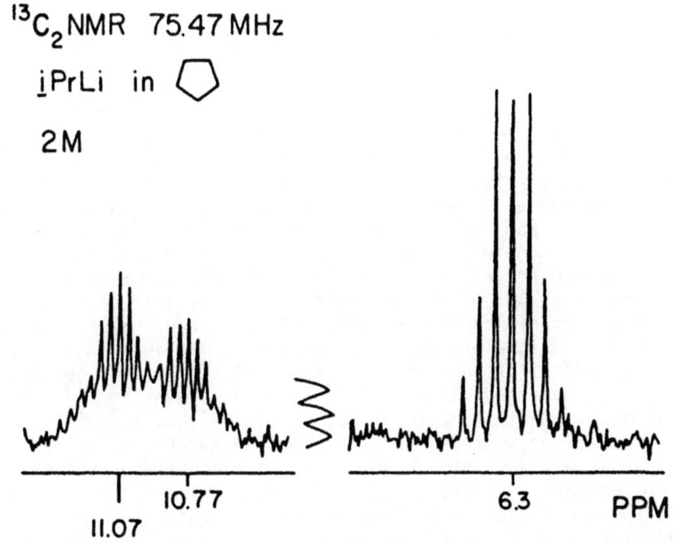

FIG. 8. ^{13}C NMR, 67.89 MHz, isopropyllithium-^6Li, 2M in cyclopentane, 240°K.

tetramer, with small amounts of dimer,

$$2R_6Li_6 \rightleftharpoons 3R_4Li_4 \qquad (3)$$

$$\Delta H = 7.1 \text{ kcal and } \Delta S = 23 \text{ eu}$$

The $^{13}C_3$ (CH_2) NMR of hexamer shows fine structure, indicative of magnetically non-equivalent diastereomeric aggregates, implying slow inversion at C_2, carbon bonded to lithium. There are four C_3 hexamer peaks, due to four fluxional diastereomers - three enantiomeric pairs $R_6 = S_6$, $R_5S = RS_5$, $R_4S_2 = R_2S_4$ and R_3S_3 and their relative intensities are similar to those expected for a random distribution of equal numbers of R and S alkyls among an assembly of sites on hexamers, 1:6:15:10, respectively [23]. This implies the hexamers have similar free energies.

Above 240°K averaging of the ^{13}C, ^6Li couplings as well as of the C_3 resonances indicates the onset of fast equilibrium interaggregate C,Li bond exchange.

Thus two typical secondary alkyllithiums exist as equilibrium mixtures of hexamers and tetramers. The hexamers undergo fast intra-agregate carbon-lithium bond exchange down to 173°K whereas in tetramers this process is slow, relative to the 6 sec^{-1} time scale associated with the ^{13}C, ^6Li coupling constant. Inversion at carbon bonded to lithium is always slow over the temperature range 183° - 320°K, relative to the NMR time scale [23].

NMR data for unsolvated primary alkyllithiums are both more interesting and more complicated than for the secondary and tertiary species. Propyllithium-1-^{13}C-1-^{6}Li (enriched in ^{13}C at C_1 to improve the ^{13}C NMR signal/noise ratio) in cyclopentane solution at 183°K gave rise to five ^{13}C, resonances all split by ^{13}C,^{6}Li coupling [15] Figure 2, above. These were shown to come from an equilibrium mixture of hexamers, octamer and three kinds of nonamers. The assignments were made on the basis that a) the C_1 resonance prevailing at room temperature represented hexamer as known from molecular weight measurements and b) the intensity ratios of this multiplet are only consistent with a C, Li bond-fluxional hexamer. Assuming all $^{13}C_1$ multiplets represent different fluxional aggregates, the different observed splittings are the result of the nearest neighbor ^{13}C,^{6}Li coupling averaged with small or zero couplings from $^{13}C_1$ to more distant lithiums. On this basis it was concluded that the mixture contained hexamer, octamer and three kinds of nonamers, of as yet unkown structures. These association numbers led to a consistent set of equilibrium constants as well as heats of interconversion of aggregates. Their distribution changes with temperature such that on warming to room temperature the equilibrium shifts entirely to hexamer, the smallest aggregate present.

The results, described above, and conclusions derived therefrom, are entirely consistent with ^{6}Li NMR data obtained for 1-propyllithium-1-^{6}Li (with ^{13}C natural abundance) in cyclopentane. Since 99% of the ^{6}Li is bonded to ^{12}C no ^{13}C, ^{6}Li coupling is observed in these spectra. The spectra consist at 183°K of five sharp lines 0.2 Hz wide over a range of half a ppm (22 Hz), which, appear by their relative intensities, appear to correspond to the ^{13}C resonances for carbon bonded to lithium [15].

Above 220° averaging takes place of the ^{13}C,^{6}Li couplings in the ^{13}C NMR of propyllithium -^{6}Li and of the ^{6}Li shifts in the ^{6}Li NMR, the result of fast carbon lithium bond exchange among aggregates. These line-shapes could be closely reproduced by assuming that carbon-lithium bond exchange takes place during collisions between pairs of aggregates and incorporating this mechanism into the equations required to draw the NMR line-shapes. Interestingly the ^{6}Li and ^{13}C line-shape analysis yielded similar rate constants confirming that the same mechanism is responsible for changes in both kinds of spectra. Activation parameters, for bi-aggregate carbon lithium bond exchange among 1-propyllithium aggregates, show a low $\Delta H^{\ddagger}$ (4.3 kcal) and a large negative $\Delta S^{\ddagger}$ (-36 eu). The latter is consistent with C, Li bond exchange taking place via a bimolecular superaggregate transition state perhaps preceded by an intermediate

$$2R_6Li_6 \rightleftharpoons R_{12}Li_{12}^{\ddagger} \rightleftharpoons 2R_6Li_6^{*} \qquad (4)$$

and less likely as a result of dissociation steps.

$$R_6Li_6 \rightleftharpoons R_2Li_2 + R_4Li_4 \qquad (5)$$

How exactly the biaggregate exchange takes place is not yet understood. Yet, the carbon-lithium bond scrambling proposed in a superaggregate has its parallel in the very fast C,Li bond exchange observed to take place inside all alkyllithium hexamers. That process may well be the result of propyls jumping between occupied and unoccupied binding sites, the latter consisting of a triangular plane bounded by three lithiums, Figure 9. So

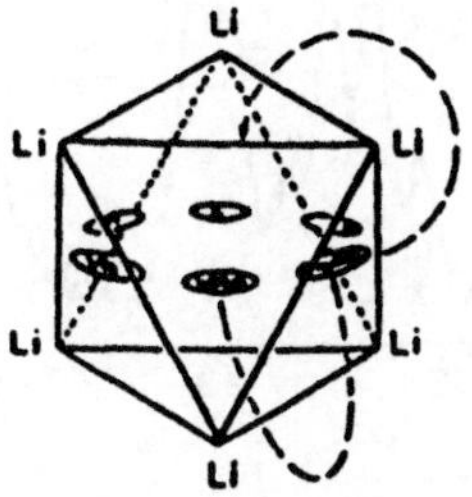

FIG. 9. Rearranging alkyllithium octahedral hexamer.

the propyls might chase each other over the surface of an octahedral matrix of lithiums. Something similar may happen in superaggregate intermediates, should they contain empty alkyl binding sites. Thus the intra- and inter-aggregate C,Li bond exchange processes may very well be part and parcel of the same thing.

In preliminary results ethyllithium exhibits NMR spectral behavior reminiscent of 1-propyllithium, as described above, though far more complicated. Figure 10 (top) shows the ^{13}C, NMR absorption of ethyllithium-^{13}C,-^{6}Li in cyclopentane solution at 200°K indicating the existence of at least six different as yet unidentified species. The different resonances show splittings due to ^{13}C, ^{6}Li coupling as amply confirmed by the results of a lithium-6 decoupling experiment, see Figure 10 (bottom). Again as observed before, the distribution of species changes with temperature and by 300°K only the highest field resonance, seen at low temperatures, is observed.

Summarizing the results for unsolvated alkyllithium compounds: Typically they exist as equilibrium mixtures containing a variety of $(RLi)_n$ aggregates. Their distribution varies with temperature, smaller aggregates prevailing at higher temperatures. Primary alkyllithiums form at least hexamers, octamers and three kinds of nonamers and as yet other detected but unidentified species; for secondary $(RLi)_n$, n equals 4 and 6 and in one case dimers have been observed; tertiary alkyllithiums form tetramers exclusively.

Hexamers and larger aggregates exhibit fast intra-aggregate carbon lithium bond exchange at least down to 173°K if not lower. In tetramers this intra-aggregate exchange is much slower compared to hexamers and the rate can be measured using NMR line-shape techniques.

All $(RLi)_n$ species exhibit NMR signal averaging by 300°K symptomatic of faster inter-aggregate C,Li bond exchange. This has been extensively

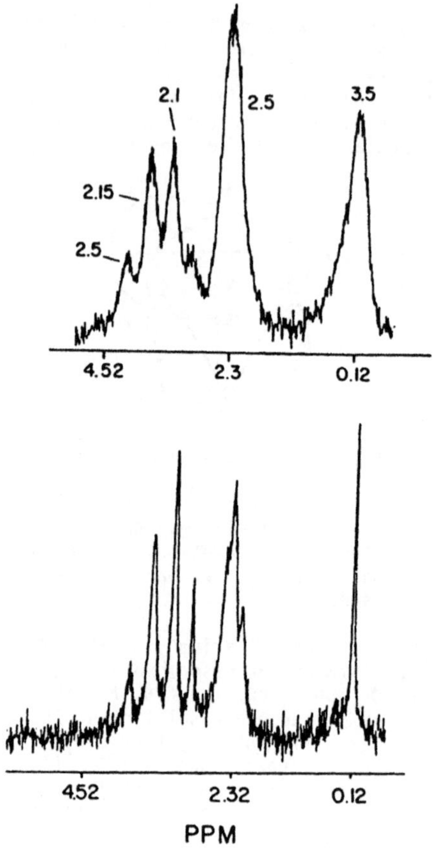

FIG. 10. ^{13}C NMR 75.47 MHz of C_1 of ethyllithium-1-^{13}C-1-^{6}Li, 0.9M in cyclopentane at 183°K top ^{13}C,^{6}Li coupled, splittings due to ^{13}C,^{6}Li coupling listed by multiplets, bottom ^{6}Li decoupled.

investigated in one case and the dynamic process appears to result from bimolecular exchanges among aggregates.

SOLVATED ORGANOLITHIUM COMPOUNDS

Many organolithium compounds, eg R = vinyl or allyl, are insoluble in alkanes and only dissolve in the presence of ethers or tertiary amines. As discussed below organolithium compounds complexed to ethers or tertiary amines form characteristic aggregates whose structures and modes of behavior often differ markedly from their corresponding unsolvated forms. This section discusses the behavior of a simple model system for investigation of solvation effects, 1-lithio-3,3-dimethyl-1-butyne.

It is common practice to teach undergraduates that alkynyl metal compounds are ionic [24] due to

$$-C{\equiv}\ddot{C}^{-}\ M^{+} \quad \mathbf{6}$$

the high degree of "s" character associated with the non-bonding electrons on carbon. Actually nothing is further from the truth. Our studies show a typical alkynyllithium compound is no more ionic than all other solvated "RLi" species [25]; it behaves just like them, hence, constitutes our stable model system, see below.

Reaction of t-butylacetylene with 1-butyllithium in pentane at room temperature precipitates the insoluble 1-lithio-3,3-dimethyl-1-butyne, **7** [25].

$$(CH_3)_3CC{\equiv}CLi \quad \mathbf{7}$$

Acetylide **7** is unreactive to dioxygen and readily hydrolyzed back to acetylene. A solution of this lithium acetylide in THF-D_8 at 173°K gave the ^{13}C NMR spectrum, triple bond region only, shown in Figure 11. Whereas the reactant alkyne shows C_1 and C_2 ^{13}C shifts at, 92 and 68 ppm, respectively, the THF-solvated acetylide has several C≡C resonances between 114 and 122 ppm. Here we see two overlapping heptets due to differently solvated tetramers centered at 114 ppm, $J(^{13}C,^{6}Li)$ = 5.7 and

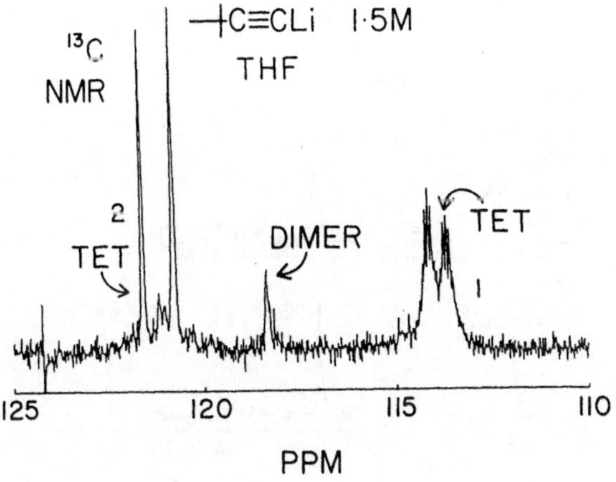

FIG. 11. ^{13}C NMR, 75.46 MHz, of C_1 and C_2 of 1-lithio-3,3-dimethyl-1-butyne-^{6}Li, 7, 0.9M in THF, 173°K.

5.8 Hz and the two corresponding resonances for C_2 of tetramers, in correct ratio, at ca. 121 ppm. In the center of the spectrum at 118 ppm are the C_1 and C_2 resonances from bridged dimer, the singlet for C_2 superposed on the quintet, $J(^{13}C, {}^{6}Li) = 8.3$ Hz for C_1.

The observation of two differently solvated tetramers implies that one is bound at least at one lithium to alkoxide impurities in the butyllithium starting material. With increasing temperature first dimer disappears then the $^{13}C,^{6}Li$ couplings average and finally so also the shifts among tetramers. By 300°K the triple bond spectrum is just two sharp lines, see Figure 12. These effects are reversible with temperature and clearly establish a) the existence of a dynamic equilibrium between tetramers and dimers and b) that above 200°K there is fast interaggregate exchange of lithiums and alkynyls among aggregates.

The equilibriuim is best written as tetramer with each Li bonded to one THF taking up four THF molecules to give two dimers in which lithium is disolvated by THF. This is in accord with all published crystallographic data on solvated organolithium compounds and preserves

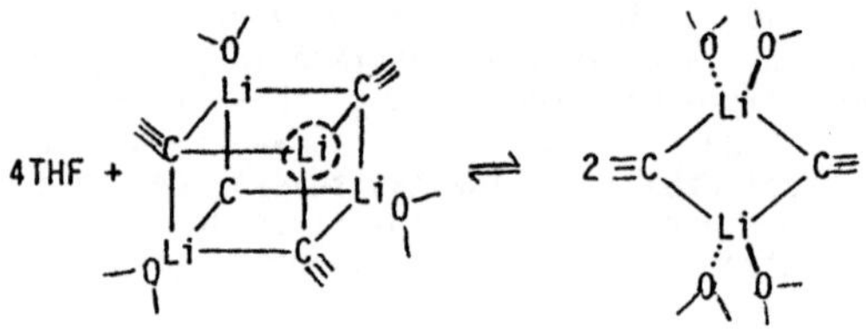

(6)

the tetrahedral bonding arrangement about lithium [26]. A second similar equilibrium involves the less abundant tetramer with one lithium bound to

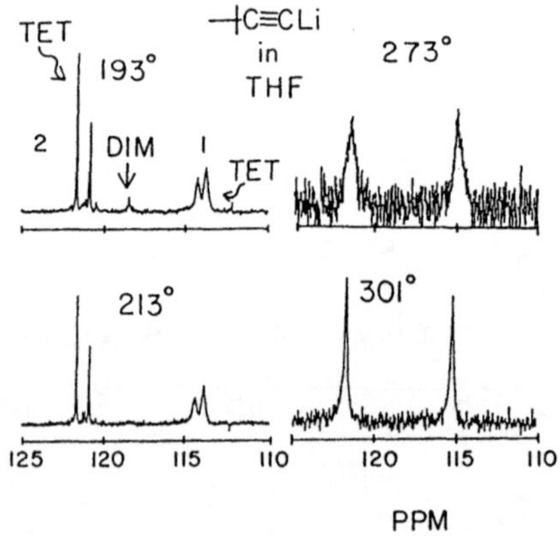

FIG. 12. ^{13}C NMR, C_1 and C_2, 75.46 MHz, compound **7**, 0.9M in THF, different temperatures, °K.

alkoxide and the rest complexed to THF. Since only one kind of dimer is detected we must assume that dimer contains no alkoxide and is only solvated by THF. The observed exothermicity of the equilibrium in the dimer direction is also consistent with the extra solvation of lithium by THF proposed here. This exothermic contribution to ΔH must exceed the positive terms from bond breaking of tetramer. Effects like these have been noted before - that solvated organolithium compounds dissociate to smaller but more highly solvated species at lower temperatures, see below.

Before inquiring too deeply into the nature of the mechanism responsible for carbon-lithium bond exchange in these THF solutions of acetylide some qualitative observations must first be mentioned. Acetylide is insoluble in cyclopentane but dissolves on addition of one equivalent of THF. The triple bond carbon resonance of this solution, 180°K, Figure 13, resembles the spectrum of **7** in pure THF at the same temperature with two important exceptions. Dimer is not detected in the spectrum of the cyclopentane solution, as expected from the much lower concentration of THF, and the multiplet splitting of the C_1 resonance for tetramers is much deeper and better defined. This reflects a much slower carbon-lithium bond exchange rate in the poorly solvated sample of **7** compared to the THF solution. Then, dimer, THF and tetramer must all be involved in the slow step for carbon lithium bond exchange. This conclusion is supported by the observation that in the THF solution of **7**

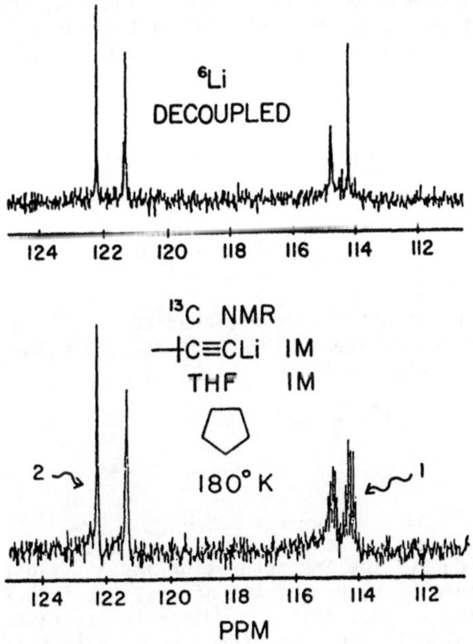

FIG. 13. ^{13}C NMR, 75.46 MHz, C_1 and C_2 of **7**, 1.0M in cyclopentane containing THF, 1.0M, 180°K, Top ^{6}Li decoupled; Bottom normal spectrum.

dimer and tetramer resonances broaden equally at the same temperature. One can imagine two most likely kinds of exchange mechanisms which involve dimer and tetamer, degenerate exchange, (7)

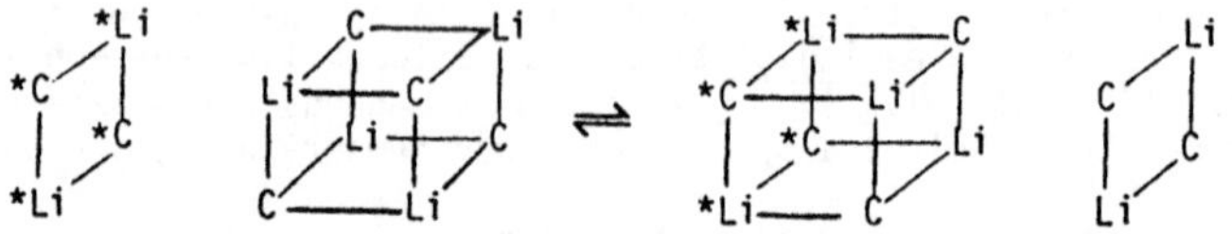

(7)

where dimer collides with a face of tetramer knocking out a new dimer and new tetramer and alternatively direct dissociation of tetramers to dimers as in (6) above.

It is instructive to outline how NMR line-shape analysis [19] is carried out of the triple bond carbon resonance of acetylide **7** in THF solution undergoing fast interconversion between dimers and tetramers. The technique does often distinguish among mechanisms of exchange. Below we show how the C_1 resonance is obtained, the more complicated problem, then briefly indicate that for C_2.

As far as NMR is concerned tetramer can be simulated as a pseudospecies in which one ^{13}C is coupled to three ^{6}Li's but not to the fourth, as $^{13}C(^{6}Li)_3{}^{6}Li$ and dimer as $^{13}C^{6}Li_2$. The $^{13}C_1$ NMR of **7** with ^{13}C in natural abundance is almost entirely due to tetramers containing only one $^{13}C_1$. A dissociation recombination is written, (8), as

$$^{13}C(^{6}Li_3)Li^* \rightleftharpoons {}^{13}C^{6}Li_2 + LiLi^*$$

(8)

$$T \rightleftharpoons D + D.$$

Since $^{6}Li^*$ in tetramer is not observable coupled to ^{13}C the final equations required to calculate the NMR line-shape are the same whether or not Li* is included. This simplifies (8) to (9)

$$^{13}C^{6}Li_3 \rightleftharpoons {}^{13}C^{6}Li_2 + {}^{6}Li$$

(9)

$$T \rightleftharpoons D + D.$$

We wish to first calculate the NMR absorption for $^{13}C_1$ in an equilibrium interconverting mixture of tetramers and dimers. It is best to use the product representation so that states of dimer and tetramer are written as products of individual spin functions (Product representation)

$$\emptyset_T = clmn, \ \emptyset_D = clm \tag{10}$$

where c,l,m,n are spin functions for ^{13}C and the two or three lithiums, respectively. Notice how the factoring of the wavefunction parallels exactly the chemical description of the molecular species [19],

$$\text{clmn} \text{ -----} \ ^{13}C_1Li_1Li_2Li_3.$$

The ^{13}C transitions of interest are

$\alpha lmn \leftarrow \beta lmn$ for tetramers

$\alpha lm \leftarrow \beta lm$ for dimer

where the states of ^6Li are depicted by their M_z eigenvalues, +1 or -1 or 0, written in simplified fashion as +, -, or 0. Notice that the product functions connected by a $^{13}C_1$ transition in dimer or tetramer are always diagonal in the states of ^6Li's. Since each lithium spin can be +, - or 0 there are altogether 27 product states lmn and 9 of lm giving 27 transitions for each tetramer and 9 for dimer.

The ^{13}C NMR absorption is obtained by summing all elements of the density matrices $\rho^{sp}_{i,j}$ of dimers and tetramers, where i,j are two states connected by a ^{13}C transition, each term weighted by the concentration of species, sp, and then the whole plotted as a function of the frequency, ω, radians/sec.

$$\text{Abs} = \sum_{lmn} \left[(T_1)\rho^{T_1}_{\alpha lmn,\beta lmn} + (T_2)\rho^{T_2}_{\alpha lmn,\beta lmn} + (D)\,\rho^{D}_{\alpha lm,\beta lm} \right] \tag{11}$$

The individual elements of the density matrices are obtained by solving the first order coupled equations which result by taking all elements of the density matrix equations where the states

$$\dot{\rho}^{sp}_{i,j} = i[\bar{H}, \rho^{sp}]_{i,j} + (E\rho^{sp})_{i,j} + \left(\frac{\rho^{sp}}{T}\right)_{i,j} \tag{12}$$

i,j are connected by a ^{13}C transition, $\bar{H}$ is the nuclear spin hamiltonian in the rotating frame, E is the exchange operator and 1/T is the phenomenological line width parameter.

The exchange term for tetramer is

$$(E\rho^T)_{i,j} = \left(\frac{\rho^T(ae) - \rho^T}{\tau_T}\right)_{i,j} \tag{13}$$

with τ_T the mean lifetime between successive exchanges and $\rho(ae)$ is the form the density matrix takes after exchange. The $\rho^T(ae)$ element can be derived by inspection of the exchange process and accompanying states, Ø.

$$^{13}C^{6}Li_3 \rightleftharpoons {}^{13}C^{6}Li_2 + Li \qquad (9)$$

$$\begin{array}{llll} \emptyset & \alpha lmn & \alpha lm & n \\ \emptyset' & \beta lmn & \beta lm & n \end{array} \qquad (14)$$

Taking account of the altogether three ways tetramer can slice up into dimers, $\rho^T(ae)$ comes out as

$$(\rho^T(ae))_{\alpha lmn,\beta lmn} = \frac{1}{3}(\rho^D_{\alpha lm,\beta lm} + \rho^D_{\alpha ln,\beta ln} + \rho^D_{\alpha mn,\beta mn}) \qquad (15)$$

making use of the fact that the diagonal element (one ^6Li) is

$$\rho^D_{n,n} = \frac{1}{3} \qquad (16)$$

One can see how exchange converts elements of ρ^T into elements of ρ^D and brings about mixing of transitions.

Using our PI method, described before,[19] the $\rho^D(ae)$ elements are obtained as

$$(\rho^D(ae))_{\alpha lm,\beta lm} = \sum_n(\rho^T_{\alpha lmn,\beta lmn} + \rho^T_{\alpha lnm,\beta lnm} + \rho^T_{\alpha nlm,\beta nlm}). \qquad (17)$$

We have applied the above procedures to calculate the ^{13}C, NMR line-shape for two tetramers, T_1 and T_2, in rapid equilibrium with dimer, D.

$$T_1 \rightleftharpoons 2D \qquad \begin{array}{l} k_{T_1} \rightarrow \\ k_{DT_1} \leftarrow \end{array} \qquad (18)$$

$$T_2 \rightleftharpoons 2D \qquad \begin{array}{l} k_{T_2} \rightarrow \\ k_{DT_2} \leftarrow \end{array} \qquad (19)$$

Taking all elements of the density matrix equations for the two tetramers and dimer generates the sixty three coupled first order inhomogeneous equations in the required density matrix elements. Two examples of these equations show the general form they take.

$$\langle \alpha+- | \dot{\rho}^D | \beta+- \rangle = 0 \tag{20}$$

$$\left[i(\omega - \omega_D) - \frac{1}{T} - \frac{1}{\tau_{DT_1}} - \frac{1}{\tau_{DT_2}} \right] \rho^D_{\alpha+-,\beta+-}$$

$$+ \sum_{T=T_1,T_2} \left[\sum_n \frac{1}{\tau_T} \rho^T_{\alpha n+-,\beta n+-} + \rho^T_{\alpha+-n,\alpha+-n}\right.$$

$$\left. + \rho^T_{\alpha+n-,\beta+n-} \right] = \frac{iC}{2}\left(\frac{1}{18}\right) \tag{21}$$

$$\langle \alpha+++ | \dot{\rho}^{T_1} | +++ \rangle = 0 \tag{22}$$

$$\left[i(\omega - \omega_{T_1} - 3J_{T_1} - \frac{1}{T} - \frac{1}{\tau_{T_1 D}} \right] \rho^{T_1}_{\alpha+++,\beta+++}$$

$$+ \frac{1}{3\tau_{T_1 D}} \rho^D_{\alpha++,\beta++} = \frac{iC}{2}\left(\frac{1}{54}\right) \tag{23}$$

Actually by judicial use of group theory one can reduce the number of variables to nineteen, the seven each for the tetramers and five for the dimer.

The absorption for the $^{13}C_2$'s of tetramers and dimer is calculated separately as a three-half spin uncoupled exchanging system [19], and then added to the absorption for $^{13}C_1$.

A similar treatment to that described above derives ^{13}C, NMR line-shape equations for 7 undergoing C,Li bond exchange via the degenerate mechanism, see above.

$$T_1 + D \rightleftharpoons D^* + T_1^* \quad k_{11} \tag{24}$$

$$T_1 + D \rightleftharpoons D^* + T_2 \quad k_{12}, k_{2,1} \tag{25}$$

$$T_2 + D \rightleftharpoons D^* + T_2 \quad k_{22} \tag{26}$$

Interestingly the line-shape equations are identical to those obtained for **7** undergoing reversible dissociation. This need not be surprising considering that the two mechanisms differ in timing rather than kind. The two mechanisms can be distinguished on the basis of their kinetics.

Recalling that τ is defined as,

$$\frac{1}{\tau_{sp}} \equiv \frac{\text{rate law}}{(sp)} \tag{27}$$

we find for degenerate exchange,

$$\frac{1}{\tau_{T_1}} = \left(k_{11} + k_{12}\right)(D) \tag{28}$$

$$\frac{1}{\tau_{T_2}} = \left(k_{22} + k_{21}\right)(D) \tag{29}$$

$$\frac{1}{\tau_D} = k_{11}(T_1) + k_{12}(T_1) + k_{22}(T_2) + k_{21}(T_2) \tag{30}$$

whereas for dissociation,

$$\frac{1}{\tau_{T_1}} = k_{T_1} \tag{31}$$

$$\frac{1}{\tau_{T_2}} = k_{T_2} \tag{32}$$

$$\frac{1}{\tau_D} = k_{DT_1} + k_{DT_2}\,(D) \tag{33}$$

Comparison of experimental NMR spectra of $^{13}C_1$ and $^{13}C_2$ of **7** in THF solution with line shapes calculated for different values of the τ's yields experimental mean lifetimes for the different species. Preliminary evidence indicates the τ_T values to be independent of concentrations of all components. This strongly implicates the operation of the dissociation mechanism. At 193°K the rate constants extricated from our data are:

$$k_{T_1} = 2.2 \text{ sec}^{-1} \qquad k_{T_2} = 1.9 \text{ sec}^{-1}$$

$$(k_{DT_1} + k_{DT_2}) = 152 \text{ M}^{-1} \text{ sec}^{-1}.$$

The principle conclusion to be derived from the experiments with lithium acetylide in THF are the existence of dimer and tetramers, exclusively and that carbon lithium bond exchange takes place via the reversible dissociation of tetramers to dimers.

The interaction of lithium acetylide **7** with a wide variety of ethers and tertiary amines has also been investigated using NMR methods. These results greatly resemble those obtained for the THF solutions. Resonances are observed only at shifts previously assigned to tetramers and dimers in the THF solutions; the accompanying ^{13}C,Li coupling constants are likewise similar to those reported above. No other species could be detected.

An equimolar solution of acetylide **7** and N-methylpiperidine in cyclopentane, both 0.75M, at 233°K gave rise to ^{13}C resonances for one tetramer only, with C_1 at 115.6 ppm, $J(^{13}C,^{6}Li)$=6Hz and C_2 at 122 ppm. At least four tetramers are observed in the ^{13}C NMR of a 1:1 molar mixture of **7** and triethylamine both 0.75M in cyclopentane solution, 183°K, yielding C_1 at 112-114 ppm, J = 5.3, and C_2 at 120.5, 120.7, 121.4 and 121.6 ppm, respectively.

Carbon NMR of acetylide **7**, 0.75M, in THF solution containing one equivalent of N,N,N',N'-tetramethyl-1,2-diaminoethane, TMEDA, at 183°K shows predominantly dimer resonances, C_1 and C_2 at 118.3 ppm, $J(^{13}C,^{6}Li)$=9.1 Hz, and weaker broad tetramer peaks at 114 ppm, (C_1), and 121 ppm, C_2. With increasing temperature, the ratio of tetramer to dimer increases so that by 301°K only tetramer is observed.

Another diamine, noted to be an excellent ligand for lithium, 1,2-dipiperidinoethane, DPE, when treated with acetylide **7** in THF (1:1 diamine:Li) exhibited only dimer resonances between 183° and 273°K, with C_1 and C_2 at 118 ppm, $(^{13}C,^{6}Li)$ = 8.8 Hz. When this experiment was carried out using DPE, 0.75, M in cyclopentane with an equivalent of acetylide **7**, dimer was observed exclusively at all temperatures from 183° to 301°K with C_1 at 117.5 ppm, $J(^{13}C,^{6}Li)$ = 8.7 Hz, and C_2 at 120 ppm. Furthermore, this sample exhibited sharp splitting of the C_1 quintet showing unusually slow carbon-lithium bond exchange at all these temperatures, see Figure 14.

It might be expected that a vicinal diamine constrained to be cis as 1,2-bis(dimethylamino)cyclopentane, would favor formation of bidentately complexed dimer. In fact, cyclopentane solution 0.75M in both acetylide and cis diamine, 8, showed mainly dimer resonance up to 253°K. The

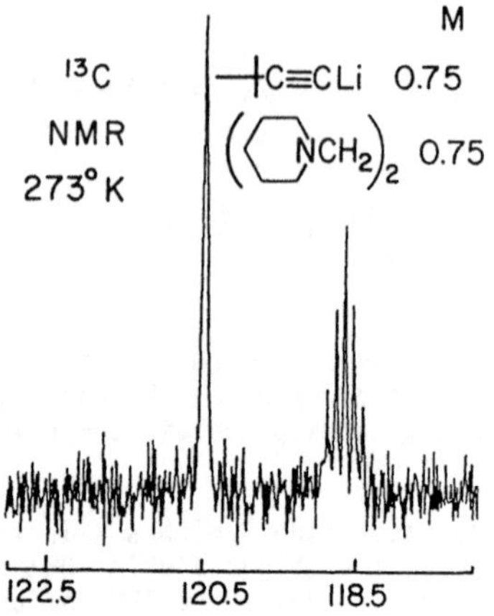

FIG. 14. ^{13}C NMR, 75.46 MHz, C_1 and C_2 of **7** with 1,2-dipiperidinoethane, both 0.75M in cyclopentane, 273°K.

cisoid diamine is clearly a superior ligand to TMEDA for favoring acetylide dimer.

Thus, the salient feature of solvated acetylide **7** is that so far it is found to always consist of dimers and tetramers; no other species have been detected.

The preference of vicinal diamines to solvate acetylide in the dimer structure can be understood by comparing two equilibria between tetramers and dimers one with tertiary monoamine and the other with tertiary vicinal diamine, e.g. TMEDA. In solvated tetramer each lithium is bound to one

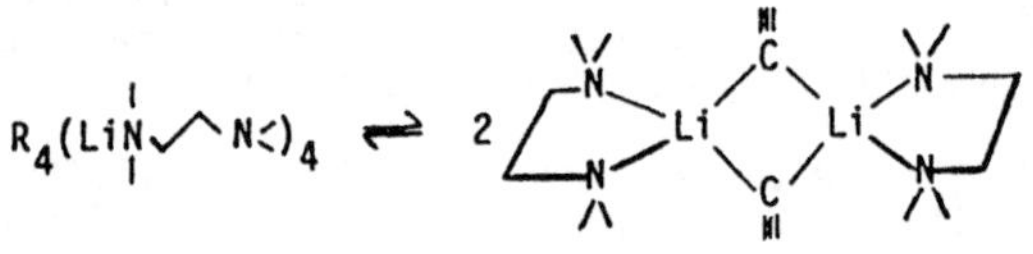

(34)

(35)

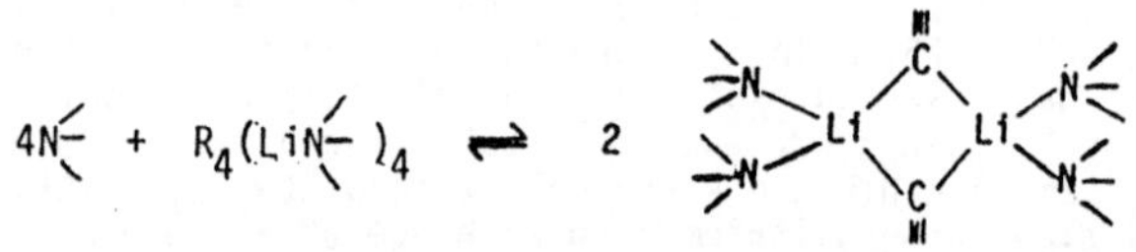

amine nitrogen either a monotertiary amine or one nitrogen of a TMEDA molecule. In the latter case, the second nitrogen of each TMEDA would be uncomplexed because TMEDA is not long enough to bind to two lithiums in the same tetramer. In dimer each lithium is complexed to two nitrogens to preserve the tetrahedral coordination around lithium, either two nitrogens of the same TMEDA molecule or to two monoamines. One can reasonably assume that the heat of amination of lithium is the same in both equilibria. Then, dimer formation will be favored among the TMEDA solvated aggregates since it is accompanied by the more positive translational entropy of reaction.

The dimer tetramer equilibrium which characterizes the behavior of solvated lithium acetylide **7** also applies to several other typical organolithium compounds. Thus, butyllithium-1-^{6}Li in THF [27] and vinyllithium-^{6}Li in diethyl ether [28] were shown from ^{13}C NMR to exist of equilibrium mixtures of tetramers and dimers. In similar fashion phenyllithium-^{6}Li and phenyllithium-^{7}Li in diethyl ether [28] with added TMEDA were shown to be mainly TMEDA complexed dimers with 10% tetramer at 198°K as determined from the ipso ^{13}C NMR. The directly bonded coupling constants in dimer are $J(^{13}C,^{7}Li)$=20 Hz and $J(^{13}C,^{6}Li)$=8 Hz. The tetramer dimer ratio increases with temperature and by 273°K resonances for dimer and tetramer are signal-averaged. Interestingly, phenyllithium crystallizes from diethyl ether solution as a distorted cubic etherate [29].

Finally, in diethylether both 1-butyllithium and ethyllithium associate as tetramers [12]. This information comes from the ^{13}C,^{7}Li coupling constants.

Schleyer, Weiss, et.al., determined that crystals formed from THF solutions of lithium tertiary butyl acetylide consisted of solvated tetramers [30]. Freezing point determinations of these solutions gave association numbers of between two and four confirming the above proposed for a tetramer/dimer equilibrium [31]. Taken together all the above data strongly support the contention that solvated $(RLi)_n$ species are mixtures of tetramers and dimers.

EXPERIMENTAL

Preparations of organolithium compounds described herein have been reported previously [15,23,25,28] as have the details of NMR instrumentation. All ^{13}C NMR and ^{6}Li NMR spectra were obtained with proton decoupling. Selected ^{13}C spectra were carried out with ^{6}Li and proton decoupling, as noted in the Figures.

ACKNOWLEDGMENT

The research described above would not have been possible without the talented efforts of my coworkers, Biing Ming Su, Berhan Tecle´, Hsipai Hsu, Michael Stier, Pradik Pramanik Howard Rosenzweig and Betsy Sprague, and our collaborators at Eastman Kodak, Mark Henrichs and Michael Hewitt who let us use their equipment in the early part of this work.

The work was supported by the National Science Foundation which also provided partial funding for purchase of the NMR equipment used. We are especially grateful to Dr. Charles Cottrell of the Chemical Instrumentation Center, Ohio State University, for technical assistance.

REFERENCES

1. B. J. Wakefield, "The Chemistry of Organolithium Compounds," Pergamon Press, New York, 1974.
2. G. Fraenkel, D. G. Adams and R. R. Dean, J. Phys. Chem., 72, 944 (1968); G. Fraenkel, S. Kobayashi and S. Dayagi, ibid. 72, 953 (1968).
3. P. v. R. Schleyer, Pure App. Chem. 55, 355 (1983); W. N. Setzner and P. v. R. Schleyer, Adv. Organometallic Chem., 24, 354 (1985); A. Streitwieser, Jr., J. E. Williams, Jr., S. Alexandratos, and J. M. McKelvey, J. Am. Chem. Soc., 98, 4478 (1976); J. Klein, D. Kost, G. W. Schriver, and A. Streitwieser, Jr., Proc. Nat. Acadm. Sci., USA, 79, 3922 (1982); A. Streitwieser, Jr., Acc. Chem. Res., 17, 353 (1984); A. E. Reed, R. B. Weinstock, and F. Weinhold, J. Chem. Phys., 83, 735 (1985).
4. R. Strozier, Goodyear Tire and Rubber Company, Private Communication, 1985.
5. M. Weiner, G. Vogel and R. West, Inorg. Chem. 1, 654 (1962).
6. D. Margerison and J. R. Newport, Trans Faraday Soc. 59, 2058 (1963); T. L. Brown, Adv. Organometallic Chem., 3, 365 (1965).
7. J. B. Collins, J. D. Dill, E. B. Jemmis, Y. Appeloig, P. v. R. Schleyer, R. Seeger and J. A. Pople, J. Am. Chem. Soc., 98, 5419 (1976); W. D. Laidig and H. F. Schaefer, III, J. Am. Chem. Soc., 100, 5972 (1978).
8. Y. Apeloig, P. V. R. Schleyer, J. S. Binkley, J. A. Pople and W. L. Jorgensen, Tetrahedron Lett, 3923 (1976).
9. Y. Apeloig, P. V. R. Schleyer, J. S. Binkley and J. A. Pople, J. Am. Chem. Soc., 98, 4332 (1976); W. D. Laidig and H. F. Schaefer, III, ibid., 101, 7184 (1979).
10. G. Fraenkel, A. M. Fraenkel, M. J. Geckle and F. Schloss, J. Am. Chem. Soc., 101, 4745-4747 (1979).
11. D. Seebach, H. Siegel, J. Gabriel and R. Hässig, Helv. Chem. Acta, 63, 2046 (1980); D. Seebach, R. Hassig and J. Gabriel, ibid., 66, 308 (1983); M. J. Goldstein and T. Wenzel, Helv. Chem. Acta, 67, 2029 (1984); A. Rajca and L. Tolbert, J. Am. Chem. Soc. 107, 2969 (1985).
12. L. D. McKeever, R. Waack, M. A. Doran and E. B. Baker, J. Am. Chem. Soc., 90, 3244 (1968); ibid., 91, 1057 (1969).
13. F. W. Wehrli, J. Magnetic Resonance, 30, 193-209 (1978).
14. D. Seebach, J. Gabriel and R. Hassig, Helv. Chem. Acta, 67, 1083 (1984).
15. G. Fraenkel, M. Henrichs, J. M. Hewitt, B. M. Su and M. J. Geckle, J. Am. Chem. Soc., 102, 3345 (1980).
16. A. Abragam, "The Principles of Nuclear Magnetism," Oxford University Press, London, 1961, p. 315.
17. J. I. Kaplan and G. Fraenkel, "NMR of Chemically Exchanging Systems," Academic Press, New York, 1980, p. 50.
18. G. Fraenkel, W. E. Beckenbough and P. P. Yang, J. Am. Chem. Soc., 88, 6878 (1976).
19. Reference 17, Chapter 6.
20. Kashi Ram, Chemistry Department, Ohio State University, 1983; R. D. Thomas, M. T. Clark, R. M. Jenson and T. C. Young, Organometallics, 5, 1857 (1986).
21. V. M. Sergutin, H. G. Antonov, V. W. Zgonnik, and K. K. Kalninsh, J. Organometal. Chem., 145, 265 (1978).

22. W. E. Beckenbough, Western Electric Company, as described in Reference 18.
23. G. Fraenkel, M. Henrichs, J. M. Hewitt, and B. M. Su, J. Am. Chem. Soc., 106, 255 (1984).
24. D. J. Cram, "Fundamentals of Carbanion Chemistry," Academic Press, New York, 1965, p. 49; A. Streitwieser, Jr., J. E. Williams, Jr., and S. Alexandratos, J. Am. Chem. Soc., 98, 4778 (1976).
25. G. Fraenkel and P. Pramanik, J. Chem. Soc., Chem. Commun., 1527 (1983); M. Stier, Regional Meeting, American Chemical Society, Akron, June 4, 1985.
26. R. Hassig and D. Seebach, Helv. Chem. Acta, 66, 2269 (1983); B. Schubert and E. Weiss, Angew. Chem., 95, 499 (1983).
27. J. Heinzer, J. F. M. Oth and D. Seebach, Helv. Chem. Acta, 68, 1848-1862 (1985).
28. G. Fraenkel, H. Hsu and B. M. Su, in R. Bach, Ed., "Lithium Current Applications in Science Medicine and Technology," John Wiley, New York, 1985, p. 273.
29. P. P. Power and H. Hope, J. Am. Chem. Soc., 55, 5320 (1983).
30. M. Geissler, J. Kopf, B. Schubert, E. Weiss, W. Neugebauer and P. v. R. Schleyer, J. Organometallic Chem., Submitted for Publication.
31. W. Bauer and D. Seebach, Helv. Chem. Acta, 67, 1972 (1984).

SPECTROSCOPIC STUDIES OF SOME CRYPTATED LITHIUM CARBANIONS AND RELATED SPECIES

S. HUBERT,* C. MOMTAZ,* P. HEMERY,* S. BOILEAU*§ and J.P. KINTZINGER**
* Collège de France, 11 place Marcelin Berthelot, 75231 Paris Cédex 05, France; ** UA 422 Institut Le Bel, Université Louis Pasteur, 4 rue Blaise Pascal, 67000 Strasbourg, France.

INTRODUCTION

Complexation of alkali cations by macroheterobicyclic ligands leads to a now well-known class of complexes, called cryptates, in which the cation is included within the central molecular cavity of the ligand [1]. Solutions of these alkali metal cryptates have been studied by different techniques [2], however, only few data exist concerning their physicochemical properties in low polarity aprotic solvents [3]. This information is needed in order to elucidate the mechanism and the kinetics of the anionic polymerization of heterocyclic and vinyl monomers with cryptates as counterions [4]. Thus we have examined the behavior and the structure of some lithium ion pairs and of their complexes with cryptands, namely the [211], in THF and in toluene by two complementary techniques : UV and visible spectrophotometry, and by ^1H and ^7Li NMR.

EXPERIMENTAL

All the operations were made in glass apparatus sealed under high vacuum according to processes usually employed in our laboratory. THF and toluene were dried over calcium hydride and several sodium mirrors [3]. Lithium hexafluoroarsenate, AsF_6Li, (Ventron) was dried before use and a sample of purified $Ø_4BLi$ was kindly supplied by Pr Van Beylen (Belgium). Fluoronyl lithium (FlLi), carbazyllithium (CbzLi), 1,1-diphenylhexyllithium (nDPLi) and 3,3-dimethyl 1,1-diphenylbutyllithium (tDPLi) were prepared by reaction of n-BuLi or t-BuLi with an excess of fluorene, carbazole and 1,1-diphenylethylene respectively. The "kryptofix" ligands (Merck) [211] and [221] were redistilled in small quartz apparatus under high vacuum, put into vials fitted with break-seals and rinsed with pure THF [5].

UV and visible spectra were recorded by means of a Perkin Elmer 554 spectrophotometer using a 0.5 cm optical vacuum cell provided with a spacer. ^1H and ^7Li NMR spectra of lithium salts solutions in deuterated toluene or THF ($0.3\ 10^{-3} < C < 5\ 10^{-3}$ M) were recorded at different temperatures, using a BRUKER WP 200 SY spectrometer.

RESULTS AND DISCUSSION

UV and visible Spectrophotometry.

Changes in the optical spectra of the ion pairs in solution can be used in some cases to study the interaction of complexing agents with the cation. Thus we examined the behavior of lithium salts which form stable and soluble complexes with the specific cryptand [211] in low polarity solvents such as THF, toluene or benzene and which present a characteristic UV or visible absorption spectrum. The studied salts are $Ø_4BLi$, fluorenyllithium (FlLi) and carbazyllithium (CbzLi) in THF as well as lithium salts of carbanions derived from 1,1-diphenylethylene : 1,1-diphenylhexyllithium (nDPLi) and 3,3-dimethyl-1,1-

Recent Advances in Anionic Polymerization
T.E. Hogen-Esch and J. Smid, Editors

diphenylbutyllithium (tDPLi), in THF and in benzene and toluene. Those last salts were found to give soluble and rather stable complexes with [211] in aromatic hydrocarbons whereas it is not the case for $Ø_4$BLi, FlLi or CbzLi.

The results of UV and visible spectrophotometric studies made on those non complexed and cryptated lithium salts are summarized in Tables I and II together with some conductance data.

Molar extinction coefficients (εmax) corresponding to the maxima of nDPLi and tDPLi were determined from the value found for FlLi in THF at 25°C [6] by adding an excess of fluorene to a THF solution of nDPLi or tDPLi. The values of ε max for the cryptated salts were measured after addition of a slight excess of [211] to the non complexed lithium salts solutions. The value of ε max = 15,000 at 372 nm found for FlLi + [211] is in good agreement with that determined in the case of fluorenylstrontium complexed by the [222] cryptand as well as in the case of fluorenylsodium complexed by dibenzo-18crown-6 in THF (ε max = 14,400) [9]. However the values of ε max and λ max (ε max = 19,000 at 415 nm) found for nDPLi in benzene are quite different from those found by Evans and George [10]. This comes from the fact that nDPLi was prepared by reaction of a slight excess of 1,1-diphenylethylene with nBuLi in benzene in the present case whereas Evans and George used a large excess of nBuLi. Waack and Doran [11] first mentioned a value of 426 nm for λ max of nDPLi in benzene before they clearly demonstrated the strong influence of an excess of nBuLi on λ max of nDPLi which is located at 415 nm in benzene when stoechiometric amounts of reagents are used [12].

The observed spectral changes for a given salt can be used to study the interaction of complexing agents with the cation. The bathochromic shift of the maxima noticed on adding the [211] to CbzLi in THF is due to the increase of the size of the counterion, indicating the change from tight ion pairs to cryptated ion pairs. A similar shift was observed on adding the [222] to a carbazylpotassium solution in THF whereas a large increase of the dissociation constants K_D was noticed (from 8.5 10^{-9} for carbazylpotassium to 1.3 10^{-5} on adding the [222], at 20°C) [12].

For the other salts in THF, a small hypsochromic shift is observed which corresponds to a change from loose ion pairs to cryptated ion pairs, the size of Li^+ + [211] being slightly smaller than that of solvated Li^+. This is confirmed by conductance measurements made on $Ø_4$BLi + [211] in THF : K_D is slightly smaller for the cryptated salt than for $Ø_4$BLi alone [8].

A very important bathochromic shift (80-90 nm) of the maximum is observed between benzene or toluene and THF in the case of tDPLi and nDPLi (Table II). This is due to the fact that the main ionic species are aggregates in hydrocarbons whereas they are loose ion pairs in THF as shown by Waack et al. for nDPLi [11]. Again, a small hypsochromic shift is noticed on adding the [211] to the THF solutions of nDPLi or tDPLi which corresponds to the change from loose to cryptated ion pairs.

The addition of [211] to a benzene or toluene solution of tDPLi or nDPLi causes a pronounced change in the position of λmax. The maximum observed when the ratio [211]/[Li^+] is higher than 1 is located at 498 nm, in the case of tDPLi in toluene (Table II), which is very close to the value of λmax observed for cryptated ion pairs in THF (496 nm). Thus it can be concluded that the corresponding species are cryptated ion pairs. It must be mentioned that this is the first example of a cryptated lithium salt which is soluble and quite stable in benzene.

TABLE I. Spectrophotometric studies of cryptated lithium salts in THF at room temperature.

Anion	Counterion	λ max(nm)	ε max	Main Ionic Species[a)]
$Ø_4B^-$ b)	Li^+	274 [7]	2,200	l.i.p.
$Ø_4B^-$ b)	Li^+ + [211]	273	-	cr.i.p.
Fl^- b)	Li^+	373	9,600 [6]	[l.i.p.]/[t.i.p.] = 0.80 [6]
Fl^-	Li^+ + [211]	372	15,000	cr.i.p.
Cbz^-	Li^+	371;390	-	t.i.p.
Cbz^-	Li^+ + [211]	397;419	-	cr.i.p.

a) l.i.p. : loose ion pairs; cr.i.p. : cryptated ion pairs; t.i.p. : tight ion pairs; b) K_D ($Ø_4$BLi) = 5.3 10^{-5} [8]; K_D ($Ø_4$BLi + [211]) = 3.9 10^{-5} [8]; K_D (FlLi) = 4.4 10^{-6} [6], at 20°C.

TABLE II. Spectrophotometric studies of cryptated lithium carbanions derived from 1,1-diphenylethylene in THF and in aromatic hydrocarbons, at room temperature.

Anion	Counterion	Solvent	λ max(nm)	ε max	Main Ionic Species[a)]
nDP^-	Li^+	THF	496	26,000	l.i.p.
nDP^-	Li^+ + [211]	THF	488	27,500	cr.i.p.
nDP^-	Li^+	Benzene	415	19,000	aggregates
nDP^-	Li^+ + [211]	Benzene	495	26,000	cr.i.p.
tDP^-	Li^+	THF	500	25,000	l.i.p.
tDP^-	Li^+ + [211]	THF	496	25,000	cr.i.p.
tDP^-	Li^+	Toluene	412	19,000	aggregates
tDP^-	Li^+ + [211]	Toluene	498	25,000	cr.i.p.

a) l.i.p. : loose ion pairs ; cr.i.p. : cryptated ion pairs.

NMR spectroscopy.

During the past decade there has been a very rapid expansion of the use of multinuclear NMR for the studies of thermodynamics and kinetics of reactions in solution. To a very large extent, this expansion is the result of progress in NMR instrumentation. Chemical shifts of alkali ions are very sensitive to the immediate environment of an alkali ion in solution; thus the alkali metal NMR is a very sensitive technique for the detection of ion-ion, ion-solvent and ion-ligand interactions [13]. However, in a general manner, all the ^{7}Li NMR data found in the literature correspond to relatively high concentrations of lithium salts ($>10^{-2}$ M) so that the nature of involved ionic species is not known. Popov et al. studied in detail some lithium cryptates in various solvents by ^{7}Li NMR [14]. Their results are summarized in Table III.

TABLE III. ^{7}Li NMR study of [211] lithium complexes in various solvents at +30°C [14].

Anion	$[Li^+]^{a)}$ (M)	Solvent	$\delta^{b)}$ (ppm)
I_3^-	0.14	CH_3NO_2	-0.37
I_3^-	0.10	THF	-0.36
I_3^-	0.15	$CHCl_3$	-0.38
ClO_4^-	0.15	CH_3NO_2	-0.41
ClO_4^-	0.25	PC c)	-0.38
ClO_4^-	0.25	DMF	-0.42
ClO_4^-	0.25	Formamide	-0.38
ClO_4^-	0.20	DMSO	-0.38
I^-	0.14	CH_3NO_2	-0.42
I^-	0.14	DMSO	-0.39
Cl^-	0.11	DMSO	-0.39
Cl^-	0.13	CH_3NO_2	-0.41
$\varnothing_4B^-$	0.10	DMSO	-0.42

a)[211]/[Li^+] = 1 or 0.5; b)vs. aqueous $LiClO_4$ at infinite dilution (corrected for magnetic susceptibility); c)Propylene carbonate.

These authors showed that in all solvents the addition of the ligand [211] in less than stoechiometric amounts results in two resonances corresponding to the free and the complexed lithium ion. Moreover, they found that the chemical shift of Li^+ + [211] is the same whatever the nature of the solvent and or the anion : $\delta \simeq -0.4$ ppm (Table III). It must be mentioned that the anions studied by Popov et al. did not contain an aromatic ring except $\varnothing_4BLi$ in DMSO. Thus these authors concluded that "the lithium ion is completely encaged by the ligand [211] and, since the ^{7}Li chemical shits are dependant almost exclusively on the nearest neighbors of the lithium ion, they are insensitive to either the solvent molecules surrounding the cryptate or to counterion in cases where a low dielectric constant of the solvent (such as THF or chloroform) would lead to ion pair formation " [14]. Moreover, Cambillau et al. [15] found a sharp signal at -0.4 ppm for Li^+ + [211] associated with the following anion :

$$[(Me\text{-}\underset{O}{C}\text{-}CH\text{-}\underset{O}{C}\text{-}OEt)_2Li^+]^-$$

in DMSO and in methylene chloride. This result is in good agreement with the findings of Popov et al.

^{7}Li NMR studies were performed on lithium salts in THF and in toluene with different cryptand/Li^+ mole ratios, in the same concentrations range as used for the UV and visible spectrophotometric studies ($0.3\ 10^{-3} < C < 5\ 10^{-3}$ M), by means of a 200 MHz spectrometer.

First of all, the behavior of AsF_6Li in THF was examined in the presence of different amounts of [211] and of [221]. Typical spectra are shown in Figure 1.

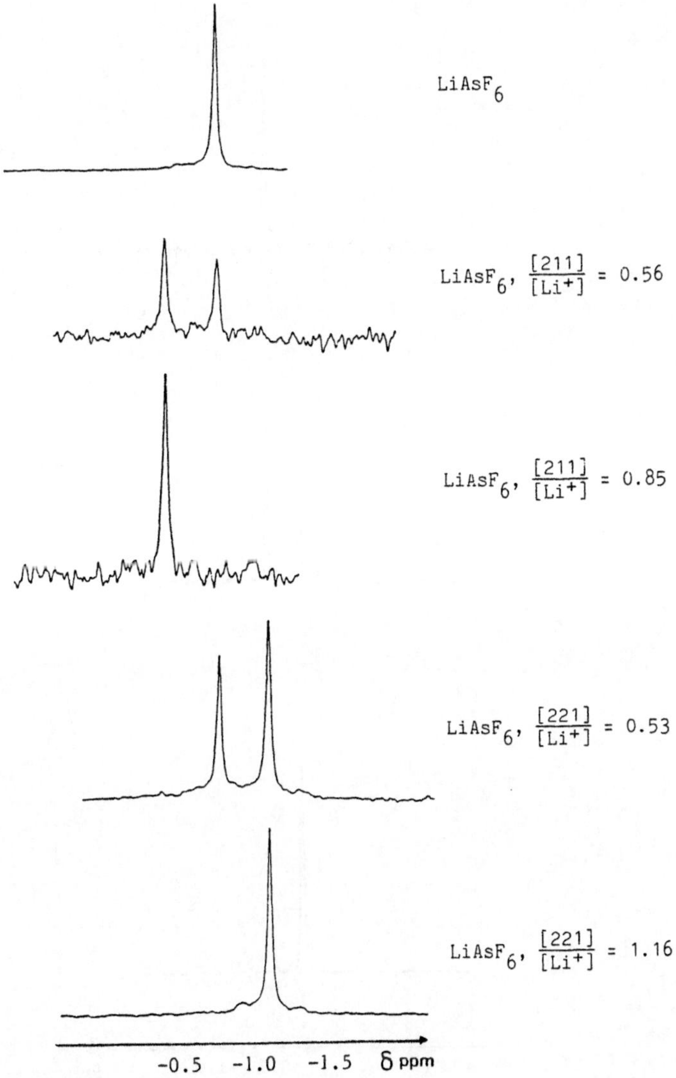

FIG. 1. ^{7}Li NMR spectra of AsF_6Li solutions in THF with and without cryptands at room temperature.

These cryptands form stable complexes with the lithium cation and two resonances corresponding to the non complexed and the cryptated Li^+ are observed for solutions containing an excess of the Li^+ ion. Moreover the chemical shift of Li^+ + [211] is equal to -0.43 ppm at 20°C. These results are in good agreement with those of Popov et al. [14]. However these authors observed a fast exchange on the NMR time scale for [221] lithium complexes in polar solvents, whereas a slow exchange is observed in THF.

The ^7Li NMR spectra of solutions of $Ø_4BLi$ with increasing amounts of [211] are shown in Figure 2.

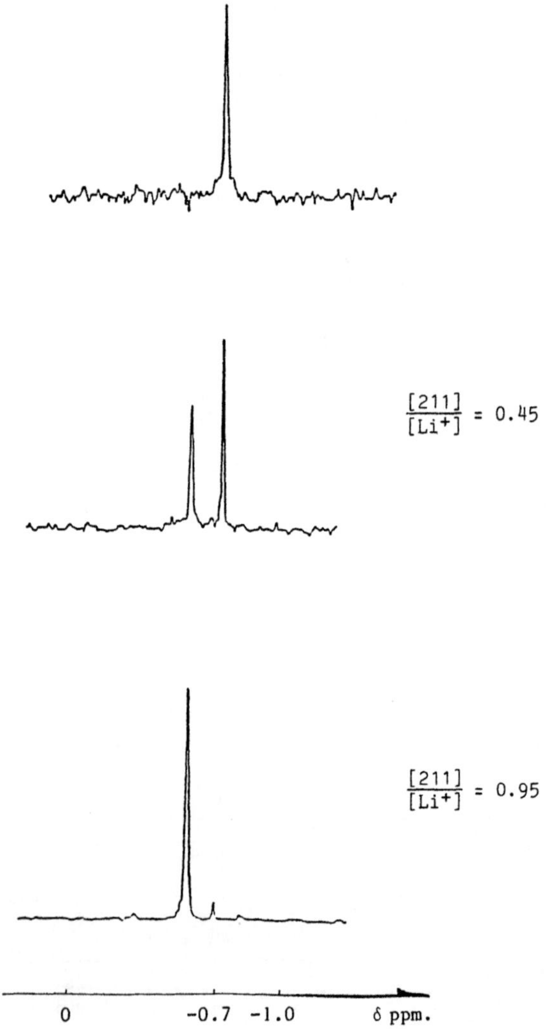

FIG. 2. ^7Li NMR spectra of $Ø_4BLi$ in TDF with and without [211] at room temperature. ([C] = 1.2 10^{-3} M).

A slow exchange between non complexed and cryptated Li^+ is observed. However, in contrast with the previous results, the chemical shift of Li^+ + [211] is not located at -0.4 ppm but at -0.57 ppm at 20°C. Thus we suspected a ring current effect of the aromatic anion upon the chemical shift of the cryptated lithium cation though this effect was not observed by Popov [14] in the case of $Ø_4BLi$ + [211] in DMSO. This can be explained by the fact that the main ionic species in THF are cryptated ion pairs whereas the proportion of free ions is predominant in DMSO : the influence of the anion is much greater in THF than in DMSO.

Other aromatic organolithium derivatives were studied such as FlLi, CbzLi and tDPLi. The ^7Li chemical shifts of these salts and of [211] lithium complexes are listed in Tables IV and V.

TABLE IV. ^7Li chemical shifts of lithium salts in THF.

Anion	[salt] (10^3 M)	δ(20°C)a) (ppm)	$\Delta\nu_{1/2}$ (Hz)	δ(-20°C)a) (ppm)	$\Delta\nu_{1/2}$ (Hz)
AsF_6^-	2.7	-0.76	$\leqslant$ 1	-0.66	$\leqslant$ 1
$Ø_4B^-$	0.7	-0.70	$\leqslant$ 1	-0.64b)	$\leqslant$ 1
tDP^-	0.6	-0.71	3.5	-0.65	2.5
Fl^-	0.6	-1.09	13	-0.90	7

a)vs $LiClO_4$ $5x10^{-2}$ M in H_2O/D_2O (90/10); b)-30°C.

TABLE V. ^7Li chemical shifts of [211] lithium complexes in THF and in toluene.

Anion	[Li^+ + (211)] (10^3 M)	Solvent	δ(20°C)a) (ppm)	δ(-20°C)a) (ppm)
AsF_6^-	5.8	THF	-0.43	-0.45
$Ø_4B^-$	0.5	THF	-0.57	-0.59b)
tDP^-	1.0	THF	-0.56	-0.60
Cbz^-	1.0	THF	-0.92	-0.96b)
Fl^-	1.0	THF	-1.03	-1.08
tDP^-	0.3	Toluene	-1.25	-1.32

a)vs $LiClO_4$ $5x10^{-2}$ M in H_2O/D_2O (90/10); $\Delta\nu_{1/2} \leq 1$Hz; corrected for the change in lock signal; b)-30°C.

A narrow signal at $\sim$-0.7 ppm is observed for Li^+ alone in THF for the examined compounds except in the case of FlLi (δ = -1.09 ppm and $\Delta\nu_{1/2}$ = 13 Hz at 20°C) which has presumably a "sandwich" structure [16]. The chemical shifts of Li^+ + [211] strongly depend on the nature of the anion in contrast to the findings of Popov et al. [14]. Ring current effects are observed for all the salts except AsF_6Li (Table V) and are more pronounced in toluene than in THF. Moreover, a deshielding effect is noticed for the chemical shift of Li^+ alone on lowering the temperature which can be explained by a larger dissociation of ion pairs in THF whereas the reverse is observed for the chemical shift of Li^+ + [211]. For all cryptated salts, a narrow signal is observed since $\Delta\nu_{1/2}$ is always lower than 1Hz even in toluene.

Finally ^{1}H NMR spectra were recorded on solutions of [211] alone and of some lithium cryptates containing an excess of Li^+. The results are listed in Table VI.

TABLE VI. ^{1}H chemical shifts of [211] in solutions of cryptated lithium salts in TDF and in deuterated toluene : δ ppm/ δTMS [ref : $-OCD_2-$ in TDF (δ_o = 3.58 ppm) or $-CD_3$ in toluene (δ_o = 2.03 ppm)].

Solution	Y_4	XX'	AA'	Y	X	B	A
TDF [17]	3.60	3.42	2.65	3.53		2.65	
AsF_6Li,TDF[a)]	3.65	3.55	2.71	3.65		2.87	2.58
Ø$_4$BLi,TDF	b)	3.44	2.59	b)		2.68	2.48
tDPLi,TDF+THF	b)	3.44	2.58	3.44		2.71	2.38
$C_6D_5CD_3$	3.49	3.22	2.45	3.53	3.43	2.59	2.48
tDPLi,$C_6D_5CD_3$	2.73	2.58	1.75	2.86		2.12	1.71

a) recorded in air; b) masked by the solvent.

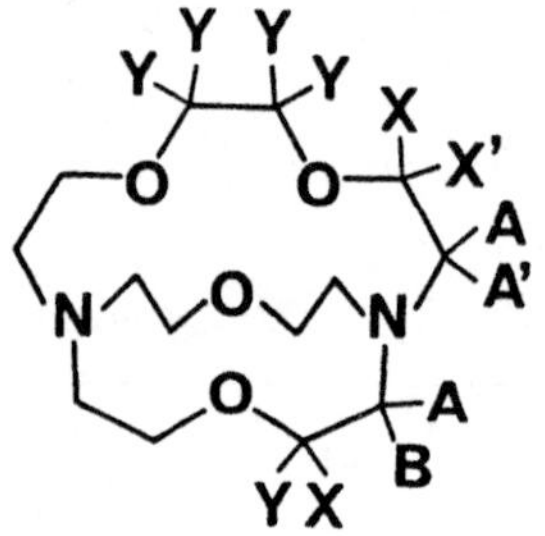

The spectra are different if the ligand is free or if it contains the cation inside the cavity. The two methylenic protons A and B close to the nitrogen in the branch containing only one oxygen are non equivalent. This was already observed for instance in the case of LiCl + [211] in $CDCl_3$ [18]. Moreover a slight deshielding of the protons of the [211] is noticed for AsF_6Li + [211] compared to the protons of the ligand alone as in the case of LiSCN + [211] in $CDCl_3$ [18]. However the reverse effect is observed for Ø$_4$BLi and tDPLi and the shielding in more pronounced in toluene than in THF. This can be explained by ring current effects of the anions.

CONCLUSION.

In conclusion, the behavior of several lithium salts such as AsF_6Li, Ø$_4$BLi, fluorenyllithium, carbazyllithium and of the two carbanions derived from the reaction of 1,1-diphenylethylene with nBuLi and tBuLi has been examined in THF, and in toluene for the last two salts. Addition of the specific ligand [211] leads to the formation of cryptated ion pairs (major species at 10^{-3} M) in which the cation is encaged by the ligand. This was shown by UV-visible spectrophotometry, and conductance measurements as well as by ^{1}H and ^{7}Li NMR. The chemical shift of ^{7}Li + [211] markedly depends on the anion and ring current effects are observed which are more prominent in toluene than in THF.

REFERENCES

§ Laboratoire de Chimie Macromoléculaire associé au CNRS; UA 24.

1. B. Dietrich, J.M. Lehn and J.P. Sauvage, Tetrahedron Letters, 2885 and 2889 (1969).
2. A.I. Popov and J.M. Lehn, Physicochemical studies of crown and cryptate complexes in Coordination Chemistry of Macrocyclic Compounds, (Plenum, New-York 1979) pp.537-602.
3. S. Boileau, P. Hémery and J.C. Justice, J. Sol. Chem. 4, 873 (1975).
4. S. Boileau, Use of cryptates in anionic polymerization of heterocyclic compounds in Anionic Polymerization, Kinetics, Mechanism and Synthesis, A.C.S. Symp. Series n°166, (Washington 1981) pp.283-305; and Anionic polymerization of cyclosiloxanes with cryptates as counterions in Ring Opening Polymerization : Kinetics, Mechanisms and Synthesis, A.C.S. Symp. Series n°286, (Washington 1985) pp.23-35.
5. P. Hémery, Thèse de Doctorat d'Etat, Paris (1976).
6. T.E. Hogen-Esch and J. Smid, J. Am. Chem. Soc. 88, 307 (1966) (a) and 88, 318 (1966) (b).
7. D. Nicholls, S. Sutphen and M. Szwarc, J. Phys. Chem. 72, 102 (1968).
8. H. Porte, Thèse de Docteur Ingénieur, Paris (1980).
9. T.E. Hogen-Esch and J. Smid, J. Phys. Chem. 79, 233 (1975).
10. R. Waack and M.A. Doran, J. Phys. Chem. 67, 148 (1963).
11. R. Waack, M.A. Doran and P.E. Stevenson, J. Am. Chem. Soc. 88, 2109 (1966).
12. B. Vidal, Thèse de Doctorat de 3° cycle, Paris (1975).
13. A.I. Popov, Pure and Appl. Chem. 51, 101 (1979).
14. Y.M. Cahen, J.L. Dye and A.I. Popov, J. Phys. Chem. 79, 1289 (1975).
15. C. Cambillau et M. Ourevitch, Chem. Comm. 996 (1981).
16. T.E. Hogen-Esch and J. Smid, J. Am. Chem. Soc. 89, 2764 (1967).
17. E. Kauffmann, private communication.
18. J.P. Sauvage, Thèse de Doctorat d'Etat, Strasbourg (1971).

NEW ASPECTS OF CARBANION CHEMISTRY. STRUCTURE OF PENTADIENYL ANIONS AND BUTENEDIYL DIANIONS AND THEIR ROLES IN ORGANIC AND INORGANIC SYNTHESES

HAJIME YASUDA, KAZUYUKI TATSUMI, AND AKIRA NAKAMURA
Department of Macromolecular Science, Faculty of Science, Osaka University, Toyonaka, Osaka 560, Mapan

ABSTRACT

Pentadienylpotassium exhibits the U-shaped (Z,Z)-geometry in ethereal solution as confirmed by trimethylsilylation and by the NMR measurement while solvated pentadienyllithium assumes the W-shaped (E,E)-geoemtry. The EHMO calculation on the pentadienylpotassium indicates the substantial stability of the U-shaped geometry. A variety of transmetallation of pentadienylpotassium is summarized briefly. A novel preparative method of 2-butene-1,4-diylmagnesium, one of the allylic dianions, is also described. This bifunctional magnesium compound serves as a versatile reagent in organic and inorganic syntheses.

1. INTRODUCTION

The major interest in carbanion chemistry has been extended from simple alkyl, alkenyl or allyl compounds to carbanions with highly conjugated and hyperconjugated systems. These systems display rich chemistry since their stereo- and regiochemistries in many cases change with the metal, reaction medium, alkyl substituent(s) and substrate. Here we describe recent advances in the chemistry of pentadienyl anions, especially, their geometrical change with counter cation and alkyl substituent(s) in addition to the versatile utility of these carbanions in organic and inorganic syntheses. The structural elucidation has been made based on the spectral and MO studies together with transmetalation reactions. As an example of alkanediyl dianions, the chemistry of 2-butene-1,4-diylmagnesium was investigated. This paper describes a facile method of preparation of a series of 2-butene-1,4-diylmagnesium compounds and synthetically useful novel transmetalations upon these substances.

2. STRUCTURE AND CHEMICAL BEHAVIOR OF PENTADIENYL ANIONS

2.1 Structural Analysis

The structural studies on the pentadienyl anions has long been focused to penetrate into the remarkable behavior of these intermediates in alkali metal catalyzed polymerization of 1,3-pentadienes, base-catalyzed isomerization of 1,3- and 1,4-dienes or in nucleophilic substitution of aromatics [1]. More recently these carbanions have attracted attention as fascinating synthetic reagents as shown in the present article.

A series of 2,4-pentadienyl metals (Li,Na,K,Rb,Cs) could be prepared easily by direct metalation of substituted and non-substituted 1,3- or 1,4-pentadienes with alkali metals in the presence of a tertiary amine [2]. The addition of a tertiary amine, for example triethylamine, completely supresses polymerization and the resulting pentadienyl metals could be isolated as crystals. The NMR studies on the parent pentadienylpotassium, -rubidium, and -cesium revealed that the carbanions assume the U-shaped (Z,Z)-geometry [3] in solution while the lithium derivative assumes the W-shaped (E,E)-structure [4].

Trimethylsilylation of pentadienyl anions with chlorotrimethylsilane

Recent Advances in Anionic Polymerization
T.E. Hogen-Esch and J. Smid, Editors

provides the most facile method of determining the geometry of carbanions [4,5]. Silylation of the U-shaped pentadienylpotassium and -cesium (1) in tetrahydrofuran, 1,2-dimethoxyethane or 1,3-dioxane gave (Z)-2,4-pentadienyltrimethylsilane in 98% regioselectivity (99% yield) while the W-shaped pentadienyllithium (2) gave the (E)-isomer in quantitative yield (eq.1 and eq.2). In the case of pentadienylsodium, the silylation led to a mixture of (E)/(Z)-isomers in 6:3 ratio. These drastic changes can be interpreted by supposing

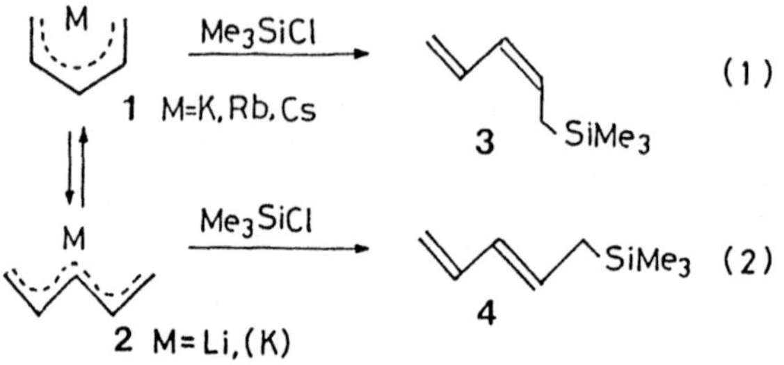

that potassium and cesium derivatives exist as contact ion pairs while lithium analogs exist as solvent-separated ion pairs. Actually, tetrahydrofuran molecules are bound so weakly to pentadienylpotassium and -cesium that they can be removed easily by heating the compounds in vacuo or even by washing them with hexane, whereas pentadienyllithium will not release coordinated ethers even under vigorous conditions.

Surprising is the formation of (E)-2,4-pentadienyltrimethylsilane (4) upon trimethylsilylation of pentadienylpotassium in the solid state (as a hexane suspension) with chlorotrimethylsilane, in sharp contrast to the exclusive formation of the (Z)-isomer (3) in ethereal solutions [5]. Thus, the change of medium is capable of changing geometry. However, the corresponding pentadienyllithium is insensitive to the medium effect. Transmetalation of any of the above pentadienyl metals with MCl_2 (M=Be,Mg,Zn) is found to be useless in determining the original geometry of the carbanions because the resulting pentadienylmetal halides or bis(pentadienyl)metals are immediately converted to the thermally more stable (E)-isomer through a rapid 1,3-rearrangement at ambient temperature [6,7] . In practice, the TMEDA complex of 2,4-pentadienylzinc chloride prepared from (Z,Z)-pentadienylpotassium assumes the zig-zag like (E)-2,4-pentadienylmetal structure as revealed by the X-ray analysis (Fig. 1) [7].

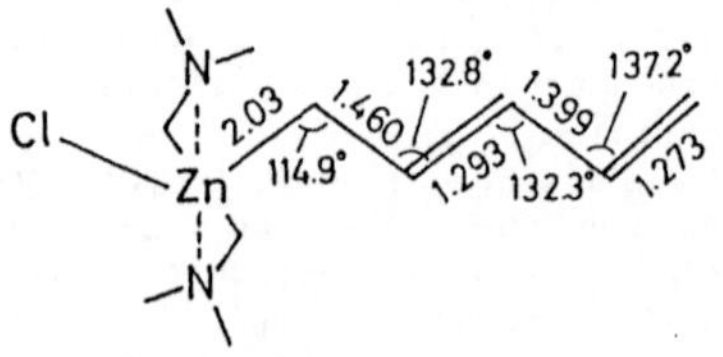

Fig. 1. X-ray structural parameters of $ZnCl(C_5H_7)\cdot$TMEDA

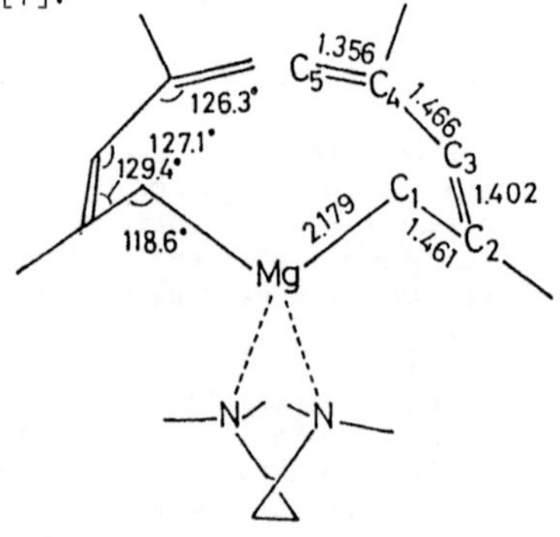

Fig. 2. $Mg(C_5H_7)_2\cdot$TMEDA

2-Methylpentadienylpotassium consists of a single isomer which may possess either the U- or S-shaped structure as indicated by its ^1H NMR spectrum. Trimethylsilylation finally confirmed the structure to be U-shaped [3] since it gave a mixture of (Z)-4-methyl-2,4-pentadienyltriemthylsilane (5a) and (Z)-2-methyl-2,4-pentadienyltrimethylsilane(5b) in 11:9 ratio, in good accord with the result of the boration/oxidation sequence of Schlosser [8]. Silylation in the solid state, however, gave a mixture of (E)-4-methyl-2,4-pentadienyltrimethylsilane(5c) and (Z)-2-methyl-2,4-pentadienyltrimethylsilane(5b) in 46:54 ratio. This result suggests that 2-methylpentadienylpotassium exists in the solid state as a single isomer of

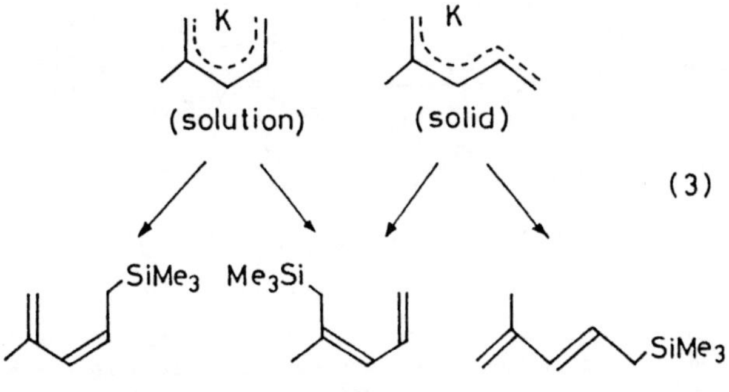

(3)

5a 5b 5c

exo-S geometry (eq.3). Despite the existence of hexadienylpotassium in THF as a mixture of exo-U- and exo-S-shaped dienyl anions, trimethylsilylation gave only (E,Z)-2,4-hexadienyltrimethylsilane due to the selective silylation of both compounds at their terminal CH_2 site. No medium effect was observed in this case.

2,4-Dimethylpentadienylpotassium as well as cycloheptadienylpotassium and cyclooctadienylpotassium always assume the U-shaped geometry regardless of the medium [3]. The severe steric repulsion between the two methyl groups in 2,4-dimethylpentadienyl anions will hamper the formation of the W-shaped structure. Actually, transmetalated compounds such as bis(2,4-dimethyl-2,4-pentadienyl)magnesium·TMEDA [6] and 2,4-dimethylpentadienyllithium [3,4] retain this property and exist as the U-shaped single isomer as confirmed by X-ray and NMR analyses. In the case of the above $Mg(C_7H_{11})_2$·TMEDA complex, the five carbons slightly deviate from coplanarity; the dihedral angle between C_1-C_2-C_3 and C_3-C_4-C_5 planes is 22.5° (Fig.2).

By repetition of the trimethylsilylation-metalation sequence, the regioselective multiple trimethylsilylation of pentadienyl anion has been realized (Scheme 1). The resulting 1,5-disilylated pentadienyl anions(7) assume the (E,E,E,E)-geometry irrespective of the counter cations, Li or K. Interestingly, 1,3,5-tris(trimethylsilyl)pentadienyllithium(8) was found to exist as an S-shaped (E,E,Z,E)-carbanion, the first example of the S-shaped open-chain pentadienyl anion, as revealed by variable temperature NMR studies. Further silylation of the 1,3,5-triply silylated carbanion 8 in THF afforded 1,1,3,5-tetrakis(trimethylsilyl)-2,4-pentadiene exclusively (99% regioselectivity) while reaction in hexane gave a mixture of 1,1,3,5-tetrakis(trimethylsilyl)-2,4-pentadiene and 1,3,3,5-tetrakis(trimethylsilyl)-1,4-pentadiene in 55:45 ratio. Alkylation of 1-(trimethylsilyl)pentadienyllithium(6) and 1,5-disilylated pentadienyllithium(7) with t-butyl bromide and benzyl bromide occurs selectively(99-100%) at the central carbon of the pentadienyl group. The reason why t-butyl bromide alkylates 6 at the central carbon rather than at the ends of pentadienyl group as Me_3SiCl may be because of a change in reaction mechanism from S_N2 with Me_3SiCl to single electron transfer to form a radical pair with Me_3CBr [5].

The σ,π-interaction as well as pπ-dπ interaction present in the silylated pentadienyl anion systems may be reflected in the electronic spectra as

Scheme 1.

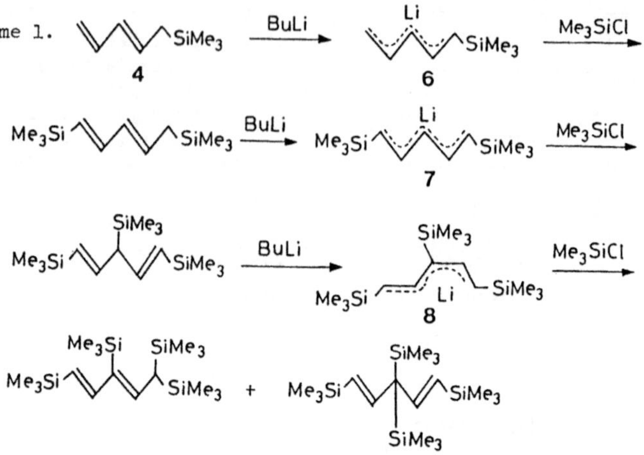

it will cause the extensive delocalization of electrons over the pentadienyl group. As shown in Fig.3, the absorption maximum shifts to the higher wave length in increasing the silyl substitution. Thus, expected extensive σ,π-conjugation in C-C-Si bonds has been detected using a series of oligo-silylated pentadienyl anions.

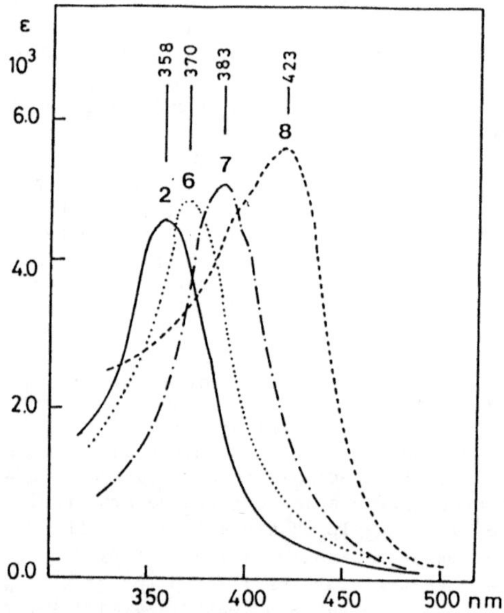

Fig. 3. Absorption spectra of silylated pentadienyllithium compounds in THF(20°)

2; C_5H_7Li
6; $C_5H_6(SiMe_3)Li$
7; $C_5H_5(SiMe_3)_2Li$
8; $C_5H_4(SiMe_3)_3Li$

We have performed ab initio calculations on a series of pentadienyl anions and their Li^+ and K^+ complexes under the STO-3G level [9]. The three limiting geometries of the metal-free anion, U-, S-, and W-shapes, were fully optimized by the gradient technique to find that planarity was retained in each case with short-long-long-short C-C bonds. Evidently the important stabilization occurs through a π-electron delocalization in the planar carbon skeleton. As far as π interactions go, the U-shape should be most appropriate [10]. However, according to our calculations the sterically less crowded

W-shape was found to be more stable than the S or U shape by 1.7 or 4.9 Kcal/mol, respectively. The MNDO calculations by Dewar et al. gave nearly the same energetics, in favor of W [11].

Figure 4 summarizes the fully-optimized structures of Li^+ and K^+ complexes of pentadienyl in the three conformations. When coordinated to these naked metal ions, pentadienyl tends to assume a U-shape as its most

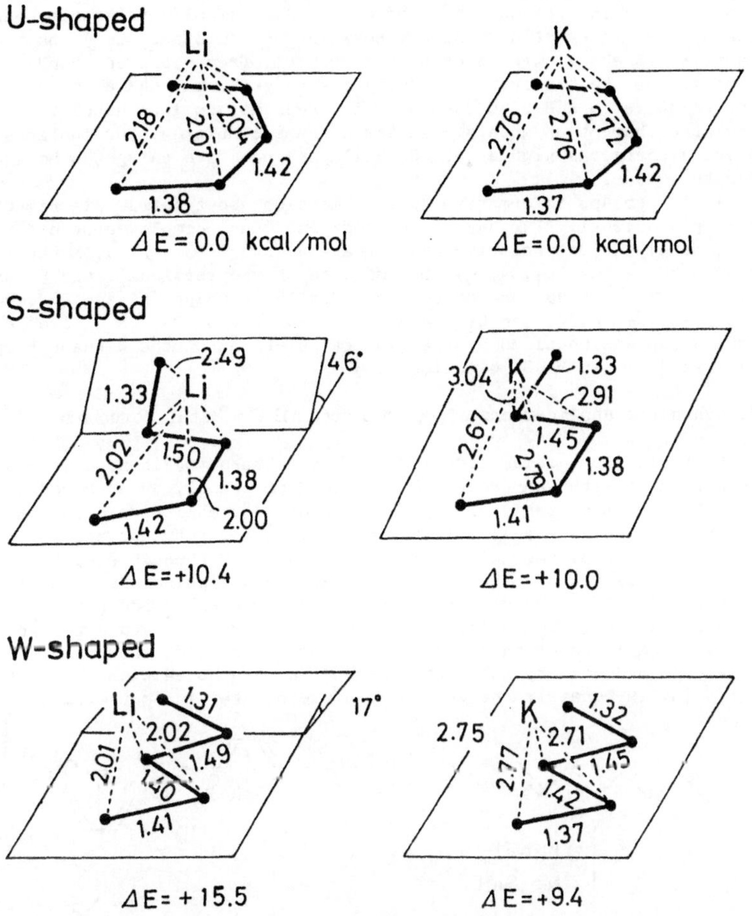

Fig. 4. Optimized STO-3G calculation on C_5H_7Li and C_5H_7K

stable form. The carbon chain remains fairly flat, and the ions settle themselves at positions approximately equidistant from the carbon atoms; Li-C=2.04-2.18Å and K-C=2.72-2.76Å. Thus even a small Li^+ ion does not move to an η^3-coordination site, and instead the symmetrical η^5-coordination geometry is preferred. It should be noted here that our calculations do not include solvation effects. The W-shaped structure was suggested experimentally for the Li^+ complex and this might be a consequence of solvent ligation to Li^+.

For the less stable S and W conformations, a metal ion is in principle bound to pentadienyl in an η^3 manner, giving rise to a localized double bond character at the terminal C-C bond. The resulting structures are, however, appreciably different between the K^+ and Li^+ complexes. In the case of K^+, the planarity of the pentadienyl portion is again retained, while for Li^+ this does not hold true. The terminal carbon in the Li^+ compelx moves toward the metal for the S-shape or moves away from the metal for the W-shape. The deviation from the planarity is more significant in the former conformation [12]. The presence of such an additional Li-C bonding interaction must be the reason the S-shape is 4.9 Kcal/mol more stable than the W-shape. The interesting contrast between the nonplanar (Li^+) and the planar (K^+) S-shape structures arises probably from different bonding capability of the two cations, namely a stronger ionic character of the K-C bonds relative to the Li-C bonds. This trend may be incarnated in the calculated charges on metals. In the S-shape complexes, for instance, K carries a charge as high as +0.62, while the positive charge on Li is moderate +0.06e.

It is perhaps informative here to mention about charge distribution within pentadienyl. For any of the conformations, complexed or not with alkali metal cation, negative charges are always accumulated on the 1, 3, and 5 carbon atoms (C_1,C_3,C_5), but more so on the terminal 1 and 5 carbons (C_1,C_5). Exemplification may be those for the U-shaped Li(and K) complexes, be they -0.16e(-0.24e) on $C_{1,5}$ and -0.14e(-0.22e) on C_3. A similar charge distribution was obtained by the CNDO calculations on the W-shape complex of Li solvated by two Me_2O molecules [13].

2.2 Synthetic Applications of Pentadienyl Alkali Metal Compounds

Pentadienylpotassium gave 1,3,7,9-decatetraene(9) in good selectivity by treating it with anhydrous CuBr or $NiCl_2$ presumably through reductive elimination from pentadienylcopper or nickel [3], while isolated bis(2,4-pentadienyl)zinc gave a 1:1 mixture of 1,3,7,9-decatetraene and 3-vinyl-1,5,7-octatriene by thermolysis in benzene at 50°C through radical coupling [7]. In the similar procedure, cyclooctadienylpotassium afforded bis-(cyclooctadienyl). The resulting tetraenes serve as a good precursor for semiconducting organometallic compounds. Tricarbonyliron and tricarbonyl-ruthenium complexes of these tetraenes, for example 10, show fairly good electrical conductivity as a compressed disk when doped with iodine (eq.4) [14]. 1,3,7,9-Decatetraene serves as a useful starting material for

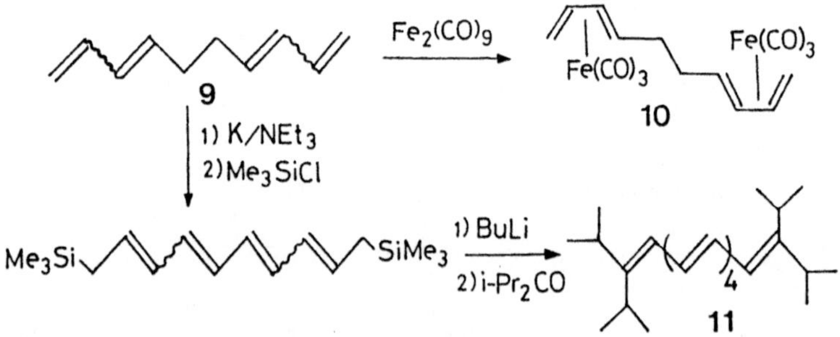

(4)

(5)

synthesis of highly conjugated polyenes. The tetraene was converted by lithiation/trimethylsilylation sequences to (E,E,E,E)-1,10-bis(trimethylsilyl)-2,4,6,8-decatetraene and the product was led to "all-trans" hexaene (11) by further lithiation followed by the addition of a bulky ketone, which undergoes the Peterson elimination (eq.5) [15].

Transmetalation of pentadienylalkaline metals has attracted recent attention due to its great utility in both organic and inorganic syntheses. A typical example of the transmetalation is summarized in Scheme 2. 2,4-

Pentadienylsilane or -stannane(12) undergoes the addition reaction with carbonyl compounds selectively at the C-Si bond by the assistance of a Lewis acid catalyst to give 3,5-hexadienol derivatives upon hydrolysis [16]. On the contrary, pentadienyl magnesium, -zinc and -beryllium(13) led to 2-vinyl-3-butenol derivatives as a result of the preferential attack of the

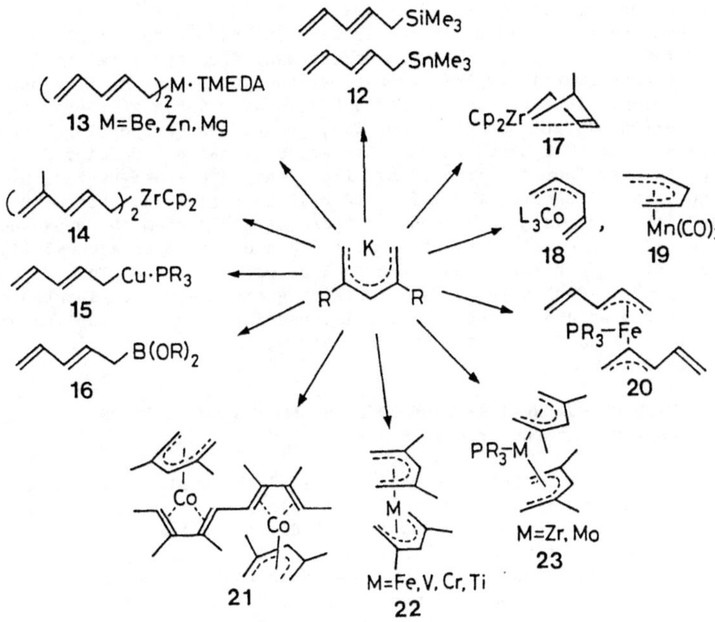

Scheme 2. Typical transmetalation reaction of pentadienylpotassium and its derivatives

central carbon of the pentadienyl group to the carbonyl compound. Thus, a high regio-control has been achieved by an appropriate choice of metals. It should be noted that pentadienylzinc compounds readily react with carbonyl compounds at 25°C whereas dialkylzincs are generally completely inert even at 70°C. Oxidation of pentadienylborane(16) with H_2O_2 provides a good route to obtain a series of 2,4-pentadien-1-ol [8]. Despite their great thermal instability, organocopper compounds are frequently used in modern organic synthesis because of their high stereo- and regio-selectivity [17]. Pentadienylcopper·PBu_3 complex(15) was found to exhibit an enhanced thermal stability and hence its ^1H NMR spectrum was readily available. The complex was found to prefer the (E,E)-geometry while undergoing rapid 1,3-rearrangement at 30°C [9].

An extensive series of pentadienyl transition metal compounds as well as pentadiene complexes have been synthesized starting from pentadienyl anions in the last five years. Hydrozirconation of pentadienylpotassium and its higher analogs with $ZrCl(H)Cp_2$ afforded a series of s-cis- or s-trans-pentadien-zirconium complexes(17) [18], which undergoes the double insertion of aldehydes or ketones at the diene termini to lead to (E)-3-hexene-1,6-diol derivatives, while the reaction with $ZrCl_2Cp_2$ gave $ZrCp_2$(pentadienyl)$_2$ (14) of (E)-geometry.

A variety of open metallocenes have been prepared by Ernst et al. stating from 2,4-dimethylpentadienylpotassium, $C_7H_{11}K$ [19]. The geometry of $M(C_7H_{11})_2$(M=Fe,V,Cr), 22, is very close to the molecular structure of the corresponging ferreocene, vanadocene, or chromocene. Their two pentadienyl units rotates ca. 50-80° (cis-eclipsed orientation defines 0° angle of rotation and the anti-eclipsed orientation 180° rotation). $M(C_7H_{11})_2 \cdot PR_3$(Mn,Zr,Mo),23, however, preferred the cis-eclipsed orientation [20].

Similar η^5-coordinated 2,4-dimethylpentadienylcobalt complex, (η^5-C_7H_{11})Co(PEt_3)$_2$, is also available [21] but the use of PMe_3 instead of PEt_3 resulted in the formation of [(η^3-anti-C_7H_{11})Co(PMe_3)$_3$]$^+$(18). The five carbon atoms are not coplanar in the latter complex and the vinyl moiety is bent out of the plane of the allylic moiety away from the metal by 37°. A η^3-syn-pentadienylmetal complex was also synthesized starting from pentadienylpotassium. The reaction of $FeCl_2(PMe_3)_3$ with C_5H_7K gave syn,syn-bis-(η^3-C_5H_7)Fe·(PMe_3)$_2$(20) where the two C_5H_7 groups assume the staggered orientation [22]. The C_1,C_2,C_3,C_4 atoms are coplanar but C_5 lies 0.26Å out of the plane away from the iron atom. The S-shaped η^3-anti-pentadienyl structure is also reported for Mn(CO_3)(C_5H_7)(19) [23]. The mode of the reaction with 2 equivalents of $CoCl_2$ differs markedly from the above noted reactions [24]. A coupling reaction of the 2,4-dimethylpentadienyl ligand occurred during the reaction and gave a binuclear cobalt complex, Co_2(η^5-C_7H_{11})$_2$(2-5-η^4, 6-9-η^4-$C_{14}H_{22}$)(21), containing bis(2,4-dimethylpentadienyl). This will serve as a good model of the intermediate for the coupling reaction of pentadienyl ligand catalyzed by CuBr, $NiCl_2$, or $CoCl_2$ affording 1,3,7,9-decatetraene.

3. PREPARATION OF BUTENE-1,4-DIYLMAGNESIUM AND ITS APPLICATION

3.1. Preparation and the structure

2-Butene-1,4-diylmetal compound and its higher analogs display one of the most interesting organometallic functionalities. We have succeeded in the preparation of a series of 2-butene-1,4-diylmagnesium by activation of magnesium using a mixture of an alkyl halide and a diene [25]. The procedure reported in the patent literature by Ramsden is useless for the present purpose because it forms magnesium-diene 1:2 and 1:3 adducts as major product [26]. More recently we have developed a more convenient method for preparation of the magnesium-diene 1:1 adduct in pure form. The typical procedure is as follows. Commercially available magnesium powder, flake or chips are treated with 5 mol% of phenyliodide in THF at 25°C and the resulted solution and suspension containing PhMgBr was removed by a syringe under an argon atmosphere. The residual gray magnesium turnings are activated by this procedure. After the addition of a THF solution of butadiene, isoprene, 2,3-dimethylbutadiene or 1,4-diphenylbutadiene (0.3 mol per magnesium), the mixture is stirred at 10-15°C for 3-7 days until the yellow or orange 1:1 adduct precipitates. Temperature control as well as the use of thoroughly dried THF is essential to obtain the pure 1:1 adduct. Raising the temperature to 25°C always brings about contamination by the 1:2 or 1:3 adduct. Free dienes remaining in the system are removed by trap-to-trap distillation. The resulting Mg-butadiene 1:1 adduct is sparingly soluble in THF but isoprene, 2,3-dimethylbutadiene and 1,4-diphenylbutadiene adducts are soluble in THF. Repeated precipitation by pouring them into a mixture of hexane and THF gives pure 2-butene-1,4-diylmagnesium derivatives in good yield. As an alternative method of activation of magnesium, Vogdanovic et al. reported the efficiency of anthracene as an activator [27] but such a procedure does not contribute effectively in acceleration of the reaction in this case. The rate of reaction is almost independent of the shape of the magnesium turnings. To evaluate the extent of the activity of the magnesium, the reaction of the magnesium with cyclopropylmethyl bromide followed by the addition of i-butanal was examined. Highly active magnesium prepared by condensation of magnesium vapor in THF is already known to react readily at low temperatures with cyclopropylmethyl bromide without ring opening of the cyclopropyl group, while magnesium turnings activated by the conventional method with iodine always result in the ring opening during the reaction [28]. The magnesium turnings prepared by the present procedure react with cyclopropylmethyl bromide at -40°C to give, after the addition of i-butanal, 3-methyl-1-cyclopropylbutan-2-ol(24) in 40% selectivity (eq.6). Thus the magnesium activated by the present method was found to possess an intermediate reactivity between the metal vapor condensed magnesium and the conventional magnesium turnings.

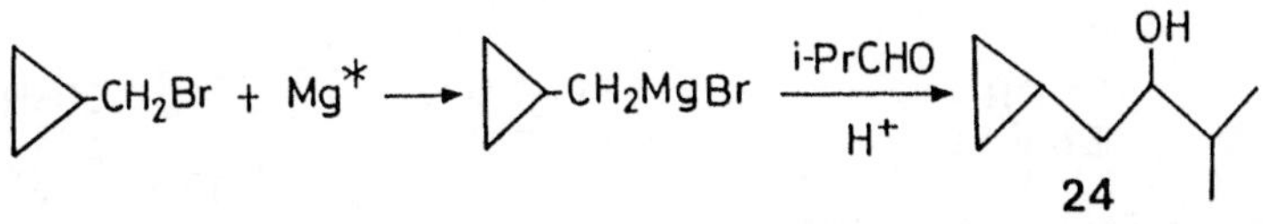

(6)

The molecular weight of 2-methyl-2-butene-1,4-diylmagnesium, $Mg(C_5H_8)\cdot(THF)_2$ determined ebulioscopically in THF together with NMR studies indicates the presence of the adduct as a cyclic pentamer(25). For the magnesium-isoprene 1:2 adduct, $Mg(C_{10}H_{16})(THF)_2$, prepared from the reaction between magnesium and isoprene (1:2 ratio) at 45°C, a cyclic trimer structure(26) has been proposed [25a]. A similar cyclic structure has been reported for o-xylenediylmagnesium [29] and butane-1,4-diylmagnesium [30]. In sharp contrast to the macrocyclic structure observed for the alkyl substituted 2-butene-1,4-diylmagnesium, $Mg(1,4\text{-diphenylbutadiene})(THF)_3$ prefers the monomeric structure as confirmed by the X-ray analysis (Fig.5) [31a]. The metal is five coordinated and the 2-butene moiety has the s-cis gemetry. The corresponding lithium derivative, $(1,4\text{-diphenylbutadiene})Li_2(TMEDA)_2$, involves two lithium atoms which are displaced to opposite sides of the s-cis-diene plane [31b].

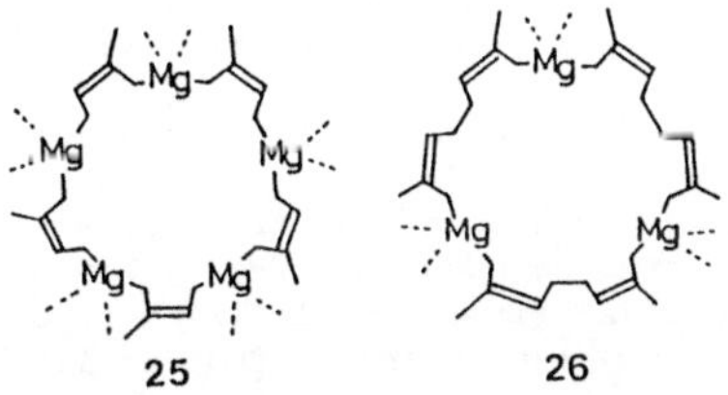

25 26

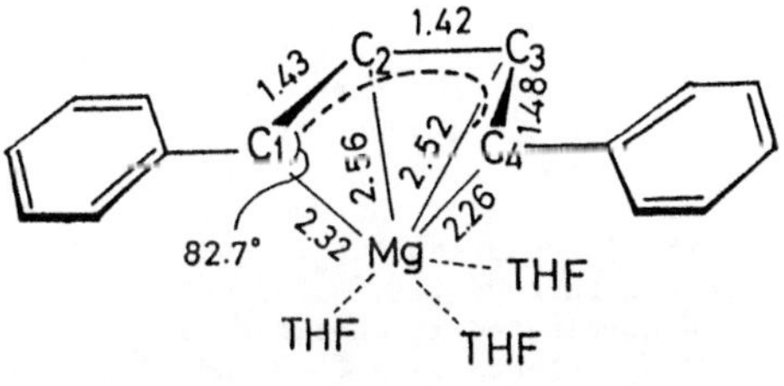

Fig. 5. X-ray parameters of $Mg(PhCH{=}CHCH{=}CHPh)\cdot(THF)_3$

3.2. Synthetic Applications of Butenediylmagnesium

2-Methyl-2-butene-1,4-diylmagnesium shows a high catalytic activity in the oligomerization of isoprene while dialkylmagnesium compounds show relatively low catalytic activity. It is worth noting that the degree of oligomerization parallels the initial molar ratio of isoprene to the 2-methylbutenediylmagnesium until the ratio reaches to 15 [25]. The oligomer involves the 3,4-polyisoprene unit exclusively. Addition of 100 mol of isoprene to the catalyst resulted in the formation of 3,4-polyisoprene having ca. 126 monomer units. An attempt to prepare macrocyclic 3,4-polyisoprene by treatment of the resulting magnesium-polyisoprene adduct with $NiBr_2(PR_3)_2$ or other metal halides has failed at present because the product comprises a complex mixture.

Trimethylsilylation of 2-methyl-2-butene-1,4-diylmagnesium with one equivalent of chlorotrimethylsilane gave 4-silylated-2-methyl-2-butenylmagnesium chloride(27) in 86% regioselectivity. The selectivity could be improved by the use of t-butyldimethylchlorosilane up to 95% [29]. The subsequent reaction with aldehydes, ketones, oxiranes and CO_2 undergoes the insertion reaction at the C_3 position to give unsaturated alcohols or an acid in good yield (Scheme 3).

Scheme. 3.

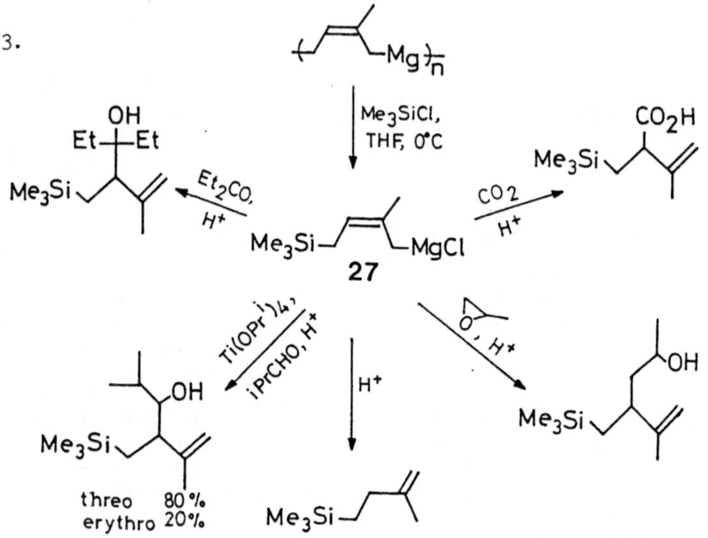

Transmetalation of butenediylmagnesium opened a new avenue in the area of early transistion metal chemistry. Thus a variety of diene complexes of group 3(Th,U) [32], group 4(Ti,Zr,Hf) [33,34], group 5(Nb,Ta) [35], group 6 (W) [36], group 7(Mn) [37], and group 8(Fe) [38], have been synthesized using butenediylmagnesium (Scheme 4). The diene complexes of group 3-5 elements with the formulation MCp_2(diene) exhibit unique bent s-cis-metallacyclo-3-pentene structure (28,29,33) or novel s-trans-diene structure(28b). The s-trans-coordinated mononuclear diene structure was first revealed with $ZrCp_2$-(1,4-diphenylbutadiene [39] and $ZrCp_2$(butadiene) [40] prepared from Mg-1,4-diphenylbutadiene 1:1 adduct or a photochemical reaction. X-ray analysis clearly confirms the geometrical change of the 1,4-diphenylbutadiene unit from s-cis to s-trans during transmetalation from Mg to Zr(Fig.6). Compare with the molecular structure of the starting material(Fig.5). Especially noteworthy is the high regio- and stereoselectivity observed in the reaction of $ZrCp_2$(s-cis-isoprene), which attacks aldehydes, ketones, esters, nitriles, CO_2 and oxirans at the C_1 position of isoprene unit, whereas it reacts with alkenes, dienes and alkynes at the C_4 atom selectively [33]. Three and five membered ring phosphines(34) were first prepared by using 2-butene-1,4-diylmagnesium [41]. Metallacyclo-3-pentenes containing Si(31) [42] and B(32) [43] have also been prepared starting from the butenediylmagnesium.

4. Concluding Remarks

By the advent of a variety of pure pentadienyl anions and butenediylmagnesium compounds, a series of organic and organometallic compounds containing pentadienyl or butenediyl units becomes readily accessible. Structural elucidation of these precursors by spectral and MO studies together with transmetalation sequences has clarified the selective reaction pathways.

Scheme. 4

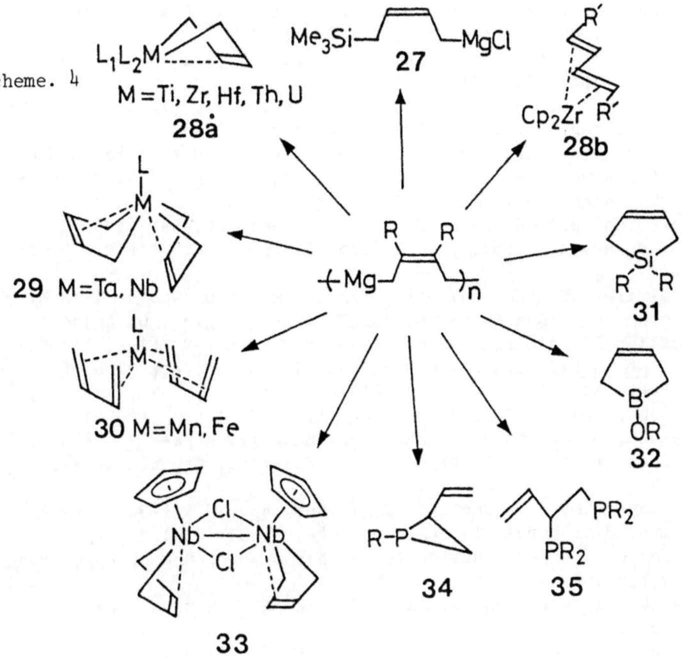

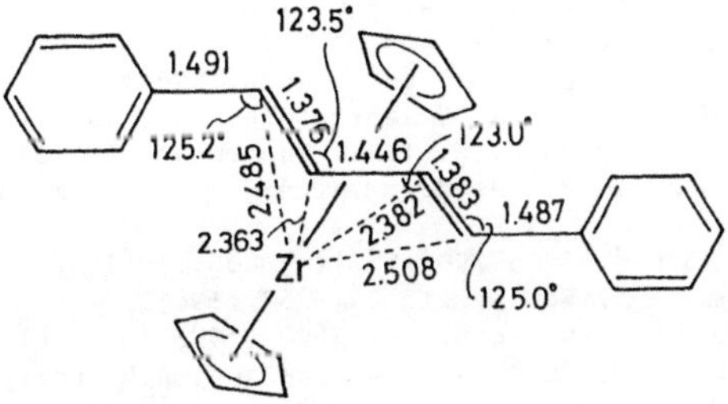

Fig. 6. X-ray structural data of $ZrCp_2(PhCH=CHCH=CHPh)$

Acknowledgement

Molecular orbital calculations were done at the Computer Center of IMS. We are grateful to Professor T. Fueno of Osaka University for valuable discussions.

REFERENCES AND NOTES

1. a) R.B. Bates, C.A. Ogle, Carbanian Chemistry, Springer-Verlag, Heidelberg, 1983. b) D. Devaprabhakara, C.G. Cardenas, P.D. Gardner, J. Am. Chem. Soc. 85, 1553 (1963).
2. H. Yasuda, Y. Ohnuma, M. Yamauchi, H. Tani, Bull. Chem. Soc. Jpn. 52, 2036 (1979).
3. H. Yasuda, M. Yamauchi, Y. Ohnuma, H. Tani, Bull. Chem. Soc. Jpn. 54, 1481 (1981).
4. R.B. Bates, D.W. Gosselink, J. A. Kaczynski, Tetrahedron Lett. 1967, 199 and 205.
5. H. Yasuda, R. Nishi, A. Nakamura, Organometallics 2, 21 (1983).

6. H. Yasuda, M. Yamauchi, A. Nakamura, T. Sei, Y. Kai, N. Yasuoka, N. Kasai, Bull. Chem. Soc. Jpn. 53, 1089 (1980).
7. H. Yasuda, Y. Ohnuma, A. Nakamura, Y. Kai, N. Yasuoka, N. Kasai, Bull. Chem. Soc. Jpn. 53, 1101 (1980).
8. M. Schlosser, G. Rauchschwalbe, J. Am. Chem. Soc. 100, 3258 (1978).
9. The IMS library package "IMSPAK" (Gaussian 70) was used for the calculations. The basis functions for Li, C, and H are standard ones and those for K were taken from, W.J. Pietro, B.A. Levi, W. J. Hehre, R.F. Stewart, Inorg. Chem. 19, 2225 (1980).
10. R. Hoffmann, R.A. Olofson, J. Am. Chem. Soc. 88, 943 (1966).
11. M.J.S. Dewar, M.A. Fox, D.J. Nelson, J. Organomet. Chem. 185, 157 (1980).
12. a) A similar, but not quite the same, distortion was derived by ab initio calculations on the Li complex. A. Bongini, G. Cainelli, G. Calrdillo, P. Palmieri, A. Umani-Ronchi, ibid. 92, C1-C3 (1975).
b) For ab initio calculations on the Li complex, see also, R.J. Bushby, A.S. Patterson, Ibid, 132, 163 (1976).
13. J.F. Sebastian, B. Hsu, J.R. Brunwell, ibid. 105, 1 (1976).
14. H. Yasuda, S. Miyanaga, A. Nakamura, Macromolecules 17, 2453 (1984).
15. H. Yasuda, T. Nishi, S. Miyanaga, A. Nakamura, Organometallics 4, 359 (1985).
16. a) D. Seyferth, J. Pornet, J. Org. Chem. 45, 1721 (1980). b) A. Hosomi, M. Saito, H. Sakurai, Tetrahedron Lett. 3783 (1980).
17. G.H. Posner, An Introduction to Synthesis Using Organocopper Reagents, John Wiley & Sons, 1980.
18. H. Yasuda, K. Nagasuna, M. Akita K. Lee, A. Nakamura, Organometallics 3, 1470 (1980).
19. R.D. Ernst, Struct. Bonding (Berlin), 57, 1 (1984).
20. L. Stahl, J.P. Hutchison, D.R. Wilson, R.D. Ernst, J. Am. Chem. Soc. 107, 5016 (1985).
21. J.R. Bleeke, W.-J. Peng, Organometallics 5, 635 (1986).
22. R. Bleeke, M.D. Hays, Organometallics 3, 506 (1984).
23. M.A. Paz-sondval, P. Powell, Organometallics 4, 1026 (1984).
24. D.R. Wilson, R.D. Ernst, S. K. Michael, Organometallics 3, 1442 (1984).
25. a) H. Yasuda, Y. Nakano, K. Natsukawa, H. Tani, Macromolecules 11, 586 (1978). b) K. Fujita, Y. Ohnuma, H. Yasuda, H. Tani, J. Organometal. Chem. 113, 201 (1976).
26. H.E. Ramsden, U.S.pat. 3 35 646, 3 35 190(1967), 3 388 179(1968).
27. B. Bogdonovic, Angew. Chem. Int. Ed. Engl. 24, 262 (1985).
28. E.P. Kundig, C. Perret, Helv. Chim. Acta. 64, 2606 (1981).
29. M.F. Lappert, T.R. Martin, C. L. Raston, B. W. Skelton, A.H. White, J. Chem. Soc. Dalton, 1959 (1982).
30. H.C. Holtkamp, G. Schat, L. Blomberg, F. Bickelhaupt, J. Organometal. Chem. 249, 1 and 217 (1982).
31. a) H. Kai, N. Kanehisa, K. Miki, N. Kasai, K. Mashima, H. Yasuda, A. Nakamura, Chem. Lett. 1277 (1982). b) A.J. Kos, P. Stein, von R. Schleyer, J. Organometal. Chem. 280, C5 (1985).
32. G.M. Smith, H. Suzuki, D.C. Sonnenberg, V.W. Day, T.J. Marks, Organometallics 5, 549 (1986).
33. H. Yasuda, K. Tatsumi, A. Nakamura, Acc. Chem. Res. 18, 120 (1985).
34. G. Erker, C. Krüger G. Muller Adv. Organometal. Chem. 24, 1 (1985).
35. H. Yasuda, K. Tatsumi, T. Okamoto, K. Mashima, K. Lee, A. Nakamura, Y. Kai, N. Kanehisea, N. Kasai, J. Am. Chem. Soc. 107, 2410 (1985).
36. R. Benn, G. Schroth, J. Organometal. Chem. 228, 71 (1982).
37. R.J. Harlow, K.J. Krusic, R.J. McKinney, S.S. Wreford, Organometallics 1, 1506 (1982).
38. J.M. McCall, J.R. Martin, K.F. Prestone, Organometallics 3, 238 (1984).
39. Y. Kai, N. Kanehisa, K. Miki, N. Kasai, K. Mashima, K. Nagasuna, H. Yasuda, A. Nakamura, J. Chem. Soc. Chem. Commun., 191 (1982).
40. G. Erker, J. Wicher, K. Engel, F. Rosenfeldt, W. Dietrich, C. Krüger,

J. Am. Chem. Soc. 102, 6346 (1980).
41. W.J. Richter, Chem. Ber. 116, 3239 (1983).
42. W.J. Richter, Angew. Chem. Int. Ed. Engl. 21, 292 (1982).
43. G.E. Herberich, B. Hessner, D. Soenen, J. Organometal. Chem. 233, C37 (1982).

ION PAIR STRUCTURE AND PHOTOCHEMISTRY OF PARTIALLY METALLATED POLY(2-VINYLFLUORENE) IN SOLUTION

MINORU MATSUDA* and AKIRA WATANABE[1]
Chemical Research Institute of Non-aqueous Solutions, Tohoku University, Katahira 2-Chome, 980 Japan

ABSTRACT

The ion pair structure of partially metallated poly(2-vinylfluorene) is studied by means of electronic absorption and fluorescence spectroscopy. The degree of metallation along the polymer chain influences the ion pair structure effectively. In the ground state the conversion of contact ion pairs into solvent-separated ion pairs along the polymer chain is depressed by electrostatic interaction between the ion pairs. The contact ion pairs with a bulky counter cation form aligned pendant fluorenide chromophores. In the excited state the steric effect of counter cation in the polymer chain hinders the rearrangement of the fluorenide chromophores after Franck-Condon transition and this effect is more remarkable with large counter cations. It is suggested that these polymer effects, electrostatic and steric effects, influence the conformation of the polymer chain. An intramolecular energy migration along the polymer chain is also studied.

INTRODUCTION

The reactivity of alkali metal salts of carbanions in solution is governed mainly by the structure of the ion pairs as shown clearly in the anionic polymerization of vinyl monomers and in electron transfer reactions in ethereal solutions (2-4). The ion pair structure in the ground state is determined by the alkali metal counter cation, solvent and temperature, as demonstrated by Smid et al. from electronic absorption spectra (5) and conductance (6) data for the alkali metal salts of fluorenyl carbanion in non-aqueous solutions. The ion pair structure is also greatly affected by the nature of the neighboring group in the carbanion (7-9). When carbanion pairs are bound to a polymer chain, their structure appears to depend on the degree of metallation along the polymer chain (10).

Smid et al. (11) studied the metallation of poly(2-vinylfluorene) in THF and 1,2-dimethoxyethane (DME) and suggested that the ion pair structure of carbanion pairs existing along the polymer chain is significantly influenced by electrostatic and steric effects in the salts. This may be understood from a representative illustration (Fig. 1). The solvent-separated ion pairs exist in equilibrium with the contact ion pairs along the polymer chain. The interactions between the different types of ion pair, and hence the shape of the polymer chain, is expected to depend on the degree of metallation along the polymer chain (Scheme 1).

The behavior of the excited state of the alkali metal salts of carbanions has been studied for fluorenyl carbanions by Hogen-Esch and Plodinec (12-14). The ion pair structure in the first excited state is also affected by the alkali metal counter cation, solvent, and temperature , in a similar manner as that of the ground state (5).

In the present work we have determined the effect of the degree of metallation of the repeating units in a poly(2-vinylfluorene) chain (Scheme 1) on ion pair structure in the ground (15) and excited states (16), and an intramolecular energy migration along the polymer chain (17) will be discussed.

Recent Advances in Anionic Polymerization
T.E. Hogen-Esch and J. Smid, Editors

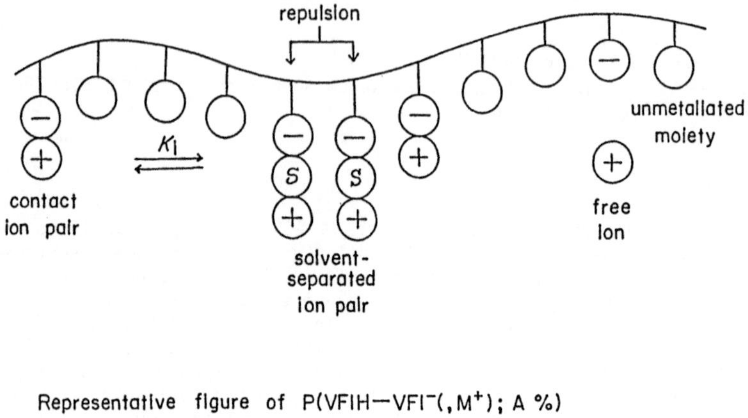

Fig. 1 A representative figure of partially metallated poly(2-vinylfluorene) in non-aqueous solution.

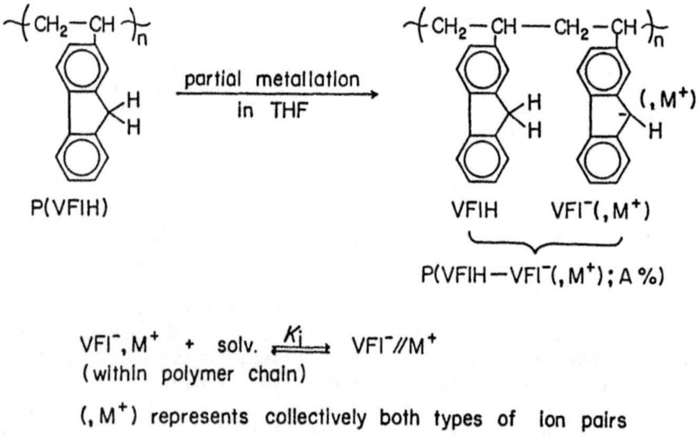

Scheme 1 Equilibrium between contact and solvent-separated ion pair in partially metallated poly(2-vinylfluorene).

EXPERIMENTAL

Poly(2-vinylfluorene) P(VFlH) was prepared by cationic polymerization of 2-vinylfluorene (18) with borontrifluoride etherate as catalyst in dichloromethane at -30 °C (19). The polymer was purified by several precipitations from benzene with petroleum ether. The molecular weight and its distribution were determined by gel permeation chromatography GPC using a Toyo Soda HLC-802 UR instrument equipped with a low angle light scattering photometer (LALS, Toyo Soda LS-8); $\bar{M}n$ = 89000 and $\bar{M}w/\bar{M}n$ = 1.43.

The alkali metal salts of vinylfluorene units $(VFl^-(,M^+))$ along the P(VFlH) chain were prepared in purified THF under vacuum (10^{-5} torr) by treatment with the alkali metal salts of 1,1,4,4-tetraphenylbutane dianion $(M^+,)^-DD^-(,M^+)$ or by metallation with alkali metals. The proportion of remaining unmetallated VFlH units (100 - A %) in the polymer was determined as follows. In the presence of excess dianion salt $(M^+,)^-DD^-(,M^+)$ the remaining VFlH units are converted into ion pairs yielding $P(VFl^-(,M^+)$; 100 %) and H-DD-H. The amount of $(M^+,)^-DD^-(,M^+)$ consumed is then determined from its known extinction coefficient (20) and hence A % is found. Smid et al. (11) reported that the extinction coefficient of $P(VFlH\text{-}VFl^-(,M^+)$; A %) decreases with increasing A %, and suggested that this is most probably the result of hypochromism caused by the close proximity of the fluorenyl chromophores. In our experiments the decrease in ε was less than 10 % when A % less than 30 %. The following values of ε ($dm^3 mol^{-1} cm^{-1}$) were used; Li^+ salt (solvent-separated), ε_{375} = 11000; Na^+ salt, ε_{360} = 11000, K^+ salt, ε_{365} = 10000; Cs^+ salt, ε_{365} = 10000. When A % was higher than 30 %, ε values were determined individually from the decrease in concentration of the dianion.

Electronic absorption spectra were measured on a Cary-14 spectrophotometer. Fluorescence emission spectra of $P(VFlH\text{-}VFl^-(,M^+)$; A %) were recorded on a Shimadzu 501 spectrophotometer. All solutions were degassed and excited at 305 nm. The quantum yields for both the monomer and excimer emissions were determined using a 0.01 M quinine sulfate aqueous solution as a standard.

RESULTS AND DISCUSSION

1. Electronic Absorption Spectra

The electronic absorption spectra in THF for a partially sodium metallated poly(2-vinylfluorene) and its monomer model, i.e., the sodium fluorenide $Fl^-(,Na^+)$, are depicted in Fig, 2. The shapes of the spectra for the polymer ion pairs are entirely similar to those of the monomeric ion pairs. The bands appearing at 375 nm were assigned to the solvent-separated ion pair and the bands at 360 nm to the contact ion pair by analogy with the spectrum of $Fl^-(,Na^+)$ in THF (5). On lowering the temperature from -35 °C, the absorption band due to the solvent-separated ion pair increases relative to that due to the contact ion pair. The equilibrium constant K_i between the contact ion pair and the solvent-separated ion pair can be determined for polymers of varying degrees of metallation from the ratio of the height of the absorption bands.

$$P(VFlH\text{-}VFl^-,Na^+;\ A\ \%) \xrightleftharpoons{K_i} P(VFlH\text{-}VFl^-/\!/\,Na^+;\ A\ \%)$$

An isosbestic point appears in Fig. 2. However, the spectrum at -80 °C did not pass through on the isosbestic point. This is due to an increase of extinction coefficient at very low temperature. In the case of the sodium salt, the positions of the absorption bands do not shift with increasing carbanion percentage A (%), although in Fig. 2 the spectra are shown only for a 9 % metallated polymer carbanion. The K_i values estimated for the

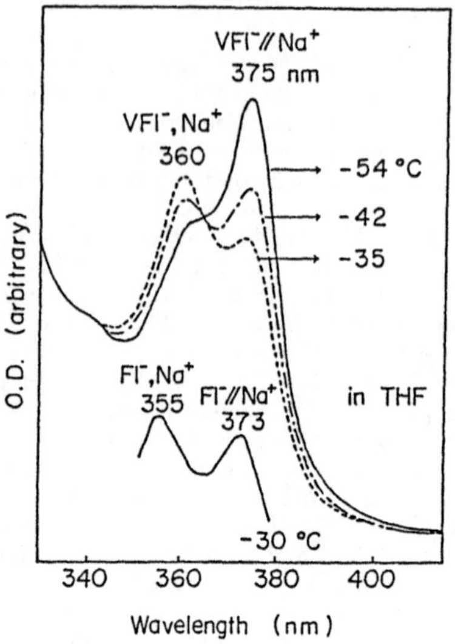

Fig. 2 Electronic absorption spectra for 9 % sodium metallated poly(2-vinylfluorene) and the monomer model (fluorenyl sodium) in THF.

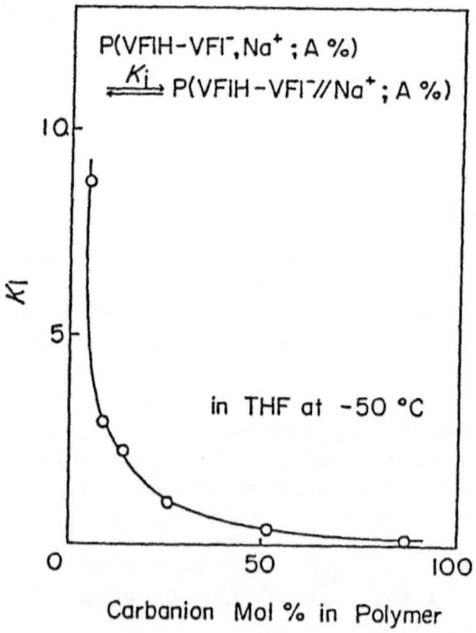

Fig. 3 The K_i values for the sodium salt as a function of carbanion mol % in the polymer in THF at -50 °C.

sodium salt in THF at -50 °C are plotted as a function of carbanion mol % in Fig. 3; the K_i values decrease strongly with increasing carbanion %. This indicates that the conversion of a contact ion pair into a solvent-separated ion pair along the polymer chain is influenced by the mutual interaction between the solvent-separated ion pairs. The enthalpy changes $-\Delta H_i^\circ$ calculated from the slopes of the straight lines in a Van't Hoff plot (Fig. 4) are plotted in Fig. 5 as a function of A %, for this and several other cation/solvent systems. In every case, $-\Delta H_i^\circ$ decreases sharply with increasing A % to ca. 25 %. The trend of the decrease becomes small above ca. 25 %. It must be attributable to a chain expansion leading to an increase of a distance between chromophores along the polymer chain. These results may be ascribed to destabilization of the solvent-separated ion pairs by an electrostatic repulsive force when they are concentrated along the polymer chain. There is another explanation for the decrease of $-\Delta H_i^\circ$ that intramolecular aggregation suppresses the conversion from contact ion pair to solvent-separated ion pair. In such a case, the higher A % the larger the intramolecular interaction (aggregation), i.e., it should have decreased $-\Delta H_i^\circ$. The reverse trend appears in Fig. 5.

Entropy changes on varying A % show a similar trend as the enthalpy changes (15); the changes in $-\Delta S_i^\circ$ lessens as A % increases. This may be explained by a disordering of the orientation of the solvent molecules surrounding the alkali metal cation. The disorder of the orientation results from electrostatic repulsions among solvent-separated ion pairs along the polymer chain. The isokinetic relation (21, 22), i.e., the relationship between $-\Delta H_i^\circ$ and $-\Delta S_i^\circ$ is plotted in Fig. 6 (the values in Fig. 6 are of A %). Linear relationships are obtained in all cases, indicating that the changes in thermodynamic parameters observed with three polymer anion systems are attributable to a common origin, namely the electrostatic repulsion between the solvent-separated ion pairs in the ground state. The slopes of the straight lines give the special temperature T_c, at which all the carbanions in the polymer chain become contact ion pairs irrespective of A %. In our polymer systems, Gibbs' free energy must be expressed by $\Delta G_i^\circ = \Delta G^o + \Delta G(A)$, where ΔG^o is intrinsic free energy to fluorenyl carbanion unit and $\Delta G(A)$ stands for the change of the free energy caused by intramolecular interaction at carbanion percentage A. Similarly, $\Delta H_i^\circ = \Delta H^o + \Delta H(A)$, $\Delta S_i^\circ = \Delta S^o + \Delta S(A)$, and thus $\Delta G(A) = \Delta H(A) - T\,S(A)$. Slopes in Fig. 6 are expressed by $d\Delta H_i^o/d\Delta S_i^o = d\Delta H(A)/d\Delta S(A) = (d\Delta G(A)/d\Delta S(A)) + T = T_c$. Therefore, the following equations are derived, $d\Delta G(A) = (T_c - T)d\Delta S(A)$, and at $T = T_c$, $\Delta G_i^\circ = \Delta G^o = RT \ln K_c$ = constant for any carbanion percentage A (%). For the P(VFlH-VFl$^-$(,Na$^+$); A %) in THF, T_c is calculated to be 34 °C, for P(VFlH-VFl$^-$(,Na$^+$); A %) in THF : DME = 1 : 2, T_c = 63 °C, and for P(VFlH-VFl$^-$(,Li$^+$); A %) in THF, T_c = 87 °C.

In the case of the potassium salts P(VFlH-VFl$^-$(,K$^+$); A %), λ_{max} shifts to a shorter wavelength when A % increases from 5 to 74 % (Fig. 7). A broadening of the absorption band is also observed. These spectra were measured in THF at 22 °C; at this temperature only contact ion pairs were present. This is because of the larger ionic radius of K^+ compared with Na^+ (5). A similar behavior was observed for P(VFlH-VFl$^-$,Cs$^+$; A %). Therefore, the shift and the broadening of absorption bands of contact ion pairs may be attributed to the bulkiness of the counter alkali metal cation. As expected, this effect does not appear with smaller lithium or sodium cation, even though the relative proportion of solvent-separated ion pairs changes with A % as for the sodium salt (Fig. 3). To show this more clearly, λ_{max} are plotted as a function of A % in Fig. 8 for the potassium, sodium, and lithium salts.

The association of carbanion salts is known to occur most effectively with lithium as counter ion as has been demonstrated by Van Beylen (23). However, the lithium salt of the fluorenyl polymer carbanion does not exhibit a shift (15) in the position of λ_{max} with increasing A % (Fig. 8). Therefore, the blue shift of λ_{max} with increasing A % for the potassium salt can not be attributed to the formation of associated dimers of pendant

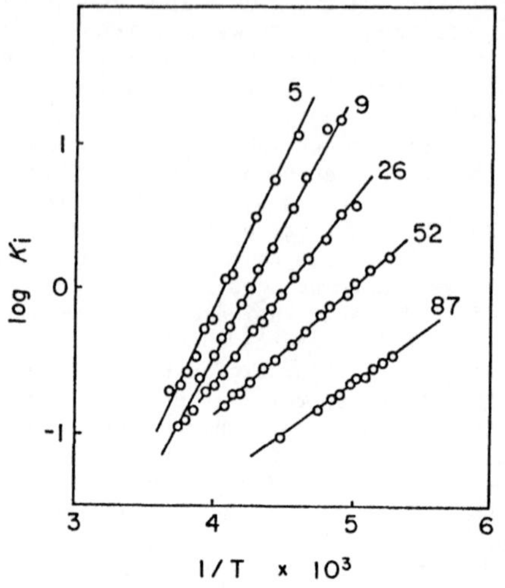

Fig. 4 Van't Hoff plots of the K_i values determined for $P(VFlH\text{-}VFl^-(,Na^+);\ A\ \%)$. The values refer to carbanion percentage A (%).

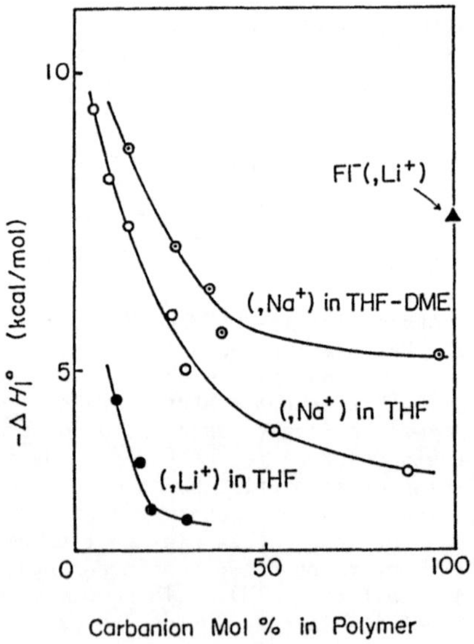

Fig. 5 The enthalpy change as a function of carbanion percentage A (%) in the polymer.

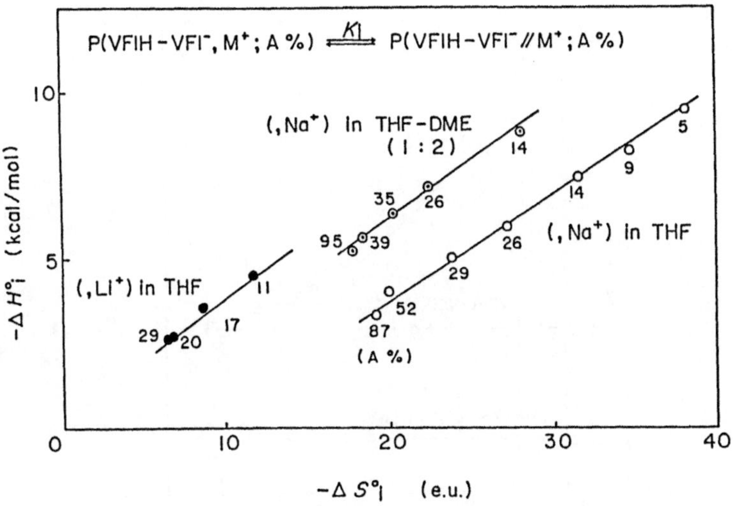

Fig. 6 The isokinetic relation. The values refer to carbanion percentage A (%).

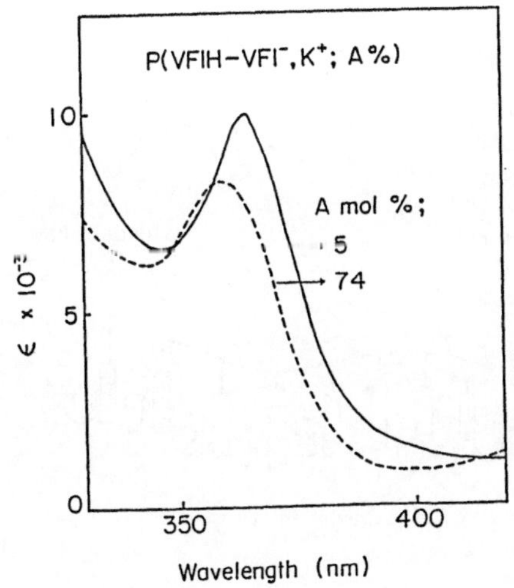

Fig. 7 Electronic absorption spectra observed for the potassium salts.

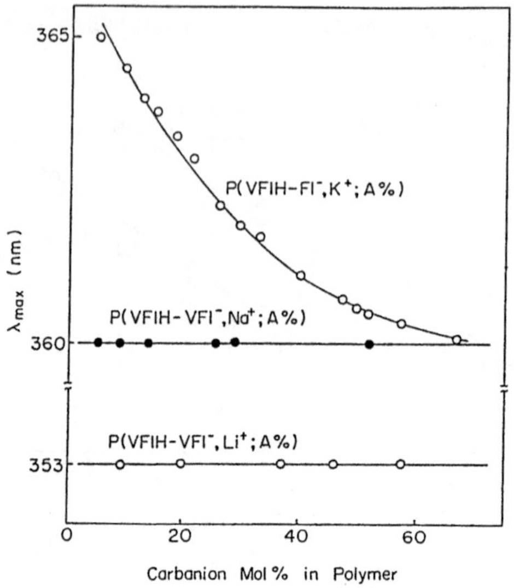

Fig. 8 The position of the absorption maxima as a function of carbanion percentage A (%) in the polymer.

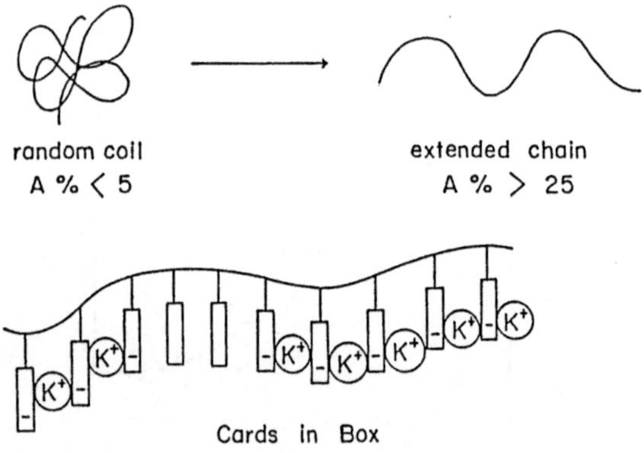

Scheme 2 A representative figure for an extended polymer with a bulky potassium cation.

potassium fluorenides along the polymer chain. The other possibility is that, when a large alkali metal cation exists in the polymer solution, the chromophores, i.e., the pendant fluorenides, are more aligned. In this case, the potassium cation is sandwiched between two neighboring chromophores. Colloquially speaking, this is something like a pack of cards in a box (24) as depicted in Scheme 2. It is known that such a configuration of chromophores, cards in box, leads to a blue shift in λ_{max} to shorter wavelength (24, 25). The shift of λ_{max} for the cesium cation (365 → 361 nm by metallation to 59 %)is not so extensive compared with that for potassium cation (365 → 360 nm by metallation from to 60 %). This result may suggest that Cs^+ cation is too large to form a sandwiched structure between neighboring chromophores; such a conformation of P(VFlH-$VFl^-(,Cs^+)$; A %) was checked by using CPK models.

In conclusion, the conversion of contact ion pairs into solvent-separated ion pairs along the polymer chain is influenced by electrostatic repulsion between the ion pairs; the higher the carbanion % in the polymer chain the more difficult is the conversion of contact ion pairs into solvent-separated ion pairs. In a low carbanion % region, the polymer anion may form a randomly coiled shape. With increasing fraction of solvent-separated ion pairs, the polymer anion may be forced into a more extended (elongated) shape. The second conclusion is that the fluorenyl chromophores are better aligned for contact ion pairs with a bulky counter cation.

2. Fluorescence Spectra

The fluorescence spectra for partially metallated poly(2-vinylfluorene)'s, P(VFlH-$VFl^-(,M^+)$; A %), in THF appear over a wavelength range of 480 - 700 nm accompanying a vibration structure. A mirror image relation between the emission and absorption spectrum is not observed. The emission of monomer models of P(VFlH-$VFl^-(,M^+)$; A %), i.e., the alkali metal salts of the fluorenyl carbanion $Fl^-(,M^+)$, have been studied in low dielectric constant solvents in detail (12 - 14). In comparison with the spectra for P(VFlH-$VFl^-(,M^+)$; A %), $Fl^-(,M^+)$ displays a rather simple emission spectrum the shape of which is not changed significantly by a change in counter cation, solvent, or temperature. The difference in spectra between $Fl^-(,M^+)$ and P(VFlH-$VFl^-(,M^+)$; A %) becomes significant on increasing the counter cation radius and A %. In $Fl^-(,M^+)$, the emission maximum of the solvent-separated $Fl^-//M^+$ ion pairs appears at a lower wavelength than that of the contact ion pairs, and a blue shift has been observed on increasing the counter cation radius. This trend is the reverse of what is observed in the absorption spectrum for $Fl^-(,M^+)$ and can be understood by considering the Franck-Condon principle and the stabilization effect of the counter cation in the equilibrium nuclei arrangement in the first excited state (26). The ease of formation of solvent-separated ion pairs in the first excited state for $Fl^-(,M^+)$ has been interpreted on the basis of calculations of the charge distributions using the semiempirical π-electron Variable Electronegativity SCF method by Velthorst et. al (27).

The emission spectra observed for partially lithium metallated P(VFlH-$VFl^-(,Li^+)$; A %) with 10 and 84 mol % of carbanion in THF at 22 °C are depicted in Fig. 9 together with the visible absorption spectrum. The spectra are not affected by a change in A % or in the temperature, and are very similar to that observed for $Fl^-(,Li^+)$ in THF. These results strongly suggest the exclusive formation of solvent-separated P(VFlH-$VFl^-//Li^+$; A %) ion pairs in the first excited state over a temperature range of -80 to 22 °C. In the ground state the lithium salt exists as a mixture of the contact P(VFlH-VFl^-,Li^+; A %) ion pair and the solvent-separated P(VFlH-$VFl^-//Li^+$; A %) ion pair, and the ratio depends on A %. The higher the A % the higher the fraction of contact ion pairs. Therefore, our results demonstrate that also in the polymer the solvent-separated fluorenyl ion pair is favored to form in the excited state. The spectra display two sharp peaks at 532 nm

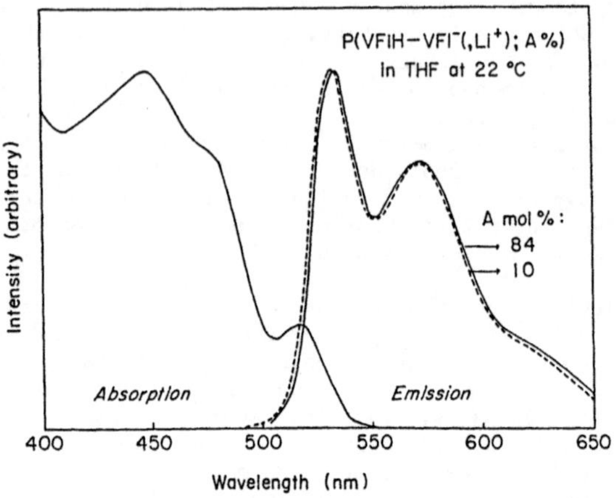

Fig. 9 Absorption and Fluorescence spectra of P(VF1H-VF1$^-$ (,Li$^+$); A %) in THF. $[P(VF1H)]_o = 6.60 \times 10^{-4}$ M.

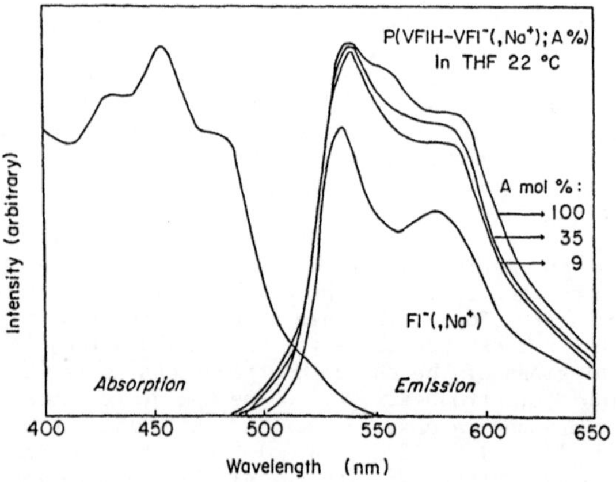

Fig. 10 Absorption and fluorescence spectra of P(VF1H-VF1$^-$ (,Na$^+$); A %) and F1$^-$(,Na$^+$) in THF. $[P(VF1H)]_o = 10^{-3} \sim 10^{-4}$ M.

(0→0 transition) and 575 nm (0→1 transition) and one broad 575 nm (0→2 transition). It is unlikely that the spectrum represents three different species since a similar pattern appears for all salts examined. Moreover, the relative intensities of the three peaks are independent of the temperature. For this reason, it is concluded that the three peaks are assigned to the vibrational structure of the fluorescence. The Stokes shift for P(VFlH-VFl$^-$(,Li$^+$); A %) is estimated to be 0.55 x 10^3 cm^{-1} in Fig. 9 and this value is rather large.

The emission spectra for the sodium salts, P(VFlH-VFl$^-$(,Na$^+$); A %), in THF at 22 °C are depicted in Fig. 10. With increasing A % the spectra broaden, and P(VFlH-VFl$^-$(,Na$^+$); 100 %) displays a new peak at 560 nm in addition to the peaks at 537 and 575 nm. On the other hand, with increasing concentration of Fl$^-$(,Na$^+$), the emission spectrum of Fl$^-$(,Na$^+$) changes distinctly and in a manner different from that of P(VFlH-VFl$^-$(,Na$^+$); A %); in the former carbanion salts, the change of emission band, specially the disappearance of 0→0 fluorescence transition which is overlapping upon the visible band, was observed, and it could be attributed to reabsorption of fluorescence. At the higher concentration of Fl$^-$(,Na$^+$) than 3 x 10^{-2} M, no 0-0 band appears in the fluorescence spectrum (16). Therefore, the change of emission spectra with increasing A % in the polymer systems shown in Fig. 10 should be ascribed to a change in ion pair structures.

The potassium salts P(VFlH-VFl$^-$(,K$^+$); 14 %) display a similar tendency as the sodium salts (16). The change of emission spectra with A % in the polymer becomes more complex in going from sodium metal counter cation to cesium cation. The emission spectra of P(VFlH-VFl$^-$(,Cs$^+$); A %) in THF at 22 °C are depicted in Fig. 11; the ratios of emission intensity for three different peaks change with A %. The difference of the spectral shape between P(VFlH-VFl$^-$(,Cs$^+$); A %) and Fl$^-$(,Cs$^+$) is more significant compared to the cases of sodium and potassium salts; the spectral shape of Fl$^-$(,Cs$^+$) is almost similar to that of other Fl$^-$(,M$^+$) as shown in Fig. 10. In Fig. 11, emission peak maximum drastically shifts to longer wavelength with increasing A %. In the ground state the cesium salt exists as contact ion pairs as shown from the electronic absorption spectrum (5). In the excited state it is most likely that the cesium salt is also composed of only contact P(VFlH-VFl$^-$(,Cs$^+$); A %) ion pairs. The change in the ratios of emission intensity for three different peaks (the vibrational structure of the fluorescence) could be ascribed to a change of the nuclear geometry of fluorenide chromophore in the first excited state, possibly induced by the bulkiness of the cesium cation (16). In the excited state the steric effect of counter cation in the polymer chain may hinder the rearrangement of the fluorenide chromophores after Franck-Condon transition, and this effect is more remarkable with large counter cation.

3. Energy Migration along the Polymer Chain

The absorption and emission spectra of P(VFlH) and P(VFl-VFl$^-$(,M$^+$); A %) in THF at 22 °C are depicted in Fig. 12. The monomer fluorescence spectrum in Fig. 12 is calculated on the basis of the spectrum at -80 °C which consists of the excimer and the monomer emission and that at 22 °C which consists of only the excimer emission. In practice, the intensity of the monomer emission is negligibly small compared with that of the excimer at 22 °C. Even in a glassy solution at 77 K, the excimer emission as well as the monomer emission was fairly well observed. This implies that a considerable singlet energy migration was taking place and that the energy was trapped at the excimer formation site.

The excimer fluorescence from the unmetallated VFlH units emitting at 380 nm at 22 °C was quenched effectively by the VFl$^-$(,M$^+$) units existing within P(VFlH-VFl$^-$(,M$^+$); A %). In Fig. 13, the quenching in these polymer systems is plotted together with those of the monomer systems; as a monomer model, fluorene FlH was used as a donor and fluorenyl carbanion salt Fl$^-$(,M$^+$) as acceptor (quencher). The intensity of the monomer emission from

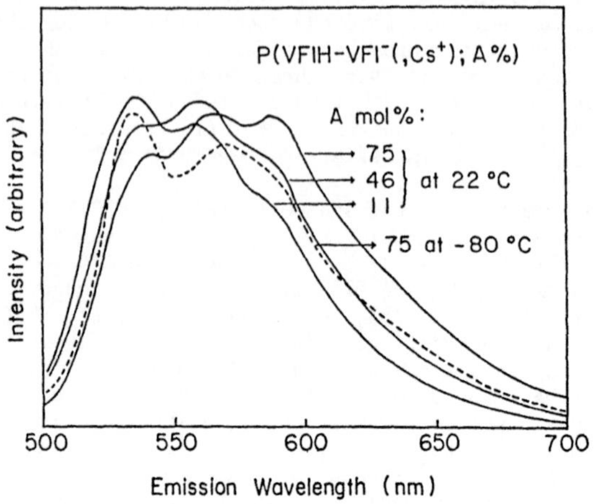

Fig. 11 Fluorescence spectra of P(VFlH-VFl$^-$(,Cs$^+$); A %). $[P(VF1H)]_o = 10^{-3} \sim 10^{-4}$ M.

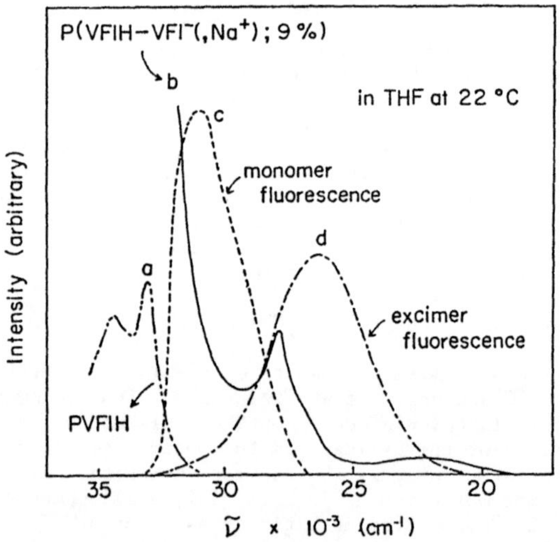

Fig. 12 The absorption and emission spectra of PVF1H and P(VF1H-VF1$^-$(,Na$^+$); A%) in THF at 22 °C.

FlH was measured at 320 nm. In the FlH + $Fl^-(,M^+)$ systems, the quenching is governed by Stern-Volmer kinetics (26, 28) as shown in Fig. 13. In the polymer systems, on the other hand, the quenchings are taking place very effectively compared to the monomer systems. Plots of $\ln I_o/I$ vs. carbanion concentration (M) gave straight lines, where I and I_o stand for the intensity of the excimer emission from P(VFlH-$VFl^-(,M^+)$; A %) or FlH + $Fl^-(,M^+)$ systems and that from P(VFlH) or FlH, respectively. The monomer emission (325 nm) at 22 °C from VFlH units in the polymer was very small, and the shape of the emission bands was little affected by A % as long as polymers containing less than 10 mol % of $VFl^-(,M^+)$ unit were used. Under these conditions, the decrease of the VFlH concentration decreases the excimer fluorescence intensity less than the decrease caused by quenching with $VFl^-(,M^+)$. The fluorescence due to $VFl^-(,M^+)$ in the polymer was negligibly small in the wavelength range between 285 and 500 nm.

The quenching scheme occurring along the P(VFlH-$VFl^-(,M^+)$; A %) chain may be written as,

$$\mathrm{VFlH} \longrightarrow \mathrm{VFlH}^* \quad (1)$$
$$\mathrm{VFlH}^* + \mathrm{VFlH} \longrightarrow \mathrm{VFlH} + \mathrm{VFlH}^* \quad (2)$$
$$\mathrm{VFlH}^* + \mathrm{VFlH} \longrightarrow (\mathrm{VFlH}\cdots\mathrm{VFlH})^* \quad (3)$$
$$\mathrm{VFlH}^* + \mathrm{VFl}^-(,\mathrm{M}^+) \longrightarrow \mathrm{VFlH} + [\mathrm{VFl}^-(,\mathrm{M}^+)]^* \quad (4)$$
$$(\mathrm{VFlH}\ \ \mathrm{VFlH})^* + \mathrm{VFl}^-(,\mathrm{M}^+) \longrightarrow 2\mathrm{VFlH} + [\mathrm{VFl}^-(,\mathrm{M}^+)]^* \quad (5)$$
$$(\mathrm{VFlH}\cdots\mathrm{VFlH})^* \longrightarrow 2\mathrm{VFlH} + h\nu \quad (6)$$

The overlap between absorption and emission bands for the polymer solution systems is seen in Fig. 12. It indicates that a non-radiative energy transfer by Förster type is taking place (26, 28). The critical distance for the energy transfer, R_o, defined as the donor-acceptor separation where 50 % of energy migration takes place, was estimated for P(VFlH-$VFl^-(,Na^+)$; A %) systems from the spectra; 22 Å for eq 2, 36 Å for eq 4, and 29 Å for eq 5. These R_o values were calculated by using the following quantum yields Φ_o of fluorescence; Φ_o = 0.98 for the monomer fluorescence of VFlH (Φ_o value of FlH) and 0.57 for the excimer fluorescence. These yields were measured in degassed THF solution. The R_o value calculated for the processes of eq 2 is long enough to expect significant energy migration.

The effective quenching observed with the P(VFlH-$VFl^-(,M^+)$; A %) systems is considered to be the result of intramolecular quenching. This conclusion is based on the following findings; (1) The quenching efficiency depends upon the degree of polymerization of the polymers used. This is clearly seen from Fig. 14; the longer the polymer chain ($\bar{P}n$ = 323) the higher the observed quenching efficiency. For intermolecular quenching such a dependency upon $\bar{P}n$ should not have been observed. (2) The quenching efficiency does not depend upon the polymer concentration. This can be seen from Fig. 15. There is no significant difference in $\ln I_o/I$ observed over a range of polymer concentrations between of 1.03 x 10^{-4} and 0.94 x 10^{-3} M. On the other hand, at higher concentration (3.6 x 10^{-3} M), an increase in $\ln I_o/I$ was observed. These results suggest that above 10^{-3} M both intermolecular and intramolecular quenchings take place. (3) The difference in the quenching behavior between the polymer and monomer systems (Fig. 13) is also evidence for intramolecular quenching, the polymer concentration being less than 10^{-3} M.

The static quenching can be confirmed from the linear relationship between $\ln I_o/I$ and quencher (carbanion) concentration, as proposed by Perrin. This type of quenching is observed in a rigid matrix solution where donor and acceptor molecules cannot alter their position (25, 27) in space relative to one another. Perrin's model predicts the relationship,

$$\ln I_o'/I' = VN\,[\,Q\,] \quad (7)$$

where I_o' is the emission intensity without quencher Q, I' that in the

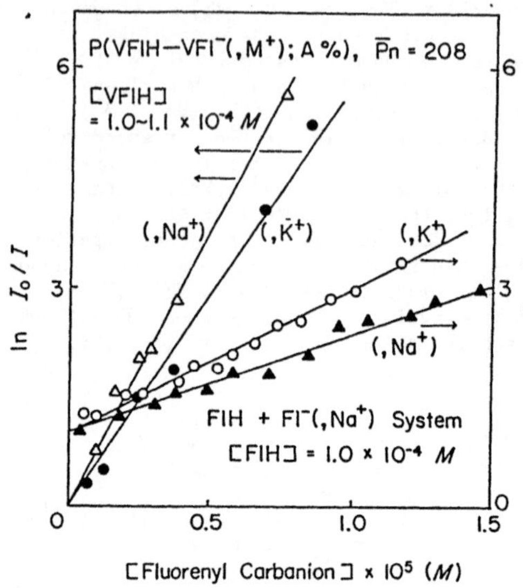

Fig. 13 Quenching plots for monomer and polymer system.

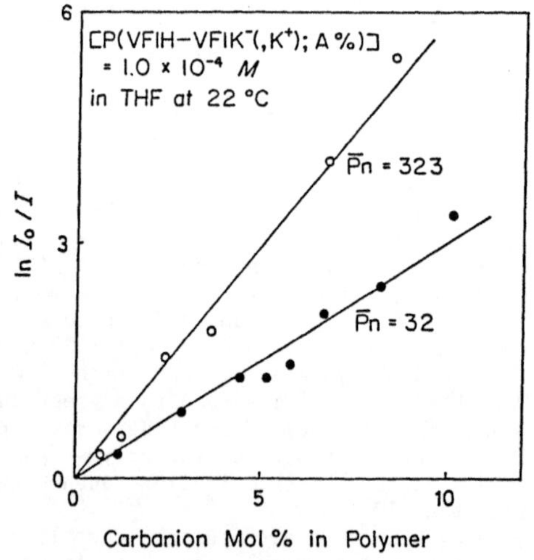

Fig. 14 Dependence of the quenching efficiency in P(VFlH-VFl$^-$(,K$^+$); A%) on degree of polymerization.

presence of quencher, V the volume of the active sphere of quenching (in cm^3), and N Avogadro number. The model applies well in the case of short-range electron exchange interactions, for example, the triplet energy quenching in a glassy solution at 77 K (28). The model, however, is a phenomenalistic one and does not contain any assumption for short-range exchange interactions. Therefore, we applied the static quenching model to our polymer quenching systems and estimated the singlet energy migration efficiency.

We assumed P(VFlH) is a one-dimensional crystal which consists of $\bar{P}n$ units, where $\bar{P}n$ is the number average degree of polymerization. If there is one quencher unit ($VFl^-(,M^+)$) in continued $\bar{n}$ units of P(VFlH), the excited states in the continued $\bar{n}$ units are quenched completely with an effective energy migration within the n units during the life time of the excited state. On the other hand, if the quencher unit is located outside of the continued $\bar{n}$ units, no quenching of excited states within $\bar{n}$ units takes place. These $\bar{n}$ values expresses the mean number of effective energy migration from directly excited unit to the nearest neighbor unit not directly excited down the polymer chain during the lifetime of the excited state. Based on the above assumptions, the following equations are derived (17),

$$\ln I_o/I = \alpha A' \quad (8)$$

$$\alpha = \bar{P}n \ln (1-\bar{n}/\bar{P}n)^{-1} \quad (9)$$

where A' is A (mol %) x 10^{-2}. The linear relationship between $\ln I_o/I$ and A' observed in Figures 18 and 19 confirms the validity of eq 8 for these systems. Thus, $\bar{n}$ can be calculated from the slope of these plots and from eq 9.

In table I, α and $\bar{n}$ values are listed. The $\bar{n}$ values lie between 31 and 76 for the P1 polymer and between 9 and 19 for the P2 polymer. Schneider (29) reported $\bar{n}$ value of about 10 for poly(N-vinylcarbazole) oligomer from fluorescence polarization measurements. The considerably larger $\bar{n}$ values obtained in our system suggest that the singlet energy migration takes place over a relatively long distance along the $P(VFl\text{-}VFl^-(,M^+); A\ \%)$ chain. These $\bar{n}$ values are a little too large because the energy migration from $VFlH^*$ to $VFl^-(,M^+)$ along the polymer chain according to the Förster type kinetics can take place with a fairly large R_o value during the energy migration. Furthermore, energy migration may also take place by skipping units other than along a polymer chain; such an effect on $\bar{n}$ value may become significant when the polymer chain is contracted in solution.

From Table I, the dependence of the counter cation of the quencher (M^+ of $VFl^-(,M^+)$ in the polymers) upon the $\bar{n}$ value can be seen. The $\bar{n}$ estimated for 1 x 10^{-3} M polymer solutions are in the order $Na^+ > K^+ > Li^+ > Cs^+$. There are two possible explanations for this counter cation dependence. One is that the $\bar{n}$ value corresponds only to the active sphere of quenching in Perrin's static quenching model, and the observed dependency upon the counter cation is attributable to the quenching efficiency of the cation. The other possibility is that an interaction between $VFl^-(,M^+)$ units alters the properties of the polymer chain in solution and this affects the efficiency of the singlet energy migration along the polymer chain. The observed order with respect to M^+ cannot be explained by a heavy atom effect (26, 30) since in that case the observed quenching order would be $Cs^+ > K^+ > Na^+ > Li^+$. For the monomer model systems as depicted in Fig. 13, the order is different from that of the polymer, i.e., $K^+ > Na^+$. Therefore, the dependence of $\bar{n}$ upon the quencher's cation may be caused by a change in the solution properties of $P(VFlH\text{-}VFl^-(,M^+); A\ \%)$. In such a case, the $\bar{n}$ values contain the effect of the solution properties of the polymer. In the quenching system with lithium salt, electrostatic repulsion between the solvent-separated $VFl^-//Li^+$ moieties in the $P(VFlH\text{-}VFl^-//Li^+; A\ \%)$ may result in the formation of more extended polymer chain and this could lower the $\bar{n}$

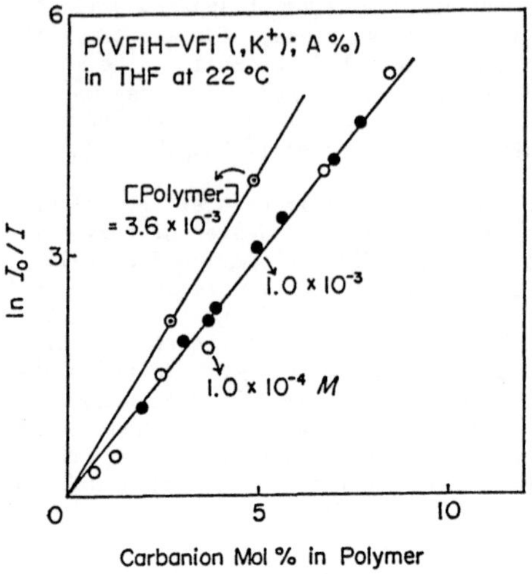

Fig. 15 Dependence of the quenching efficiency of P(VFlH-VFl$^-$(,K$^+$); A%) on the polymer concentration.

Table I The mean number for effective energy migration $\bar{n}$.

Run No	Quenching System	$\bar{P}n$	Polymer Concn (M)	α	$\bar{n}$
1	P(VFlH-VFl$^-$//Li$^+$)	323	1.08×10^{-3}	58	53
2	P(VFlH-VFl$^-$(,Na$^+$)	323	1.00×10^{-3}	86	76
3	P(VFlH-VFl$^-$(,Na$^+$)	323	1.07×10^{-4}	81	72
4	P(VFlH-VFl$^-$(,K$^+$)	323	3.61×10^{-3}	84	74
5	P(VFlH-VFl$^-$(,K$^+$)	323	0.94×10^{-3}	63	57
6	P(VFlH-VFl$^-$(,K$^+$)	323	1.03×10^{-4}	60	55
7	P(VFlH-VFl$^-$(,K$^+$)	32	1.00×10^{-4}	30	19
8	P(VFlH-VFl$^-$,Cs$^+$	323	1.71×10^{-3}	33	31
9	P(VFlH-VFl$^-$,Cs$^+$	32	1.06×10^{-4}	14	9

in THF at 22 °C.

value.

On the other hand, the sodium, potassium, and cesium salt (specially the last two) exsist mainly as contact ion pairs. In these cases, the rigidity caused by steric hindrance of the counter cation may decrease the $\bar{n}$ value. For these salts $\bar{n}$ decreases in the order $Na^+ > K^+ > Cs^+$, the reverse order as the cation radius. In the quenching system with a larger cesium cation, the steric hindrance may induce an extension of the polymer chain and reduce $\bar{n}$ value. Such an effect may not be significant for the smaller sodium ion.

REFERENCES

1. Present adress, Department of Applied Chemistry, Faculty of Engineering, Iwate University, Ueda, Morioka, 020 Japan.
2. M. Szwarc, "Carbanions, Living Polymers, and Electron Transfer Processes"; Interscience Publishers: New York, 1968.
3. M. Szwarc, "Ions and Ion pairs in Organic Reactions, Vols. 1 and 2"; Wiley-Interscience: New York, 1972.
4. A. Streitwieser Jr, Acc. Chem. Res., 17, 353 (1984).
5. T. E. Hogen-Esch and J. Smid, J. Am. Chem. Soc., 87, 669 (1965); 88, 307 (1966); 90, 4654 (1968).
6. T. E. Hogen-Esch and J. Smid, J. Am. Chem. Soc., 88 318 (1966).
7. S. Nishi and M. Matsuda, J. Am. Chem. Soc., 101, 4632 (1979).
8. S. Nishi and M. Matsuda, Phosphorus and Sulfur, 16, 181 (1983).
9. H. R. Huson, Annual Reports on the Progress of Chemistry, London; The Royal Society of Chemistry, 76, 66, (1979).
10. A. Watanabe, M. Iino, and M. Matsuda, Polymer Preprints, Japan, 30, 200 (1981); 31, 649 (1982); 31, 423, (1982)
11. K. Wong, L. Ambroz, and J. Smid, J. Polymer Bull., 8, 411 (1982).
12. M. J. Plodinec and T. E. Hogen-Esch, J. Am. Chem. Soc., 96, 5262 (1974).
13. T. E. Hogen-Esch and J. Plodinec, J. Phys. Chem., 80, 1085, 1090 (1976).
14. T. E. Hogen-Esch and J. Plodinec, J. Am. Chem. Soc., 100, 7633, (1978).
15. A. Watanabe and M. Matsuda, British Polym. J., Part 1, in press.
16. A. Watanabe and M. Matsuda, Part 2, to be published.
17. A. Watanabe and M. Matsuda, Part 3, to be published.
18. M. Minabe, M. Yoshida, M. Fujimoto, and K. J. Suzuki, J. Org. Chem., 41, 1935 (1976).
19. E. Gipstein, W. A. Hewett, and O. U. Need, J. Polym. Sci. Part B, 9, 671 (1971).
20. H. C. Wang, G. Levin, and M. Szwarc, J. Am. Chem. Soc., 100, 6137 (1978).
21. G. C. Greenacre and R. N. Young, J. Chem. Soc. Perkin II, 874 (1976).
22. H. M. Parkes and R. N. Young, J. Chem. Soc. Perkin II, 1137, (1980).
23. A. Persoons and M. Van Beylen, Pure and Applied Chem., 51, 887, (1979).
24. M. Kasha, Radiation Res., 20, 55 (1963).
25. A. S. Davydov, "Theory of Molecular Excitons"; New York, 1962.
26. M. J. Turro, "Modern Molecular Photochemistry"; Benjamin: California, 1978.
27. H. W. Vos, C. MacLean, and N. H. Velthorst, J. Chem. Soc. Faraday II, 72, 63 (1976).
28. J. B. Birks, "Photophysics of Aromatic Molecules"; Interscience: New York, 1970.
29. F. Z. Schneider, Naturforsch, 24, 863, 1969.
30. C. A. G. Brooks and K. M. C. Davis, J. Chem. Soc. Perkin II, 1649 (1972).

MECHANISMS

CLARIFICATION OF SOME PROBLEMS OF ANIONIC POLYMERIZATION

MICHAEL SZWARC
Department of Chemistry, University of California San Diego
LaJolla, CA 92093

INTRODUCTION

Ionic, and especially anionic polymerization is of great interest to polymer chemists as evident by this second International Symposium on Anionic Polymerization. Vigorous studies of anionic polymerization started in the late fifties, and progress in this field has been reported in numerous reviews published during the last 20 years [1]. The published results are, on the whole, mutually consistent and accounted for by plausible mechanisms accepted by the workers in this field. There are, nevertheless, some contradictory claims pertaining mainly to polymerizations involving Li counter-ions and proceeding in hydrocarbon solvents.

The advances in ionic polymerization stem from the discovery of living polymers that led to many important synthetic applications and to a better understanding of the nature of the active polymer end-groups. The first unequivocal proof of the feasibility of homogeneous chain addition polymerization free of termination and chain-transfer was reported by Szwarc et al. [2] in 1956. The concept of living polymers was developed in these publications and its ramifications were fully outlined. The initial work dealt with anionic polymerization of styrene and the dienes, but the underlying ideas were soon applied by other workers to a great variety of polymer systems. For the sake of illustration, examples of various kinds of presently known living polymers are given below.

The earliest observation of living polymers, although not referred to by this term, was reported by Ziegler [3]. He noted that the molecular weight of polybutadiene resulting from initiation of the reaction by metallic sodium increased when fresh monomer was added to the reactor. This work, reported in the thirties, failed to make the impact which otherwise would be expected because the heterogeneous nature of this reaction obscured its important features.

The feasibility of the living nature of anionic polymerization of ethylene oxide was fully appreciated by Flory [4] who realized that such polymers should have a Poisson molecular-weight distribution. Unfortunately, this reaction was performed in those days in solutions containing alcohol needed for solubilization of the initiator. Proton transfer from the alcohol to the growing polymers, i.e., a chain-transfer, led to their termination followed by initiation of another chain, thus reducing their anticipated molecular weight and broadening their molecular weight distribution.

These short-comings of oxirane polymerization were eliminated when new catalysts were developed. An excellent catalytic system developed by Inoue [5] allowed him to produce truly living poly-oxiranes having Poisson molecular-weight distribution and to synthesize polymers composed of 2 or 3 blocks of different polyoxiranes, each characterized by a narrow molecular-weight distribution.

Recent Advances in Anionic Polymerization
T.E. Hogen-Esch and J. Smid, Editors

Controlled hydrolysis of organometallic compounds, studied by Tsuruta [6] and by Vandenberg [7] led to a most interesting class of initiators effectively utilized by Teyssié [8] for the preparation of living poly-ε-caprolactams. Although the improved catalyst, composed of Al-Zn oxyalkoxides, contains four active centers, only one of them appears to be active in polymerization.

The anionic systems discussed so far are not unique in their ability to yield living polymers. Cationic growth of living polymers is presently well documented. The living nature of cationically propagating polytetrahydrofuran was established simultaneously and independently by three research groups, and since then many ring-opening polymerizations were found to proceed through cationic growth of their living cyclic-onium end-groups [9].

Living, cationically propagated polymerization of vinyl ethers, initiated by equimolar mixture of HI and I_2, was described by Higashimura [10]. It yielded polymers of narrow molecular weight implying a rapid initiation. Cationic polymerization of isobutene, initiated by esters such as $(CH_3)_3CO(OR)$ activated by BCl_3, was shown to yield living polyisobutene even at temperatures as high as -30^oC [11].

Polymerization of propylene propagated through a coordination mechanism involving a soluble $V(acetylacetone)_3/AlEt_2Cl$ complex was studied by Doe et al. [12]. It yielded living syndiotactic polypropylene of relatively narrow molecular weight distribution.

Entirely new kinds of living polymers were reported recently. Living group-transfer polymerization of methyl methacrylate was developed by the DuPont team, [13] and this new approach was adapted recently for polymerization of aldehydes [13a]. Spontaneous polymerization of some cyclobutane derivatives resulting in the formation of living polymers was described by Hall [14]. Finally, a living polymer growing by virtue of an olefin metathesis process involving carbene was reported recently by Grubbs [15].

The above examples demonstrate the wide range of systems capable of forming living polymers.

LONGEVITY OF LIVING POLYMERS

It was remarked in the original paper introducing the concept of living polymers [2] that "living polymers are not immortal". They are subject to destruction and this could be of considerable practical value. Since the activity of their end-groups is retained after completion of polymerization, addition of a suitable reagent may endow the final polymer with some desired functional end-group. Such a functionalization of living polymers is especially valuable when performed on polymers possessing active groups on both of their ends since it results in the formation of bi-functional polymers.

This deliberate termination of living polymers should be distinguished from spontaneous deactivation of their growth end-groups that eventually has to take place. Polymers are considered to be living if they retain their capacity of growth for a time needed for completion of a desired task, whether a synthesis, or some kinetic or spectroscopic study. Nevertheless, ultimately they have to either isomerize, or decompose, or undergo some reaction with their surroundings, thus loosing their ability to grow further. Their lifetime depends on their nature and on the conditions prevailing in the polymerized system. For example, at elevated temperatures their lifetime is shorter and indeed, some living polymers have to be kept at extremely low temperatures during and after their formation to prevent their premature termination.

It is the task of the investigator to prolong the lifetime of the living polymers he intends to study. This is achieved by choosing optimal conditions for the investigated polymerization process. The most rigorous purification of reagents, solvents, walls of the reaction vessels, etc., are the obvious requirements to be fulfilled. An interesting question arises: is the lifetime of living polymers prolonged by keeping them supplied with monomer? The answer depends on the kind of mechanism governing the propagation.

Most common propagations take place in a single step, i.e.,

$$P^*_n + M \longrightarrow P^*_{n+1}, \qquad k_p$$

where P^*_n denotes a growing polymer, and M its monomer. Simultaneously with propagation some very slow reactions may take place such as

$$P^*_n \longrightarrow D_n \text{ or } D_n + X, \qquad k_{is}, \quad (a)$$

$$P^*_n + R \longrightarrow \text{inactive species}, \qquad k_R, \quad (b)$$

Here (a) denotes some irreversible isomerization or decomposition, and (b) represents a reaction with some species present in the surroundings that deactivates the growing group. Hence the lifetime, τ, of the living polymer is given by

$$\tau = 1/\left(\Sigma k_{is} + \Sigma k_R R\right)$$

The independence of the lifetime of living polymers on the monomer concentration should not be confused with the dependence of their maximum average length on the concentration of monomer. The maximum average length increases, of course, with k_pM, but in any practical experiment it is not attained.

The problem is different when propagation involves 2 steps. For example, consider a propagation described by the scheme:

$$P^*_n \longrightarrow A_n, \qquad k_1$$

$$A_n + M \longrightarrow P^*_{n+1} \qquad k_2$$

where P^*_n denotes a dormant species inactive in propagation and protected from destructive reactions, whereas A_n is capable of growing by adding monomer, but it is also susceptible to destructive reactions causing its termination. The polymers spend a fraction $k_1/(k_1 + k_2M)$ of time in the form P^*_n, being protected from the destructive reactions, and a fraction of time $k_2M/(k_1 + k_2M)$ in the vulnerable form A_n. Hence, their lifetime increases with increasing concentration of monomer, and is reduced, perhaps to an unacceptable level, by their "starvation", i.e., by leaving them without monomer for a prolonged time. Metathetic ring opening polymerization of norbornyl, reported by Grubbs [15] and described by scheme A, provides a good illustration of this kind of behavior. It proceeds smoothly at ambient temperature in the presence of the monomer, its rate being zero order in the latter reagent because $k_2M >> k_1$. This reaction displays all the features of living polymerization as long as it is not "starved" by lack of monomer for a prolonged time. The activity of these polymers may be preserved for longer times by chilling them to a sufficiently low temperature. Their growth stops then as they "hybernate", but it resumes when the system is brought to room temperature.

SCHEME A

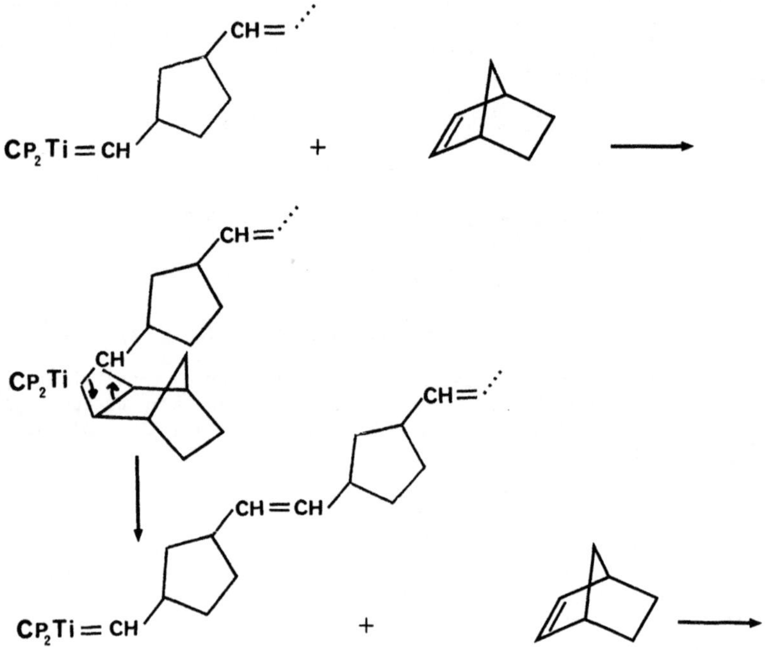

Alternatively, let us consider a polymerization involving a monomer-polymer complex,

$$P^*_n + M \underset{k_b}{\overset{k_f}{\rightleftarrows}} (P^*_n, M)$$

$$(P^*_n, M) \xrightarrow{k_d} P^*_{n+1}$$

Denote by P^*_{total} the total concentration of living polymers in whatever form. The rate of propagation, R_p, is

$$R_p = \{k_d k_f/(k_d + k_b + k_f M)\} P^*_{total} M,$$

and, therefore, the reaction is first order in monomer for $(k_d + k_b) >> k_f M$, and zero order for $(k_d + k_b) << k_f M$.

In either case the fraction of time the polymer spends in the P^*_n form is given by

$$(k_d + k_b)/(k_d + k_b + k_f M).$$

Provided that P^*_n is the labile form, susceptible to destruction, while the complex is protected from these side reactions, we conclude that the longevity of such polymers increases with monomer concentration.

OCCAM RAZOR RULE

Correlations between experimentally determined variables were, and still are the subject of many investigations. The results are fitted to some function that interrelates the values of the observed variables. For example, the rates, R, of some studied reaction are correlated often with the temperature, T, at which it proceeds through the relation (a),

$$R = \alpha \exp(\beta/T) \qquad (a),$$

involving two constants: α and β. There are other relations, often more complex, that fit the data better but which require a greater number of constants. For example, it might be necessary to introduce a third constant, γ, into the relationship (a) to get a better fit between the observed and computed values of R and T, for instance,

$$R - \gamma = \alpha \exp(\beta/T) \qquad (b).$$

Whenever a desired relation involves only two constants it is always possible to transform the variables in such a manner that the resulting relation becomes linear. For example, relation (a) becomes linear when R is substituted by exp.x and T by 1/y, i.e.,

$$x = \ln \alpha + \beta y. \qquad (c).$$

Let it be stressed that the linear relation (c) does not correlate the observed pairs of variables better than the non-linear relation (a). One gains only through this transformation a more convenient way of plotting the experimental data.

It is not uncommon to find two distinguishable functional relations, both correlating the experimental data equally well, but one involving a greater number of constants than the other. In such a case the relation involving a smaller number of constants if preferred. This is the essence of the Occam Razor Rule.

Consider now two distinct relations, both requiring the same number of constants and reproducing the experimental data equally well. In such a case we prefer the relation which is based on a physically justified model and involves meaningful constants of plausible values. Let us illustrate such a problem by considering the following example.

Smets et al. [16] investigated the anionic polymerization of lithium poly(o-methoxystyrene) in a hydrocarbon solvent. The monomeric Li-poly(o-methoxystyrene), PLi, maintained in equilibrium with its dimers, $(PLi)_2$ seems to be the only active species propagating the reaction:

$$\text{PLi} + \text{monomer} \longrightarrow \text{propagation}, \qquad k_p,$$

$$(\text{PLi})_2 \rightleftarrows 2\ \text{PLi}, \qquad K_{diss},$$

The fraction of the dissociated polymers is relatively large in this system, hence, their concentration has to be calculated from the equation:

$$[\text{PLi}]^2 + (K_{diss}/2)[\text{PLi}] - (K_{diss}/2)[\text{PLi}]_{anal} = 0,$$

where $[\text{PLi}]_{anal}$ denotes the total concentration of living polymers. Solving the equation gives

$$[\text{PLi}] = \{-K_{diss}/2 + (K_{diss}^2/4 + 2K_{diss}[\text{PLi}]_{anal})^{1/2}\}/2,$$

and the observed apparent pseudo-first order rate constant of propagation, i.e.,

$$k_{app} = (-d\ln[M]/dt)/[\text{PLi}]_{anal}$$

is

$$k_{app} = k_p[\text{PLi}].$$

The experimental data provided k_{app} and $[\text{PLi}]_{anal}$, their values being listed in Table 1. The plot of $\log(k_{app})$ vs. $\log[\text{PLi}]_{anal}$, shown in Figure 1, was found to be slightly curved. The reported k_{app}'s were well reproduced by taking $K_{diss} = 10^{-3}$ M and $k_p = 50\ M^{-1}\text{min}^{-1}$, as is evident on inspection of Table 1. Thus, the proposed mechanism implies that the percent of dissociation of the dimeric (PLi_2) varies from about 15 to 65% over the investigated concentration range of $[\text{PLi}]_{anal}$.

TABLE 1. Pseudo-first order propagation constants, k_{app}, for various total concentrations of living Li-poly(o-methoxystyrene) at 20°C. The values listed in the first 2 columns are taken from Table 2 of ref. 16.

$[\text{PLi}]_{anal}$ x 10^3M	k_{app}(exp)*	k_{app}(Smets)*	k_{app}(from linear relation)*
18.2	.141	.139	.145
10.0	.1015	.100	.100
3.59	.055	.056	.053
1.30	.029	.0297	.0284
.997	.025	.0250	.0241
.910	.022	.0235	.0236
.532	.016	.0163	.0162

* All in units of min^{-1}

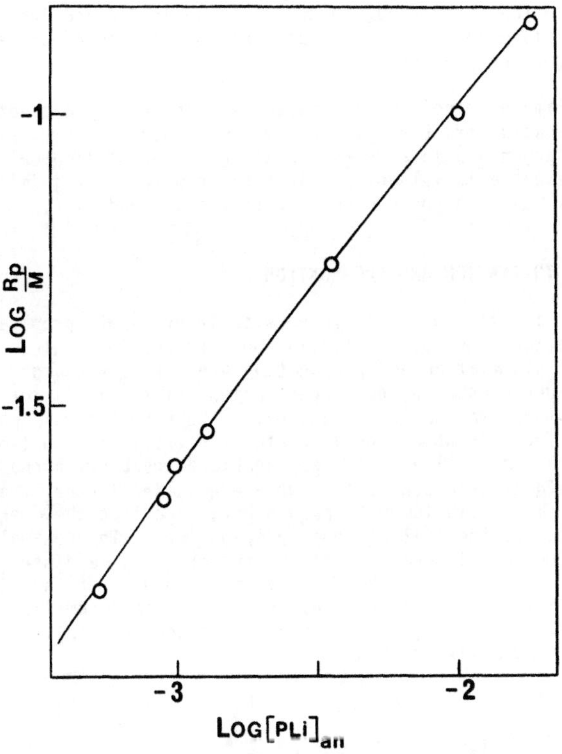

FIGURE 1. Effect of total living end concentration on the rate of anionic polymerization of o-methoxystyrene. (Taken from ref. 16)

This well documented approach was disputed in a paper [17] presented during the previous Symposium on Anionic Polymerization and again in a recent review [18], on the ground that the reported experimental pairs: $\log(k_{app})$ and $\log[PLi]_{anal}$, fit the linear relation

$$\log(k_{app}) = -1.621 + .622 \log[PLi]_{anal}$$

with an excellent correlation coefficient of .9994. The k_{app}'s calculated from this relation are tabulated in the last column of Table 1. The agreement of k_{app}'s calculated from the linear relation with the observed k_{app}'s, is slightly poorer than that of those computed on the basis of Smets' mechanism, although the differences are not significant.

Both relations, the linear one and the curved one derived from Smets'approach, involve two constants. The linear one requires a value of the slope and the intercept, whereas the curved one is characterized by the values of K_{diss} and of k_{app}. Which one should be preferred?

The advocates of the linear relation did not attribute any meaning to their constants, nor proposed a mechanism that could account for their relation. On the other hand, Smets' relation is derived from a reasonable mechanism and involves meaningful constants of plausible values. Obviously the latter is to be preferred.

These simple principles discussed here apparently are not appreciated by the advocates of the linear relation since they repeated their questionable arguments in a recent review [18], confusing only the issue. The above example is valuable. It shows how to treat similar problems whenever they arise in the course of a scientific endeavor.

SIMULTANEOUS INITIATION AND PROPAGATION

Consider a living polymer system with irreversible propagation. Such a polymerization involves initiation and propagation only. Let it be assumed that initiation and propagation are both first order in monomer, while their dependence on the concentrations of the initiator and of the growing polymers may be quite complex. The complexity is caused by the variety of forms in which the initiator and the living polymers may be present, each form being a species capable to react with the monomer with its own individual rate constant. We are assuming further that all these species are in equilibrium with each other, and that these equilibria are not perturbed by the initiation or propagation. For example, let the initiator be present in the form of a tetramer (I_4), a dimer (I_2), and a monomer (I), as well as in the form of its complex with the living polymer (IP). Similarly, the living polymers may form monomeric and dimeric species, i.e., P and P_2, and of course the complex (IP). The following equilibria are established then:

$$2I \underset{}{\overset{K_1}{\rightleftarrows}} I_2; \quad 4I \overset{K_2}{\rightleftarrows} I_4; \quad (IP) \overset{K_3}{\rightleftarrows} I + P; \quad 2P \overset{K_4}{\rightleftarrows} P_2$$

leading to 4 equations:

$$I_2 = K_1 I^2, \quad I_4 = K_2 I^4, \quad (IP) = K_3 IP, \quad \text{and } P_2 = K_4 P^2.$$

These, in conjunction with the 2 stoichiometric equations, namely:

$$I + 2I_2 + 4I_4 + (IP) = I_{total}, \text{ and } P + 2P_2 + (IP) = P_{total},$$

determine the concentrations of the 6 species: I, I_2, I_4, (IP), P, and P_2. The above result is general. Whatever the complexity of the system, the concentrations of all the individual species are uniquely determined by the pertinent equilibrium constants and the two variables: I_{total} and P_{total}, abbreviated as I_t and P_t, respectively. Since the addition of one monomer molecule to the initiator in whatever form converts one I into one P, the total concentration of living polymer, P_t, is given by

$$P_t = I_0 - I_t$$

where I_0 denotes the initial concentration of the initiator. Therefore, the rate of the initiator consumption is

$$-dI_t / dt = F(I_t, I_0 - I_t) M,$$

and the rate of monomer consumption is

$$-dM/dt = [F(I_t, I_0 - I_t) + G(I_t, I_0 - I_t)]M.$$

Hence,

$$d M/d I_t = 1 + G(I_t, I_0 - I_t)/F(I_t, I_0 - I_t).$$

Note that $d M/d I_t$ is independent of monomer concentration, provided that the previously made two assumptions remain valid, namely: the first order dependence of each elementary reaction on monomer, and the maintenance of the equilibria between all the species participating in the reaction.

It may be helpful to give an example illustrating the forms of $F(I_t, I_0 - I_t)$ and $G(I_t, I_0 - I_t)$. For the previously discussed system involving the 6 species I, I_2, I_4, (IP), P, and P_2

$$F(I_t, I_0 - I_t) = k_1 I + k_2 I_2 + k_3 I_4 + k_4(IP),$$

and

$$G(I_t, I_0 - I_t) = k_5 P + k_6 P_2 + k_4(IP),$$

where k_1, k_2, k_3, k_4, k_5, and k_6 refer to the respective reactions:

$$I + M \longrightarrow P, \qquad k_1,$$

$$I_2 + M \longrightarrow (IP) \text{ or } I + P, \qquad k_2,$$

$$I_4 + M \longrightarrow I_2 + (IP), \qquad k_3,$$

$$(IP) + M \longrightarrow 2P \text{ or } P_2, \qquad k_4,$$

$$P + M \longrightarrow P, \qquad k_5,$$

$$P_2 + M \longrightarrow P_2, \qquad k_6.$$

Some k's may be 0, e.g., k_3 and k_6 equal 0 implies that I_4 and P_2 are inert.

The independence of $d M/d I_t$ of M, and therefore of its initial concentration M_o, implies that the curves giving M as functions of I_t form a family of parallel shifted curves for a constant I_0, each curve corresponding to a different value of M_o. Such curves are depicted in Figs. 2A and 2B.

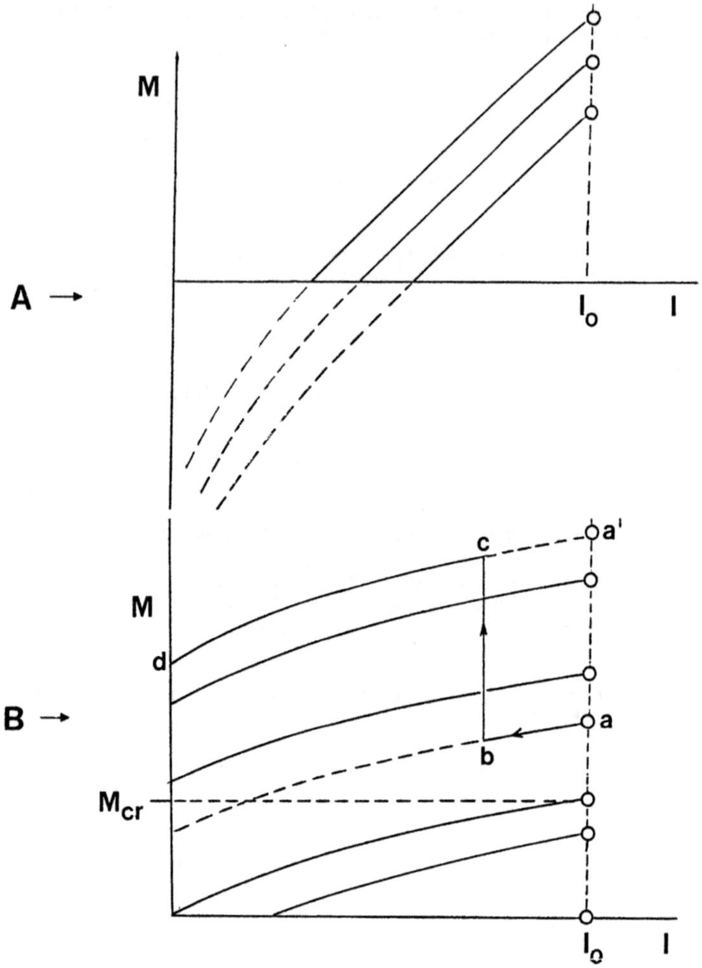

FIGURE 2A AND 2B. Curves depicting the change in monomer concentration as a function of the total initiator concentration, I, for different initial values of I_0 in systems of simultaneous initiation and propagation (for further explanation, see text).

There are two kinds of such curves. Those shown in Fig. 2A do not intersect the M axis but approach it asymptotically. They reflect the behavior of systems such as the one for which the rate of initiator consumption is proportional to I_t, being independent of $I_0 - I_t$, and the rate of propagation is proportional to $I_0 - I_t$, independently of I_t. The respective differential equation is

$$d\,M/d\,I_t = 1 + (k_p/k_i)\,(I_0 - I_t)\,/\,I_t,$$

giving the integrated form

$$M = M_0 + (k_p/k_i - 1)(I_0 - I_t) + (k_p/k_i) I_0 \ln (I_t/I_0).$$

It implies that all the monomer is eventually consumed while the concentration of the residual initiator, I_r, reaches the value determined by the equation

$$M_0 = -(k_p/k_i - 1)(I_0 - I_r) + (k_p/k_i) I_t \ln (I_t / I_r).$$

In systems when the rate of initiator consumption is given by

$$-d I_t/dt = k_i I_t^{1/n}, \text{ with } n > 1,$$

while the rate of propagation is proportional to $I_0 - I_t$, i.e., to the concentration of living polymers, then

$$d M/d I_t = 1 + (k_p/k_i)(I_0 - I_t)/I_t^{1/n},$$

and for $n > 1$ the dependence of M on I_t is given by a curve resembling the curves shown in Figure 2B, i.e., it intersects the I_t axis for $M_0 < M_{cr}$ and the M axis for $M_0 > M_{cr}$, where M_{cr} is the critical concentration of the monomer, given by

$$M_{cr} = I_0 + \{n^2/(2n - 1)(n - 1)\} (k_p/k_i) I_0^{2-1/n}.$$

Thus, starting with the initial concentration of the initiator I_t, we find all of it consumed when $M_0 > M_{cr}$, while some of it remains still unreacted when $M_0 < M_{cr}$.

A more complex system was discussed by Worsfold and Bywater [19], namely the initiation is 1/6 order in the initiator and propagation 1/2 order in living polymers. An approximate solution was reported. The results resemble qualitatively those discussed above, i.e., all the initiator is consumed when the concentration of the monomer exceeds some critical value which, in turn, depends on I_0.

It is interesting to consider the situation depicted in Figure 2B by the line described by abcd. Polymerization starts with a relatively low concentration of the monomer corresponding to point a of Figure 2B. At some stage (point b) fresh monomer is added increasing its concentration to the value corresponding to point c. The reaction continues then along the curve cd. It follows from our discussion that the polymer obtained at point d has the same degree of polymerization and the same distribution of molecular weights as the polymer produced at the respectively higher initial concentration of the monomer corresponding to point a'. Hence, the "seeding technique" recommended by some authors does not improve the molecular weight distribution. This conclusion was confirmed by experiments of Van Beylen (Macromolecules, 1987 in press).

COPOLYMERIZATION AND REACTIVITIES OF MONOMERS

For a series of compounds, A_1, A_2, ...A_n, undergoing a bimolecular reaction with X,

$$X + A_i \longrightarrow \text{products}, \quad k_{xi},$$

the concept of relative reactivity has a well defined meaning, namely A_i is considered to be more reactive than A_j if $k_{xi} > k_{xj}$. The ratio k_{xi}/k_{xj} is defined as the relative reactivity of A_i with respect to A_j for the reaction with X.

The ordering in such a series depends on the nature of X as well as on the thermodynamic conditions prevailing in the reaction. For example, there might be a series of reactions involving Y, i.e.,

$$Y + A_i \longrightarrow \text{products}, \qquad k_{yi},$$

such that $k_{yi} < k_{yj}$ although $k_{xi} > k_{xj}$.

This kind of relations applies to copolymerization. For a series of monomers, M_1, M_2,.....M_n, copolymerizing with a growing polymer, X^*, i.e., for a series of processes,

$$\sim\!\sim X^* + M_i \longrightarrow \sim\!\sim XM_i^*, \qquad k_{pxi},$$

M_i is more reactive than M_j if $k_{pxi} > k_{pxj}$. However, here again the gradation depends on $\sim\!\sim X^*$. For example, in the copolymerization of polystyryl anion with butadiene or isoprene in THF the former monomer is more reactive than the latter, while a reverse is anticipated in cationic copolymerization. This could be expected because the electron-donating tendency of the methyl group greatly reduces the rate of addition of the negative center of $\sim\!\sim CH_2CH(Ph)^-$ to carbon 1 of $CH_2{=}C(CH_3)CH{=}CH_2$, leaving the addition to carbon 4 only slightly affected. Consequently, the reactivity of isoprene in anionic copolymerization is about twice as low as that of butadiene [20], while the reactivity 2,3-dimethylbutadiene is lower by a large factor of about 70. This effect of a methyl group in anionic copolymerization is general; it makes α-methylstyrene less reactive than styrene, reduces the reactivity of methyl methacrylate relative to methyl acrylate, or of methacrylonitrile relative to that of acrylonitrile. Apparently this obvious relation is not fully understood. For example, in a recent review [21] its authors stated: it is "a bizarre assertion that 1,3-butadiene will be twice as reactive as isoprene (in anionic copolymerization) since one of the ends of isoprene, relative to butadiene, is unreactive while the other retains its normal reactivity".

Reactivities of monomers should not be gauged by comparing the rate constants of their homopropagations. This was realized long ago in the course of studies of radical copolymerization. The rates of homopropagation, say

$$\sim\!\sim A^* + A \longrightarrow \sim\!\sim AA^*, \qquad k_{pA},$$

or

$$\sim\!\sim B^* + B \longrightarrow \sim\!\sim BB^*, \qquad k_{pB},$$

are determined by the reactivities of end-groups of the growing polymers, as well as by the reactivities of the monomers. As a rule, any structural changes that increase the reactivity of the end-groups of growing polymers decrease the reactivity of the respective monomers. Hence, the outcome of a modification in the nature of the monomer on its respective rate constant of homo-propagation is unpredictable. This lack of relation between the structure of the monomer and its homopropagation rate constant is not always appreciated. For example, one finds a statement [21] such as:

"McGrath's results [22] regarding the rate constants (of anionic homo-propagation) of 1,3-butadiene and isoprene (imply) butadiene < isoprene".

An additional factor complicates the comparison of rates of homopropagation in the Li-hydrocarbon systems. The Li salts of the polybutadiene$^-$ and polyisoprene$^-$ anions form aggregates (presumably tetramers) in hydrocarbon solvents at conventional concentrations of living polymers. However, propagation, at least in aromatic hydrocarbons, seems to be carried out by the monomeric form of the polymers. Hence, the observed propagation constant is given by $(K_{diss}/n)^{1/n}k_p$ where K_{diss} denotes the dissociation constant of the aggregated polymers into their active monomeric form, while k_p is the homo-propagation constant of the active species. Even for k_p of butadiene larger than k_p of isoprene, the apparent homo-propagation constant of the former monomer may be smaller than that of the latter, provided that the K_{diss} of butadiene is much lower than that of isoprene. Still more complex relations are expected when mixed aggregates are formed. Under such conditions the simple concept of an apparent rate constant is inappropriate for the evaluation of the reactivity of the monomer. Examples of such complexities were reported in the literature, see e.g., ref. 23 and 24.

Copolymerization of styrene and butadiene in the Li-hydrocarbon system provides a classic case of such a problem. This reaction was investigated first by Korotkov who reported [25] an unexpected phenomenon. The co-polymerization starts slowly and initially consumes butadiene. When this monomer is exhausted, the reaction speeds up and styrene polymerizes then. Korotkov's explanation was based on the idea of preferential solvation of the end group of the growing polymer by butadiene. Provided that butadiene is a much better solvating agent than styrene, the growing polymer "sees" only the former monomer, and this one is then consumed by propagation, with virtual exclusion of styrene. After exhaustion of butadiene, the growing polymers become by necessity surrounded by styrene which then starts to polymerize.

This interesting hypothesis, which might be applicable for some future system, was disproved by Worsfold [26]. To ease experimentation he studied a closely similar system involving isoprene instead of butadiene. The rate of addition of styrene to Li-polyisoprene was determined in the presence of variable concentrations of isoprene and found to be constant, contrary to predictions deduced from the hypothesis of preferential solvation. Moreover, according to this hypothesis a trend in the order of the reaction should be observed. The order in butadiene monomer is expected to drift to zero as its concentration increases, while no trend was revealed even when the concentration of the monomer was varied by four powers of ten.

The correct explanation of this phenomenon was offered by O'Driscoll and Kuntz [27], namely: the addition of butadiene, or isoprene, to Li-polystyrene is much faster than the addition of styrene to Li-polybutadiene, or polyisoprene. Hence, virtually only Li-polydiene is present in the polymerized mixture as long as the diene monomer is available. This suggestion was verified by Ellis and Morton [28], and by Worsfold [26,29] who found the apparent rate constants of the dienes addition to Li-polystyrene to be extremely fast.

An interesting problem deserves to be mentioned. Consider again the addition of styrene to Li-polybutadiene. Initially the rate of the reaction is given by

$$(K_{diss},\ \text{Li-BD}/4)^{1/4}\ k_p(\text{LiBD})^{1/4}\ \text{Styrene}.$$

As the reaction proceeds, yielding Li-polystyrene, mixed aggregates are formed, and these could be more dissociated than the original (Li-polybutadiene)$_4$. As a result the reaction would speed up, provided that the mixed aggregates are more readily dissociated, or more reactive, than the original tetramers. This kind of auto-acceleration was observed by Fetters and Young [30].

The term "reversal of polymerization" used in the description of the styrene-butadiene system is misleading. In the Li-hydrocarbon systems the dienes are genuinely more reactive than styrene. Under such conditions they add faster than styrene whether to Li-polystyrene or Li-polydienes [21], although the reverse is observed in the Li-THF system [31]. It happens, perhaps not surprisingly, that homopropagation of Li-polystyrene is faster than that of Li-polydienes. However, as stated earlier, the rates of homopropagations are unpredictable. Nonetheless, it is interesting to inquire why the dienes are more reactive than styrene in the Li-hydrocarbon systems, while styrene is more reactive in the Li-THF system, or in hydrocarbon solutions with other alkali ions. It seems that the dienes, by virtue of their geometry, strongly interact with the small, unsolvated Li^+ ion in the transition state of the addition. This interaction speeds up their addition. Let it be stressed again that such an interaction should be distinguished from the formation of a complex between dienes and the growing polymer; formation of such a complex is disproved by the work of Worsfold [26]. Here we encounter an example showing the importance of thermodynamic conditions on the order of the monomer reactivity. A change of solvent leads to a change in the order of reactivities.

REFERENCES

1. (a) S. Bywater; Adv. Polymer Sci., 4, 66 (1956).
   (b) R.C. Cubbon and D. Margerisson; Prog. Reaction Kinetics, 3, 403 (1965).
   (c) M. Szwarc; "Carbanions, Living Polymers, and Electron Transfer Processes", John Wiley, New York (1968).
   (d) H.L. Hsieh and W.H. Glaze; Rubber Chem., 43, 22 (1970).
   (e) S. Bywater; Prog. Polymer Sci., 4, 27 (1974).
   (f) M. Szwarc; in "Ions and Ion-Pairs in Organic Reactions" M. Szwarc, ed., Vol. II, 375 (1974).
   (g) D.H. Richards; in "Developments in Polymerization", R.N. Haward, ed., Vol. I, (1977).
   (h) J. Smid; in "Structure and Mechanism in Vinyl Polymerization", T. Tsuruta and K.F. O'Driscoll, Eds., Marcell Dekker, Inc. New York, Ch. 11, (1969).
   (i) S. Bywater; in "Chemical Kinetics", C.H. Bamford and C.F.H. Tipper, ed, Vol. 15, (1976).
   (j) M. Szwarc; Adv. Polymer Sci., 49, 1 (1983).
   (k) M. Morton; "Anionic Polymerization", Academic Press (1983).
   (l) S. Bywater; in "Encyclopedia of Polymer Science", H. Mark et al., eds., Vol. II, 1 1984).
2. (a) M. Szwarc, M. Levy, and R. Milkovich; J. Amer. Chem. Soc., 78, 2656 1956).
   (b) M. Szwarc; Nature, 178, 1168 (1956).
3. K. Ziegler et al.; Ann., 473, 36 (1929), 479, 150 (1930), 511, 13, 45, 64 (1934).
4. P.J. Flory; J. Amer. Chem. Soc., 62, 1561 (1940).
5. S. Inoue et al.; Macromolecules, 14, 1162, 1166 (1981).
6. R. Sagata and T. Tsuruta; Makromol. Chem., 40, 64 (1960).
7. E.J. Vandenberg; J. Polymer Sci., 47, 486 (1967).

8. Ph. Teyssie et al. Macromolecules, 6, 651, (1973), ACS Symp. Series 59, 165 (1977).
9. See for review: S. Penczek, P. Kubisa, and K. Matyjaszewski; Adv. Polymer Sci. 37, 1 (1980).
10. T. Higashimura et al.; Macromolecules, 18, 611 (1985) and ref. therein.
11. J.P. Kennedy et al.; J. Polymer Sci., in press.
12. Y. Doi et al.; Macromolecules, 12, 814 (1979).
13. O.W. Webster et al.; J. Amer. Chem. Soc., 105, 5706 (1983), and the following papers.
14. K. Hall, J. Amer. Chem. Soc. (1987) in press.
15. R.H. Grubbs and L.R. Grilion; J. Amer. Chem. Soc.; 108, (1986).
16. (a) G. Geerts, M. Van Beylen, and G. Smets, J. Polymer Sci., A7, 2859 (1969).
(b) G. Smets and M. Van Beylen, 23rd Congress of Pure and Appl. Chem., 4, 405, Butterworth, Pub. (1971).
17. L.J. Fetters and R.N. Young; in ACS Symps. Series, 166, 100 (1981).
18. R.N. Young, R.P. Quirk and L.J. Fetters, Adv. Polymer Sci., 56, 33-34 (1984).
19. J.D. Worsfold and S. Bywater, Can. J. Chem. 38, 1891 (1960).
20. (a) M. Szwarc, ref. 1a, pp. 539-540.
(b) M. Shima, J. Smid and M. Szwarc; J. Polymer Sci. B, 2, 735 (1964).
21. Ref. 18, p. 62.
22. J.E. McGrath, et al., ACS Symp. Series, 166, 529 (1981).
23. (a) Z. Laita and M. Szwarc; Macromolecules, 2, 412 (1969).
(b) A. Yamagishi, et al.; ibid, 11, 504 (1978).
(c) A. Yamagishi, et al.; ibid, 11, 607 (1978).
24. A. Yamagishi and M. Szwarc; ibid, 11, 1091 (1978).
25. A.A. Korotkov, Ang. Chem., 70, 85 (1958).
26. D.J. Worsfold; J. Polymer Sci., A5, 2783, (1967).
27. K.F. O'Driscoll and I. Kuntz; J. Polymer Sci., 61, 19 (1962).
28. M. Morton and F.R. Ellis; J. Polymer Sci., 61, 25 (1962).
29. A.F. Johnson and D.J. Worsfold; Makromolek. Chem. 85, 273 (1965).
30. R.N. Young and L.J. Fetters; Macromolecules, 11, 899 (1978).
31. D. Kelley and A.V. Tobolsky; J. Amer. Chem. Soc., 81, 1597 1959).

THE EFFECT OF MOLECULAR WEIGHT ON ASSOCIATION IN ANIONIC POLYMERIZATION

S. Bywater and D.J. Worsfold
Division of Chemistry
National Research Council of Canada
Ottawa, Canada, K1A 0R9

ABSTRACT

The effect of molecular weight on the equilibrium between the tetrameric and dimeric associates of the active chain-ends in the polymerization of isoprene by lithium based initiators has been studied by UV spectroscopy. This equilibrium is equivalent to the association of two polymer chains linked by active groups centrally placed on the chains where large changes of equilibrium constant are found in going from low molecular analogs to polymers in other systems. Large effects are found in this case also in both cyclohexane and benzene solution. The dissociation constant in benzene solution with high molecular weight species is sufficiently high that it modifies the kinetics of the propagation reaction. The kinetic order changes from one quarter to one half with dilution to follow the change in equilibrium, with parallel changes in activation energy. Also the propagation rate differs for very low molecular weight species.

INTRODUCTION

Several theoretical studies on the effect of the polymer chain on reactions between active groups substituted on separate polymer molecules have dwelt on the reversible formation of the activated complex, in non-reversible reactions, as being the important factor. [1,2,3] The subsequent unimolecular decomposition of the activated complex to give products being considered of less importance.

The treatments of this problem by Morawetz, [1] by Kozlov, [2] and by Iwata, [3] have started from the theory of absolute reaction rates where the second-order rate constant is defined as

$$k_2 = \kappa\ kT/h \cdot Q^{*}_{1,2}/\ Q_1 \cdot Q_2 \exp(-\Delta E/kT)$$

The Q factors are the partition functions of the reagents and the activated complex and were considered to differ in polymers because of the excluded volume effect. The transmission constant κ and the activation energy ΔE are suggested to remain unchanged from the small molecule reagents to the polymeric reagents.

When considering the equilibrium constant in reversible reactions the equivalent form is

$$A + B \rightleftharpoons AB$$
$$K = Q_{AB}/\ Q_A \cdot Q_B \cdot \exp(-\Delta E/kT)$$

If the same assumption is made that the partition coefficients are modified by the excluded volume effects, it is possible to predict that changes should occur in equilibrium constants between reactive groups when they are attached to polymers.

Published 1987 by Elsevier Science Publishing Co., Inc.
Recent Advances in Anionic Polymerization
T.E. Hogen-Esch and J. Smid, Editors

Measurements confirming these effects have been made for both irreversible [4] and reversible [5] reactions by comparing polymeric systems with comparable low molecular weight model systems. In the latter case equilibrium constants were measured between polystyrenes substituted with a single tertiary amine group placed at selected positions along the chain and the di-(dinitrohydroquinone) ester of either adipic or sebacic acids. The esters are bidentate acids, because of the two nitrophenol groups, and form complexes with two amine groups, in effect coupling two chains together.

$$\mathrm{HOROH + PSMe_2N \overset{K_1}{\rightleftharpoons} PSMe_2\overset{+}{H}N.\overset{-}{O}ROH}$$

$$\mathrm{PSMe_2\overset{+}{H}N.\overset{-}{O}ROH + PSMe_2N \overset{K_2}{\rightleftharpoons} PSMe_2\overset{+}{H}N.\overset{-}{O}R\overset{-}{O}.\overset{+}{N}HMe_2PS}$$

HOROH is 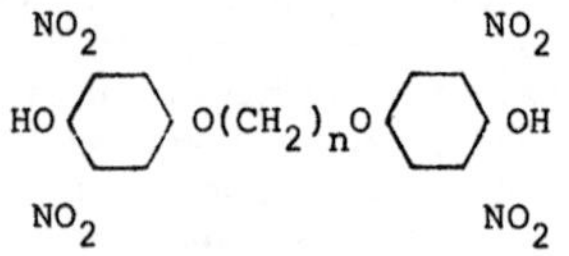 , PS is Polystyrene

$K_1 >> K_2$, and K_2 is MW sensitive

TABLE 1. Equilibrium Constants Ratios for Interpolymer Association Polystyrene MW 180,000, 30°

Position[1]	K_{pol}/K_{model}	MW x 10^{-3}	K_{pol}/K_{model}
0	0.69	0.3	0.35
0.1	0.41	0.4	0.34
0.2	0.35	0.5	0.35

[1] Fraction of chain length from end of chain that amine group is substituted.

TABLE 2. Equilibrium Constants Ratios for Interpolymer Association[1]

Polystyrene		Polyisoprene	
MW x 10^{-3}	K_{pol}/K_{model}	MW x 10^{-3}	K_{pol}/K_{model}
90	0.37	34	.27
180	0.35	155	.18
300	0.27		

[1] For centrally substituted chains.

Table 1 summarizes the results obtained for group position, Table 2 for molecular weight and polymer. Although the effect is clearly demonstrated, the magnitude is only about half that predicted by the theoretical calculations. It is also quite clear that, as predicted by theory, when the groups are attached to chain ends only a modest effect is seen, compared to central substitution.

Equilibria between macromolecules in anionic polymerization are important in several contexts. Of particular interest are the aggregates formed in the lithium based polymerizations in hydrocarbon solvents, and triple ion formation in more polar solvents at high ionic concentrations. It is likely that in the latter case only the triple ion composed of two polymeric living chain ends coupled to one cation is important in this sense, [6] and would only involve end to end chain association of polymers where excluded volume effects on equilibrium constants are small. The same would apply to the dimeric associations found for polystyryllithium in hydrocarbon solvents. But in the case of diene polymerization in hydro carbon solvents, tetrameric association occurs and larger effects could be predicted, and in fact may be diagnostic of this type of association. It is likely that the higher aggregates found for lithium organic compounds in hydrocarbon solvents are based on dimeric structures [7] which self aggregate to form tetramers, and in some cases higher aggregates. If this is the case the latter aggregates would entail the association of the basic dimer in what is in effect a mid-chain interaction, and the effect of excluded volume could be expected to be greater than in dimer/monomer association.

The equilibrium between such dimeric and tetrameric species has been illustrated in two ways, first by light-scattering [8] measurements on living polymer systems, and secondly by UV [9] measurements of the active chain-end which show differences between different aggregates.

Light-scattering experiments were made on living polymerization systems simplified for the measurements by using polystyrene as the base polymer in cyclohexane solution at 35°, the Θ temperature. Then by adding a few units of diene the effective chain-end and degree of aggregation is changed to that typical of say polybutadienyllithium or isoprenyllithium, but retaining for the light-scattering measurements the solution properties of the polystyrene. Such measurements have shown that whereas the polystyryllithium showed a two-fold association, the butadiene capped polymer four-fold association, the polyisoprenyllithium end-capped polymer showed an intermediate association. At higher concentrations the association was four-fold, but as the concentration decreased this fell and approached two-fold. It was possible to calculate an equilibrium constant for a tetramer-dimer equilibrium.

The UV spectrum of isoprenyllithium in hydrocarbon solvents shows changes on dilution that are not seen in the spectrum of butadienyllithium which is unchanged in the measurable range of 10^{-2} to 10^{-6}M. These changes were attributed to the same tetramer-dimer equilibrium found in the light-scattering experiments. Making some assumptions concerning the final spectrum, an equilibrium constant was calculated, and this was found to be somewhat solvent dependent in aliphatic solvents and greatly changed in benzene solution. In cyclohexane, however, making allowance for the temperature differences, between the two sets of experiments the equilibrium constants diverged more than expected (x4). At the time the reason for this was not clear, but when considered in terms of the excluded volume effect above, it is not unreasonable because in the light scattering experiments the base polymer was a high molecular weight polystyrene, but the spectroscopic experiments used a low molecular weight polyisoprene.

To confirm that the interpretation of the above experiments in light-scattering and UV spectra was in fact correct, further UV studies have been made to confirm that high molecular weight systems have different association constants. Predictable effects on kinetics and energetics of the propagation step would also be expected assuming a dissociative mechanism. In this way if the magnitude of the effect is sufficient, it would be confirmation of the higher degree of association for diene systems than for styrene [13] which is sometimes disputed on the basis of concentrated viscosity [10] measurements, whose validity themselves have been questioned. [11]

EXPERIMENTAL

The techniques used in these experiments have all been used before. [12] All experiments were made in glass and quartz apparatuses under vacuum with the vessels prewashed with butyllithium solution.

The UV experiments were made in apparatuses fitted with two UV cells with suitable movable quartz spacers so that the optical path could be varied from .01 to 10 cm. The apparatuses were designed so that solutions could be diluted or concentrated by distillation of the solvent from one part to another. Living end concentrations were found from an isosbestic point (290 nm) used in a previous study, using the same extinction coefficient. The concentrations of the two different associates were measured from the ratio of the optical densities at 290 nm and 320 nm. sec-Butyllithium was prepared and distilled under vacuum before diluting and subdividing into a suitable solvent. The propagation rate was measured by UV spectroscopy. Only the later stages of the reaction were followed after the polymer chain had mostly formed. The reaction in benzene was followed in the window of the benzene spectrum near 220-230 nm using a .01 cm UV cell.

RESULTS

The UV measurements on higher molecular weight living polyisoprenes showed a far greater change of spectra at similar concentrations than for lower molecular weight polymers comparable to those used in the previous study, [8] (Figures 1 and 2). That dissociation was in fact complete for the most dilute solutions was seen from the asymptotic behaviour of the ratio of the absorption at 320 and 290 nm. It was found possible in cyclohexane to determine the final spectrum of the presumed dimer which had not been possible before. Hence more reliable dissociation constants were obtained as only an assumed spectrum for the dimer was used previously which was found to be inadequate. Table 3 shows the various dissociation constants found for differing molecular weights and two solvents. The rapid increase from low molecular weights is shown, but any smaller variations at higher molecular weights are lost in the experimental error.

TABLE 3. Tetramer/Dimer Equilibrium Constants for Isoprenyllithium at 31°

Solvent	MW x 10^{-4}	K_D x 10^6
c-Hexane	.05	1.6
"	.25	13
"	.95	40
"	1.25	50
"	1.8	35
"	12.5	45
"	41.0	37
Benzene	.05	60
"	1.1	350

The temperature coefficient of the dissociation was measured over a 30° range and the heat of association was found to be 9100 cals. in benzene solution. Propagation rate measurements were made over a wide range of active chain end concentrations in benzene solution covering the region from where the association is primarily tetrameric to the dilute region where it is dimeric. The results are shown in Figure 3, and demonstrate a break at the mid range. The line drawn at higher concentrations is of slope 0.25, at lower 0.5. The minimum molecular weight at which the rate

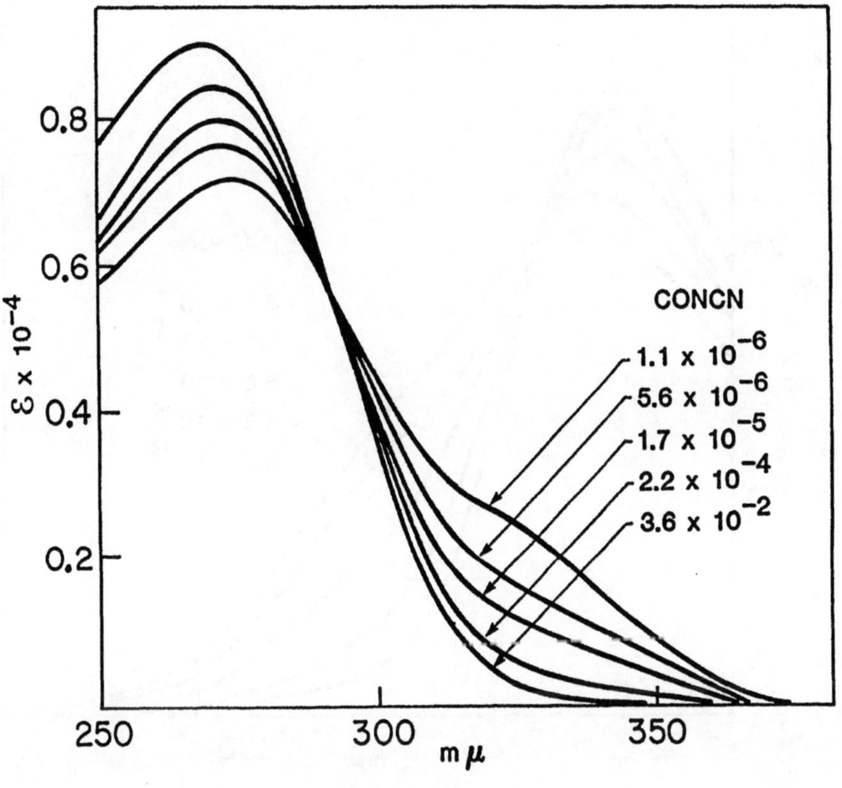

Figure 1. U.V. Spectra of Low Mol. Wt. Polyisoprenyllithium, cyclohexane, 31°.

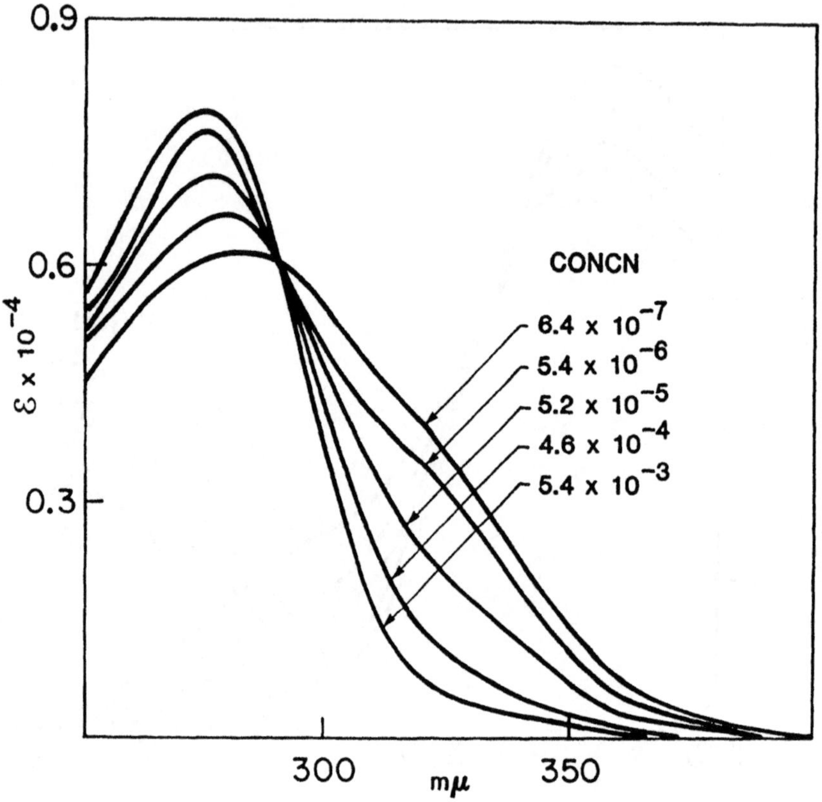

Figure 2. U.V. Spectra of High Mol. Wt. Polyisoprenyllithium, cyclohexane, 31°.

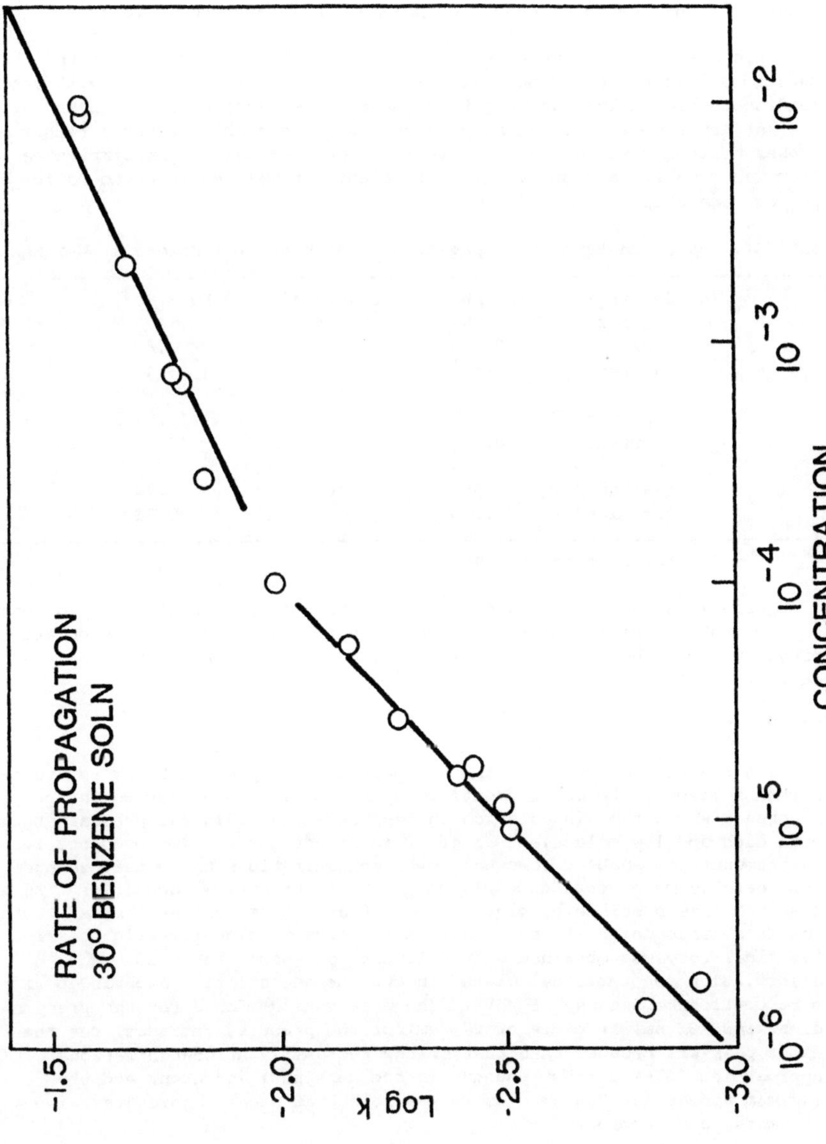

Figure 3. Log/log plot of k_p v. [Li] for isoprene in benzene solution 31°.

was measured was 10,000, a limit imposed by the desire to be in the range of the higher dissociation constants, but still be in the dilute polymer solution range at the highest used polymer chain-end concentrations.

Propagation rates were measured covering the final stages of the reactions leading to low molecular weight polymers and compared with those leading to high molecular weight polymers. The results are shown in Table 4. The lower molecular weight polyisoprene in cyclohexane has a rather higher molecular weight than desirable to ensure that the initiation reaction was complete and still have sufficient monomer remaining to follow the propagation step.

TABLE 4. Rate Constant of Propagation at High and Low Molecular Weights

Solvent/ Monomer	MW	Conc x 10^4 M	Kp x 10^3
c-Hexane/	2,000	1.2	3.4
Isoprene	High[1]	1.2	5.1
Benzene/	500	7.0	11.6
Isoprene	20,000	7.0	16.7
c-Hexane/	500	10	1.74
Butadiene	26,000	10	2.38

[1] Literature data ref. 12.

Also measured was the apparent activation energy at high and low chain end concentrations in the regions where tetrameric or dimeric association predominates. The values found were 17,200 cal. at 8 x 10^{-3}M, and 14,400 cal. at 8 x 10^{-6}M.

DISCUSSION

In the earlier study of the UV spectra of isoprenyllithium in hydrocarbon solvents, the measurements were made on low molecular weight polyisoprenes where the dissociation is less extensive. In the present study, both high and low molecular weights were investigated. Improvements in instrumentation enabled somewhat lower concentrations to be reached and considerably more confidence placed in the lower optical densities. As a result it was possible by plotting the OD at 320 nm against that at 290 nm, the isosbestic point, to show that in the higher molecular weight series the final form was obtained at the lowest concentrations used. The dissociation constants calculated in cyclohexane solution are viewed with more confidence because of this. The peak maximum found for the presumed dimer was 282 nm, in place of 273 nm for the presumed tetramer, but the dimer peak was broader than that of the tetramer. At concentrations approaching 10^{-2}M further changes in the peak position occur and the maximum appears at 268 nm. It is possible that higher aggregates, i.e., hexamers, are forming.

The change in the equilibrium constant on going from the low molecular weight material to the high is very substantial, a factor of nearly thirty in cyclohexane and six in benzene solution. The magnitude of the effect is so much greater than that found for a supposed monomeric chain-end association to dimer, that the only reasonable explanation is that mid-chain association of dimeric associates to higher aggregates is occurring.

The propagation reaction, in the case when isoprene is polymerized with lithium based initiators in hydrocarbon solvents, has been found to have a kinetic order of 0.25, and this has been explained by the tetrameric association of the active chain ends. [13] The rate expression derived from the scheme:

$$P_4 \overset{K}{\rightleftharpoons} 4.P_1$$

$$-\frac{dM}{dT} = k_p[P_1][M]$$

$$= k_p K^{1/4}[P_4]^{1/4},$$

assumes only inconsiderable dissociation to P_1 the monomeric chain end. Intermediate degrees of association were not excluded however so that if, the actual scheme is

$$P_4 \overset{K_1}{\rightleftharpoons} 2.P_2$$

$$P_2 \overset{K_2}{\rightleftharpoons} 2.P_1$$

Then $-\frac{d[M]}{dt} = k_p[P_1][M]$

and $= k_p K_1^{1/4} K_2^{1/2}[P_4]^{1/4}$ at high concentrations

or $= k_p K_2^{1/2}[P_2]^{1/2}$ at low concentrations.

where P_4 and P_2 are the major species at those concentrations.

If appreciable dissociation of the tetramer can take place over the measurable concentration range, the reaction order would vary between .25 order at high concentrations, where the tetramer is prevalent, to .5 in dilute regions where most of the active chain ends are dimeric. In aliphatic solvents the primary aggregate is tetrameric over the whole measurable concentration range, in agreement with the .25 order generally found.

In benzene solution the range where the association change takes place is at the lower end of the readily measurable concentration range for kinetic measurements. Literature values [14,15] cover only the range down to 10^{-4}M in active chain end concentration, and the order found is .25. The measurements made in this study cover also the lower concentration range and demonstrate that the anticipated change in kinetic order occurs.

It would be expected that as the propagation reaction at high concentrations involves dissociation of the tetramer to the dimer, whereas at low concentrations it does not, the apparent activation energy would differ at these extremes. The apparent activation energy will involve terms for the true activation energy of the reaction of the monomeric chain end with the monomer, plus half the heat of association of the dimer to monomer at low active chain end concentrations. But at high concentrations it will include also a term which is one quarter of the heat of association of the tetramer to the dimer. The measured heat of this latter association is 9,100 cal. in benzene solution, hence the expected activation energy difference between high and low concentrations would be 2,300 cal. The measured difference is from 17,200 to 14,400 cal, i.e., 2,800 cal, close enough to support the suggestion that the dissociation measured by spectroscopic experiments is from tetramer to dimer. If the dissociation measured in the spectroscopic experiments was from a dimeric to a monomeric form of active chain end, the calculated change in activation energy would be half of 9,100, or 4,550 cal. in much worse agreement with experiment. Moreover the change in kinetic order from ¼ to ½, would, in the latter case, be difficult to interpret in terms of a dimer-monomer equilibrium.

Because the dissociation constants of the aggregates appear in the kinetic equation governing the propagation reaction, the rate of propagation should be molecular weight dependent, following the molecular weight dependence of the dissociation constant. This is demonstrated in Table 4. The equilibrium constant K_1 appears to the .25 power in the kinetic equation, at the concentrations chosen, so a large effect cannot be expected. For isoprene in cyclohexane the ratio of rates leads to a ratio of dissociation constants near 5, about correct for a molecular weight of 2,000 compared with a high molecular weight value. In benzene the corresponding ratio is 4.3 compared with the expected value of 6, but within the probable experimental error because less than a 5% variation in each rate would bring the measured ratio to the expected value. Table 4 also lists the corresponding values for butadiene. Although there is no measurable aggregate dissociation over the experimentally accessible concentration range in the case of butadienyllithium, the same dissociation constant appears in the kinetic equation and would still be sensitive to excluded volume effects and hence to the molecular weight of the polymeric species. The found ratio of rates corresponds to a ratio of dissociation constants near 3.5, lower than for isoprenyllithium but nevertheless still in the region found in the previous study. [5]

Thus in summary, it has been confirmed by spectroscopic measurement that an aggregate dissociation process occurs on dilution of polyisoprenyllithium in hydrocarbon solvents. It has further been shown that this process is molecular weight and solvent dependent. The molecular weight dependence produces a related effect on the propagation rate as expected of a dissociative mechanism. In benzene, this primary dissociation process is so extensive that for active center concentrations much below 10^{-4}M it is essentially complete. The expected changes in kinetic order and activation energy of the propagation step between experiments carried out around 10^{-3}M and around 10^{-5}M active center concentrations can best be attributed to a tetrameric association of the active chain ends which on dilution breaks down to a dimeric association. The changes in the association constant with molecular weight are caused by the excluded volume effect introducing an extra entropy factor opposing the close association of the polymer chains in dilute solution. The magnitude of the effect is such that it is best explained in terms of association between groups centrally placed on the polymer chains as would occur when end to end associates in dimeric isoprenyllithium associate further to tetramers.

REFERENCES

1. H. Morawetz, J.R. Cho, and P.J. Gans, Macromolecules, 6, 624 (1973).
2. S.V. Kozlov, Eur. Polym. J., 16, 1241 (1980).
3. H. Iwata and Y. Ikada, Makromol. Chem., 181, 517 (1980).
4. P.E. Black and D.J. Worsfold, J. Polym. Sci. Polym. Chem. Ed., 19, 1841 (1981).
5. D.J. Worsfold, J. Polym. Sci. Polym. Chem. Ed., 21, 2271 (1983).
6. D.N. Bhattacharyya, C.L. Lee, J. Smid and M. Szwarc, J. Phys. Chem., 69, 612 (1965).
7. M.Y. Darensbourg, B.Y. Kimura, G.E. Hartwell, and T.E. Brown, J. Amer. Chem. Soc., 92, 1236 (1970).
8. D.J. Worsfold and S. Bywater, Macromolecules, 5, 393 (1972).

9. J.E.L. Roovers and S. Bywater, Polymer, 14, 594 (1973).
10. M. Morton, L.J. Fetters, R.A. Pett, and J.F. Meier, Macromolecules, 3, 327 (1970).
11. D.J. Worsfold, J. Polym. Sci. Polym. Phys. Ed., 20, 99 (1982).
12. S. Bywater and D.J. Worsfold, Can. J. Chem., 38, 1891 (1960).
13. S. Bywater and D.J. Worsfold, Can. J. Chem., 42, 2884 (1964).
14. J. Feller, Thesis Univ. of Akron (1967), University Microfilms. Ann Arbor, Michigan.
15. D. Margerison, D.M. Bishop, G.C. East and P. McBride, Trans. Farad. Soc., 64, 1872 (1968).

ANIONIC POLYMERIZATION OF 4'-VINYLBENZOCROWN ETHERS

FREDDY GILS and MARCEL VAN BEYLEN
Laboratory for Macromolecular and Physical Organic Chemistry, University of Leuven, Celestijnenlaan 200 F, B-3030 Leuven (Heverlee), Belgium

and

JOHANNES SMID
Polymer Research Institute, Chemistry Department, College of Environmental Science and Forestry, State University of New York, Syracuse, New York 13210, USA

ABSTRACT

The mechanism of anionic polymerization of vinylbenzocrown ethers in tetrahydrofuran below -50°C was studied by conductance, optical spectroscopy and kinetic measurements. Dissociation constants of potassium and cesium polystyryl carbanions derived from 4' -vinylbenzo-15-crown-5 and 4' -vinylbenzo-18-crown-6 were determined and compared with those obtained for corresponding salts of polystyryl carbanion itself, their complexes with monomeric crown ethers (e.g., benzo-15-crown-5 and benzo-18-crown-6), and the salts of poly(p-methoxystyryl) and poly (3,4-dimethoxystyryl). Combined with spectral data, information was obtained on the ion pair structure of the carbanion salts such as the formation of intra- versus inter-molecular complexes between the counter cations and the polymer-bound crown ether ligands. From the dissociation constants and the observed rate constants at different carbanion salt concentrations the ion pair and free ion propagation rate constants for styrene, p-methoxystyrene and 4 -vinylbenzo-15-crown-5 with potassium as counter ion were obtained in THF at -70°C.

INTRODUCTION

The cation-chelating properties of neutral macroheterocyclic ionophores such as crown ethers (1,2), cryptands (3) and spherands (4) have been extensively investigated over the past twenty years. Their usefulness in applications such as phase transfer catalysts, ion-selective electrodes or ion-separation processes are often greatly enhanced by immobilizing the ligands on a polymeric matrix. The anchoring of these ligands to macromolecules can alter their binding characteristics (5). For example, comonomer substituents may interfere with or can enhance the binding of ionic species to the ligands (5,6). Also, when the polymer-bound ligands are in close proximity, formation of 2:1 ligand-cation complexes are favored in systems where such complexes are stable, e.g., complexes of K^+ or Ba^{2+} with 15-crown-5 type ligands and complexes of Cs^+ with 18-crown-6 type compounds (7-10). Studies on polymers derived from vinylbenzocrown ethers and other vinylcrown ether monomers have demonstrated that ion selectivities of these poly(crown ether)s are indeed different from those of the corresponding monomeric ligands (5). They can be further modified by incorporation of comonomer units which alters the distance between the

Recent Advances in Anionic Polymerization
T.E. Hogen-Esch and J. Smid, Editors

pendant ligands (5,8).

The vinylbenzocrown ethers are useful monomers for synthesizing poly(crown ether)s since radical or anionic initiators can convert them into high molecular weight homopolymers (11). Moreover, a variety of vinyl monomers can be used to obtain random copolymers or, when using anionic initiators, block copolymers. Addition of cross-linking agents yields network-type poly(crown ether)s. In spite of their frequent use, little attention has been given to the polymerization mechanism of these monomers. This should be especially of interest in anionic polymerization since the cation-binding ligands on monomer and polymer molecules will affect the degree of dissociation of the living polystyryl carbanion pairs in the type of solvents generally employed in such polymerizations, e.g., toluene or ethereal solvents. Since the rate and stereospecificity of the polymerization both depend on the structure and degree of dissociation of the propagating ion pairs, the presence of the crown ether ligands are expected to profoundly affect the polymerization process.

This paper deals with the propagation mechanism of anionic polymerization of vinylbenzocrown ethers. For comparative purposes some polystyryl and substituted polystyryl salts are included in the investigation. Special emphasis is placed on the conductance behavior of the propagating carbanionic species since the rate of polymerization is dependent on the extent of dissociation of the ion pairs. The instability of the polystyryl-type carbanion pairs derived from vinylbenzocrown ethers in solvents such as tetrahydrofuran made it necessary to carry out the conductance experiments at temperatures between -70° and $-55^{o}C$, while rates were measured only at $-70^{o}C$.

EXPERIMENTAL

Materials. Procedures for synthesizing vinylbenzo-15-crown-5 (VB15C5) and vinylbenzo-18-crown-6 (VB18C6) have been reported (11-13). The purity of the monomers was checked by melting point, TLC and NMR. The white crystals were stored under vacuum.

Styrene, p-methoxystyrene and 3,4 dimethoxystyrene (Aldrich Europe) were dried over CaH_2, then distilled under vacuum and again dried on sodium potassium alloy until the orange polystyryl carbanion color persisted. Following vacuum distillation the monomers were stored in ampoules at $-25^{o}C$. Tetrahydrofuran (THF) was dried by standard procedures and eventually distilled in vacuo from the polystyryl potassium salt and stored in ampoules. Benzo-18-crown-6 (B18C6) was synthesized as reported (14), while 15-crown-5 (15C5), 18-crown-6 (18C6) and benzo-15-crown-5 (B15C5) were Merck products. The crown ethers were purified on an Al_2O_3 (activity V) column and recrystallized from dry petroleum ether ($40-60^{o}C$).

The polymer carbanion salts were synthesized under vacuum at $-70^{o}C$ in an all glass-sealed apparatus using cumyl potassium or cesium as initiator. To minimize termination of the reactive carbanions the conductivity cell was already attached to the reaction vessel. The full details of these procedures will be published elsewhere.

Measurements. Immediately following preparation of the living polymers, their conductivities were measured by means of a Wayne Kerr universal bridge B642. Carbanion concentrations were determined in attached optical cells using a Cary 17 spectrophotometer. A quartz dewar with optical windows was placed in the cell compartment. It contained methanol the temperature of which could be lowered to $-70^{o}C$ by means of a

Colera KT90S thermostat. Kinetic data were obtained spectrophotometrically by monitoring the disappearance of one of the absorption bands of the monomer, i.e., styrene at 291 nm, VB15C5 at 296 nm and p-methoxystyrene at 302 nm. A description of the all-glass sealed apparatus and other details of the measurements are reported elsewhere (15), but the general procedures to obtain conductance and kinetic data for living polymers are well known (16,17)

RESULTS AND DISCUSSION

Vinylbenzocrown ethers can be polymerized anionically to high molecular weight polymers (11) by an anion living polymer-type mechanism. The two polymeric carbanions derived from VB15C5 and VB18C6 have been abbreviated as $P15C5^-$ and $P18C6^-$, respectively, and their structures are depicted below.

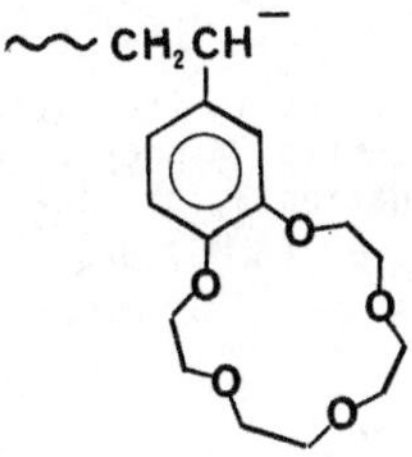

$P15C5^-$

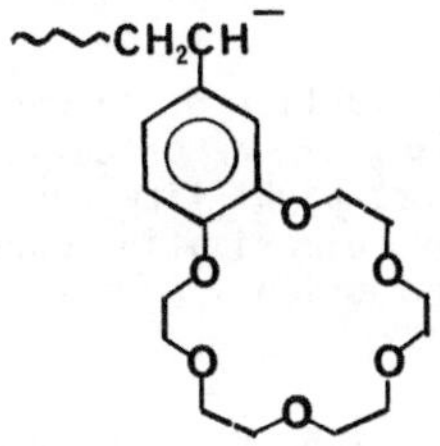

$P18C6^-$

The two poly (crown ether) carbanions are much less stable than the polystyryl carbanion. Generally, termination of the carbanions is faster the smaller the alkali counterion. For example, at 70°C the carbanion concentration of a 3×10^{-4} M solution of $P15C5^-K^+$ in THF decreases 8% in 18 hours, and at -60°C the decrease is 30% in 15 hours. With Li^+ as counterion $P15C5^-$ is very unstable even at -70°C (11). The sensitivity of the termination reaction to a change in counterion suggests that the cation may catalyze the reaction. This has also been observed in ring opening reactions of ethylene oxide with polystyryl (18) and fluorenyl carbanions (19), and in protonation reactions of radical anions with alcohols (20). A possible termination reaction for the poly (vinylbenzocrown ether) carbanion is the attack of the carbanion on the PhO-C bond of a terminal or nearby benzocrown ether group yielding a stable phenoxy ion.

<u>Conductance data</u>. In THF the main source of conducting species for polystyryl-type salts is the dissociation of ion pairs into free ions. However, depending on experimental conditions and the type of carbanion salt, triple ion formation ($A^-M^+ + M^+ \rightleftharpoons M_2^{2+}A^-$ and $A^-M^+ + A^- \rightleftharpoons M^+A_2^{2+}$) and dimerization of ion pairs ($2\,A^-M^+ \rightleftharpoons [A^-M^+]_2$) can also produce conducting species or affect their concentrations. Therefore, the conductance data were analyzed by means of the Kraus-Bray equation 1 (26), the Fuoss relationship 2 derived for systems where two types of triple ions are formed ($M^+A_2^{2-}$ and $M_2^{2+}A^-$) (22) and the Wooster expression 3 which is applicable when only one kind of triple ion is present (23).

$$1/\Lambda = 1/\Lambda_o + C\Lambda/\Lambda_o^2K_d \qquad (1)$$

$$\Lambda = (\Lambda_oK_d^{1/2}) / C^{1/2} + (\lambda_oK_d^{1/2}K_t)C^{1/2} \qquad (2)$$

$$\Lambda^2C = \Lambda_o^2K_d + (2\Lambda\lambda_o - \Lambda_o^2)K_dK_tC \qquad (3)$$

In these equations C denotes the total carbanion concentration, $_o$ and $_o$ are the equivalent conductivities at infinite dilution of the free ions and triple ions, respectively, K_d the ion pair dissociation constant and K_t the formation constant for triple ions. Dimerization of ion pairs can be taken into account by using the modified Wooster equation 4

$$C\Lambda^2/(1-\Lambda/\Lambda_o)\alpha_2 = \Lambda_o^2K_d + 2(\Lambda_o\lambda_o-\Lambda_o^2)(1-\Lambda/\Lambda_o)K_dK_tC\alpha_2 \qquad (4)$$

where α_2 is the fraction of nonconducting species in the form of ion pairs. Values of α_2 can be calculated from equation 4 once K_t and K_d are obtained from the straight line portion of equation 3 at low carbanion concentration. The dimerization constant $K_{dim} = [(A^-M^+)_2]/[A^-M^+]^2$ can be derived from the expression

$$1/\alpha_2 = 1 + 2\,K_{dim}\alpha_2C(1-\Lambda/\Lambda_o) \qquad (5)$$

by plotting $1/\alpha_2$ vs. $\alpha_2C/1-\Lambda/\Lambda_o)$ (23).

One of the main problems is to obtain reliable Λ_o and λ_o values. A detailed discussion of how they were computed will be reported in a future publication. Suffice to say that for the homopolymer salts of the two vinylbenzocrown ethers ($P15C5^-M^+$ and $P18C6^-M^+$) the cation on dissociation of the ion pair is assumed to complex to a crown ligand of another poly(crown ether) molecule. However, on dissociation of the polystyryl salts of p-methoxystyrene ($PpMeOS^-M^+$) and of 3,4-dimethoxystyrene ($PDMeOS^-M^+$) the cation is not likely to bind to the weakly complexing methoxy substituents on the polymer chain. Rather, the cations are solvated by THF molecules, and for such cations Λ_o^+ at various temperatures are known (24,25). These values only change slightly when part of the THF solvation shell is replaced by a crown ether such as B18C6 or 15C5 (26). It is estimated that the accuracy of the calculated Λ_o values is about 25% and that of K_d approximately 50%.

Some of the conductivity results are depicted in Figures 1-3. For $P15C5^-K^+$ (Figure 1) and $P18C6^-K^+$ (Figure 2) as well as for $P18C6^-Cs^+$ the Kraus-Bray plots are linear. For these systems a slope of -0.5 is found by plotting log Λ vs. log C (for low degrees of dissociation equation 1 can be rearranged to read logΛ = -0.5 log C + constant). Slopes for the Fuoss and Wooster plots are zero. Apparently, under the chosen experimental conditions (10^{-3}-10^{-5}M salt concentration) formation of triple ions or ion pair association to dimers can be neglected, and conducting species only arise from ion pair dissociation into free ions. The respective K_d values are listed in Table I. They were computed from the slopes of the respective Kraus-Bray plots. The calculated Λ_o values for the three salts were 6.4 (-70°), 7.0 (-65°), 7.8 (-60°) and 8.6^o(-55°).

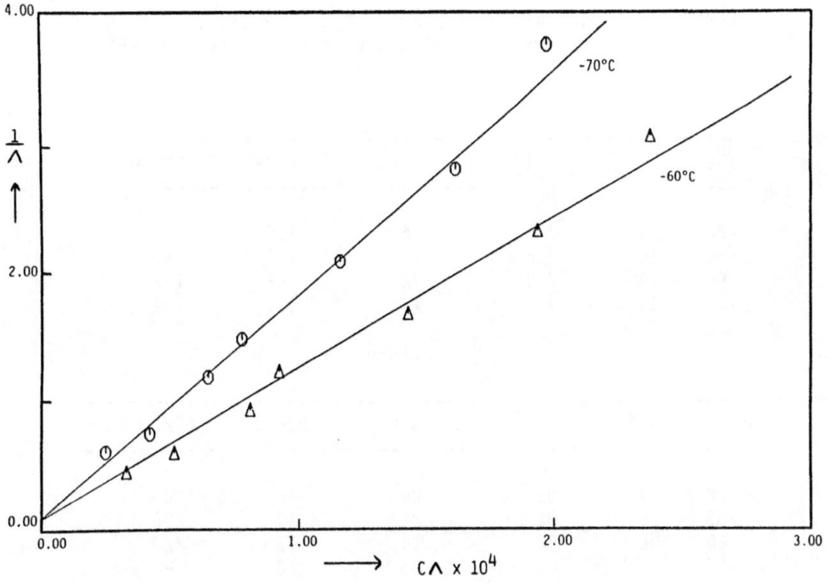

FIGURE 1 Kraus-Bray plots for $P15C5^{-}K^{+}$ at -60°C (△) and -70°C (○).

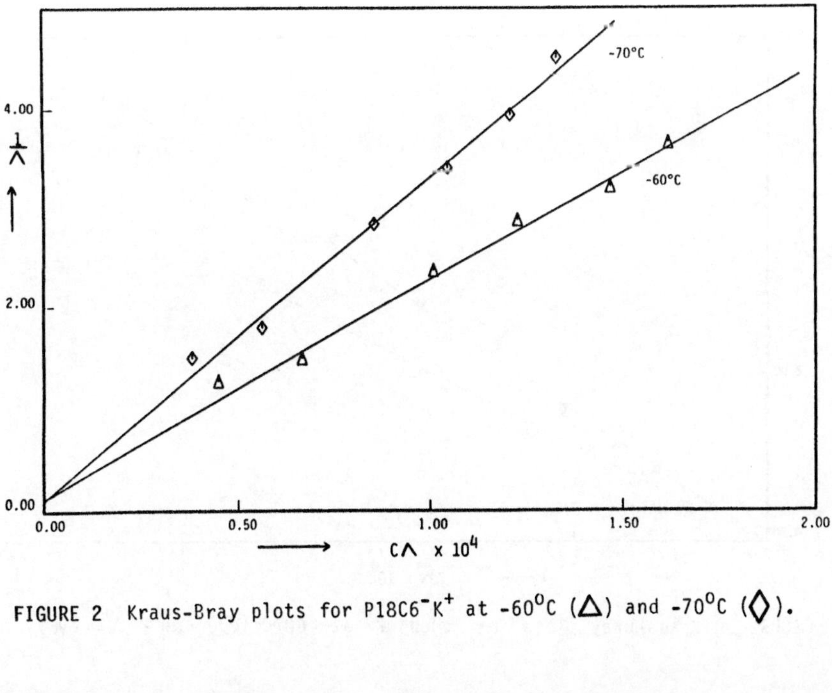

FIGURE 2 Kraus-Bray plots for $P18C6^{-}K^{+}$ at -60°C (△) and -70°C (◇).

TABLE 1. Dissociation Constants, K_d, of Potassium Salts of Polystyryl and Related Carbanion Salts in THF[a]

$K_d \times 10^7$,M

T^oC	$P15C5^-K^+$	$P18C6^-K^+$	$P18C6^-Cs$	$PpMeOS^-K$	$PDMeOS^-K^+$
-70	14	7.5	66	1.4	1.2
-65	14	7.6	74	1.3	1.2
-60	14	7.4	72	1.2	1.2
-55	14	7.6	73	1.1	1.2

PS^-K^+/Ligand

	15C5	B15C5	18C6	B18C6	C222
-70	74	91	26	13	200
-65	70	96	26	13	220
-60	68	98	24	12	220
-55		105	23	12	--

(a) $K_d \times 10^7$,M for PS^-K^+ is 0.74 at -70°C and 0.52 at -50°C, and for PS^-Cs^+ it is 0.09 at -70°C.

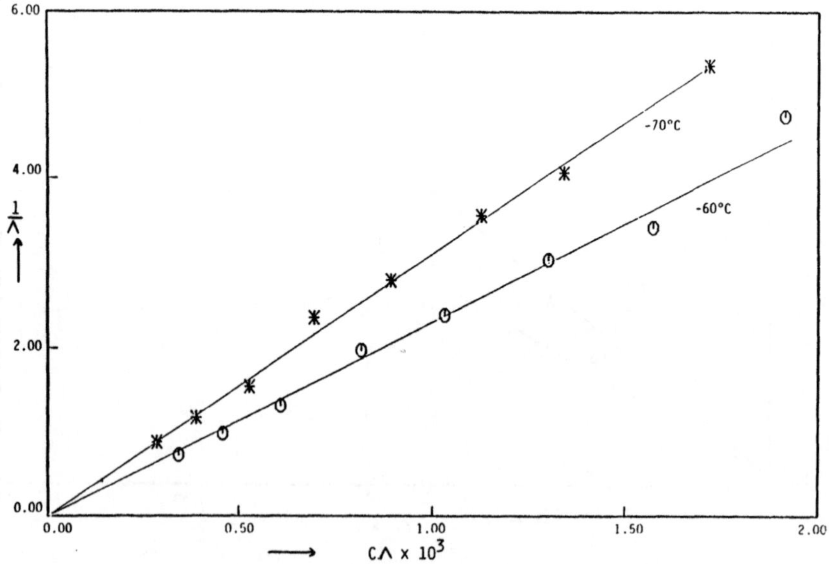

FIGURE 3 Kraus-Bray plots for $PpMeOS^-K^+$ at -60°C (○) and -70°C (✱).

The dissociation constants of PS^-K^+, PS^-Cs^+, $PpMeOS^-K^+$ and $PDMeOS^-K^+$ were measured in order to determine the effect of meta and para methoxy substituents. The Kraus-Bray plots of these polystyryl salts are all linear (see Figure 3 for $PpMeOS^-K^+$) and under the experimental conditions no measurable formation of triple ions or dimerization can be detected. The respective K_d values are also listed in Table 1. For the three potassium salts the Λ_0 values are 15.2 (-70^0), 17.0 (-65^0), 18.7 (-60^0) and 20.5 (-55^0) (15).

Inspection of the K_d values of the various substituted polystyryl salts reveals the following. A methoxy group at the para position ($PpMeOS^-M^+$) increases K_d by a factor 2 relative to the K_d of PS^-K^+, while an additional meta OCH_3 substituent ($PDMeOS^-K^+$) has little or no effect (Table I). The increase in K_d appears to contradict published data by Favier et al. (27). Their $K_d \times 10^9$ (M) values for the cesium salts in THF at 0^0C are 6.6 for PS^-, 3.2 for $PpMeS^-$ and 1.6 for $PpMeOS^-$. A decrease seems reasonable since the electron-donating effect of the substituents should increase the charge density on the carbanion and make ion pair dissociation more difficult. However, in studies with substituted benzyl and styryl carbanions Van Beylen et al (28) have found that the carbanion causes the p-methoxy substituent to rotate out of the plane of the benzene ring. This decreases the electron-donating mesomeric effect of the methoxy group and its electron-withdrawing inductive effect becomes dominant. This in turn weakens the ionic interaction and could explain the higher K_d for $PpMeOS^-K^+$.

It is obvious that the "methoxy" effect does not explain the high K_d values for the poly (crown ether) salts. Relative to $PDMeOS^-K^+$ the K_d for $P15C5^-K^+$ is about twelve times higher, and replacing PS^-Cs^+ with $P18C6^-Cs^+$ increases K_d by a factor 700. The potassium ion of the $P15C5^-K^+$ ion pair is most likely intramolecularly bound to a crown ligand of the penultimate unit (or to one that can closely approach the counterion) rather than to the terminal crown unit since this would require a much larger stretching of the interionic ion pair distance. It can be argued that intramolecular complexation, while stretching the ionic bond of the tight ion pair, would nevertheless hinder the release of the counterion. However, the energy needed to break the complex is more than compensated for by the complexation of the dissociated cation with the crown ligands of another P15C5 chain, especially because the free K^+ ion forms a stable 2:1 crown-cation complex. A similar compensating effect is expected on dissociating $P18C6^-K^+$, except that in this system the carbanion pair and the free K^+ ion both form 1:1 complexes with the B18C6 ligand. The small difference between the K_d's for $P15C5^-K^+$ and $P18C6^-K^+$ can have a number of reasons, e.g., differences in interionic ion pair distances or in the complexation energies of the various complexes.

The ten times higher K_d of $P18C6^-Cs^+$ relative to the potassium salt is remarkable since at -70^0 in THF the K_d for PS^-Cs^+ is eight times less than that of PS^-K^+. In the latter salts dissociation is from tight ion pairs, but even if the interionic ion pair distance in PS^-Cs^+ would be larger, stronger interaction of THF molecules with the free K^+ ion favors the dissociation of PS^-K^+ (24,29). On the other hand, P18C6 is an excellent complexer of Cs^+ since this cation forms 2:1 complexes with the pendant B18C6 ligands, while K^+ only forms 1:1 complexes (5). This facilitates the dissociation of $P18C6^-Cs^+$, where the interionic ion pair distance may also be larger than in the potassium salt.

Further insight in the behavior of poly(crown ether) carbanion salts was gained from conductance measurements on PS^-K^+ in the presence of a large excess of the cation-binding ligands 15C5, B15C5, 18C6, B18C6, and C222 (cryptand 2.2.2). It was assumed that under our experimental conditions PS^-K^+ and the free K^+ cation are all complexed to one or two ligand molecules. The conductivity data were analyzed by a curve-fitting procedure in which through optimizing K_d and K_t the calculated values matched the experimental Kraus Bray plot. The latter plot exhibited a curvature at higher salt concentration indicating triple ion formation. The K_d values for the various system are listed in Table 1.

The highest dissociation is found for PS^-K^+/C222. Since the K^+ ion forms an inclusion complex with the cryptand, the interionic ion pair distance is large. This is also the case with 15C5 and B15C5 both of which form sandwich-type 2:1 crown-cation complexes with K^+. For example, B15C5 added to fluorenyl potassium in THF yields a 2:1 crown-separated ion pair (30) although with a more localized carbanion salt such as PS^-K^+ a 1:1 complex may exist as well. Note that the K_d for PS^-K^+/B15C5 exceeds that of $P15C5^-K^+$ by a factor 7 in spite of the fact that the proximity of B15C5 ligands on a P15C5 chain and the formation of 2:1 complexes of K^+ with this ligand make P15C5 a much better complexer for K^+ than B15C5 itself (5). However, B15C5 converts PS^-K^+ into a 2:1 crown-separated ion pair, while $P15C5^-K^+$ is a 1:1 intramolecularly complexed tight ion pair. The difference in K_d between PS^-K^+/B18C6 and $P18C6^-K^+$ is much smaller since in both cases the ion pair as well as the free cation form 1:1 crown complexes and the lengths of the ionic bonds in the two ion pair complexes are not expected to differ by much.

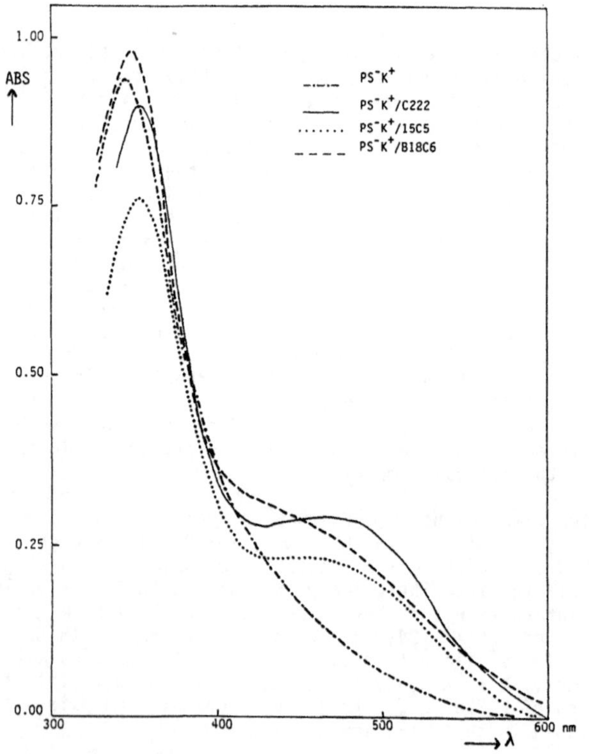

FIGURE 4 Optical spectra of PS^-K^+, PS^-K^+/C222, PS^-K^+/15C5 and PS^-K^+/B18C6, all in THF at -70°C.

More direct information on interionic ion pair distances in the crown complexes can be derived from spectral data. The conversion of a tight to a loose (solvent- or ligand-separated) ion pair in carbanion salts is often accompanied by significant changes in their optical spectra (30,31). Although spectral changes in polystyryl salts are less dramatic than in fluorenyl carbanion salts, optical shifts associated with ion pair equilibria have been reported for poly(α-methylstyryl) and polystyryl salts (32,33). For polystyryl-type carbanions the main absorption band around 340 nm shifts to about 350 nm for a separated ion pair while a new band becomes visible at 460 nm. The extent to which the latter band is developed is related to the fraction of loose ion pairs, or to the extent to which the length of the PS^-K^+ bond is increased on adding a ligand. Figure 4 shows that the 460 nm band is clearly visible for PS^-K^+/C222 in THF at 70°. It is somewhat less developed but still clearly visible for PS^-K^+/15C5 and PS^-K^+/B15C5. For PS^-K^+/B18C6 and PS^-K^+/18C6 the spectrum differs from PS^-K^+ but the 460 nm band is only a shoulder, indicating a predominance of 1:1 externally-complexed tight ion pairs. Interestingly, very little change and almost no 460 band or shoulder are found in the spectra of $P15C5^-K^+$ and $P18C6^-K^+$ (Figure 5), confirming the earlier conclusion that these ion pairs are intramolecularly, externally complexed tight ion pairs.

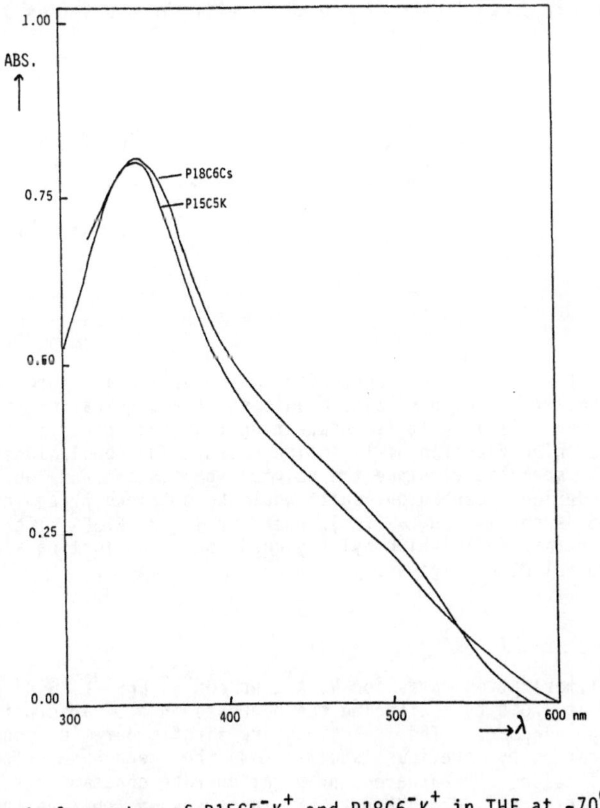

FIGURE 5 Optical spectra of $P15C5^-K^+$ and $P18C6^-K^+$ in THF at -70^oC.

An attempt was made to investigate carbanion salts in which only the terminal or the last two units of the polymer chain contained a crown ether ligand. This could shed light on the role of these ligands in determining the structure of the ion pair. Starting from PS^-K^+ an equivalent amount of VB15C5 or twice the amount was slowly added to the polystyryl salt. It is argued in the next section that the addition of PS^-K^+ to VB15C5 is probably fast relative to the homopolymerization step $P15C5^-K^+$ + VB15C5. Hence, the main species on adding equivalent amounts of VB15C5 is a polystyrene chain with a terminal VB15C5 unit (abbreviated as $PS15C5^-K^+$). Whether the rate of addition of a second VB15C5 molecule will differ from that of the homopolymerization step cannot be predicted, but we have assumed formation of chains with two VB15C5 units realizing that it is more likely a mixture of chains with one, two, or more VB15C5 units (abbreviated as $PS(15C5)_2^-K^+$).

The conductance of the two salts shows a rather complex behavior, and a curve fitting procedure indicates a rather high dimerization constant in the order of $10^3 M^{-1}$ (15). Although dimerization of crown-complexed ion pairs may be surprising at first sight, it is not without precedent. B18C6-separated ion pairs of fluorenyl sodium in ethereal solvents including THF at room temperature were reported to dimerize rather extensively, while the externally complexed tight ion pairs dimerize much less (34). Dimerization of $PS15C5^-K^+$ could result from an interaction of the terminal crown ligand of one chain with the counterion of another chain (I)

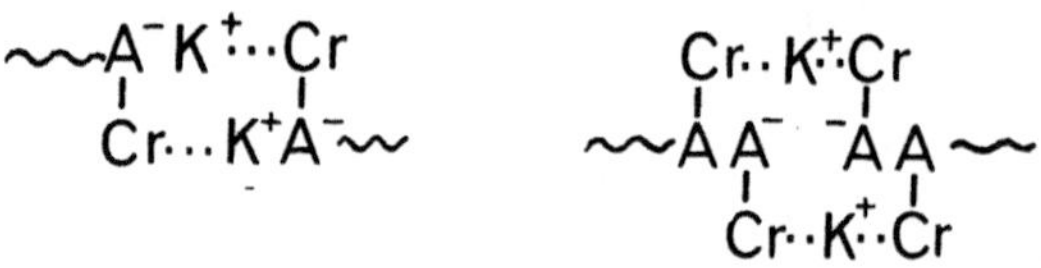

I II

With $PS(15C5)_2^-K^+$ the counterion may conceivably form a 2:1 complex involving its own penultimate unit and the terminal crown ligand of a second chain (II). It is interesting that the spectrum of $PS(15C5)_2^-K^+$ shows a high fraction of loose ion pairs. The conclusions of course are somewhat speculative since the polymer species are not well defined. A better defined carbanion salt would be obtained by adding to PS^-K^+ a compound such as $CH_2 = C(Cr)_2$ where CR is a B15C5 ligand. Only one molecule (as in 1,1-diphenylethylene) would add forming a carbanion with two terminal B15C5 ligands.

Kinetic Studies.

Polymerization rates for PS^-K^+, $PpMeOS^-K^+$ and $P15C5^-K^+$ were determined in THF at -70°C by monitoring the decrease in a UV absorption band of the reacting monomer. The reactions are first order in monomer (M). As demonstrated by previous studies on the mechanism of anionic vinyl polymerization, the apparent propagation rate constant, k_p, computed from the expression $(-dM/dt)/M = k_p[LE]$ increases on lowering the concentration of living ends, LE (16,17,31,35). This dependence of k_p on [LE] results

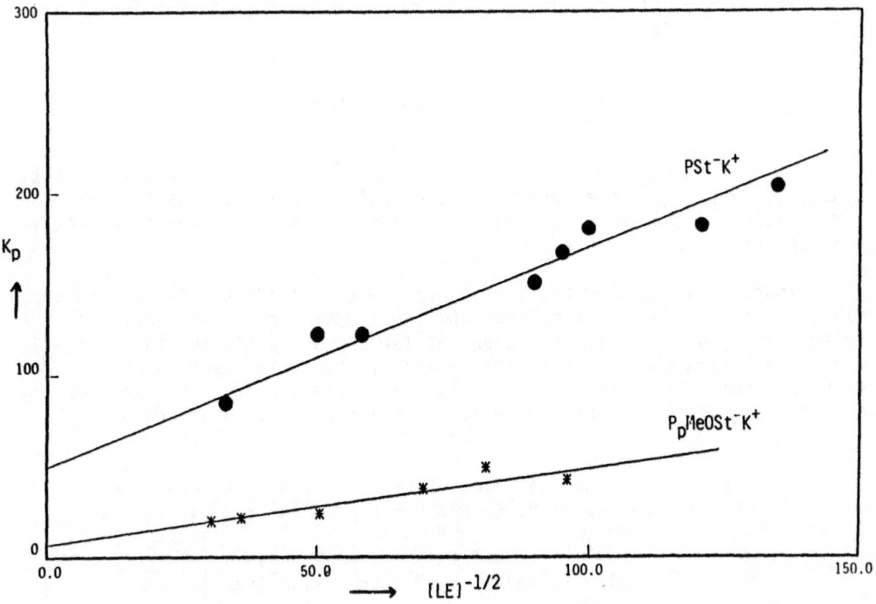

FIGURE 6 Plot of apparent homopropagation rate constant, k_p, vs. [living ends]$^{-1/2}$ in THF at -70°C for PS^-K^+ (O) and $PpMeOS^-K^+$ (✱).

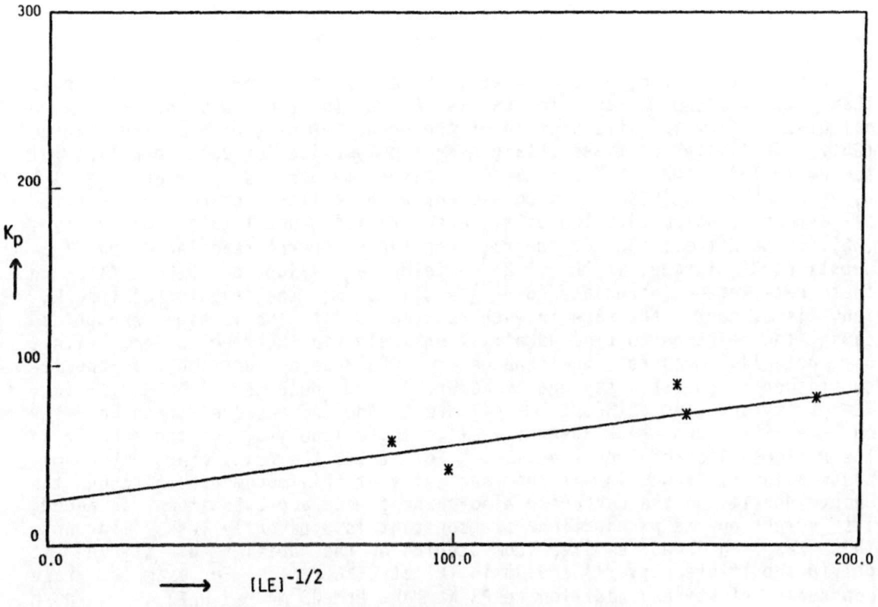

FIGURE 7 Plot of apparent homopropagation rate constant, k_p, vs. [living ends]$^{-1/2}$ in THF at -70°C for $P15C5^-K^+$ (✱).

from the high reactivity of free ions relative to those of the ion pairs, and is given by the expression

$$k_p = k_i + (k_f - k_i)K_d^{1/2}[LE]^{-1/2} \quad (4)$$

where k_i and k_f are the respective ion pair and free ion propagation rate constants and K_d is the ion pair dissociation constant. As long as the dissociation of ion pairs is limited, [LE] may be taken as the total active end concentration.

Figures 6 and 7 show plots of k_p vs. $[LE]^{-1/2}$ for the three monomers. Values for k_i and k_f computed from the slopes and intercepts of the respective plots and the K_d values of Table 1 are collected in Table 2. The long extrapolation and scattering of the data renders the k_i for $P15C5^-K^+$ somewhat uncertain. The k_f values are subject to experimental errors in K_d as the slopes from which they are calculated depend on $K_d^{1/2}$.

TABLE 2. Propagation Rate Constants of Ion Pairs and Free Anions of PS^-K^+, $PpMeOS^-K^+$ and $P15C5^-K^+$ in THF at -70°C.

Living Polymer	$k_f, M^{-1}sec$	$k_i, M^{-1}sec^{-1}$
PS^-K^+	4000	56
$PpMeOS^-K^+$	950	9
$P15C5^-K^+$	270	26

In spite of the uncertainties in k_f and k_i it is gratifying to note that $k_f = 4000\ M^{-1}sec^{-1}$ for PS^- is close to the value of about 4500 calculated from an extrapolation of the Arrhenius plot of the free anion (35). The latter plot was obtained from polymerization data on polystyryl sodium in THF between 25° and -50°C. Also, the decreasing order $k_f(PS^-) > k_f(PpMeOS^-) > k_f(P15C5^-)$ is consistent with earlier findings on the rates of anionic copolymerization of styrene with its p-substituted derivatives (36). The Hammett plot of the rate constants for the reaction PS^-Na^+ + p-substituted styrene in THF at 25°C yielded a value of 5.0. Although these data were obtained at $[PS^-Na^+] = 3 \times 10^{-3}M$, the fraction of free PS^- ions is of course the same in each reaction. It also is high enough to cause the reaction to proceed almost entirely via the free anion, (i.e., one actually measures the change in $k_f(PS^-)$ as a function of the p-substituent on the styrene monomer. The decrease in k_f for p-methoxystyrene and VB15C5 is mainly due to the increased electron density on the vinyl bond which adversely affects the reactivity of the monomer in the nucleophilic addition reaction. For VB15C5 the bulky crown ether may cause a further reduction in the reactivity of this monomer. Although the charge density on the carbanion also changes when a p-substituent is added, its effect on the propagation rate constant is apparently less important. This was concluded earlier from studies on the addition of styrene to poly(p-substituted styryl) sodium in THF at 25°C (36). For example, rate constants of styrene addition to PS^-Na^+ and $PpMeOS^-Na^+$ at $[LE] = 3 \times 10^{-3}M$ are 950 and 1150 $M^{-1}sec^{-1}$, respectively. Although at $3 \times 10^{-3}M$ the fractions of free ions for the two carbanion salts are not identical, the difference in free ion concentrations are not more than a factor 1.4 since the K_d values differ by only a factor 2 (Table 1). Hence, the reactivities of PS^- and $PpMeOS^-$ for styrene addition can not differ by much, since at [LE] = 3 x 10^{-3}M the bulk of the reaction again proceeds via the free ion. A similar conclusion was arrived at by Hirohara and Ise (37), but Favier et al. obtained different results for the reaction of PS^-Cs^+ and $PpMeOS^-Cs^+$ with 1,1-diphenylethylene (27). Their k_f value for PS^- in THF at 0° is ten times that for $PpMeOS^-$. As noted earlier, their K_d for the methoxy salt is four times lower than for PS^-Cs^+, contrary to our results obtained for the potassium salts at -70°C. These apparent discrepancies may result from the difficulty in obtaining accurate dissociation and rate constants in these rather complex and sensitive systems. It also could reflect on our inadequate knowledge on the role of counterions, carbanion structure and other variables on the rate processes in these systems in spite of the considerable body of information presently at our disposal.

The ion pair rate constant k_i = 56 $M^{-1}sec^{-1}$ for PS^-K^+ is five times less than k_i for PS^-Na^+ (16). This rate constant is a composite given by the expression $k_i = fk_{is} + (1-f)k_{it}$, were f is the fraction of loose ion pairs with rate constant k_{is} and (1-f) the fraction of tight ion pairs with rate constant k_{it} (31,38). The equilibrium between tight and loose (solvent or ligand-separated) ion pairs for carbanions like PS^-M^+ or fluorenyl salt is exothermic, and in solvents such as THF the fraction, f, generally increases in the order Cs^+<K^+<Na^+. Also, in anionic vinyl polymerization k_{is} is usually much larger than k_{it}. At -70°C the fraction f for PS^-Na^+ is considerably larger than for PS^-K^+ and high enough so that fk_{is}>>(1-f)k_{it}. At 25°C the k_i value for PS^-NA^+ and PS^-K^+ do not differ much since f is so small that the term (1-f)k_{it} also contributes to k_i. The rate constant for tight ion pairs of PS^-M^+ salts increases with the size of the counterion as shown from data obtained in dioxane (17), hence, $k_{it}(PS^-Na^+) < k_{it}(PS^-K^+)$.

The six times lower value of k_i for $PpMeOS^-K^+$ relative to that of PS^-K^+ is consistent with the trend in the k_- values and the lower reactivity of the p-methoxystyrene monomer. Again it should be pointed out that comparison of the two k_i values should also take into consideration differences in the f value for the two carbanion salts. However, since at -70°C the two ion pair dissociation constants differ by only a factor 2, the fractions of loose ion pairs formed by partial separation of the two ions is not likely to be very different for PS^-K^+ and $PpMeOS^-K^+$. The situation is different for $P15C5^-K^+$. The k_i for VB15C5 is higher than for p-methoxystyrene while the reverse is true for k_f. As argued earlier, spectral and conductivity data suggest an increased interionic ion pair distance for $P15C5^-K^+$ as a result of intramolecular complexation between counterion and a crown ether ligand. This increases the reactivity of the ion pair. The same argument was used in rationalizing the high reactivity of ion pairs of sodium poly(2-vinylpyridinyl) (39) and of poly(o-methoxystyryl) (40). An interesting difference between $P15C5^-M^+$ and PS^-M^+ is that for PS^-M^+ the fraction of loose ion pairs rapidly increases on lowering the temperature, at least for PS^-Na^+. This can give rise to negative apparent activation energies as found for k_i for PS^-Na^+ in THF (31,35,38). Because of its intramolecular complexation the ion pair structure of $P15C5^-M^+$ may not significantly change with temperature. This means that the activation energy for ion pair propagation may be close to that of the free ion. In such a case the k_i for $P15C5^-K^+$ at room temperature (if it could be measured) is expected to be much higher than for PS^-K^+.

Another interesting question is whether prior to addition, the crown ligand of the monomer complexes with the counterion, especially in systems where a 2:1 crown-cation complex can exist. This may not play a significant role in the propagation of $P15C5^-K^+$ since the K^+ is already complexed intramoleculary to a crown ligand and after addition of a VB15C5 molecule the counterion is probably not bound to the terminal crown ligand because it requires too large a separation between the two ions. However, in the reaction of PS^-M^+ with VB15C5 the monomer, through its crown ligand, can form a complex with the counterion similar to those formed with a B15C5 ligand. This probably means a high rate constant for the reaction PS^-M^+ + VB15C5, and it would be interesting to study this reaction, both in the absence and presence of a salt such as $M^+BPh_4^-$. The excess M^+ ions would complex with the monomer and prevent its complexation with PS^-M^+. This in turn may drastically affect the rate of addition of VB15C5.

Acknowledgment. This research was supported in part by a grant from NATO, by the Belgian Ministry of Scientific Programming and by the Polymers Program of the National Science Foundation.

REFERENCES

1. F. de Jong and D.N. Reinhoudt, Adv. Phys. Org. Chem. 17, 279 (1980).
2. R.M. Izatt, J.S. Bradshaw, S.A. Nielsen, J.D. Lamb and J.J. Christensen, Chem. Rev. 85, 271 (1985).
3. J.M. Lehn, Acc. Chem. Res. 11, 49 (1978).
4. D.J. Cram, Science 219, 1177 (1983).
5. J. Smid, Topics Curr. Chem. 121, 105 (1984).
6. Ph. Gramain and Y. Frere, Macromolecules 12, 1038 (1979).
7. S. Kopolow, Z. Machacek, U. Takaki and J. Smid, J. Macromol. Sci. Chem. A7, 1015 (1973).
8. S.C. Shah, S. Kopolow and J. Smid, Polymer 21, 189 (1980).
9. K. Kimura, T. Maeda and T. Shono, Anal. Lett. A11, 821 (1978).
10. U. Takaki and J. Smid, J. Am. Chem. Soc. 96, 2588 (1974).
11. S. Kopolow, T.E. Hogen Esch and J. Smid, Macromolecules 6, 133 (1973).
12. J. Smid, B. El Haj, T. Majewicz, A. Nonni and R. Sinta, Org. Prep. Proc. Int. 8, 193 (1976).
13. K. Kikukawa, K. Nagira and T. Matsuda, Bull. Chem. Soc. Japan 50, 2270 (1977).
14. C. J. Pedersen, J. Am. Chem. Soc., 89, 7017 (1967).
15. F. Gils, Ph.D. Thesis, Kath. Univ. of Leuven, 1983.
16. D.N. Bhattacharyya, C. L. Lee, J. Smid and M. Szwarc, J. Phys. Chem. 69, 612 (1965).
17. M. Szwarc, "Carbanions, Living Polymers and Electron Transfer Processes" Interscience, New York, 1968.
18. A.A. Solovyanov and K.S. Karanskii, Vysokomol. Soyed. A16, 595 (1974).
19. C.J. Chang, R.F. Kiesel and T.E. Hogen Esch, J. Am. Chem. Soc. 95, 8446 (1973).
20. A. Rainis, R. Tung and M. Szwarc, J. Am. Chem. Soc. 95, 659 (1973).
21. C.A. Kraus and J. Bray, J. Am. Chem. Soc. 35, 1315 (1913).
22. R.M. Fuoss and C.A. Kraus, J. Am. Chem. Soc. 55, 2387 (1933).
23. C.B. Wooster, J. Am. Chem. Soc. 59, 377 (1937).
24. C. Carvajal, K.J. Tölle, J. Smid and M. Szwarc, J. Am. Chem. Soc. 87, 5548 (1965).
25. S. Boileau and P. Hemery, Electrochim. Acta 21, 647 (1976).
26. R. Ungaro, B. El Haj and J. Smid, J. Am. Chem. Soc. 98, 5198 (1976).
27. J.C. Favier, P. Sigwalt and M. Fontanille, J. Polym. Sci., Polym. Chem. Edit. 15, 2373 (1977).
28. a. G. Van Ermen, S. Toppet, P. Geerlings and M. Van Beylen, J. Chem. Soc. Perk. Trans. II, 699 (1986).
    b. G. Van Ermen, M. Van Beylen and P. Geerlings. J. Phys. Chem. 90, 603 (1986).
29. T. Ellingsen and J. Smid, J. Phys. Chem. 73. 2712 (1969).

30. U. Takaki, T.E. Hogen Esch and J. Smid, J. Am. Chem. Soc. 93, 6760 (1971).
31. Ions and Ion Pairs in Organic Reactions, Vol. II., M. Szwarc, Ed. Wiley Interscience, 1974.
32. J.M. Ginn and K.J. Ivin, Makromol. Chem. 139, 57 (1970).
33. F. Peeters, F. Gils and M. Van Beylen, Proceedings 28th Macro Symposium, Amherst, 104 (1982).
34. U. Takaki and J. Smid, J. Phys. Chem. 76, 2152 (1972).
35. G.V. Schulz, L.L. Bohm, M. Chmelir, G. Lohr, and B.J. Schmitt, Adv. Polym. Sci. 9, 1 (1972).
36. M. Shima, D.N. Bhattacharyya, J. Smid and M. Szwarc, J. Am. Chem. Soc. 85, 1306 (1963).
37. H. Hirohara, N. Nakayama and N.Ise, J. Chem. Soc. Faraday I 68, 58 (1972).
38. T. Shimomura, K.J. Tolle, J. Smid and M. Szwarc, J. Am. Chem. Soc. 89, 796 (1967).
39. M. Fisher and M. Szwarc, Macromolecules 3, 23 (1970).
40. J. Geerts, M. Van Beylen and G. Smets, J. Polym. Sci. A1, 2859 (1969).

KINETICS AND THERMODYNAMICS OF ANIONIC POLYMERIZATION OF 2-ISOPROPENYL PYRIDINE IN TOLUENE

ADIB ABOUDALLE, Laboratoire de Recherches sur les Macromolecules, Universite de Paris Nord, 93430 Villetaneuse, France
ALAIN SOUM*, MICHEL FONTANILLE, Laboratoire de Chimie des Polymeres Organiques, Institut du Pin, 33405 Talence Cedex, France and
THIEO E. HOGEN-ESCH, Center for Macromolecular Science and Engineering and Department of Chemistry, University of Florida, Gainesville, FL 32611, USA

INTRODUCTION

The thermodynamics and kinetics of anionic vinyl polymerization has been the subject of several investigations over the last few decades. [1-9] Of these, the anionic polymerization of alpha-substituted vinyl monomers [3-9] was most readily studied because of their relatively low ceiling temperature. In these studies, both the polymerization rate constants and equilibrium constants have been determined but the rate of depolymerization has not been measured directly, due to a number of reasons such as the lack of stability of the propagating carbanions and the rapid rates of depolymerization under conditions of interest. We now wish to report the direct determination of both the rates of polymerization and depolymerization of 2-isopropenyl pyridine (2-IP) in toluene with Li^+ as counter-ion, and the comparison of the thermodynamic results with the kinetic data.

The present system was chosen since the rates of polymerization and depolymerization are slow enough to allow determination by ^1H NMR. Furthermore, under the reaction conditions, the occurrence of side-reactions could be kept minimal, in contrast to the case of 2-vinyl pyridine (2-VP) in THF [10,11] and alpha-methyl styrene (α-MS) in benzene and in bulk [12,13] that are both similar to 2-IP.

EXPERIMENTAL

The synthesis of 2-IP, purification of reactants and polymerizations were carried out as previously reported [9,14].

Kinetic studies of polymerization were performed as described for 2-VP [14], by measuring the disappearance of the ^1H NMR signal of the vinylic proton of 2-IP (δ_{TMS} in toluene = 5,3 ppm) with time in a polymerization initiated by living oligomers ($DP_n \cong 10$). These oligomers were prepared by reacting 2-IP with 1,1-diphenyl hexyl lithium in toluene at 20°C.

The kinetics of depolymerization was studied by following, by ^1H NMR, the appearance of the vinylic proton of the monomer with time. The living polymer was first cooled to -35 °C for complete polymerization and then rapidly heated to the temperature chosen for the depolymerization studies. In all cases, the thermal equilibrium is reached rapidly compared to the rate of depolymerization. Depending on the temperature range, the equilibrium concentration $[M_e]$ as a function of temperature was measured by ^1H NMR, gas chromatography (GC) (using as internal standard hexamethyldisiloxane for NMR and n-octane for GC) or by UV spectroscopy. In NMR and GC experiments the values of $[M_e]$ were determined by comparison with standard monomer solutions. In UV studies the values of $[M_e]$ were directly calculated from the optical density of monomer absorption at 290 nm (ε_{290} monomer = 2600 $l.mol^{-1}.cm^{-1}$) after substracting the absorption of carbanionic species at the same wavelength (ε_{290} = 15000 $l.mol^{-1}.cm^{-1}$).

Recent Advances in Anionic Polymerization
T.E. Hogen-Esch and J. Smid, Editors

NMR experiments were performed on a WP 60 DS BRUKER NMR spectrometer; UV determinations were carried out on a CARY 219 spectrophotometer and GC measurements were done on a HEWLETT PACKARD 5740G gas chromatograph.

RESULTS

Poly (2-IP) prepared by anionic polymerization in THF and toluene, with Li^+, K^+, Cs^+ or Mg^{2+} as counter-ion, was previously studied [9] by NMR and size exclusion chromatography (SEC). The number average molecular weight (determined from the monomer/initiator ratio) was in good agreement with the experimentally determined value and the molecular weight distribution was found to be narrow ($M_w/M_n \cong 1.07$). This indicates that the initiation is fast and complete and that propagation occurs without perceptible side reactions.

A detailed study of the stability of the active centers, carried out by UV spectroscopy, demonstrated that the absorption of the carbanionic species remained unchanged over several days at temperatures ranging from 20°C to 60°C. At 80°C, the decrease in concentration is relatively small (about 6% after 20 hours of reaction) (Figure 1). The disappearance of the living polymer follows first order kinetics and the apparent rate constant of degradation, k_d, is equal to : 1.5 (± 0.05) 10^{-6} s^{-1} at 80°C. Thus, the rate of degradation is much lower than that of the polymerization and depolymerization reactions and, therefore, kinetic and thermodynamic experiments could reliably be performed below this temperature.

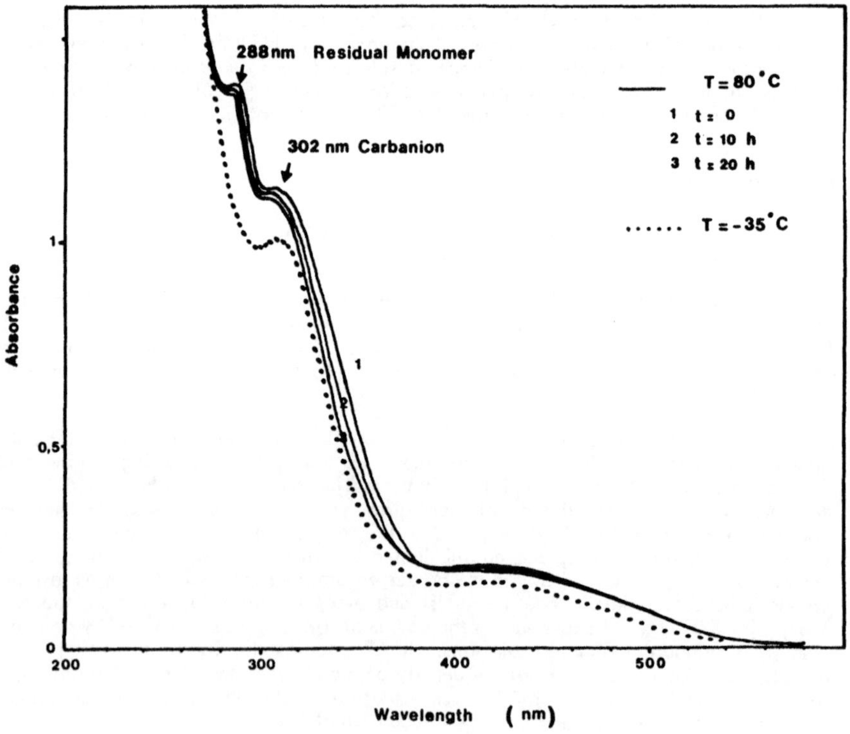

FIG. 1 (···) UV absorption of the lithium salt of living 2-poly(2-isopropenylpyridine)

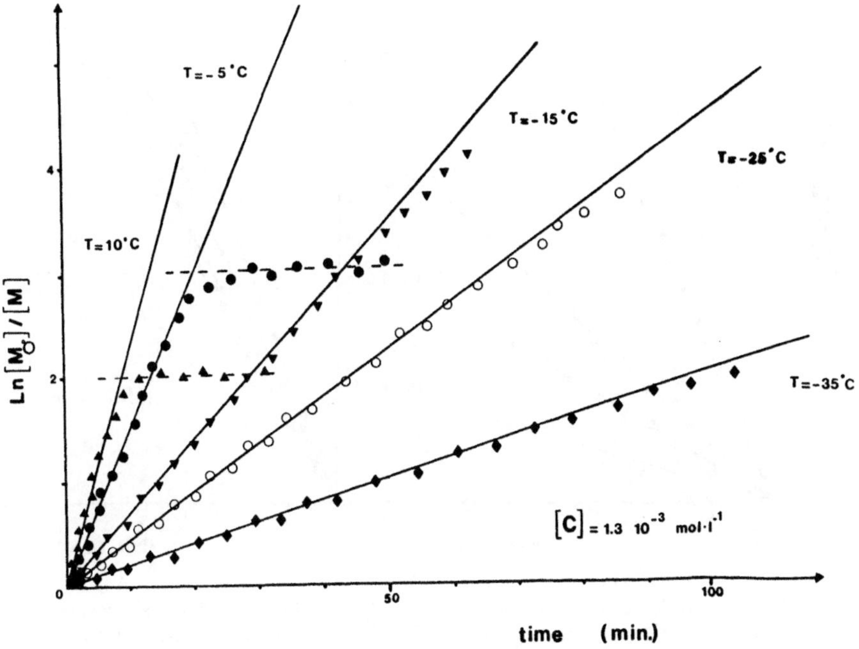

FIG. 2. Pseudo first-order plots of monomer disappearance in the anionic polymerization of 2-IP in toluene initiated by 1,1-diphenyl hexyl lithium at 10, -5, -15, -25, -35°C.

Kinetic Studies

The rates of polymerization of 2-IP initiated by living oligomers in toluene are first order in carbanion (5.10^{-4} mol.l^{-1} < [C] < 5.10^{-3} mol.l^{-1}) and monomer (0.015 mol.l^{-1} < [M] < 0.5 mol.l^{-1}) concentrations at temperatures varying from -50°C to +30°C. Figure 2 shows typical examples for several temperatures. The reactions are first order in monomer up to about 75% conversion. At higher temperatures the monomer concentration deviates from first order behavior due to the influence of the polymerization-depolymerization equilibrium below -5°C, the rate of depolymerization is negligible compared to the rate of polymerization). However the relationship : $\ln([M_o] - [M_e])/([M]-[M_e]) = k_p [C]t$ (Figure 3) established for equilibrium polymerization (4b, 6a), is linear over the entire temperature range. Figure 4 shows a linear plot of ln k_p versus 1/T. This linearity suggests that only one ionic species is contributing to the polymerization reaction. Furthermore, because of the very low values of the ion pair dissociation constants of 2-pyridine-stabilized carbanions in THF [15], there appears to be little doubt that the polymerization and depolymerization proceed through ion pairs.

Between 15 °C and +25° C the initial rate of depolymerization of poly (2-IP) lithium in toluene was also found to be first order in carbanion concentration so that $\ln([M_e]/([M_e] -[M])) = k_{-p}[C]\, t$ (Figure 3).

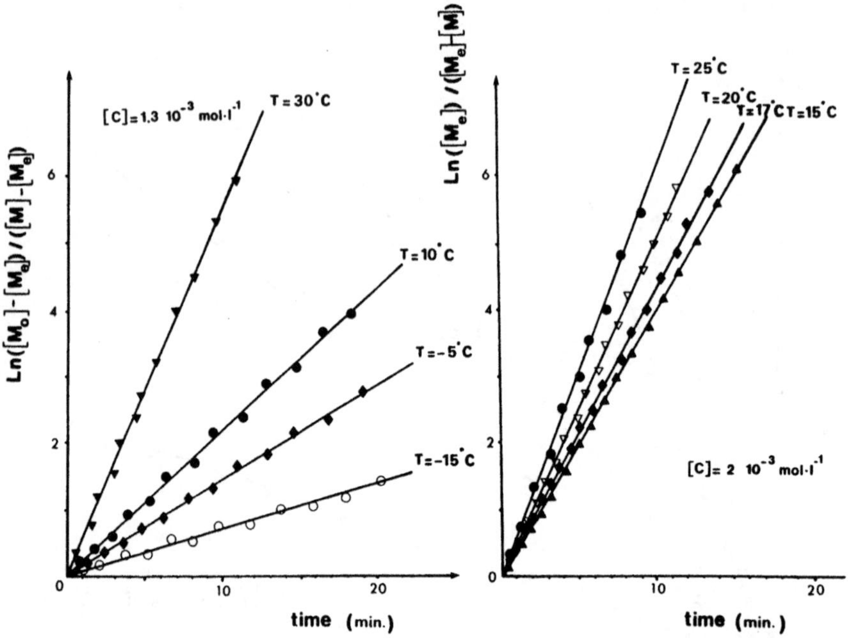

FIG. 3. First-order plots of monomer concentration in the anionic polymerization and depolymerization of poly(2-IP) lithium in toluene at various temperatures.

The initial slopes $(R_{-p})_o$ of the [M] versus time plots could also be determined and are equal to:

$$(R_{-p})_0 = \frac{d[M]}{dt} = k_{-p}[C]_o - k_p[C]_o[M]_o$$

For the depolymerization $[M]_o \cong 0$, therefore, the values of k_{-p} could be directly measured from the depolymerization reactions (Figure 5).

Figure 4 also shows the temperature dependence of the rate constants of depolymerization k_{-p}, measured by this method. As was the case for propagation, linear kinetics is observed. The activation enthalpies and entropies of the forward and backward reactions are given in Tables 1 and 2, along with similar data obtained by other authors on the equilibrium polymerization of α-MS and 2-IP.

Equilibrium Studies

Figure 6 shows the temperature dependence of the equilibrium constant K_e of monomer addition, calculated from the equilibrium monomer concentration using $K = 1/[Me]$ for $DP_n \cong 10$ [3].

The equilibrium monomer concentration was determined by ^{1}H NMR, GC and UV spectroscopy (UV was used to measure low values of $[M_e]$).

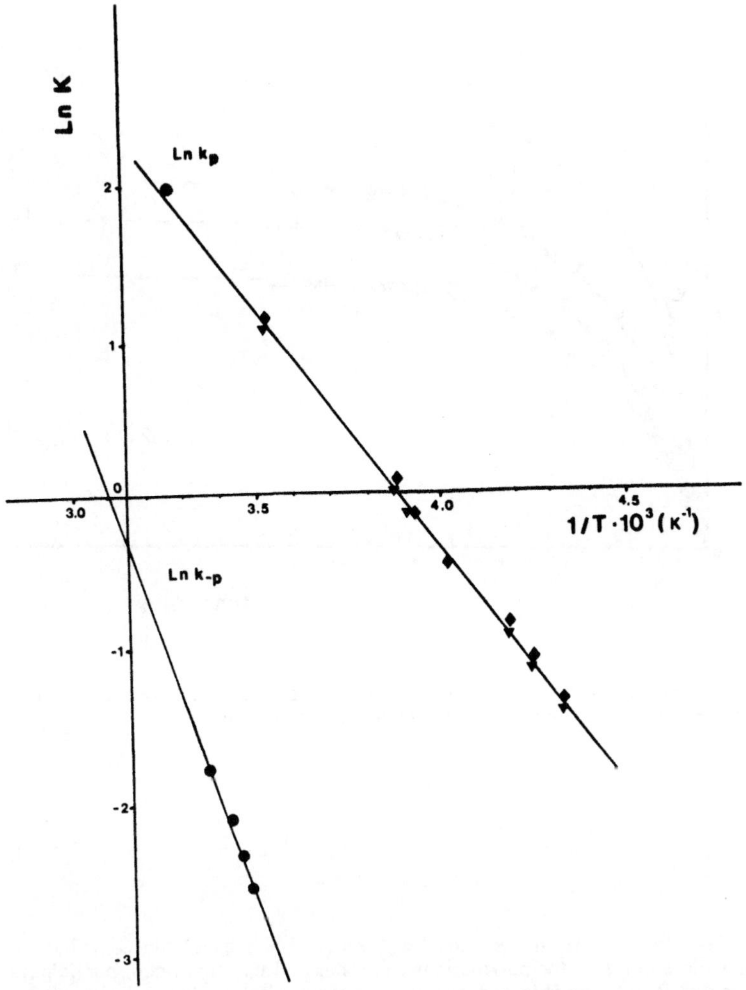

FIG. 4. Plots of $\ln K_p$ and $\ln k_{-p}$ as a function of temperature 3.0×10^{-3}M; 1.3×10^{-3}M.

Values for the equilibrium monomer concentration measured directly at different temperatures, and those computed from kinetic data ($[M_e] = k_{-p}/k_p$) together with the corresponding K_e values are collected in Table 3. The agreement between calculated and experimental values is satisfactory. We believe that this is the first case in which the propagation and depolymerization rate constants of vinyl polymerization have been directly determined along with the equilibrium constant.

DISCUSSION

The direct determination of the rate constant of depolymerization is an *uncommon feature in this research* and is made possible by the considerable stability of the 2-pyridyl-stabilized carbanions in the required temperature range.

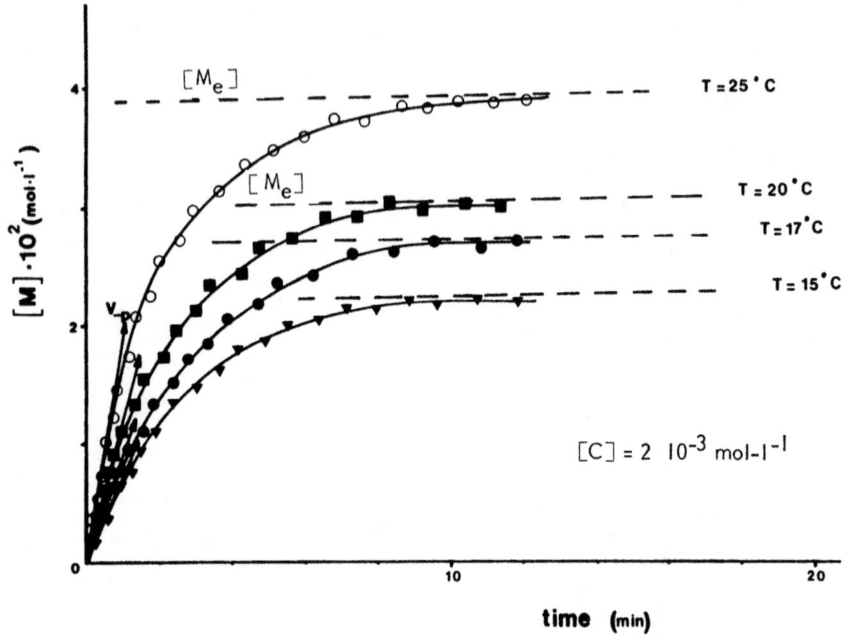

FIG. 5. Variation of monomer concentration in the anionic depolymerization of 2-IP in toluene with Li^+ as counter-ion, at several temperatures.

The linear plots of $\ln k_p$ and $\ln k_{-p}$ versus 1/T suggest the contribution of a single ionic species to the polymerization process. Thus, significant participation of free anions is not probable and this is consistent with the low polarity of the medium. The participation in the reaction of dimers or higher aggregates is also inconsistent with the observed first order dependence on the carbanion concentration. Therefore, it is most likely that the reaction proceeds through ion pairs. These results differ from those found in the anionic polymerization of styrene [16, 17], α-MS [1,2] and in THF or THP with Li^+ or Na^+ as counterions, or in the bulk polymerization of α-MS [13]. In that case pronounced deviations from linearity were found and the participation of several ionic species in the reaction was demonstrated.

However, the polymerization of 2VP initiated by organomagnesium compounds in toluene was previously found to be first order in active centers [14]. In this case, the lack of aggregation was demonstrated by viscosity measurements and attributed to the existence of an intramolecular coordination of the metal counter-ion by the neighboring pyridine groups. This complexation apparently takes precedence over the formation of intermolecular aggregates. The present results appear to be consistent with this explanation.

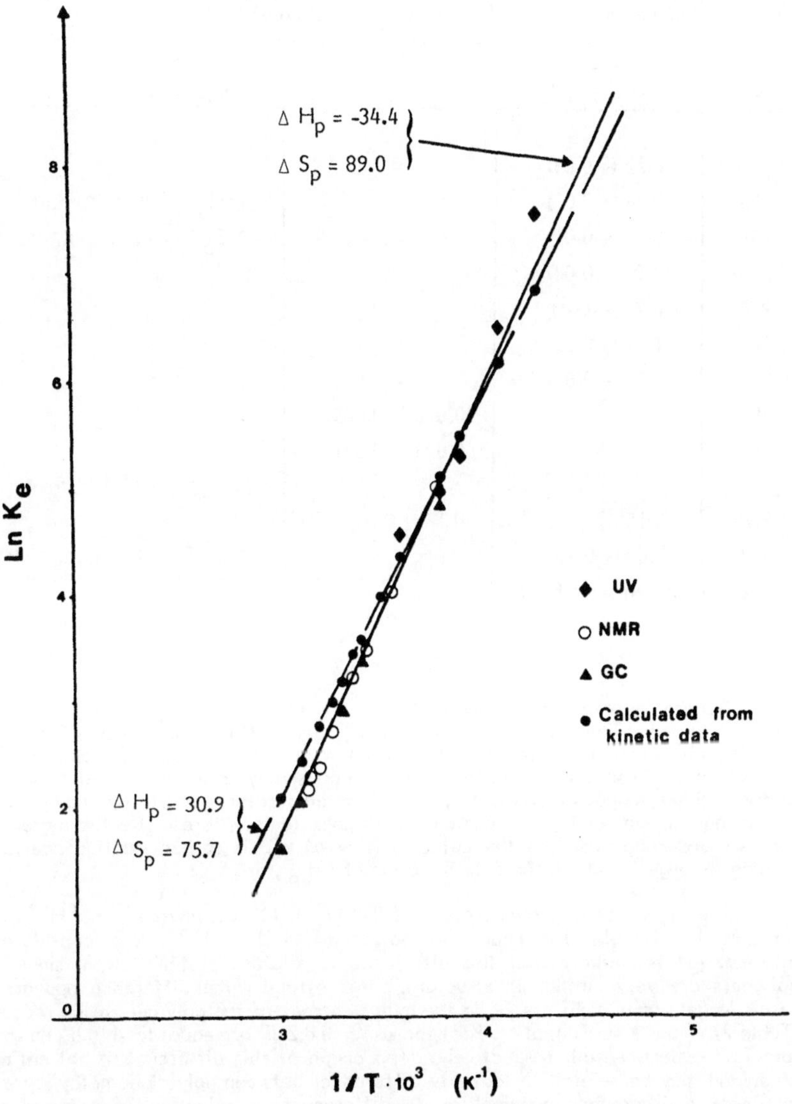

FIG. 6. Plots of $\ln K_e$ versus 1/T for equilibrium polymerization of 2-IP in toluene: (-●-) Calculated from kinetic data, (___) Determined from measurements of residual monomer concentration.

Compared with α-MS and styrene, the anionic polymerization of 2-IP in toluene proceeds at a fairly rapid rate at all temperatures. This is probably due to the fact that the poly(2-IP)$^-$, Li^+ in toluene is not aggregated. This in turn may be related to the increased polarity of the carbanion environment as a result of the presence of the polar pyridine groups.

TABLE I. Values of the absolute rate constant of propagation $k_p^{\pm}$ and depolymerization $k_{-p}^{\pm}$ versus temperature, for the polymerization of 2-IP in toluene.

T °K	$k_p^{\pm}$ $l.mol^{-1}.s^{-1}$	$k_{-p}^{\pm}$ s^{-1}	Thermodynamic Parameters
228	0.22 (± 0.03)	-	
233	0.26 (± 0.01)	-	$\Delta H_p^{\neq} = + 26.8\ KJ\cdot mol^{-1}$
238	0.30 (± 0.01)	-	$\Delta S_p^{\neq} = - 143\ J\cdot mol^{-1}\cdot K^{-1}$
248	0.56 (± 0.02)	-	
258	0.9 (± 0.01)	-	
268	1.84 (± 0.05)	-	
283	2.78 (± 0.01)	-	
288		0.076 (± 0.005)	
290	-	0.095 (± 0.001)	
			$\Delta H_{-p}^{\neq} = + 57.7\ KJ\cdot mol^{-1}$
298	5.15 a)	0.17 (± 0.01)	
303	7.00 (± 0 2)	-	$\Delta S_{-p}^{\neq} = - 67.3\ j.\ mol^{-1}.K^{-1}$

a) Extrapolated from Figure 4

General agreement is observed between our $\Delta H_p°$ and $\Delta S_p°$ values and those reported by other investigators for 2-IP [8] and α-MS [3-6] (Table 2). Some deviations between calculated and determined values of $\Delta H_p°$ and $\Delta S_p°$ exist for the 2-IP system. These deviations may have been caused by errors in the determination of the residual monomer concentration at low temperatures and/or by uncertainties in the equilibrium and rate constants at higher temperatures. Furthermore, the temperature range used in the determination of k_{-p} is rather small conceivably leading to larger errors in the determination of $\Delta H_{-p}^{\neq}$, $\Delta S_{-p}^{\neq}$.

The higher ceiling temperature of 2-IP($T_c \cong 120°C$, where $T_c = \Delta H_p° / \Delta S_p°$ for $[M_e] = 1\ mol.l^{-1}$) compared to that of α-MS ($T_c \cong 10°C$) is of considerable interest. It is unlikely that this difference is related to steric strain since the polymers are very similar in structure. Interestingly, the difference appears to result largely from a difference in the molar entropy of the polymerization reaction (Table 2). Thus $\Delta S_p°$ ($J.mol^{-1}.K^{-1}$) for α-MS and 2-IP are equal to -110.8 and -89.0 (or -75.7 experimental), respectively. The origin of this difference is not entirely clear and may be related to tacticity differences between poly(alphamethylstyrene) and poly (2-isopropenylpyridine) or to differences in solvation of polymer and monomer for the two systems. Another possibility would be the greater number of pyridine orientations arising from the asymmetrically located pyridine dipole.

The origin of the rather substantial difference between the $\Delta H_p°$ and $\Delta S_p°$ values for the polymerization of 2-IP in the presence of Li^+ in toluene and in the presence of Na^+ in THF is not quite clear. It is possible that this is due to the fact that in the work of Y. Iwakura and Toda [8], the equilibrium monomer concentrations were measured for a two-ended living system. It is also plausible that the stability of the 2-IP carbanion is somewhat less than in toluene so that side reactions may have occurred in the Na/THF system. Differences in $\Delta H_p°$ and $\Delta S_p°$

TABLE 2. Kinetic and Thermodynamic Parameters for the Anionic Polymerization of Alpha-Substituted Monomers

Monomer / Solvent / Parameters M$^+$	2IP[a] Tol. Li$^+$	2-IP THF Na$^+$	α MS THF Na$^+$	α MS THF Na$^+$	α MS THF Na$^+$	α MS Diox. Na$^+$
$\Delta H_p^{\neq,o}$ (KJ.mol^{-1})	26.8	-	-	-	8-12	32
$\Delta S_p^{\neq,o}$ (J.mol^{-1}K^{-1})	-143.0	-	-	-	-155.5	-
$\Delta H^{\neq,o}_{-p}$ (K.J.mol^{-1})	57.7					
$\Delta S^{\neq,o}_{-p}$ (J.mol^{-1}K^{-1})	-67.3					
kp (l.mol^{-1}.sec^{-1})[b]	5.15	60[c]	17		65	.02
k_{-p} (sec^{-1})	.17		.05[d]			
ΔHp^o (K.J.mol^{-1})	-34.4 (-30.9)	-25.9	-31.4	-29.3	-33.4	-35.3
ΔSp^o (J.mol^{-1}K^{-1})	-89.0 (-75.7)	-69.4	-110.8	-103.7	-120.4	-123.3
Tc (°C)	113.5 (135.2)	100.2	10.4	9.5	4.4	13
Reference	This work	[8]	[3]	[5]	[4]	[6]

a) From Ln M_{e} vs. 1/T; bracketed values from kinetic data, b) at 25°C, c) apparent rate constant, d) determined for dimer-trimer equilibrium.

are also observed for the anionic polymerization of alphamethylstyrene in THF and dioxane (Table 2).

The comparatively high stability of the propagating carbanions, combined with the relatively low ceiling temperature should make it possible to synthesize polymers possessing a single 2-IP end group. For instance, heating a living poly(butadiene)-co-poly(2-IP) anion, followed by protonation, is expected to form a poly(butadiene) containing a single pyridine end group. Studies along these lines are in progress.

TABLE 3. Experimental and calculated values of equilibrium monomer concentration and equilibrium constant for the polymerization of 2-isopropenylpyridine, in toluene, with Li^+ as counter-ion.

T°K	Experimental[a] $[M_e]\cdot 10^2$	K_e[a] experimental	Calculated[b] $[M_e]\cdot 10^2$	K_e Calculated[b]
233	0.05	1998	0.11	909
243	0.15	665	0.21	476
256	0.49	204	0.42	238
263	0.70	143	0.60	167
263	0.77	129	-	-
278	1.02	97	1.20	
278	1.30	76	-	-
293	3.2	31	2.9	38.1
298	3.3	26	3.3	30.0
303	4.1	24	4.2	23.8
303	5.3	19	-	-
308	6.6	15	5.1	19.6
313	9.0	11	6.2	16.1
318	12.5	8.0	8.7	11.5
333	19.8	5.1	12.0	8.3

[a]Determined by UV, GC and NMR : $K_e = 1/[M_e]$, [b]Calculated from kinetic data : $[M_e] = k_{-p}/k_p = 1/K_e$

ACKNOWLEDGEMENTS

Support for this research by the NSF-CNRS US-FRANCE Scientific Exchange Program and NSF-DMR, Polymers Program, is gratefully acknowledged.

REFERENCES

1. M. Szwarc, Carbanions Living Polymers and Electron Transfer Processes (Intersciene, New York (1968), Chapter 3.
2. S. Bywater, Comprehensive Chemical Kinetics, CH. Bamford, CFH. Tipper, eds., (Elsevier, Amsterdam), Chapter 1.
3. A. Vrancken, J. Smid, M. Szwarc, Trans. Far. Soc., 58, 2036 (1962)
4. a) D. J. Worsfold, S. Bywater, J. Polym. Sci., 26, 299 (1957); b) D.J. Worsfold, S. Bywater, Can. J. Chemistry, 36, 1141 (1958).
5. H. C. McCormick, J. Polym. Sci., 25, 488 (1957).
6. a) F. S. Dainton, K. J. Ivin, Quat. Rev., 12, 61 (1958); b) F. S. Dainton, G. A. Harpell, K. J. Ivin, Eur. Polym. J., 5, 395 (1969).
7. F. S. Dainton, K. M. Hui, K. J. Ivin, Eur. Polym. J., 5, 387 (1969).
8. Y. Iwakura, F. Toda, J. Macromol. Sci., Chem. (A9), 6, 1063 (1975).
9. A. H. Soum, C. F. Tien, T. E. Hogen-Esch, N. B. D'Accorso, M. Fontanille, Makromol. Chem., Rapid Commun., 4, 243 (1983).
10. a) M. Fontanille, P. Sigwalt, Bull. Soc. Chim. Fr., 4087 (1967); b) A. Dondos, P. Rempp, C. R. Hebd. Sceances, Acad. Ser., C264, 869 (1967).
11. C. C. Meverden, T. E. Hogen-Esch, Makromol. Chem., Rapid Commun., 5, 749 (1984).
12. D. P. Wyman, I. H. Song, Makromol. Chem., 115, 64 (1968).
13. D. Ades, M. Fontanille, J. Leonard, Can. J. Chem., 60, 564 (1982).
14. A. Soum, M. Fontanille, Makromol. Chem., 182, 1743 (1981).
15. T. E. Hogen-Esch, W. L. Jenkins, J. Am. Chem. Soc., 103, 3666 (1981).
16. T. Shimomura, K. J. Tolle, J. Smid, M. Szwarc, J. Am. Chem. Soc., 89, 796 (1967).
17. T. Shimomura, J. Smid, M. Szwarc, J. Am. Chem. Soc., 89, 5743 (1967).

MECHANISTIC ASPECTS OF THE REVERSIBILITY AND IRREVERSIBILITY OF THE ANIONIC POLYMERIZATION OF α-METHYLSTYRENE

J.-Y. Bergeron and J. Léonard
Département de chimie et Centre de recherche
en sciences et ingénierie des macromolécules (CERSIM)
Université Laval, Québec, Canada, G1K 7P4

ABSTRACT

The anionic polymerization of α-methylstyrene is examined in terms of the reversibility and the irreversibility of its propagation. The conditions for strict reversibility of a polymerization are defined. Deviation from strict reversibility for polymerization in bulk and in non-polar solvent is discussed together with its implication for thermodynamic studies based on data obtained from the equilibrium polymerization. In the case of the bulk polymerization initiated with tert-butyllithium in presence of N,N,N',N'-tetramethylethylenediamine (TMEDA), the reversible propagating species formed at low temperature can be transformed into irreversible propagating species at high temperature. Transfer to the monomer also takes place and the presence of short active chains and irreversible species perturbs the equilibrium position of the polymerization. For the polymerization in a non-polar solvent, the specific case of the polymerization initiated with potassium in toluene is considered. Reversible propagating species are formed. However, because the propagation is very slow, transfer to the solvent and monomer and transformation of the reversible species into irreversible ones occur before the polymerization can reach the equilibrium position predicted by thermodynamics.

INTRODUCTION

The anionic polymerization of α-methylstyrene is often considered as a typical example of a reversible polymerization. Indeed, most of the early studies published on the subject show that the polymerization is reversible and its equilibrium has been thoroughly investigated [1-6]. However most of that work has been carried out in polar solvents and it has been observed that departure from these conditions leads to deviations from strict reversibility which cannot be attributed to a simple deactivation of the active centers. The present paper deals with the extent of the reversibility of the anionic polymerization of α-methylstyrene and the implication it has for the thermodynamics of these polymerization systems.

STRICT REVERSIBILITY

In order to study the thermodynamics of a polymerization through the use of data obtained from an equilibrium polymerization, a certain number of mechanistic requirements are to be met. These are: a rapid initiation, in order to avoid problems linked with a broad molar mass distribution; the existence of a polymerization-depolymerization reaction; the stability of the active centers over a wide range of temperatures; the stability of the active centers which is independent of the solvent and the counterion. Under these conditions, a polymerization can be said to be strictly reversible and the data can be used without any restriction for thermodynamic computation. This

Recent Advances in Anionic Polymerization
T.E. Hogen-Esch and J. Smid, Editors

appears to be the case for the studies of the equilibrium anionic polymerization of α-methylstyrene mentioned above. Upon examination these studies are found to share common features. Thus they were carried out in a polar solvent using relatively low initial monomer concentration and temperatures below 40°C. Developments in thermodynamic equations allow for a quantitative evaluation regarding the effect of the solvent and the polymer concentration on the equilibrium position of a reversible polymerization. After slight tranformation of the equation written in terms of volume fractions [7], the effect of the medium on the equilibrium monomer concentration, $[M]_e$ (mol/L), can be expressed as:

$$[M]_e = [M]_e^o + \frac{(\chi_{mp} - \beta)(V_p/V_m)}{(\beta + \chi_{mp} - 1/\phi_m^o)}[P] - \frac{(\ln([P]V_p) + 1)/nV_m}{(\beta + \chi_{mp} - 1/\phi_m^o)} \qquad (1)$$

where $\beta = \chi_{ms} - \chi_{sp}(V_m/V_s)$. In this equation n is the average degree of polymerization; χ is the Flory-Huggins interaction parameter between any two components of the system, subscripts m, s and p referring to monomer, solvent and polymer respectively, V_m and V_s are the monomer and solvent molar volumes and V_p is the volume of 1 base-mol of polymer (1 base-mol = molar mass of the polymer/n). [P] is the polymer concentration (expressed in base-mol/L) and $[M]_e^o$ is the equilibrium monomer concentration at [P] = 0 for large values of n and $\phi^o_m = [M]^o V_m$. If n is large, the third term of the right hand side of Equation 1 becomes negligible. The effect of polymer and solvent on $[M]_e$ can be evaluated through [P] and β. If n is small, the third term becomes important and $[M]_e$ is smaller that it is for large n (short chain effect). Experimental evidence of those effects can be found through the use of high initial monomer concentrations in various solvents. However irreversible propagation and deviation from the expected equilibrium position were observed for the near-bulk polymerization, that is $[M]_o > 5$ M [8]. Similar observations were made for the polymerization in solution when one or more of the following conditions were met: a relatively high initial monomer concentration, a temperature above 40°C, a non-polar solvent and the presence of a coinitiator. Under these conditions irreversible propagation, multimodal distribution and low equilibrium monomer concentration due to the presence of short active chains were observed [9, 10]. In view of the similarity of various observations made on polymerizations carried out in bulk and in solution, the case of the bulk polymerization is examined first.

BULK POLYMERIZATION

Thermodynamics of an equilibrium bulk-polymerization is much more simple than it is for the same polymerization in a solvent. In addition, the media of the polymerization being different, the value of $[M]_e$ varies considerably from one type of system to another [11]. For these reasons studies of equilibrium bulk-polymerization are of the highest interest. Unfortunately the anionic bulk polymerization of α-methylstyrene shows a certain number of anomalies. It is quite a common experience to find that pure α-methylstyrene lying on a sodium mirror can be converted spontaneously into a dark red polymer which does not depolymerize, while retaining its dark red color, when heated at temperature above the ceiling temperature. However it has been shown [12] that if the polymerization is initiated at low temperature with tert-butyllithium (t-BuLi) in presence of N,N,N',N'-tetramethylethylenediamine (TMEDA), polymerization-depolymerization reactions

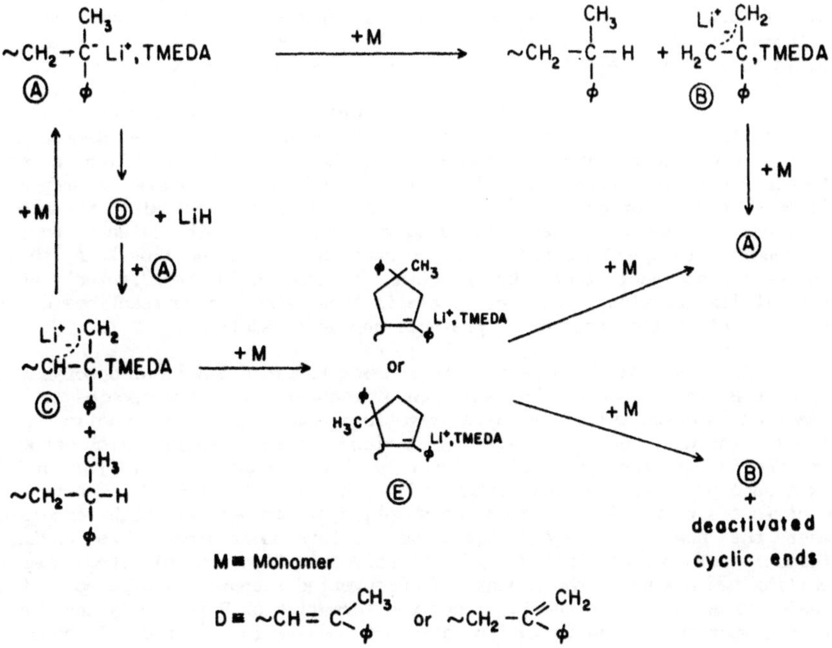

FIG 1. Propagation mechanisms for the bulk polymerization of α-methylstyrene initiated with tert-butyllithium.

can take place. Polymerization-depolymerization reactions can be carried out a limited number of times on a given sample by decreasing and increasing the temperature up to a certain point where the polymerization becomes irreversible. On the other hand initiation of the polymerization at temperatures above 25°C leads to irreversible polymerization. It was observed that, although deactivation of active centers took place to a certain extent, a large fraction of the chains remained active and polymerization resumed upon addition of fresh monomer. It was also found that a high yield of low molar mass polymer was obtained when the polymerization was initiated around the ceiling temperature.

The bulk polymerization was recently reexamined [13] and the main conclusions are summarized in Figure 1. At temperatures below 0°C species A is formed. This is the expected active species which leads to reversible polymerization with a maximum absorption at 330 nm. The stability of this species is a function of the ratio r = [TMEDA]/[t-BuLi] and temperature. In the absence of TMEDA, the species formed has a maximum absorption at 321,5 nm and the half-life time is about 5h at 25°C and a few minutes at 60°C. Under these conditions considerable deactivation takes place, the reaction does not reach equilibrium and there is no reversibility of the polymerization. For values of r ranging from 0,5 to 25, species A are found to be quite stable over a long period of time at 25°C. For temperatures around 25°C and higher, species B and C with absorptions at 390 and 424 nm respectively are observed. Upon aging of the reacting mixture, species C becomes the most abundant species.

As shown in Figure 1, species A may undergo deactivation forming an unsaturated polymer end group D (absorptions at 304 and 314 nm). Molecules D may be deprotonate by A and yield a saturated polymer and the allylic anion

C. Evidence for the presence of allylic species has been shown by various authors [14-16]. If monomer M is available, molecules C may react with it to yield again species A or species E which contains a cyclic structure. Evidence for the existence of species similar to E, which is believed to absorb at 330 nm, has been found [17]. Species E may add monomer M to give A. All these reactions lead to the formation of A which can depropagate. However upon increasing the temperature, the rate of deactivation increases together with the proportion of species C and E which cannot depropagate. Since C and E can only add monomer, the latter is eventually exhausted so that only species C and deactivated polymer remain. If the rates of addition of monomer to species C and E are the same as for the addition to A, then it would be conceivable that the reacting system would be "frozen" at its equilibrium position. However little is known about these reactions and the presence of cyclic structure E prevents any depropagation.

Species A may also react with monomer to yield deactivated polymer and species B, which upon further addition of monomer gives back species A. This reaction is essentially a transfer reaction to monomer. This reaction can be traced through GPC (or size exclusion chromatography) which shows a broadening of the peak at moderately high temperature (35°C) and the existence of a bimodal distribution in the vicinity of 60°C. The reaction contributes to the irreversible consumption of monomer at high temperature where the new active short chains may in turn undergo reactions leading to formation of species C and E. If the polymerization is initiated near the ceiling temperature, the amount of unconverted monomer is high so that the chains thus formed are short. Since the formation of B is being enhanced at high temperature, the formation of short chains is favored. It should be pointed out that in the case of bulk equilibrium polymerization, polymer with low molar mass is bound to be formed above the ceiling temperature [18].

POLYMERIZATION IN A NON-POLAR SOLVENT

Equilibrium anionic polymerization studies of α-methylstyrene in non-polar solvents are of a great interest from the point of view of thermodynamics. It has been shown that the effect of solvent on the equilibrium monomer concentration according to Equation 1 is related to χ_{ms}, the Flory-Huggins interaction parameter between monomer and solvent. A recent study [19] shows that for a change of χ_{ms} from -0.35 for the α-methylstyrene-tetrahydrofuran system [20] to 0,5 for the α-methylstyrene-cyclohexane system [19], the equilibrium monomer concentration should decrease approximately by a factor of two. However experimental results for this type of polymerization systems are scarce. An equilibrium monomer concentration of 3,53 M is reported at 40°C for the polymerization of α-methylstyrene in cyclohexane initiated with sec-butyllithium [21]. However the polymerization is carried out in presence of tetrahydrofuran or diethyl ether and it is difficult to estimate its effect on the kinetics and thermodynamics of the polymerization. In the absence of tetrahydrofuran, the polymerization is much slower and the equilibrium monomer concentration tends to be lower although in this case the equilibrium position is not really attained [10]. In none of the studies has the stability of the active centers been monitored.

Because polymerizations of α-methylstyrene in non-polar solvents bear much resemblance with the bulk polymerization, recently the study of the anionic polymerization of this monomer initiated with potassium in toluene, has been undertaken [22]. From results obtained with bulk polymerization, the active centers appear to be more stable with potassium than with lithium and this polymerization system allows comparison with both the bulk polymerization and the polymerization in cyclohexane.

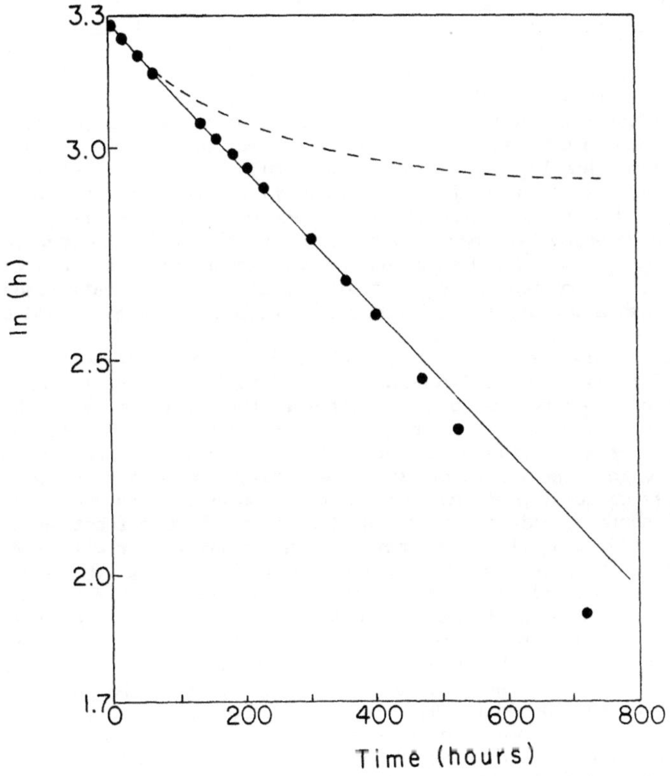

FIG 2. Variation of ln h, the level of the meniscus, with time for the anionic polymerization of α-methylstyrene in toluene; broken line, polymerization with deactivation of the active centers.

The polymerizations in toluene are followed through dilatometry, using high vacuum techniques. The initial monomer concentrations range from 1,1 M to 2,2, M. Figure 2 shows a typical example of the variation of ℓn h with time at - 3,0°C, h being the level of the meniscus in a dilatometer with respect to a reference mark corresponding to a reference volume. It can be shown that within a few percent of error, h is directly proportional to [M], the monomer concentration. A linear variation of ℓn h is observed up to 280 hrs where deviation from linearity is noted. Addition of slight amounts of tetrahydrofuran (about 0,5% V/V) does not modify the general trend of the polymerization except that deviation from linearity seems to occur at a much longer time (450 hrs).

The change in the rate constant of the polymerization may be attributed to a modification of the nature of the active centers; the active centers are simply deactivated or they are transformed into new species having a different reactivity [13]. Deactivation of the active centers has to be ruled out. The broken curve plotted in Figure 2 is obtained assuming k_p/k_d = 1000 where k_p is the propagation rate constant and k_d, the deactivation rate constant. Transformation of the active centers is also supported by UV-visible spectrophotometry and GPC analysis. The initial polymerization mixture shows a maximum absorption at 329 nm which identifies the expected active centers found in reversible polymerizations. After about 50 days an absorption at 530 and 570 nm is observed and after 150 days a strong absorption band is observed at 432 nm. This result can be related to the slow transformation of the reversible species (329 nm) to presumably irreversible species (432 nm) similar to species C found in the bulk polymerization (424 nm). Another similarity with the bulk polymerization is revealed through GPC analysis. After a period of about 50 days, GPC chromatograms show bimodal distributions similar to the ones obtained for the bulk polymerization at high temperature which indicates important transfer reactions. However this transfer is not accompanied by observation of a peak around 390 nm which has been identified in the bulk polymerization with species B. This would indicate that transfer does not occur through a direct transfer to the monomer. One possible explaination is that transfer takes place through a mechanism which would involve both the monomer and toluene.

Apparently active species which lead to reversible polymerization are formed initially. The active centers would be considered to be stable for most anionic polymerizations considering the required time to get to equilibrium. Since the propagation is very slow, it is not possible to determine whether equilibrium is attained before the active centers are transformed into species which would propagate in an irreversible fashion. Under these circumstances, measured monomer concentrations cannot be used for the usual thermodynamic computations since thermodynamic functions associated with the addition of a monomer to an active center will probably vary according to the nature of the active center. Assuming the polymerization system gets to equilibrium before the active centers are transformed, the equilibrium would be disturbed by transfer reactions. As mentioned earlier, according to Equation 1, the equilibrium between monomer and short active polymer chains is not the same as for the equilibrium involving long active polymer chains. The presence of deactivated polymer chains in equilibrium has also been considered [5, 9]. In the case of short chains, it has been shown [9] that the presence of short deactivated chains does not interfere with the equilibrium. However if the short chains are active, then they have a specific effect which tends to lower the equilibrium monomer concentrations. Because each chain possesses two active centers, most of the chains remain active after a transfer reaction and the equilibrium would involve a mixture of long and short active chains together with deactivated polymer. In short, even if the propagation is reversible and the active centers are relatively stable, due to the fact that the propagation is very slow, the attainment of equilibrium in the anionic polymerization in toluene is highly problematic.

CONCLUSION

The anionic polymerization of α-methylstyrene in bulk and in non-polar solvents seems to lead to the formation of active centers which are different from the active species found with relatively low initial monomer concentration in polar solvents. Since the ultimate monomer units thus formed are different from the usual ones, the thermodynamic parameters computed for the polymerization of α-methylstyrene cannot be applied here. The presence of such species may lead to a new equilibrium. If the species cannot depropagate, then the thermodynamic equilibrium cannot be attained if the monomer concentration happens to be under its equilibrium value.

ACKNOWLEDGEMENTS

The authors wish to acknowledge financial support from the Natural Sciences and Engineering Research Council of Canada. J.Y.B. wants to thank the Minister of Education (Quebec) for a fellowship.

REFERENCES

1. H.W. McCormick, J. Polym. Sci. 25, 488, (1957).
2. D.J. Worsfold and S. Bywater, J. Polym. Sci. 26, 299, (1957).
3. A. Vrancken, J. Smid and M. Szwarc, Trans. Faraday Soc. 58, 2036, (1962).
4. H. Hopff and H. Lussi, Makromol. Chem. 62, 31, (1963).
5. K.J. Ivin and J. Leonard, Eur. Polym. J. 6, 331, (1970).
6. J. Leonard and S.L. Malhotra, J. Polym. Sci. A-1, 9, 1983, (1971).
7. J. Leonard, Macromolecules, 2, 661, (1969).
8. J. Leonard and S.L. Malhotra, J. Macromol. Sci.-Chem. A 11, 1867, (1977).
9. J. Leonard and S.L. Malhotra, Macromol. Sci.-Chem. A 10, 1279, (1976).
10. S.L. Malhotra, J. Leonard and M. Thomas, Macromol. Sci. A 11, 2213, (1977).
11. J. Leonard and D. Maheux, J. Macromol. Sci.-Chem. A 7, 1421, (1973).
12. M. Thomas, Ph.D. Thesis, Univ. Laval, Québec, 1975.
13. D. Adès, M. Fontanille and J. Leonard, Can. J. Chem. 60, 564, (1982).
14. T. Kaufmann, Angew. Chem. Int. Ed. Engl. 13, 627, (1974).
15. G.F. Luteri and W.T. Ford, J. Orgamomet. Chem. 105, 139, (1976).
16. G.F. Luteri and W.T. Ford, J. Org. Chem. 42, 820, (1977).
17. J. Comyn and M.D. Glasse, J. Polym. Sci.: Polym. Letters 18, 703, (1980).
18. R. Binet and J. Leonard, Polymer 14, 355, (1973).
19. V.T. Bui and J. Leonard, To be published.
20. V.T. Bui and J. Leonard, J. Chem. Soc., Faraday Trans. 1, 81, 1745, (1985).
21. R.E. Cunningham, Polymer 19, 729, (1978).
22. J.Y. Bergeron, J. Leonard and J.C. Favier, To be published.

DONOR-ACCEPTOR INTERACTIONS OF THE ACTIVE CENTRES OF CHAIN PROPAGATION IN THE ANIONIC POLYMERIZATION OF ACRYLONITRILE AND METHACRYLONITRILE AND SOME OTHER POLAR VINYL MONOMERS

C.B.TSVETANOV, D.T.DOTCHEVA, D.K.DIMOV, E.B.PETROVA, AND I.M.PANAYOTOV
Central Laboratory for Polymers, Bulgarian Academy of Sciences,
1040 Sofia, Bulgaria

INTRODUCTION

Studies of the anionic polymerization of polar vinyl monomers, although very intensive, still remain within the frames of mainly qualitative and semi-quantitative evidence and conclusions. Much of the published data on principle parameters are contradictory and insufficiently reliable. The only exceptions are the polymerizations of methyl methacrylate [1] and the vinyl pyridines [2] .

The main reason for the different degree of knowledge of anionic polymerization mechanisms of polar and non-polar monomers is the presence of a reactive functional group in the molecule of polar monomers. This is involved in a number of unwanted processes with both the initiator and the growing polymer chain end. The polar group is often a strong electron pair donor and is capable of competing with the solvent in interaction with the counterion. Coordination of the counterion to the polar group is usually accepted to be intramolecular and is related to an essential influence on the structure and reactivity of the growing centres [3] . A familiar example is Cram's cycle, postulated to explain the stereospecific polymerization of methyl methacrylate in non-polar media [4] .

Direct evidence for the existence of the penultimate effect in the anionic polymerization of a series of polar vinyl monomers is the formation of cyclic products in the early stages of the process [5] . Generation of six-membered rings is preceded by the formation of an intermediate complex as illustrated for the trimer of acrylonitrile on the basis of CNDO/2 calculations [6] (Fig. 1).

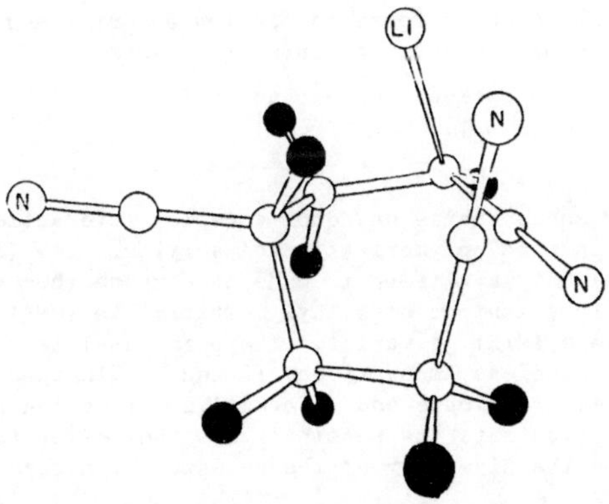

FIG. 1. Cyclic complex conformation of the trimer of "living" polyacrylonitrile lithium on the basis of CNDO/2 calculations [6] .

Recent Advances in Anionic Polymerization
T.E. Hogen-Esch and J. Smid, Editors

Polar vinyl monomers are further known to form complexes with a number of inorganic salts in non-aqueous solvents [7] . This suggests that specific interaction between monomer and counterion could possibly precede the step of monomer incorporation into the polymer chain when the reaction is performed in non-polar medium. No experimental data have been presented to prove complex formation between monomer and active centres; numerous results from kinetic studies can, however, be interpreted only in terms of intermediate complex formation. According to Erussalimsky [8] two complexation modes are possible:

$$\sim M^-,Mt^+ \quad + \quad M_1 \rightleftharpoons \sim M^-,Mt^+...M_1 \longrightarrow \sim MM_1^-,\ Mt^+$$

$$\sim M^-,Mt^+...M_1 \quad + \quad M_2 \rightleftharpoons \sim \underset{\searrow\ \swarrow \atop M_2}{M^-,Mt^+}...M_1 \longrightarrow \sim MM_2^-,Mt^+...M_1$$

It is obvious that the anionic polymerization of polar vinyl monomers is accompanied by competitive donor-acceptor interactions to be summarized as follows:

A. Active centre - solvent:

Peripheral solvation of counterion

B. Active centre - polar group of monomer unit:

1. Influence on the structure of polymer: Cram's cycle, Bernoullian distribution, Markov's chains
2. Formation of E and Z structures, ion triples and higher associates, "living" oligomers of different reactivity
3. Side reactions: inactivation of the active centres by the formation of six-membered rings, formation of cyclic products

C. Active centre - monomer:

1. Intermediate complex formation in low polarity medium - effect on kinetics and structure of polymer
2. Side reactions between monomer and initiator

The present paper concentrates on donor-acceptor interactions of the active centres in the anionic polymerization of acrylonitrile (AN) and methacrylonitrile (MAN) but an attempt is made to compare them with similar interactions in the polymerization of methyl methacrylate (MMA) and methyl isopropenyl ketone (MIK). A variety of experimental techniques can be employed in the studies. Among these, IR and UV VIS spectroscopy and conductivity studies are simple and informative. It is the aim of this communication to illustrate the possibilities they offer to reveal some of the features in the diversity of the anionic polymerization of polar vinyl monomers.

IR SPECTROSCOPY AS A METHOD IN STUDIES OF DONOR-ACCEPTOR INTERACTIONS IN THE ANIONIC POLYMERIZATION OF ACRYLONITRILE AND METHACRYLONITRILE

The active centres (AC's) in the anionic polymerization of AN and MAN are salts of ambident anions, the conjugation of the carbanion to the cyano group reducing the absorption frequency of the latter by 20 to 200 cm^{-1} [9] . Similar bathochromic shifts are observed for the carbonyl groups of the AC's in the anionic polymerization of MMA [10] and MIK [11] :

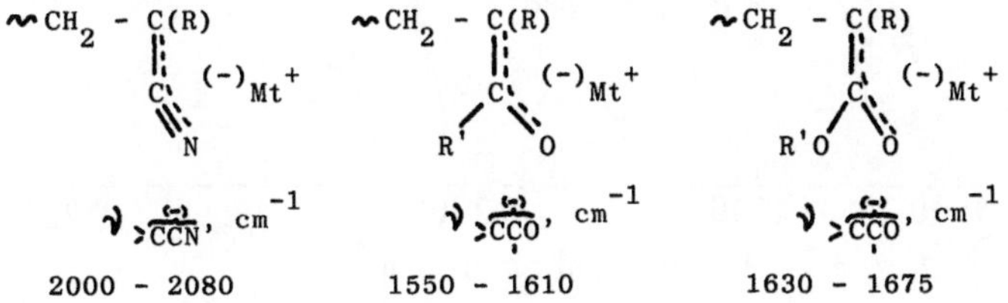

The wave lengths of the nitrile absorption bands are strongly dependent on the nature of the counterion and solvent and the presence of donor or acceptor additives, but remain unchanged within the whole temperature range between 20°C and -60°C in THF.

Following the procedure of Lochmann [12] isobutyronitrile lithium ($P_1^{CN}Li$) and 1-Li-1,3,3-trimethyl glutaronitrile ($P_2^{CN}Li$) were prepared. Their IR spectra in THF are shown on Fig. 2. The complex character of the band of the dimer is of interest and should be explained by participation of the second cyano group in both intra- and intermolecular interactions with the lithium counterion.

On increasing the mass number of the counterion the band which is characteristic for the ν >CCN(-) group shifts to the smaller wave numbers. A linear correlation is being observed depending on the reciprocal of the crystallographic radius of the counterion (Fig. 3). The structure of AC's with divalent counterions, $>CC\overset{(-)}{N},Mt^{++},\overset{(-)}{N}CC<$, resembles that of the ion triple of the $[ABA]^-$ type, or the structure of "ate" complexes [13] . Indeed, "ate" complexes of isobutyronitrile lithium with lithium alkoxides [12] , diethyl zinc [14] , and triethyl aluminium exhibit characteristic IR absorption bands for the ν >CCN(-) group at higher frequencies (Fig.3).

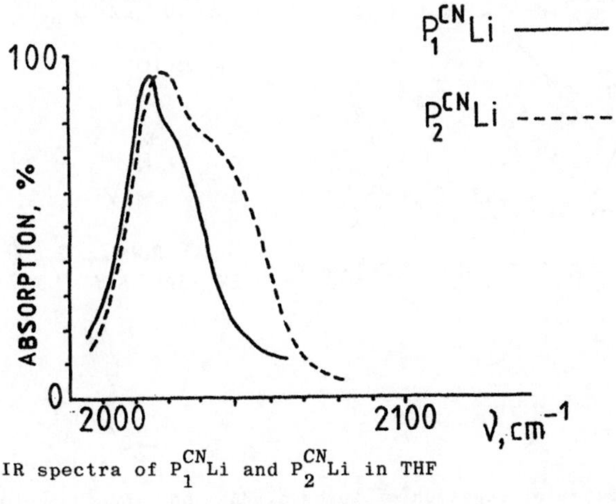

FIG. 2. IR spectra of $P_1^{CN}Li$ and $P_2^{CN}Li$ in THF

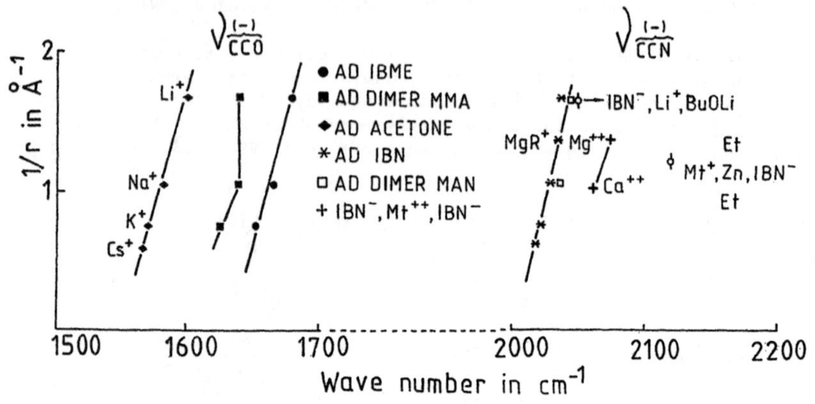

FIG. 3. Dependence of the wave number of absorption bands of the alkali derivatives (AD) of model compounds of MMA and MAN in THF
IBME: methyl isobutyrate; IBN: isobutyronitrile

With the aid of IR spectroscopy it is also possible to study the interaction of anionic AC's with the cyano group of the monomer. It was found that in non-polar solvents the AC's of MAN with lithium or magnesium as counterions form complexes with nitriles having characteristic bands in the range 2250 - 2285 cm^{-1} [13,15] , as shown on Fig. 4. Furthermore, some evidence is in favour of the possibility to experimentally follow both the complex formation and the polymer chain propagation. Preliminary results [13] showed the polymerization of MAN to be initiated by ether-free t-Bu_2Mg at -30°C, but not at -78°C. IR spectra recorded in this temperature range could give the ratio of the individual bands for the AC's with one and two polymer chains linked to the Mg cation, $\sim M^-, Mg^{++}, {}^-R$ and $\sim M^-, Mg^{++}, {}^-M\sim$ respectively, and the band for the complex (Fig.4).

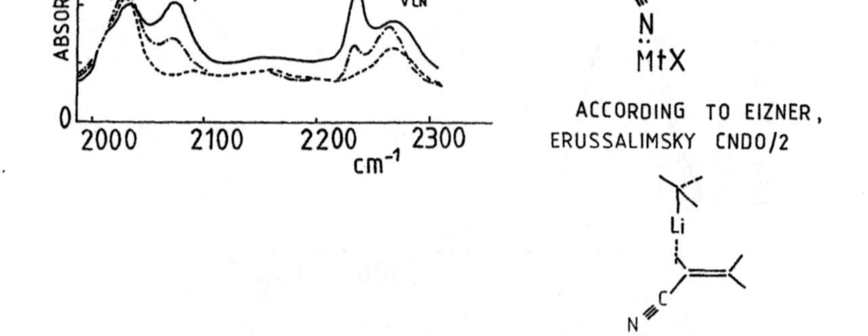

FIG. 4. IR spectra of reaction mixtures of MAN and t-Bu_2Mg in toluene

The position of the ν_{CN} band of the complex is indicative of its structure, i.e. the formation of coordinate bonds between the cation and the lone electron pair of the nitrogen atom [7] . This is, however, in contradiction with the energetically most favourable structure as computed by Eizner and Erussalimsky [16] (Fig.4).

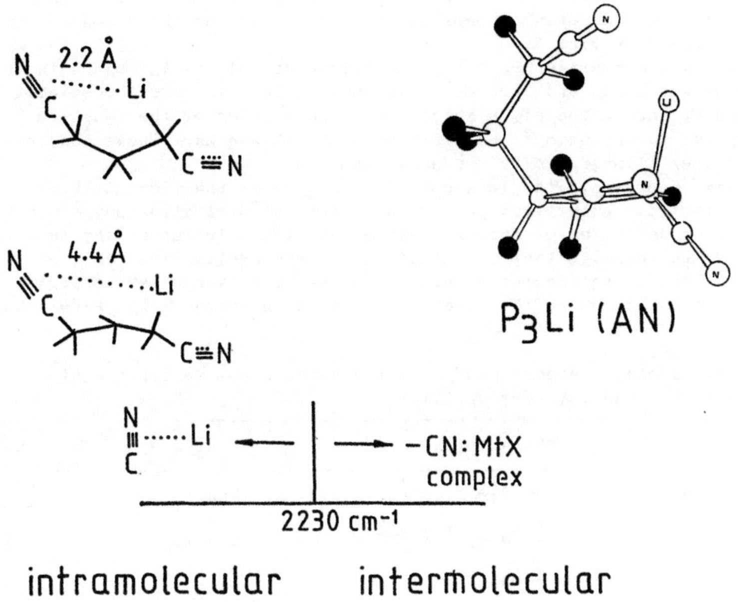

FIG. 5. Complexation modes for "living" polyacrylonitrile lithium

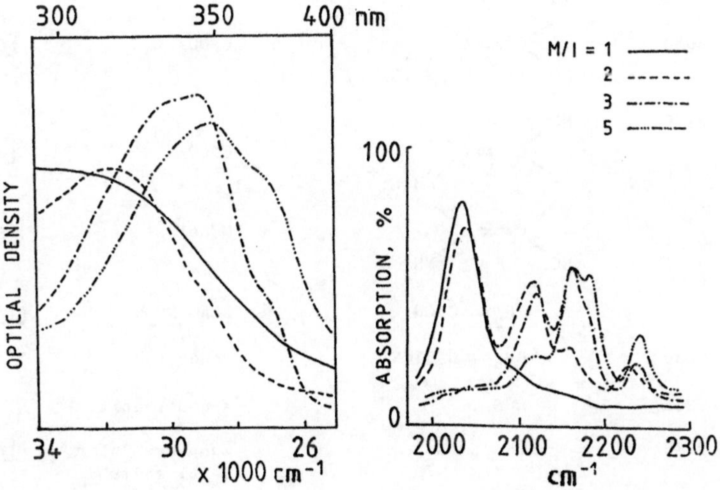

FIG. 6. IR and electronic spectra of reaction mixtures of AN and polystyryl potassium in THF at -40°C

The intramolecular coordination of the lithium counterion with nitrile groups of the neighbouring monomer units occurs with the π-electrons of the cyano group (Fig. 5). According to quantum chemical calculations such an interaction should cause a decrease of the order of the cyanic bond to result in a new band at lower frequency [6] . The geometry of the energetically most favourable conformer of the trimer (P_3Li) of polyacrilonitrile lithium where chelate bonds are formed between the lithium counterion and the cyano groups of the second and the third monomer unit is also shown on Fig. 5.

Table I summarizes data of the assignments of the ν_{CN} bands in the range between 2000 and 2300 cm^{-1}. It demonstrates the great possibilities offered by the method of IR spectroscopy in studies of the reaction mechanisms in the anionic polymerization of AN and MAN. These possibilities are further illustrated by the next examples.

Fig. 6 presents the IR and UV VIS spectra of reaction mixtures acrylonitrile - polystyryl potassium in the polymerization performed by gradual condensation of monomer vapour onto the solution of the initiator. The spectra indicate that the Ziegler - Thorpe cyclization reaction proceeds to a large degree resulting in the formation of aminonitrile structures as shown by the intensive absorption in the range 2160-2185 cm^{-1} [17] .

TABLE I. IR band frequencies ν_{CN} in the anionic polymerization of acrylonitrile and methacrylonitrile.

ν_{CN} cm^{-1}	Characteristic group	Proved by	Ref.
2000 - 2050	$>C{\cdots}\overset{(-)}{C}{\cdots}N$, Mt^+	Models	[17]
2060 - 2080	$>C{=}\overset{(-)}{CN}$, Mt^{++}, $\overset{(-)}{NC{=}}C<$	Models	[13]
2100	$>C{=}\overset{(-)}{CN}$, Mt^{+-} The most stable P_3Li	CNDO/2	[6]
2110 - 2130	$(NC)>CH{-}C{=}N^{(-)}$, Li^+	Model	[17]
2070 - 2080	$(NC)>C{=}C<NH^{(-)}$, Li^+	CNDO/2	[6]
2160 - 2170	$(NC)>C{=}C<NH^{(-)}$, Li^+	CNDO/2	[6]
2185	$(NC)>C{=}C<NH_2$	Model	[18]
2200	$CH_2{=}\overset{(-)}{C}.CN$	LCAO MO	[17]
2220 - 2230	$CH_2{=}C(R)CN$	Monomers	
2230 - 2245	RCN	Monomer units	
2260 - 2285	$-CN : MtX_n$	Complexes with metal salts	[7]
2100 - 2220	$C{\equiv}N \cdots Li$	Intramolecular complexes, CNDO/2	[6]

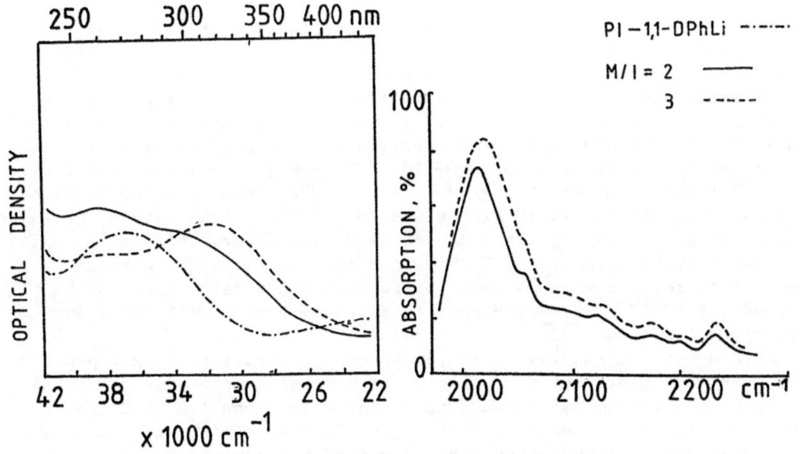

FIG. 7. IR and electronic spectra of reaction mixtures of MAN and polyisoprenyl 1,1-diphenyl lithium in heptane at -70°C

An essentially different picture is revealed (Fig. 7, Fig. 8) when the polymerization is carried out by instantaneous mixing of the solutions of polyisoprenyl 1,1-diphenyl lithium and monomer. Well pronounced side reactions are not observed in this case and the concentration of AC's remains practically unchanged in the process of chain propagation [19] . Without going into detail about the reasons for the different behaviour of the two systems cited, i.e. initiator, solvent, different mode of monomer addition, these examples clearly demonstrate the possibilities of IR spectroscopy to produce an idea of the reaction mechanisms involving anionic active centres of MAN and AN.

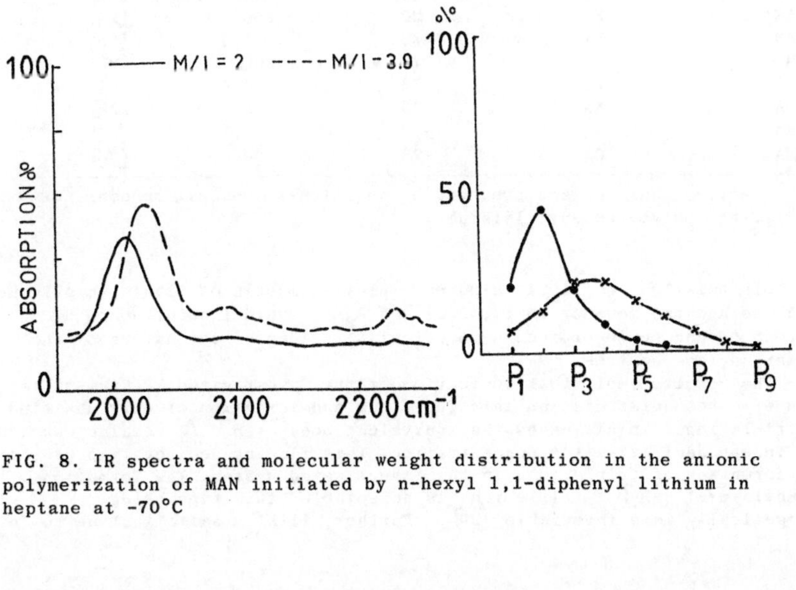

FIG. 8. IR spectra and molecular weight distribution in the anionic polymerization of MAN initiated by n-hexyl 1,1-diphenyl lithium in heptane at -70°C

COMPARATIVE INVESTIGATION OF THE DISSOCIATION EQUILIBRIA OF P_1Li AND P_2Li CONTAINING DIFFERENT POLAR SUBSTITUENTS IN THF

Determination of the dissociation constants of "living" polymer chains in solution is a method widely used in studies of the nature of active centres. A frequent drawback in this approach, however, is the occurrence of side reactions during the polymerization process which might have a substantial effect on the conductivity yielding erroneous results. A typical example is the difference between reported constants of dissociation in THF for polyisoprenyl lithium, $K_D \sim 10^{-10}$ M [20] ,and polybutadienyl lithium, $K_D \sim 10^{-8}$ M [21] . Later, Vinogradova et al. [22] found the solutions of polybutadienyl lithium in THF to be unstable at room temperature, presumably as a result of solvent cleavage and alkoxide formation. This could explain the relatively high values of K_D reported for the "living" polymers. Conductivity measurements at temperatures below 0°C gave values of K_D of the order of 10^{-10} M, i.e. the results were comparable with those for polyisoprenyl lithium [23] .

Such errors in studies of the dissociation equilibria might prove to be even more prevalent for polar vinyl monomers since side reactions in this case are often more pronounced. The data presented in Table II demonstrate the discrepancy in results obtained by different authors and the lack of correlation between the nature of the counterion and the K_D values. Probably, the real systems contain certain amounts of aminonitrile and iminonitrile anions, in the case of AN and MAN, or metal alkoxides, in solutions of "living" PMMA. These interact with the primary active centres to form mixed associates and complexes which should modify the electrochemical characteristics of the solution.

TABLE II. Dissociation constants of "living" polymers of AN, MAN and MMA in THF.

Monomer	Counterion	Temperature (°C)	K_D (10^{10} M)	Ref.
AN*	Li	20	24	[24]
AN*	Na	20	40	[24]
AN*	K	20	300	[24]
MAN*	Li	-70	1.4	[25]
MMA	Li	-78	200	[26]
MMA	Li	-78	4.4	[27]
MMA	Na	-78	35	[28]
MMA	K	-78	2.1	[27]
MMA	Cs	-78	20	[28]

* Active centres were prepared by addition of several monomer units to "living" polybutadienyl lithium.

Dissociation equilibria studies employing models of lithiated oligomers with one and two monomer units, P_1Li and P_2Li, avoid the influence of byproducts and could produce a more realistic idea of the nature of the primary active centres.

The results imply that ionization of the investigated lithium salts is due to both dissociation into free ions and formation of only one kind of triple ions. In all cases the equivalent conductance Λ remains constant and independent of salt's concentration. This finding decisevely proves the formation of triple ions in the investigated system. The formation of unilateral $[ABA]^-$ triple ions is acceptable, this type being energetically more favourable [29] . Further, lithium salts incline to form "ate" complexes [30] .

The equilibria can be described by the following equations:

$$A^-,B^+,A^- \underset{}{\overset{k_T}{\rightleftharpoons}} A^- + A^-,B^+ \qquad (1)$$

or

$$2\ A^-,B^+ \underset{}{\overset{K_T}{\rightleftharpoons}} A^-,B^+,A^- + B^+ \qquad (2)$$

According to Van Beylen and Szwarc [31] $K_T = K_D/k_T$ and can be derived from the equation:

$$\Lambda = \left\{ K_T^{1/2}/(1 + K_T^{1/2}) \right\} \lambda_o \qquad (3)$$

Inspite of the small intercepts in Wooster's plots [32] and the uncertain determination of the dissociation constants, the data are presented in Table III.

TABLE III. Dissociation constants of ion pairs K_D and ion triples k_T of lithium and sodium salts with ambident anions.

P_nMt -X	K_D (10^{12} M)	k_T (10^4 M)	$K_T^{a)}$ (10^8)	$K_T^{b)}$ (10^8)	$[ABA]^-/[A]^-$	pK_a (P_nH)	Ref.
P_1Li							
-CN	69	7	9.8	11	7	25	[33]
$-COC(CH_3)_3$	75	6.5	11	14	8	23	[34]
$-COOC_2H_5$	0.4	0.2	1.8	1.9	250	25	[34]
P_1Na							
$-COOCH_3$	7	2	3.5	4.9	25		[34]
P_2Li							
-CN	6	1	6	8	50		[33]
$-COOCH_3$	1.5	3	0.5	0.6	16		[34]

a) $K_T = K_D/k_T$ b) Calculated according to eq. (3)

It is seen from Table III that mainly tight ion pairs are formed with a minimum of the interionic distance of about 2.5 Å.

The dissociation constants of the studied salts are smaller than these of polybutadienyl lithium and polyisoprenyl lithium and diminish in the following order:

$$K_D \text{ ("living" polybutadienyl lithium)} > K_D\ (\alpha\text{-lithiated ketone}) \approx K_D\ (\alpha\text{-lithiated nitrile}) > K_D\ (\alpha\text{-lithiated ester})$$

Obviously, when C-atom in the anionic group of the AC is replaced by a heteroatom, the electrostatic interaction with the counterion is increased. Further, these salts show tendency to form higher associates similar to the alkali alkoxides. The lack of correlation between K_D and pK_a of the corresponding C-H acids indicates the specific interaction of the cation with the ambident anion to be the deciding factor for the value of K_D.

The dissociation constants of α-lithiated esters are smaller by about 2 orders of magnitude compared to these of the ketones and nitriles. This fact can be due to hindered charge distribution in the anion of the esters.

The lesser degree of anion delocalization could also explain the different shifts of the $\nu_{C\bar{C}O}$ bands in the IR spectra of α-lithiated esters [10] and lithium enolates [11] with respect to the absorption of a carbonyl group, which indicates higher electron density of the >CCO (-) group in the latter case.

Indeed, Vancea and Bywater [35] registered reduced motion freedom of the alkoxy groups in THF solutions of α-lithiated methyl isobutyrate at low temperatures by the aid of ^{13}C NMR technique which was attributed to the planar structure of the >C-·C-·O - R group (δ^-O, δ^+).

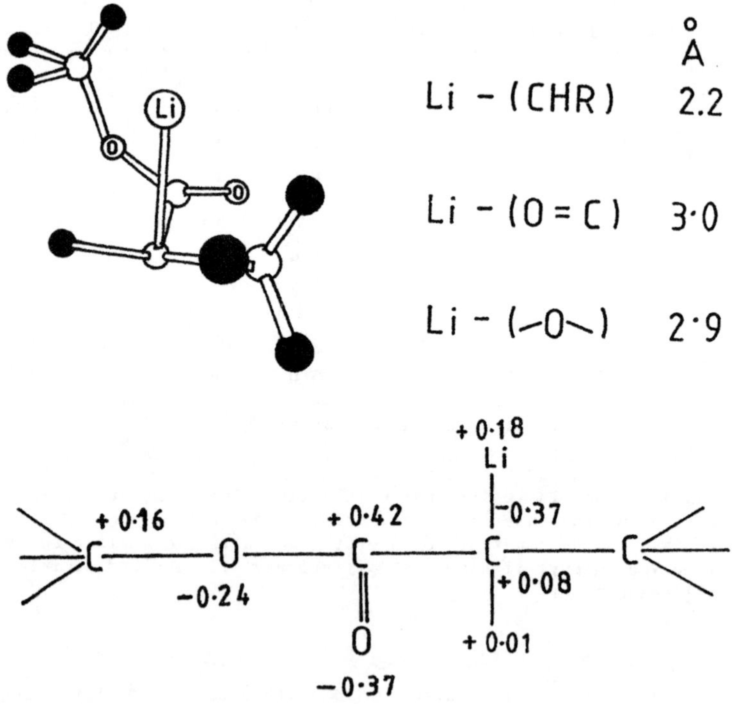

FIG. 9. Model of the active centre of methyl acrylate. (Reproduced with permission of Yu. Ye. Eizner [36])

An approximate idea of the geometry of the AC's of MMA can be obtained from the calculations of Eizner and Erussalimsky [36] (Fig.9). Contrary to the above mentioned they assume the oxygen atom of the alkoxy group to be coordinated to the lithium counterion, so that the latter resides over the plane of the nearly planar structure of the anion.

A comparison of the results for the monomer P_1Li and the dimer P_2Li of esters and nitriles is of particular interest. The data can be discussed as follows. In the case of lithium derivatives of nitriles the dissociation constant of the dimer P_2Li is smaller than the constant of the monomer P_1Li which is indicative of chelate complex formation.On the other hand, the concentration of ion triples in P_2Li solutions is higher than in solutions of P_1Li. This strongly suggests that intermolecular interaction of the cyano group of the second monomer unit with lithium cation also occurs. We consider the most probable structure of the ion triples to be as shown:

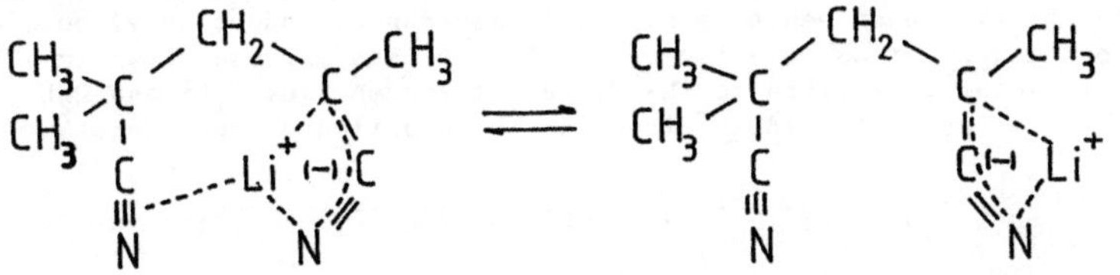

Z E

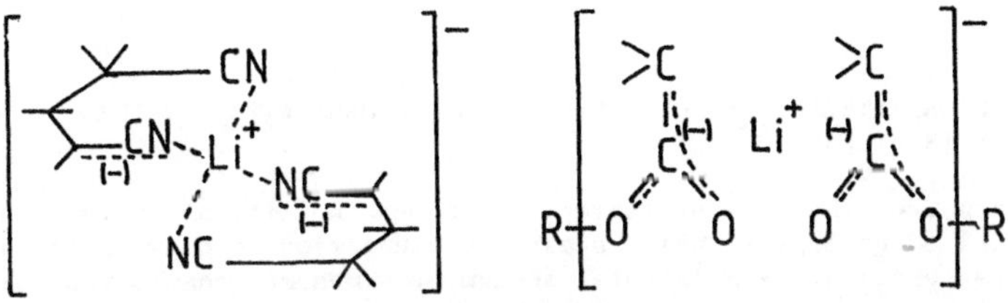

$[ABA]^-$

The data for lithium derivatives of esters are quite different. The difference between the dissociation constants of P_1Li and P_2Li is insignificant, the constant of the dimer being even slightly higher. It can be assumed that in THF solutions the equilibrium between the E and Z form is strongly shifted to the E form. NMR spectra of the lithiated dimer in THF are also not in accordance with the formation of chelate structures [37]. In contrast to the cyano derivatives, the concentration of ion triples in solutions of P_1Li is several times higher than in P_2Li solutions. In case the proposed structure of the contact ion pairs is true and the cation resides over the plane of a nearly planar >C-COO(-) group, the ion triples $[ABA]^-$ should possess a sandwich structure which is energetically more favourable for P_1Li than for P_2Li.

Hence, a principle difference in the structure of ion triples of lithiated dimers of MMA and MAN can be formulated. It consists in the assumption that the cyano group of the second monomer unit in the lithiated dimer of MAN participates in the ion triple complex by the formation of a coordinate bond, while complexation of cation with the second ester group is not likely in the case of the α-lithiated dimer of MMA.

The values of K_D and k_T of lithiated dimer of a ketone are not available at present which does not allow to complete the comparison. However, IR spectra of α-lithiated oligomers of methyl isopropenyl ketone give evidence of complex formation between the lithium cation and the carbonyl groups of the neighbouring monomer units in THF even at temperatures as high as 20°C [11] . These results make it reasonable to assign the order of polar groups in the monomer unit next to the AC of chain propagation with respect to their capacity to form coordinate bonds with the lithium counterion in THF solutions. Most effective is the keto group, the cyano group ranking second and the ester group showing the weakest coordination tendency.

The data for α-lithiated esters can be used to explain the different reactivities of P_1Li and P_2Li in the anionic polymerization of MMA [38] . It has been found that $k_{p1} > k_{p2}$ which is, however not likely to be due to the formation of a less reactive chelated form of P_2Li for it has not been confirmed by independent evidence. Several factors might be responsible for the observed phenomenon, e.g. steric hindrance, induction effect, or different degree of association of the AC's of monomer and dimer in THF. The last factor is related to the different tendency of P_1Li and P_2Li to form ion triples of the $[ABA]^-$ type. In the equation of rate constants:

$$k_p = (1 - \alpha - 2\gamma)k_p^{-+} + \alpha k_p^{-} + \gamma k_p^{-+-} \qquad (4)$$

the second term at the right hand side can be neglected in view of the extremely low values of K_D. The contribution of the third term has not been estimated. However, k_p^{-+-} could be expected to exceed k_p^{-+}, and results are in favour of a higher concentration of $[ABA]^-$ triples in solutions of P_1Li than in P_2Li solutions.

DONOR-ACCEPTOR INTERACTIONS OF POLAR VINYL MONOMERS WITH LITHIUM COUNTERIONS IN DIOXAN [39]

As seen from the data of Table III, the probability of chain propagation on free ions is negligible which shows the counterion to play an essential role in the reaction. Even in polar medium, the monomer has to replace a solvent molecule from the immediate surrounding of the active centre to form a short-lived intermediate complex [40] . Under these conditions the polymerization reaction preserves its statistical character. The situation can be thoroughly changed, however, in low polarity medium when the monomer is capable of "expelling" solvent molecules from the solvation shell of the counterion to form a stable complex. Specific interaction between monomer and counterion should have a profound effect on the reaction mechanism, especially in co-polymerization processes, and may prove to be more important than intrinsic monomer reactivity. Unfortunately, no direct experimental evidence has been reported to demonstrate it in real polymerizing systems.

Data on the electron donating capacities of the monomers might contribute to the problem so a comparative investigation of the affinity of some polar vinyl monomers for lithium picrate was performed in a solvent of low polarity, viz. dioxan. The salt was selected for several reasons. It is a consistent model of the anionic AC's of the studied monomers since the anion is similarly of ambident character and the salt forms mainly tight ion pairs in solution. Complex formation prior to monomer addition to the growing chain is most probable for lithium and magnesium counterions [41] . Lithium picrate has the advantage that it does not initiate polymerization of the monomers which is of primary significance for the studies. Finally, picrates have a characteristic absorption in the near UV region of the optical spectrum [42] which allows the use of a simple experimental technique.

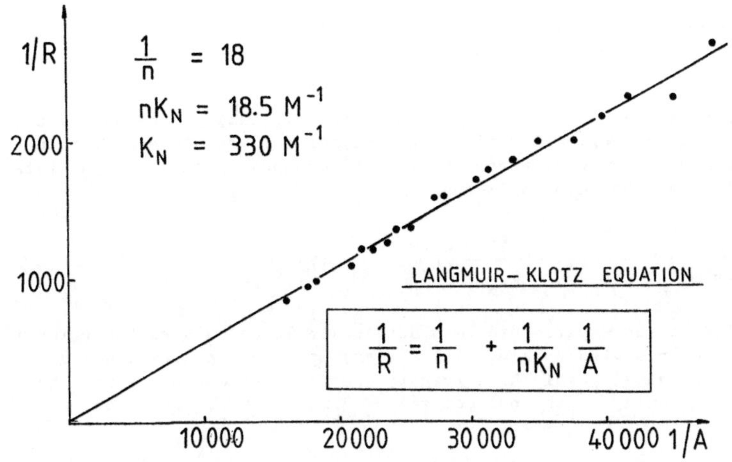

FIG. 10. Binding of lithium picrate to cross linked PEO in dioxan

The approach of Smid et al. [43] was used in the studies. It utilizes competitive binding of the ionic solute by a soluble ligand and an insoluble polymer gel to determine the constant of complex formation with the soluble ligand. An important feature of the method is the requirement for comparable binding capacities of the two ligands. For this reason, in studies of weak interactions the authors employ polymer networks with anchored polyethylene glycol chains [44].

On the basis of previous results on the coordination of alkali metal cations by linear poly(ethylene oxide) [45], cross-linked PEO was prepared by γ-irradiation of aqueous solutions of the linear polymer [46]. The resulting gels exhibit a relatively low affinity for lithium cations showing them also to be a suitable material for studies of weak interactions. Thus, the binding constant of lithium picrate to cross-linked PEO in DOX was found to be 330 M^{-1} as calculated by the equation of Langmuir-Klotz [47] (Fig. 10). Even so, the complex formation constants of the studied monomers fell beyond the acceptable limit of sensitivity of the method. A new procedure was developed to improve the sensitivity combining the approach of Smid with the equation of Langmuir-Klotz:

$$\frac{1}{R} = \frac{1}{n} + \frac{1 + K_L[L]}{nK_N}\,\frac{1}{A'} \qquad (5)$$

Here, K_L and K_N are the binding constants to the soluble ligand and the polymer gel, respectively, and 1/n is the effective number of EO units involved in complexation per lithium cation. A' denotes the overall salt concentration in the supernatant solution, i.e. the sum of the equilibrium concentrations of free salt and soluble ligand complex. When K_L is small and the soluble ligand is in a sufficiently large excess, the concentration of uncomplexed ligand $[L]$ is practically equal to the total ligand concentration $[L]_o$, which can be easily kept constant in the experiment. Under these conditions the proportionality coefficient turns into a constant:

$$[L] \approx [L]_o = \text{const.}$$

$$\frac{1}{R} = \frac{1}{n} + \frac{1}{K' A'} \qquad (6)$$

The constant K' in eqtn. (6) is the apparent binding constant of the ionic solute to a single EO unit of the polymer gel when in competition with the soluble ligand. The reciprocal 1/K' depends linearly on the total concentration of the soluble ligand:

$$\frac{1}{K'} = \frac{1}{nK_N} + \frac{K_L}{nK_N}[L]_o \qquad (7)$$

Results of the experiments with acrylonitrile calculated according to eqtn. (6) are presented on Fig. 11. As seen, upon raising the monomer concentration the slope of the straights increases which corresponds to reduced salt binding to the polymer gel:

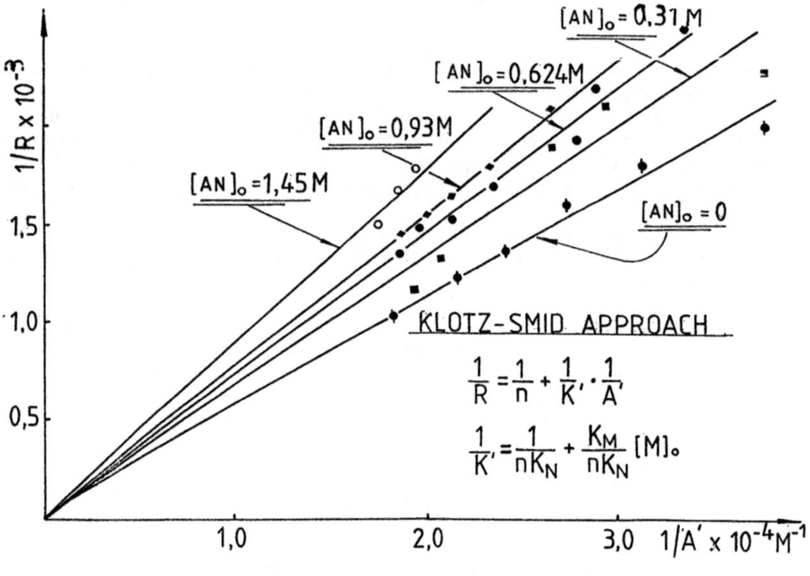

FIG. 11. Binding of lithium picrate to cross-linked PEO in dioxan in the presence of different amount of acrylonitrile in the solution

From eqtn. (7) the complex formation constants of lithium picrate with the monomers were calculated as shown on Fig. 12. It is to be noted that the proportionality coefficient contains the constant of interaction with a single EO unit nK_N, instead of the intrinsic binding constant of the polymer gel K_N as it is in the equation of Smid et al. [43] . Actually, this fact is responsible for the improvement in studies of very weak interactions by extending the method's lower limit of sensitivity by one order of magnitude.

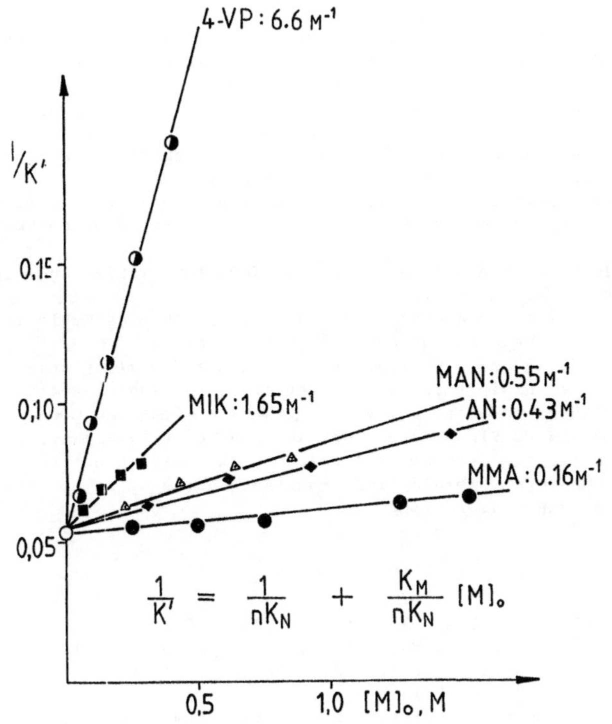

FIG. 12. Binding capacity of cross-linked PEO as a function of the total concentration of competing monomer in solution, eqtn. (7)

The final results obtained by the combined Klotz-Smid approach are presented in Table IV. As seen, there is no direct correlation between complex formation constants and dielectric susceptibilities, or dipole moments of the monomers. This suggests the observed effects to be due to specific interaction between monomer and salt. It is also to be pointed out that the polar groups of monomers exhibit the same order of descending capacities to coordinate lithium counterions as registered for the monomer units next to the active centre, viz. $>C{=}O > -C{\equiv}N > -COOR$.

TABLE IV. Constants of complex formation K_M with lithium picrate in dioxan at 25°C, dielectric constants ε (20°C), and dipole moments μ of some polar vinyl monomers.

Monomer	K_M, M^{-1}	$\varepsilon^{20°C}$	μ, D
Methyl isopropenyl ketone	1.65	-	2.8
Methacrylonitrile	0.55	-	3.69
Acrylonitrile	0.43	38	3.88
Methyl methacrylate	0.16	2.9	1.79

CONCLUSIONS

The nature of the active centres in the anionic polymerization of acrylonitrile and methacrylonitrile, and some other polar vinyl monomers, gives evidence of the important effect of the polar substituent. The marked difference in physicochemical characteristics of the active centres of polar and non-polar monomers are due to the interaction of the polar group with the counterion. Some of the important features of the anionic active centres of polar vinyl monomers are as follows:

1. Minimal participation of free ions and predominent contribution of contact ion pairs of the active centres in the reaction of polymer chain propagation.
2. Tendency of the active centres to form ion triples and higher associates in THF solutions.
3. The intimate mechanism of chain propagation presumably includes complex formation between monomer and active centres.
4. The polar groups, whether in the monomer repeating unit or in the monomer molecule, exhibit the same order of descending capacity to coordinate lithium counterions, viz. $>C=O > -C\equiv N > -COOR$.

Infrared and UV VIS spectroscopy and conductivity measurements are shown to be effective methods in studies of the anionic active centres of polar vinyl monomers combining the important merits of simplicity of operation and being informative.

REFERENCES

1. A.H.E. Müller, ACS Symp. Ser. No 166, 441 (1981).
2. C.C. Meverden, and T.E. Hogen-Esch, J. Polym. Sci., Polym. Chem. Ed. 23, 159 (1985); S. Strong Huang, T.E. Hogen-Esch, J. Polym. Sci., Polym. Chem. Ed. 23, 1203 (1985).
3. B.L. Erussalimsky, J. Polym. Sci., Polymer Symposium 62, 29 (1978).
4. D.J. Cram, and U.R. Kopecky, J. Am. Chem. Soc. 81, 2748 (1959).
5. A. Ottolenghi, and A. Zilkha, J. Polym. Sci. A 1, 687 (1963).
6. Ch.B. Tsvetanov, Yu.Ye. Eizner, and B.L. Erussalimsky, Europ. Polym. J. 16, 219 (1980).
7. S.J. Bryan, P.G. Hagget, and K. Wade, Coordination Chem. Rev. 44, 149 (1982).
8. B.L. Erussalimsky, Vysokomol. Soed. A 13, 1293 (1971).
9. I.N. Juchnovski, and I.G. Binev in: The Chemistry of Functional Groups, Suppl. C, Part 1, S.Patai and Z. Rappoport, eds. (John Wiley & Sons 1983) p. 107.
10. L. Lochmann and J. Trekoval, Makromol. Chem. 183, 1361 (1982).
11. Ch.B. Tsvetanov, D.T. Petrova, and I.M. Panayotov, Makromol. Chem. 183, 517 (1982).
12. H.J. Adler, L. Lochmann, D. Dotcheva, and Ch.B. Tsvetanov, Makromol. Chem. 187, (1986).
13. Ch.B. Tsvetanov, Europ. Polym. J. 15, 503 (1979).
14. Ch.B. Tsvetanov and V.N. Zgonnik, C.r. Acad. Bulg. Sci. 32, 55 (1979).
15. Ch.B. Tsvetanov and I.M. Panayotov, Europ. Polym. J. 11, 209 (1975).
16. Yu.Ye. Eizner and B.L. Erussalimsky, Quantum Chemistry of Polymerization Reactions, (Nauka, Leningrad 1976).
17. Ch. Tsvetanov, I. Panayotov, and B. Erussalimsky, Europ. Polym. J. 10, 557 (1974).
18. S. Baldwin, J. Org. Chem. 26, 3288 (1961).
19. Ch.B. Tsvetanov and I.M. Panayotov, Polymer Preprints 27, 147 (1986).
20. S. Bywater and P.J. Worsfold, Can. J. Chem. 45, 1821 (1967).

21. A.A. Arest-Yakubovich, A.R. Gantmakher, and S.S. Medvedev, Dokl. Akad. Nauk SSSR 167, 1069 (1966); A.Kh. Bagdasaryan, B.A. Dolgoplotsk and V.M. Frolov, Dokl. Akad. Nauk SSSR 198, 1077 (1971); D.N. Bhattacharyya, C.L. Lee, J. Smid, and M. Szwarc, J. Phys. Chem. 69, 612 (1965).
22. L.V. Vinogradova, N.I. Nikolaev, V.N. Sgonnik, B.L. Erussalimsky, G.V. Sinitsina, Ch.B. Tsvetanov, and I.M. Panayotov, Europ. Polym. J. 17, 517 (1981).
23. L.V. Vinogradova, V.N. Sgonnik, N.I. Nikolaev, and Ch.B. Tsvetanov, Europ. Polym. J. 15, 545 (1979).
24. Ch. Tsvetanov, V. Sgonnik, I. Panayotov, and B. Erussalimsky, J. Liebigs Annal. Chem. 763, 53 (1972).
25. W. Berger, B. Morgenstern, B.L. Erussalimsky, and L.V. Vinogradova, 7th International Symposium on Cationic Polymerization and Related Processes (Aug. 12-15, Jena, GDR, 1985), Abstracts.p. 75.
26. W. Fowells, C. Schuerch, F.A. Bovey, and F.P. Hood, J. Am. Chem. Soc. 89, 1396 (1967).
27. J.E. Figueruelo, Makromol. Chem. 131, 63 (1970).
28. G. Löhr and G.V. Schulz, Makromol. Chem. 172, 137 (1973); G. Lohr and G.V. Schulz, Europ. Polym. J. 10, 121 (1974).
29. D. Hounore, J.C. Kavier, P. Sigwalt, and M. Fontanille, Europ. Polym. J. 10, 425 (1974).
30. W. Tochtermann, Angew. Chem. 78, 355 (1966).
31. B. De Groof, M. Van Beylen, and M. Szwarc, Macromolecules 8, 396 (1975).
32. C.B. Wooster, J. Am. Chem. Soc. 59, 377 (1937).
33. Ch. Tsvetanov and D. Dotcheva, J. Polym. Sci., Polym. Chem. Ed. 24, 2253 (1986).
34. D. Dotcheva, L. Lochmann, and Ch. Tsvetanov, J. Polym. Sci., Polym. Chem. Ed., to be published in more detail.
35. L. Vancea and S. Bywater, Macromolecules 14, 1321 (1981).
36. Yu.Ye. Eizner and B.L. Erussalimsky, Europ. Polym. J. 12, 59 (1976).
37. L. Vancea and S. Bywater, Macromolecules 14, 1776 (1981).
38. Ch.B. Tsvetanov, A.H.E. Müller, and G.V. Schulz, Macromolecules 18, 863 (1985).
39. D.K. Dimov, E.B. Petrova, I.M. Panayotov, and Ch.B. Tsvetanov, to be published in more detail.
40. A.H.E. Müller, H. Jeuck, C. Johann, and P. Kilz, published in the same ACS Series Edition.
41. B.L. Erussalimsky, International Symp. on Macromolecular Chemistry (Budapest 1969), Main Lectures, Akademiai Kiado, p. 281.
42. M. Bourgoin, K.H. Wong, J.J. Hui, and J. Smid, J. Am. Chem. Soc. 97, 3462 (1975).
43. J. Smid and R. Sinta, in: Crown Ethers and Phase Transfer Catalysis in Polymer Science, L.J. Mathias and C.E. Carraher, Jr., eds. (Plenum Press, New York 1984) p. 329.
44. W.Y. Xu and J. Smid, J. Am. Chem. Soc. 106, 3790 (1984).
45. D.K. Dimov, I.M. Panayotov, V.N. Lazarov, and Ch.B. Tsvetanov, J. Polym. Sci., Polym. Chem. Ed. 20, 1389 (1982).
46. K.G. Vassilev, D.K. Dimov, R.T. Stamenova, R.S. Boeva, and Ch.B. Tsvetanov, J. Polym. Sci., Polym. Chem. Ed. 24, (1986).
47. I.M. Klotz, F.M. Walker, and R.B. Pivan, J. Am. Chem. Soc. 68, 1486 (1946).

PYRIDINE-MEDIATED ANIONIC HOMO-AND COPOLYMERIZATION OF ALKYLMETHACRYLATE

G. HUYNH-BA,* AND J. E. McGRATH**
*E. I. Du Pont De Nemours and Company, Experimental Station, Polymer Products Department, Bldg. 269, Room 215, Wilmington, Delaware 19898;
**Virginia Polytechnic Institute and State University, Chemistry Department, Blacksburg, Virginia 24061

ABSTRACT

Several years ago we discovered that pyridine played a significant role in the anionic homo- and copolymerization of polar monomers such as N,N-dimethylacrylamide, ε-caprolactone and alkylmethacrylate. Pyridine gives rise to monodisperse polymethylmethacrylate (PMMA) at a temperature higher than normally observed. We believe this reflects a new mechanism, involving a living polymerization, which should have profound implications for the design of PMMA copolymers via anionic techniques.

INTRODUCTION

The synthesis of AB and ABA block copolymers can be performed by using a mono- or difunctional anionic initiator (1). When A and B are nonpolar monomers such as styrene or isoprene, the synthesis of these copolymers does not pose any problems; however, when A and/or B are alkylmethacrylates, the synthesis of such copolymers may raise questions concerning the living character of the propagating chain. Data have shown that long or bulky alkyl groups (R>C4) serve to improve the living character of the chain (2, 3).

The anionic polymerization of methylmethacrylate has been widely studied (4, 5, 7, 10). One can find considerable disagreement in the literature regarding the living character of the propagating chain as measured by the molecular weight distribution (MWD). These studies were made in tetrahydrofuran (THF) or in a mixture of toluene/THF using different alkali salts (mostly lithium) as the initiator.

Both polymer and monomer contain ester groups which could be subject to undesirable reactions. It is clear now that these reactions involve nucleophilic attack of carbanion on the ester with displacement of alkoxy group:

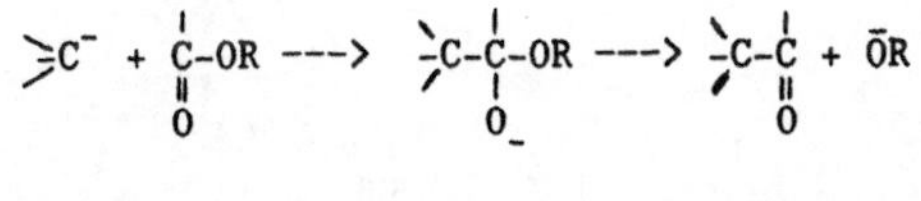

Side reactions apparently occur in the presence of the lithium counterion at temperatures above -70°C (4). The main evidence for such a reaction is the presence of lithium methoxide (5, 6). On the other hand, monodisperse PMMA was claimed to form when biphenyl lithium or biphenyl sodium were used as initiators at low temperature. However, no gel permeation chromatography (GPC) was reported on these materials (7, 8). Polymerizations carried out in mixtures of solvents such as toluene and THF yielded a broad bimodal distribution with the sodium dianion of alpha methylstyrene dimer as initiator even at low temperature (9). The most detailed studies of the kinetics of homopolymerization of MMA and related alkylmethacrylates were done by Schulz and coworkers (10, 11). Their conclusion was that the rate of the termination reaction of the carbanion (i.e., living PMMA) increased with temperature (>-70°C) and with smaller cations (i.e. Li).

Recent Advances in Anionic Polymerization
T.E. Hogen-Esch and J. Smid, Editors

The anionic block copolymerization of MMA with styrene gave a branched structure as reported by Rempp if the living polystyryl potassium anion was not capped with 1,1-diphenylethylene (DPE) (12). However, Gallot obtained linear diblock polyisoprene-PMMA under the same conditions without capping with DPE (13).

Pyridine has been used as a solvent to study the tacticity of PMMA (16). Recently, Andrews described briefly the use of pyridine and a mixture of pyridine/toluene or pyridine/THF in the anionic polymerization of MMA (17), from -78°C to as high as -20°C. Using pyridine as solvent or cosolvent and sec-butyllithium (sec-BuLi) as initiator, we are able to control the living character of the alkylmethacrylate polymerization. In this report we would like to discuss the living anionic polymerization of MMA at different temperatures with different counterions, their effect on the tacticity of PMMA and the mechanism of polymerization. Finally, we wish to report several attempts to copolymerize MMA and alkyl methacrylates with polar and nonpolar monomers.

RESULTS AND DISCUSSION

Homopolymerization of Methylmethacrylate in Pyridine - a New Mechanism

It was reported in 1962 (16) that MMA could be polymerized by BuLi in the presence of pyridine and that the resulting PMMA possessed a high syndiotactic content. However, in our case we used pyridine for two purposes:

1) to dissociate the alkyl lithium aggregates and decrease the basicity of the alkyl lithium initiator through formation of the "σ complex" (eqn. A).

2) to maintain the "σ complex" (eqn. B) (14) during the polymerization so that the side reactions (attack of the ester by carbanion) may be avoided.

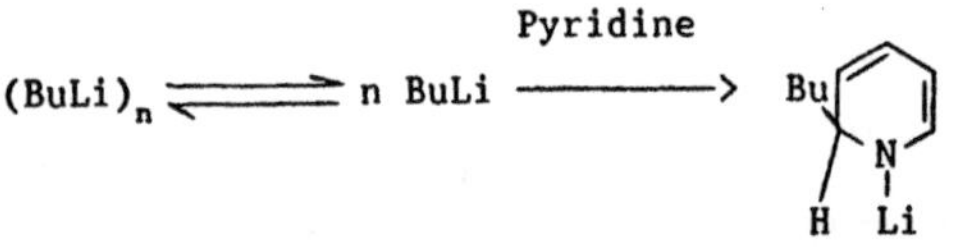

(A)

(σ complex)

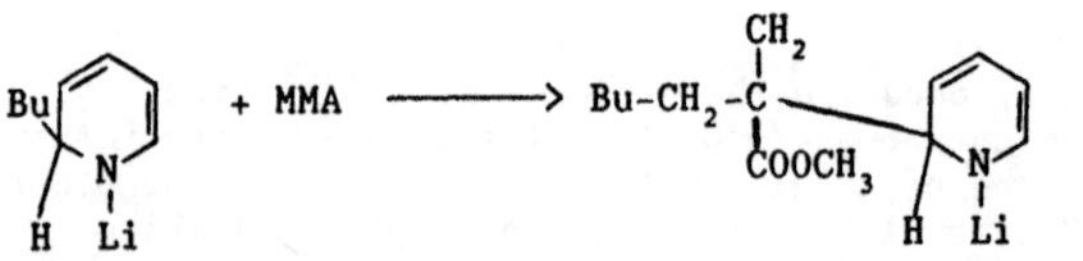

(B)

(σ complex)

Table 1 clearly shows the living character of the MMA polymerization: the polydispersity is between 1.02 - 1.03, (Case 1, and Case 3, Table 1). Moreover, the yield is quantitative and the number average molecular weight determined by viscometry (20) is almost identical, within experimental error, to the theoretical value (Table 1). In practice we are able to prepare PMMA with M_w as high as 83,000 with a

TABLE I. Homopolymerization of MMA in Pyridine initated by sec-BuLi at -45°C (a).

CASE	INITIATOR (moles)	YIELD %	M/I	M_n(theor)	(c) M_n(exp)	TOL(b) $[\eta]$25°C	GPC(d) (M_w/M_n)
1	10^{-3}	100	47	4700	4800	0.054	#1 (1.02)
2	$0.5.10^{-3}$	100	93	9300	8400	0.070	
3	$0.22.10^{-3}$	100	211	21100	21700	0.115	#2 (1.03)

(a) Reaction conditions: 4.65 gr of MMA was added to 150 ml of a sec-BuLi solution in pyridine at -45°C; termination by $CH_3OH\backslash HCl$ after 15 minutes
(b) $[\eta]^{toluene}_{25°C} = 7.8.10^{-4}.M_v^{0.5}$ (dl. gr^{-1})
(c) Because of the low dispersity we have assumed that $M_n = M_v$.
(d) Based on polystyrene standards.

polydispersity of less than 1.1. All the polymerizations are homogeneous at the polymerization temperature of -45°C. According to the literature (5, 10, 11, 17), at this temperature a "true" living polymerization (life time of living end greater than 10 minutes) is not observed when a lithium counterion is used (no quantitative yield, Mn theory < Mn exp. polydisperse MWD material). Therefore, pyridine must be involved in the polymerizations yielding monodisperse PMMA. Our kinetic study shows that the rate of polymerization is first order in monomer and initiator in pure pyridine at -45°C. The apparent rate constant is equal to 80 mol^{-1} .l. sec^{-1}. The 1,4 addition on pyridine ring was excluded because studies showed only the 1,2 addition when pyridine is reacted with organo alkali compounds (21, 22). One of the possible following polymerization mechanisms is proposed (FIG 1).

Equation 1 is proposed because we add sec-BuLi to pyridine prior to MMA addition. This σ complex is well known in the literature (18, 21, 22). During polymerization, the pyridine unit must play an active role because side reactions (such as the attack of the propagating chain on the ester groups of the monomer or of the chain) are avoided; therefore, the pyridine group may be at the living end (σ complex, eqns 2 and 3). It is also possible that pyridine protects the ester group of the monomer and/or the polymer (from side reactions), through some specific association with it. For example, there is a known specific complex formation between pyridine and the amide group (15). Our results are not consistent with ester group association.

The IR absorption of the monomer carbonyl is the same in pyridine and toluene. Therefore, in contrast to N,N'-dimethylacrylamide (14), there probably is no association of pyridine and the carbonyl of MMA. Furthermore, deactivated PMMA showed identical IR absorption wave lengths of the carbonyl in toluene, in CH_2Cl_2 and in pyridine solution. The same deactivated PMMA in $CDCl_3$ and in pyridine solution showed the same tacticity splitting and position in ^{13}C NMR of the carbonyl group. Therefore, pyridine does not appear to associate with the ester groups of the PMMA chain during the polymerization.

Since the evidence implies no association between pyridine and ester groups of monomer and of polymer, one explanation for the absence of normal side reactions in pyridine could be the formation of a σ complex at the growing end (eqns 2 and 3). This is consistent with the exothermic equilibrium of equation 4. Side reactions will occur with the enolate anion (eqn. 6) when the concentration of pyridine is low and/or the temperature is high. This can be seen in Table II where the effects of temperature and pyridine concentration on the experimental versus theoretical mass molecular weight and its polydispersity are summarized. Similar observations were also mentioned by Andrews (17).

FIGURE 1

PROPOSED MECHANISM

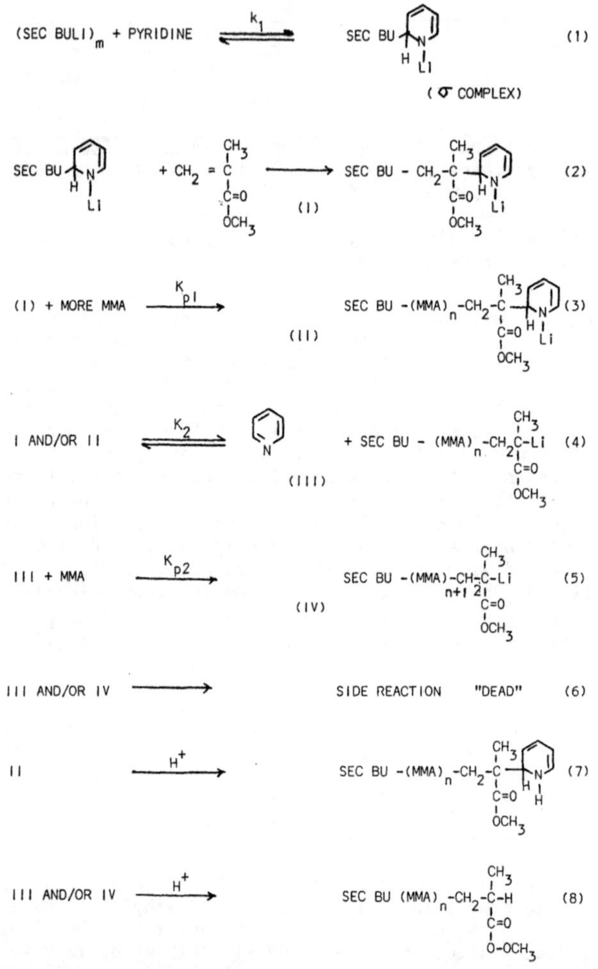

TABLE II. Effect of Temperature and Pyridine concentration on the homopolymerization of MMA initiated by sec-BuLi.

CASE	SOLVENT	TEMP (°C)	YIELD	M_n (theor)	(b) M_v (exp)	(a) GPC (M_w/M_n)	[η] TOL 25°C
1	PY	25	46	2100	20,000	1.8	0.11
2	PY=10% BZ=90%	25	7	300	6,000	>2	--
3	PY=10% BZ=90%	0	50	2300	10,000	1.6	0.078
4	PY	-45	100	4700	4,800	1.02	0.054

(a) M_w/M_n from polystyrene standards.
(b) See footnotes Table I.

The presence of the addition product of living PMMA and pyridine (eqns. 2 and 3) requires the presence of a dihydropyridine in the protonated (i.e. deactivated) polymer (eqn. 7). This was found to be the case: when PMMA obtained from sec-BuLi/Pyridine at -45°C after protonation (CH_3OH/H^+) is purified through extraction of pyridine with aqueous HCl, the UV spectrum still indicates an absorption which we attribute to the dihydropyridine viz. 240-340 nm with λ max = 260-265, 295-305 nm (Figure 2, UV #1). By contrast, neither PMMA from THF/sec - BuLi (-78°C) nor PMMA simply mixed with pyridine possessed this absorption after similar purification (Figure 2, UV #2). Additional evidence for the presence of the dihydropyridine group comes from GPC analysis of low MW PMMA (M_n = 8000). Using the UV detector (λ = 280 nm), a broader distribution is obtained than seen with a refractive index (RI) detector (Figure 3). This is consistent with the UV detection of end groups (independent of the weight of the chain) versus the RI detection of MMA units (dependent on the weight of the chain). Thus we confirm Andrews finding (17) that dihydropyridine is bound to each PMMA chain, but the living character of the chain argues for structure (E) as opposed to Andrews' structure (D). This could be confirmed via a labeling experiment as shown by equations C, which can give either structure D or E.

FIGURE 2

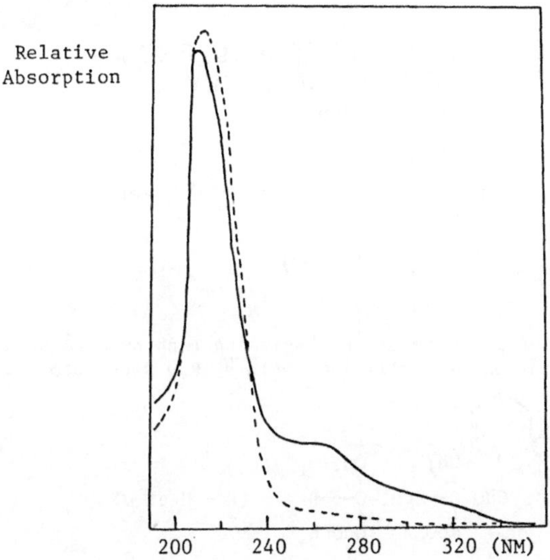

UV SPECTRUM OF PMMA PREPARED FROM PYRIDINE (#1—) AND FROM THF (#2---)

FIGURE 3

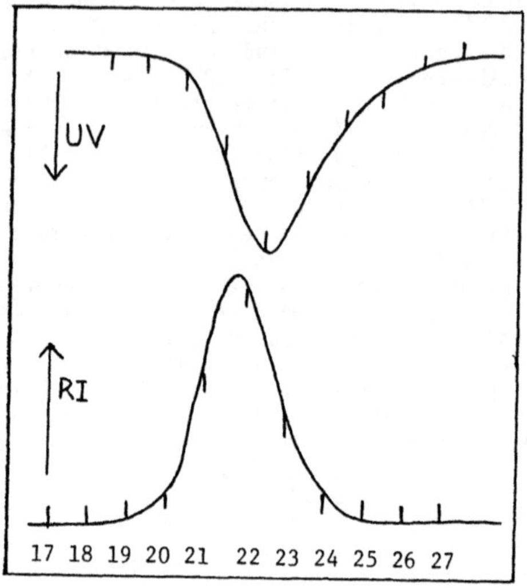

GPC OF PMMA PREPARED FROM PYRIDINE BY RI AND BY UV (λ = 280 NM)

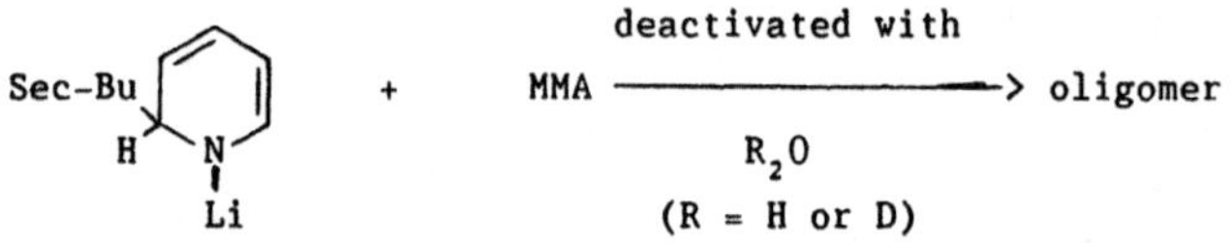

(C)

Structure D

If the dihydropyridine group is at the beginning (Andrews proposition), then, after deactiviation of the living chain by R_2O structure D will be:

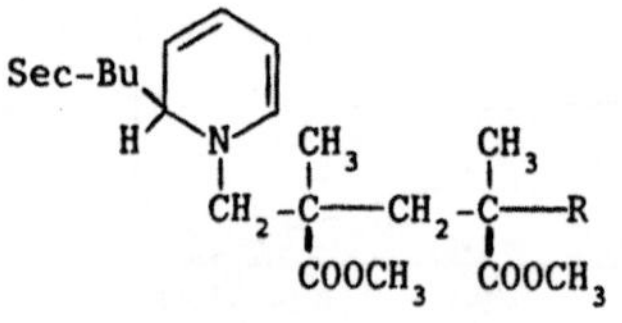

(R = H or D)

Structure E

If the dihydropyridine group is at the end (our proposition), then after deactivation of the living chain by R_2O structure E will be:

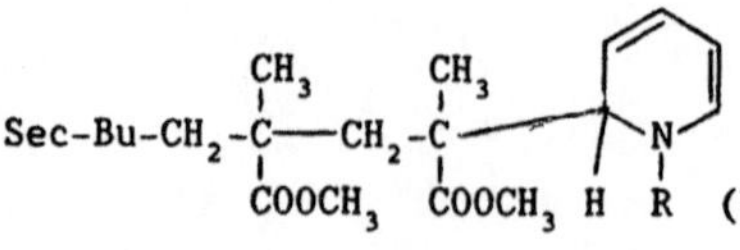

(R = H or D)

ε-caprolactone (CL) under similar homopolymerization conditions (sec Buli/pyridine as initiator - pyridine as solvent) showed no sign of the dihydropyridine group (UV detector λ = 280 nm) by GPC analysis (GPC #6, of low MW of PCL Figure 4). This suggests that the nucleophile is sec-Bu$^-$ (eqn. F) rather than $\rangle N^-$ (eqn. H) and consequently, the dihydropyridine group cannot be at the beginning of the PCL chain (eqn. G). The same reasoning can be applied to the case of PMMA.

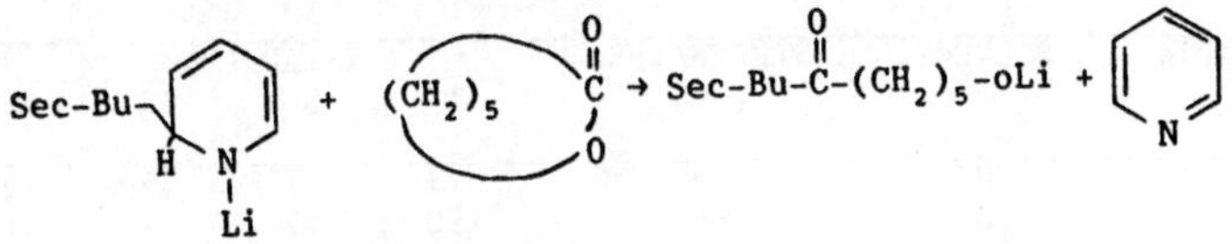

(F)

$$\text{Sec-Bu-}\overset{\overset{\text{O}}{\|}}{\text{C}}\text{-}(CH_2)_5\text{-oLi} + n(\text{CL}) \rightarrow \text{Sec-Bu }(\overset{\overset{\text{O}}{\|}}{\text{C}}\text{-}(CH_2)_5)\ \text{O})_{(N+1)}\text{-Li} \qquad \text{(G)}$$

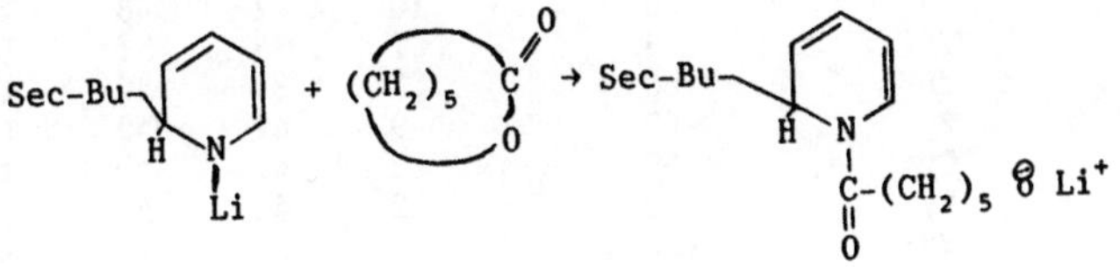

(H)

FIGURE 4

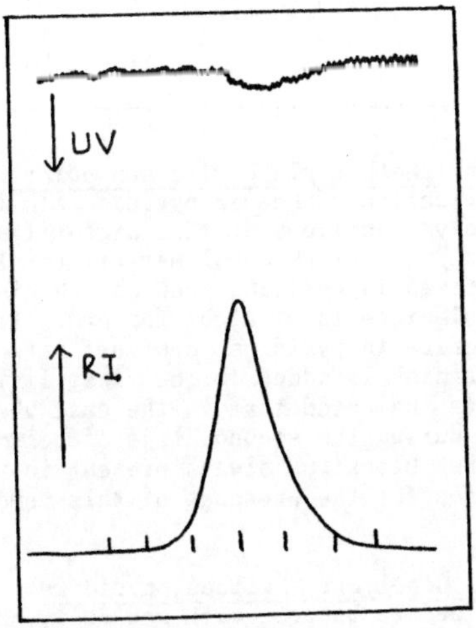

GPC OF PCL (M_N = 4000)
PREPARED FROM PYRIDINE
BY RI AND BY UV
(λ = 280 NM)

Tacticity of PMMA prepared in the presence of pyridine: Since the mechanism that we have proposed for the polymerization of MMA involves the pyridine unit at the chain-end, the tacticity should be different from that in other polar solvents such as THF. The results (Table III) show that in all cases, the syndiotactic triad content of PMMA prepared in pyridine as solvent is always higher than that prepared in THF.

TABLE III. Polymerization of MMA initiated by Alkali Metal Salts in THF and Pyridine - Dependence of Tacticity on Counterion and Temperature.

COUNTERION	SOLVENT	TEMPERATURE (°C)	TRIAD (%) I	S	H	REFERENCE
Li	THF	-78	7	57	36	(19)
Li	THF	-60	10	51	39	this work
Li	Py	-60	7	60	33	(16)
Li	Py	-60	6	68	26	this work
Li	Py	-40	8	60	32	this work
Li	Py	0	9	56	35	this work
Na	THF	-78	10	56	34	(24)
Na	THF	-60	15	38	47	this work
Na	Py	-60	10	51	39	(16)
Na	Py	-60	6	56	38	this work
Na	Py	0	10	42	48	this work
K	THF	-78	7	33	57	(25)
K	THF	-60	7	33	56	this work
K	Py	-60	9	42	49	(16)
K	Py	-60	9	41	50	this work
K	Py	0	12	41	47	this work
Rb	THF	-60	9	39	52	this work
Rb	Py	-60	8	46	46	this work
Rb	Py	0	10	41	49	this work
Cs	THF	-66	4	43	53	(11)
Cs	THF	-60	8	43	49	this work
Cs	Py	-60	7	54	39	this work
Cs	Py	0	11	50	39	this work

Copolymers of Methyl Methacrylate with non polar monomers: Since the anionic living polymerization of MMA in pyridine at -45°C has now been established block copolymerizations at this higher temperature should be possible. However, only the block copolymerization of monomers which can be anionically polymerized in pyridine such as N,N-disubstituted acrylamide, or ε-caprolactone is straight forward. For monomers which do not anionically polymerize in pyridine, one must introduce MMA as a second block after pyridine is added to the first living block. Preliminary experiments indicated that in the case where pyridine is present as co-solvent during the second stage of polymerization, the homopolymer of the first block was always present in variable proportions. The reason for the presence of this homopolymer is unclear at this time.

Polystyrene - PMMA Copolymer - without pyridine: After styrene (ST) is completely polymerized in benzene with sec Buli, MMA (or 2-ethyl-hexylmethacrylate (EHMA)) is added at -10°C; the yellow color of PST lithium disappears and the viscosity increases slightly. After a few minutes, the viscosity increases significantly. The conversion of MMA (or EHMA) is 40% (or 100%). The GPC traces using RI (for both monomers) and UV (λ = 280 nm for ST) showed the presence of some linear dimer of PST and a very broad distribution of copolymer mainly PMMA (or PEHMA) (Figure 5A). The dimer can be formed by exposure of PST lithium to air (oxygen and/or carbon dioxide) (Figure 5B).

FIGURE 5

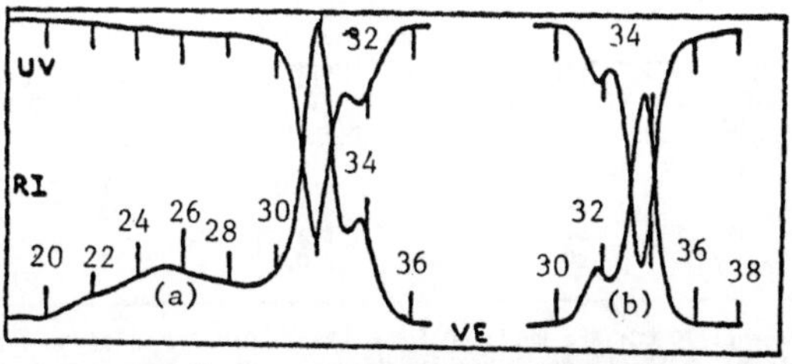

(a) PSTLi$^+$ + EHMA in Benzene -0°C
(b) PSTLi$^+$ + Air (O_2 + CO_2)

GPC OF COPOLYMER PST-PEHMA PREPARED FROM BENZENE AND HOMOPOLYMER PST EXPOSED TO AIR

Polystyrene - PMMA Copolymer in the presence of pyridine as co-solvent: After the ST is completely polymerized in toluene/sec-BuLi, pyridine is added as co-solvent (6% by volume); then at -10°C and under vacuum MMA is added. The conversion of PMMA was only 47% and a broad distribution of the copolymer PST - PMMA ($\bar{M}_w/\bar{M}_n$ > 1.5) is observed without linear dimer of PST. Therefore, pyridine is essential to avoid dimer formation. However, when the above polymerization is carried out at -50°C , the conversion of MMA is quantitative. After fractionation by addition of hexane to toluene the yield of copolymer is 71% (29% is homo PST). The copolymer has polydispersity (GPC) of 1.05 (Table IV, Case 1).

In the case where pyridine is added in larger amounts (65% by volume under vacuum), followed by addition of MMA, the conversion is quantitative and the yield of copolymer is only 54% (46% homo PST). Interestingly the distribution of the diblock is 1.05. This suggests a living character of the copolymerization (Case 2, Table IV). The presence of large amount of homo PST could be attributed to deactivation of the living chain from unknown impurities in pyridine.

Polyisoprene - PMMA Copolymer: Pyridine as Co-Solvent: Isoprene is polymerized to completion in hexane/sec BuLi under high vacuum. After addition of benzene and pyridine at -45°C, MMA is added. Again the conversion is quantitative. The yield of copolymer is 57% (43% homo PIP) after fractionation by addition of n-hexane to toluene to remove soluble PIP, and addition of methanol to toluene to remove homo PMMA. The polydispersity is a little higher: 1.1 (Table IV, Case 3). This type of polymerization could be extended to the preparation of polybutadiene - PMMA block copolymer (PB-PMMA). It is important to note that the PIP block copolymer has a 1,4 cis structure. On hydrogenation the PB and PIP could be converted to a polyethylene and poly (ethylene-propylene) copolymers, respectively. This would be a novel technique for preparing polyolefin - PMMA block copolymers.

PMMA - Poly N,N, dimethylacrylamide copolymer: Pyridine as solvent. MMA is polymerized in pyridine using sec-BuLi at -50°C. Then, N,N-dimethylacrylamide (DMA) is added. The conversion of DMA is quantitative. After fractionation with methanol, benzene and n-hexane, the yield of the copolymer is 90% (Table IV, Case 4) (10% homo PMMA). Note that it is not possible to initiate MMA with the living polymer of DMA.

TABLE IV. Anonic Copolymerization of MMA and Styrene, Isoprene and N,N Dimethylacrylamide in the presence of Pyridine.

MONOMER A/B	YIELD OF COPOLYMER (GR)	% YIELD COPOLYMER	% WT OF PMMA	[η] OF FIRST BLOCK dl. gr^{-1}	[η] OF COPOLYMER dl. gr^{-1}	GPC (e) COPOLYMER	TYPE OF BLOCK
ST/MMA (a)	11.4	71	61	--	1.59	1.05	A-B
ST/MMA (b)	11.4	54	63	0.1	0.29	1.05	A-B
IP/MMA (c)	12	57	70	(f)	--	1.10	A-B
MMA/DMA (d)	7.7	90	46	0.055	0.17	1.15	A-B

(a) ST:9.06 GR; Toluene: 70 CC; Sec BuLi: 10^{-4} Mol; Temp: 0°C, Time: 20 hrs. Pyridine: 5 ML; MMA: 7.02 GR; Temp: -50°C; Time: 10 Mn; [η] Toluene, 30°C
(b) ST: 15.86 GR; Benzene: 100 ML; Sec BuLi: 10^{-3} Mol; Temp: 0°C; Time: 20 hrs. Pyridine: 230 ML; MMA: 7.2 GR; Temp: -45°C, Time: 30 Mn; [η] Toluene, 30°C
(c) IP: 13.6 GR; Hexane: 100 ML; Sec BuLi: $0.36.10^{-3}$ Mol; Temp: 0°C; Time: 20 hrs. then Benzene: 150 CC, Pyridine: 40 CC; MMA: 8.4 GR; Temp: -40°C, Time: 30 Mn.
(d) MMA: 4.68 GR; Pyridine: 200 ML; Sec BuLi: $1.3\ 10^{-3}$ Mol, Temp: -60°C, Time: 40 Mn DMA: 4.07 GR; Time: 10 Mn, Temp: -50°C; [η] in $CH_2\ CL_2$, 25°C; DMA: CH_2=CH-CO-N$(CH_3)_2$
(e) M_w/M_n from PST standards
(f) From GPC, M_n = 38,000 - 40,000

PMMA - Poly 2-ethylhexylmethacrylate (EHMA) copolymer:

1) AB Copolymer: MMA (A) is polymerized in pyridine using sec-BuLi at -45°C. EHMA (B) is then added. The conversion is quantitative. After fractionation as in the above case PIP-PMMA, the yield of copolymer is 70% (30% homo PMMA). The polydispersity determined by GPC is about 1.05.

2) ABA Copolymer: MMA (A) is polymerized in toluene at -60°C using sec BuLi/Pyridine (2% volume). EHMA (B) is added to form the AB block, then MMA (A) is added to form the ABA copolymer. The yield is quantitative (no homopolymer of A or B) and the GPC of the ABA block is relatively narrow (Mw/Mn = 1.15, Table V, Case 2).

PMMA - Poly ε-caprolactone copolymer: Pyridine as co-solvent. MMA is polymerized in toluene using sec- BuLi in pyridine (5% volume) at -45°C. Then ε-caprolactone (CL) is added at -65°C and the polymerization is carried out at -65°C to 25°C for 18 minutes allowing temperature to rise without cooling (34% conversion of CL). After fractionation (cold methanol to remove PMMA and hot methanol to remove PCL) the yield of copolymer is 73% (27% homo PMMA). The GPC of the copolymer (GPC #14, Figure 6) is a little broader ($\bar{M}_w/\bar{M}_n$ = 1.2) compared to that of the PMMA (1st) block (GPC #13, Figure 6, Table V, Case 3).

CONCLUSION

We have demonstrated the versatility of using pyridine as co-solvent or more properly as "modifier" in the homopolymerization and copolymerization of methylmethacrylate. The polymerization is living below -45°C and the copolymerization can be performed in good yields with other polar and nonpolar vinyl monomers such as styrene, isoprene, 2-ethylhexylmethacrylate and N,N-dimethylacrylamide. The presence of a "σ complex" is probably the key to this living polymerization.

ACKNOWLEDGMENT

The authors wish to thank Professor T. E. Hogen - Esch for reviewing the manuscript and to Dow Chemical for financial support.

TABLE V. Copolymerization of MMA - EHMA or CL in the presence of Pyridine: AB and ABA type.

CASE	YIELD OF COPOLYMER (GR)	YIELD OF COPOLYMER (%)	% WT OF MMA IN COPOLYMER	GPC #	SOLVENT	TEMP °C	TYPE
1 (a)	6.3	70	30	4	Pyridine	-45	A-B
2 (b)	12.4	100	41	9	Toluene Pyridine	-60	ABA
3 (c)	8.1	73	25 (d)	12	Pyridine Toluene	-60	A-B

(a) MMA: 4.65 GR; PY: 170 ML; Sec BuLi: 10^{-3} Mol; Time: 2 Min; EHMA: 4.4 GR; Time: 10 Min.
(b) MMA: 1.68 GR; Tol: 170 ML; Sec BuLi: 10^{-3} Mol in 4 CC Pyridine; Time: 7 Mn EHMA: 7.48 GR; Time: 10 Min; MMA: 3.54 GR; Time: 15 Min $[\eta]$ Tol: 0.22 dl. gr^{-1} L.G.$^{-1}$; Composition: 13/59/28: MMA/EHMA/MMA 25°C
(c) MMA: 8.4 GR; Tol: 170 ML; Sec BuLi: 10^{-3} Mol in 4 CC Pyridine; Time: 15 Min.; CL (ε-caprolactone): 10.3 GR - only 34% conversion - Time: 12 Min. $[\eta]$ Tol PMMA: 0.1 L.G.$^{-1}$ - $[\eta]$ Tol Copolymer: 0.25 L.G.$^{-1}$
30°C 30°C
(d) % wt of CL in copolymer.

FIGURE 6

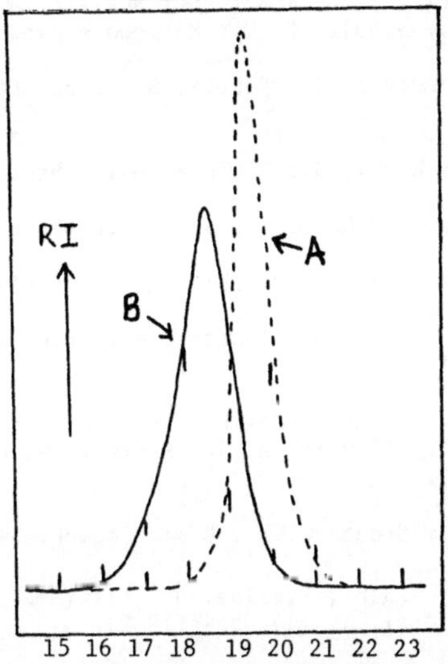

GPC OF COPOLYMER PMMA - PCL →B AND PRECURSOR PMMA ---A FROM PYRIDINE/ SEC-BULI/-45°C

BIBLIOGRAPHY

1. Noshay, A. and MC Grath, J. E.; "Block Copolymers: Overview and Critical Survey" Academic Press, New York (1977).

2. Gallot, Y.; Seow, P. K.; Skoulios, A.: Die Makromol Chem, 176, 3153 (1975).

3. Mc Grath, J. E.: Macro Iupac Dublin Preprints Meetings (1977).

4. Bywater, S.: Fortschr. Hoch Polymer Forsch. 4 (1), 66 (1965).

5. "Structure and Mechanism in Vinyl Polymerization" Ed. by Tsurita, T. and Odriscoll, K. F.; 1969, Chapt 8 - P. 223 - by Wiles, D. M., - Ed. Dekker.

6. Wiles, D. M.; Bywater, S.: Chem. Ind. (London) 1909 (1963).

7. Figueruello, J. E.; Roig, A.; Llano, E.: J. Polym. Sci. PT B-3 (3), 171 (1965).

8. Ibid, Jour. Polym. Sci. C - 16 - 4141 (1968).

9. Allen, P.E.M.; Chaplin, R. P.; Jordan, D. O.: Euro. Polym. J. 8 (2), 271 (1972).

10. a) Schulz, G. V.; Loehr, G.: Die Makromol. Chem. 172 - 137 (1973); b) Schulz, G. V.; Loehr, G.; Muller, A.H.E.; Warzelhan, V.: 175, 497 (1974); c) Schulz, G. V.; Warzelhan, V.; Loehr, G.; Hoecker, H.: 179, 2211 (1978)

11. a) Mueller, A.H.E.; Hoecker, H.; Schulz, G. V.: Macromolecules 10 (5), 1086 (1977); b) Kraft, R.; Mueller, A.H.E.; Warzelhan, V.; Hoecker, H.; Schulz, G. V.: Macromolecules 11 (6) 1093 (1978).

12. Rempp, P.; Freyss, D.; Benoit, H.: Jour. Polym. Sci. B(2) 17 (1964).

13. Gallot, B.; Rossi, J.: Die Makromol. Chem. 177, 2801 (1976).

14. Huynh-ba, G. and Mc Grath, J. E.: to be published.

15. Moriarty, R. M.; J. Org. Chem. 28, 1296 (1963).

16. Braun, D.; Herner, M.; Johnsen, U.; Kern, W.: Die Makromol Chem 51 - 15 (1962).

17. Andrews, G. D.; Anderson, B. C.; Arthur, P.; Jacobson, H. W.; Melby, L. R.; Playtis, A. J.; Sharkey, W. H.: Macromolecules 14, 1599 (1981).

18. Giam, C. S.; Stout, J. L.; Chem. Commun. 142 (1969).

19. Hatada, K.; Sugino, H.; Ise, H.; Kitayama, T.; Okamoto, Y.; Yuki, H.: Polym. Jour. 12 (1) -55- (1980).

20. Polymer Handbook: Brandrup, J. and Immergut, E. H.; 2nd edition 1975 Section IV.

21. Abramovitch, R. A. and Saha, J. G. in "Advances in Heterocyclic Chemistry" page -29- edited by Katritzky, A. R. and Boulton, A. J. Vol 6 (1966).

22. Abramovitch, R. A. and Singer, G. M. in "Pyridine and its Derivatives", Vol 14, Part 1, Page 1, (1974) edited by Abramovitch, R. A.

23. Andrews, G. D.: Personal Communication.

24. Junguera, J., Cardona, N., Figueruelo, J. E.: Die Makromol. Chem. 160, 159 (1972).

25. Pascault, J. P., Kawak, J., Gole', J., Pham, Q. T.: Europ. Polym. J. 10, 1107 (1974).

STEREOCHEMISTRY

STEREOSPECIFICITY IN ANIONIC POLYMERIZATION

S. BYWATER
Division of Chemistry
National Research Council of Canada
Ottawa, Canada, K1A 0R9

ABSTRACT

Active centers in anionic polymerization exist in cis and trans forms. Equilibrium fractions and interconversion rates depend on solvent and counter-ion. With diene polymerization in polar solvents, the isomers equilibrate rapidly down to very low temperatures if lithium is the counter-ion, but only very slowly at room temperature if it is potassium. In such solvents, the kinetic preference is trans but at equilibrium it is cis. Trans centers are more vinyl stereospecific for a given counter-ion and react faster with monomer. In vinyl polymerization, cis-trans interconversion rates are slow with all counter-ions on the reaction time scale. Counter-ion positional exchange is however rapid in THF but slows somewhat in toluene. Diad structure is not defined until the next monomer adds. The determination of 1,2 or 1,4 structure in diene polymerization is decided in the same manner, a common characteristic of anionic polymerization.

One of the most interesting features of anionic polymerization is the ability to form stereoregular polymers, principally of dienes and some vinyl monomers. Although stereoregulation is a complex process whose mechanism is often difficult to elucidate, much information can be obtained from spectroscopic studies (nmr or UV-visible) on oligomeric active species [1]. Studies of this type have shown that the active center is a planar (or near planar) delocalized system of the type:

[2,3] [4] [5]

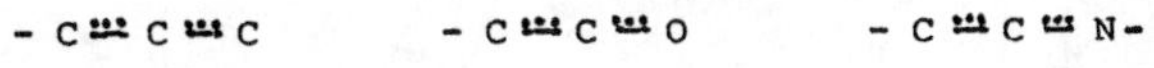

for dienes, acrylates or 2-vinyl pyridine respectively. The barrier to rotation about the β-γ bond is quite high, so that the active centers exist in cis and trans forms (Z,E) which do not necessarily rapidly interconvert.

This formulation is generally accepted to apply in polar solvents such as tetrahydrofuran (THF). Some authors [6,7] suggest that under non-polar conditions allylic lithium compounds exist primarily with lithium covalently bonded to the terminal carbon atom. Cis-trans isomerization would then proceed via a low concentration of a γ-lithium isomer. The simpler picture wherein the bonding is always ionic but with differing degrees of charge delocalization [8] is generally more attractive, particularly since allyllithium and its derivatives show typical near-UV absorption bands not normally observed with covalent compounds. The γ ^{13}C nmr shift of γ substituted allyllithium compounds are appreciably to high field (~20 ppm) of that of the corresponding 2-olefin even in hydrocarbon solvents which suggests some charge exists there [8]. Indeed the most recent ab-initio molecular orbital calculations predict ionic bonding for allyllithium even in the non-solvated state [9]. It should be noted that nmr measurements alone do not give unambiguous answers since the small concentration of a γ-lithium substituted isomer has never been observed.

Published 1987 by Elsevier Science Publishing Co., Inc.
Recent Advances in Anionic Polymerization
T.E. Hogen-Esch and J. Smid, Editors

Nevertheless whatever the form of bonding, the observation of two geometrical isomers must be important, for example in the classical lithium/isoprene/hydrocarbon solvent polymerization system first described in the Firestone laboratories [10], since a cis-center would be expected to be frozen into a cis-1,4 in-chain unit when the next monomer adds, and correspondingly for trans-centers. The high cis-1,4 stereospecificity in isoprene polymerization is caused by a kinetic preference for the formation of cis-centers and their higher reactivity towards monomer addition [11]. If monomer addition is slowed, or relaxation to the thermodynamically stable trans form is faster (as with butadiene) more trans in-chain units appear in the polymer. For dienes, therefore, a knowledge of anion-pair isomerization rates is necessary to describe the mechanism, together with individual rates of monomer addition to cis and trans forms.

In polar solvents, cis forms are in general the stable ones, but trans centers are preferentially formed in the transition state, ie. the situation is reversed in polar solvents [12]. Cis and trans forms might be expected to be irrelevant in such solvents for the polymers formed have high vinyl contents. For a given counter-ion, however, it appears that the trans species add monomer more rapidly and that they could be more vinyl stereospecific [12]. A knowledge of anion-pair isomerization and monomer addition rates is still important. Measurements were, therefore, made on neopentylallyl lithium, sodium and potassium, their β-methyl derivatives, n-butylallyl derivatives and in some cases butadiene active oligomers. Fig. 1 shows that isomerization rates (in this case the β-methyl neopentyl-allyl lithium series, models for isoprene) are solvent dependent, the biggest difference being between polar and non-polar solvents. This is not surprising as the environments are widely different, solvated in the first case, and in an aggregate in the second. Correspondingly the monomer addition rates change so that nearly all systems can have a range in which competition is important.

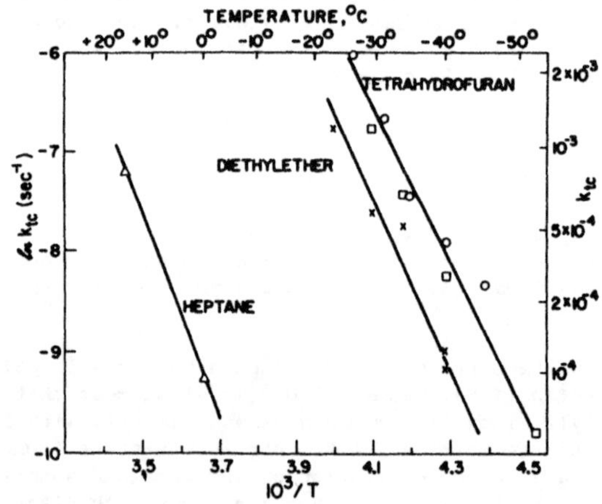

1. Temperature dependence of the first order rate constants for isomerization (k_{tc}) of β-methyl neopentylallyllithium in different solvents. Note that in heptane the spontaneous process goes cis to trans. k_{ct} will therefore be higher.

A more systematic study was made on neopentylallyl lithium, sodium and potassium in THF [13]. These compounds are relatively easy to prepare, the first by direct addition of t-butyllithium to butadiene in a non-polar solvent, the second by Li to Na exchange and the third by metallation of 5,5'-dimethylhexene-2 by trimethylsilylmethylpotassium. This latter method is interesting in that since it proceeds smoothly in THF at low temperatures and cis and trans 2-olefins are often available, then both geometrical isomers of allylic potassium compounds can be prepared in high yield. In this way the rates of butadiene addition to either can be measured. As seen in Fig. 2 it is directly confirmed that the trans form is more reactive. Unfortunately sodium based metallations do not proceed so readily and isomerization always intervenes so that it is not possible to confirm the relative reactivity in that case.

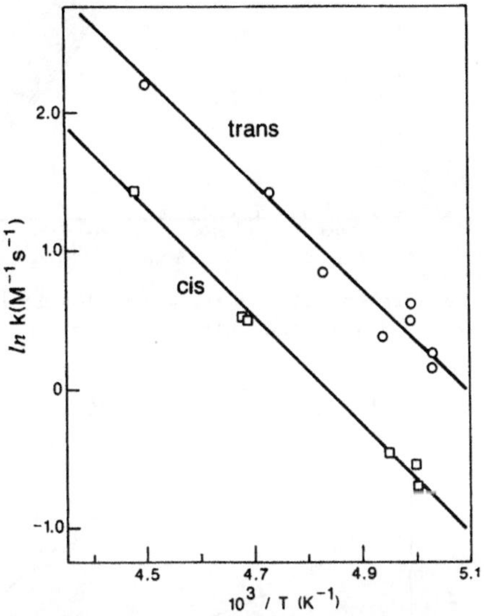

2. Temperature dependence of the ion-pair rate constant of butadiene addition to n-butylallylpotassium in THF [14].

Isomerization rates of the neopentyl compounds could be measured by nmr techniques or by following the changes in the near UV absorption band characteristic of the allylic systems. The trans species absorbs at longer wavelength (Fig. 3). Measurable rates of trans-cis isomerization ($\sim 10^{-3}$sec^{-1} to 10^{-5}sec^{-1}) could be observed in the temperature range -30 to -50° with Li^+ as counter ion but over a much higher temperature range (0° to 30°) with the K^+ ion. The sodium compound's rates were intermediate in behavior [13]. The activation energy for the process increases as counter-ion size increases rising from $13._7$ kcal per mole (Li^+) to $17._2$ kcal (Na^+) and $22._2$ kcal (K^+). It appears that stronger cation-anion interaction reduces the activation energy required. Indeed not only the barrier to isomerization increases in the series but also the barrier to rotation about the terminal methylene group follows the same trend. The lithium derivative has therefore a very labile structure while at the other extreme the potassium one is quite rigid.

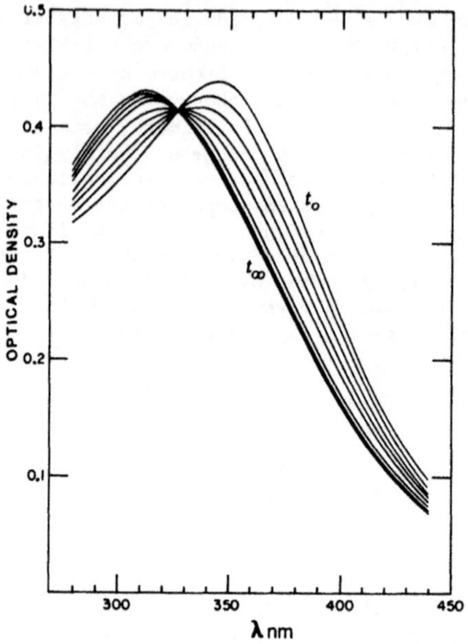

3. Spectroscopic measurement of the isomerization of trans neopentyl allylpotassium, T = 1°, Elapsed time ~4 hours.

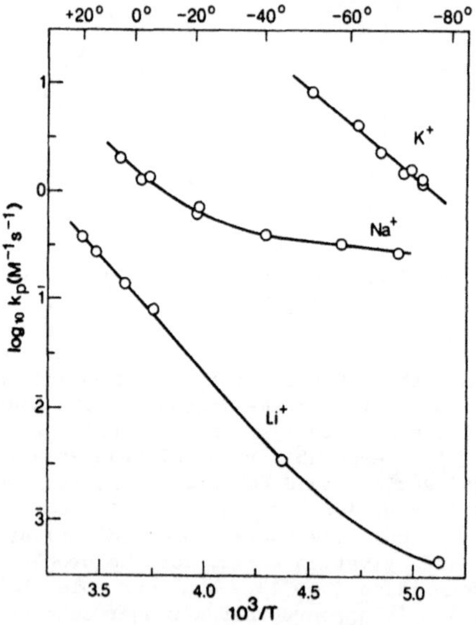

4. Ion-pair propagation rate constants ($k_p^{\pm}$) for polymerization of butadiene in THF. The lithium $k_p^{\pm}$ at -78° may depend on monomer concentration. The potassium $k_p^{\pm}$ refers to the addition of butadiene to trans n-butylallylpotassium. A few experiments with multiple butadiene additions showed that the actual propagation rate constant is only a little lower.

The lithium ion-pair under polymerization conditions is even more labile, at least as regards isomerization, than suggested by the above figures since replacement of the bulky t-butyl group by a butadiene oligomer chain produced the same range of isomerization rates mentioned above in the temperature range of -60° to -85°. Much smaller effects of substitution were observed with other counter-ions [14]. The trend of polymerization rate is in the opposite direction to that of the isomerization rates, i.e., lithium counter-ions induce slow polymerization but potassium counter-ions quite fast rates (Fig. 4). A comparison of rates leads to the conclusion that equilibration will almost always occur in THF with lithium-based polymerizations; only at ~-78°, dependent on monomer concentration, can monomer addition and isomerization rates become comparable. At the other extreme potassium based polymerizations of butadiene under any practical experimental conditions will retain the highly trans ratio of active centers produced in the transition state. Sodium is an intermediate case, active center structures will become frozen somewhat below 0°.

The isomerization experiments correlate with the observations made on the effect of polymerization temperature on polybutadiene microstructure [15]. In THF with potassium as counter-ion, no change in microstructure is observed between 0° and -78°; nor does the percentage of cis in the 1,4 units change. Lithium or sodium, however, as counter ions produce higher 1,2 contents in the polymer at -78° than at 0° and the fraction of cis units in the remaining 1,4 structures becomes predominantly trans. Again this is in agreement with the supposition that for a given counter-ion trans centers are somewhat more reactive and somewhat more likely to produce in-chain 1,2 units. Still somewhat of a problem to explain however is the reason for the generally much higher vinyl content in polybutadiene produced in THF (or ether) if one uses lithium as counter-ion than with the heavier alkali metal cations. The active center population will, as we have seen, be under most conditions largely cis and ^{13}C nmr experiments can, on the neopentyl series, be interpreted as showing most charge occurs at the α-position with lithium. A perhaps naive assumption that monomer attack occurs at the position of highest charge density would require that polymers would be highly 1,4 in structure. It has been found that protonation of allylic lithium does indeed occur preferentially at the α-position (and so does attack of trimethylsilylchloride) in THF so that the position of attack depends on the particular reaction studied. No generalizations are therefore possible.

Table 1. Microstructure of polybutadiene under polar conditions

Solvent	Counter-ion	Temperature	% Vinyl
THF	K^+	0°,-78°	67%
DEE[1]	Li^+	0°	75%
THF	Li^+	0°	87%
THF	Li^+	-78°	91%
CH[2]	Li^+.TMEDA	5°	85%[3]
CH[2]	Li^+.DIPIP	5°	99+%[3]

[1] Diethylether [2] Cyclohexane [3] Maximum attained. This usually requires more than one molecule of diamine per lithium.

The lithium/THF system at low temperatures gives ~91% 1,2 polybutadiene at -78°, but it is still not the best system for producing high vinyl polymer (Table 1). The complex lithium: dipiperidinoethane (DIPIP) in hydrocarbon solvents gives 99-100% 1,2 polybutadiene at much more useful temperatures [16]. This observation can be correlated with the fact that of all polar systems based on lithium which have been investigated, it alone seems to produce 100% trans active centers at equilibrium [17]. This would be irrelevant if in the transition state, cis forms were preferentially formed and cis-trans isomerization was slow. It is more likely that, as generally in polar systems, trans centers are preferentially formed in the monomer addition step, and that at the temperatures used (>0°) equilibration is reasonably fast. If this is so, kinetic and thermodynamic factors reinforce to keep a high trans center population, unlike the situation in the other systems discussed. In contrast the very similar complexing agent tetramethylethylenediamine gives a mixed cis/trans active center population at equilibrium and a polybutadiene containing only ~85% 1,2 units under identical conditions. All the studies discussed are in agreement with the postulate that trans active centers are the most efficient precursors for vinyl in-chain units.

Similar spectroscopic studies on vinyl polymerization offer greater difficulties in mechanism elucidation. One monomer unit models give some evidence on charge distribution and population of cis and trans centers but since stereoregulation depends on the interaction of two or more monomer units, longer chain models are required. Two unit models of the type $(CH_3)_2C(R)CH_2C^-(CH_3)R$ where R can be a phenyl, or ester group are interesting because they offer the possibility of study of cation-anion dynamics even if only in a limited way. Such alkali metal compounds can be synthesized by processes similar to those described above, i.e., by addition of a monomer unit to cumyl salts or to metallated methyl or ethyl isobutyrate. Isolation of two unit material by distillation of the protonated product followed by re-metallation may be necessary in some cases.

With planar anions, the cation is likely to be placed in the π-cloud above or below the anion plane. Whereas most models of stereoregulation are static in type, possible rapid cation flips from one side of the plane to the other should be considered. Very rapid exchanges of this type have been described [18] for radical-anions of condensed aromatics but little is known about them for less extensively delocalized systems.

If the cation "flip" rate is greater than the monomer addition rate, the structure of the active center is not defined. It will only freeze in a meso or a racemic diad with respect to its penultimate unit when it is deactivated by monomer addition. The diastereotopic gem-dimethyl groups on the two-unit models described above are in principle non-equivalent if an asymmetric center exists at the carbanion. The chemical shift difference is not predictable and could be too small to measure, but the protonated molecules should be a reasonable guide. In the α-methylstyrene [19] and methyl methacrylate [20] models the difference is 75Hz and 30Hz at 100MHz respectively, i.e., easily observable. The alkali metal ion-pairs of both, however, in THF show only a single signal attributable to each pair of methyl groups down to -70°. The counter-ions are exchanging positions at a rate greater than the chemical shift difference, i.e., the lifetime in each position is probably less than a tenth of a millisecond. Since the average individual ion-pair lifetime, for example, in methyl methacrylate polymerization in THF at -75° and 0.4 molar initial monomer concentration is ~0.03 sec (Na^+) [21] or 0.5 sec (Li^+) [22] many cation oscillations will occur between monomer addition steps. The diad structure will then be decided by the direction of approach of the next monomer molecule with respect to the two faces of the carbanion.

It would be interesting to determine if, under some other conditions there is a slowing of the cation oscillations. The most favorable conditions would appear to be where the solvent is non-polar and the monomer units are polar as in methyl methacrylate. Unfortunately its lithio dimer is not sufficiently soluble in toluene for ^{13}C nmr experiments to be made but that of ethyl methacrylate fortunately is soluble. With this compound a single sharp signal attributable to a pair of methyl groups is observed at 15° which splits to two equal but rather broad signals at -55° separated

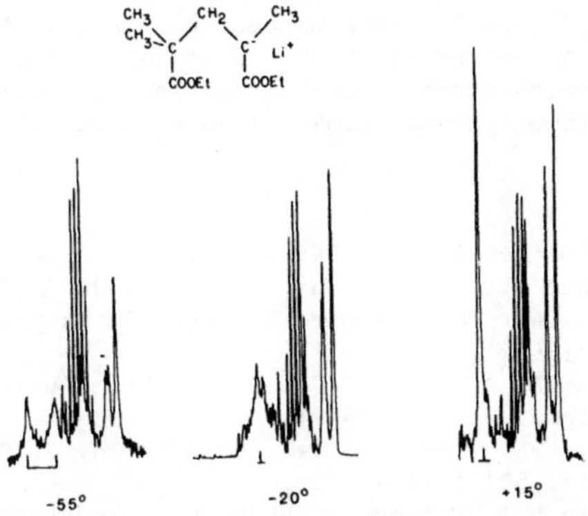

5. ^{13}C nmr spectrum of the methyl region of the lithio dimer of ethyl methacrylate in toluene. The multiplet at the center is due to deuterotoluene, it obscures CH_3-α. Signals from $(CH_3)_2$ to the left and from the ester methyls to the right.

by ~120Hz (Fig. 5). The coalescence temperature is close to -30°. The lifetime of a particular cation configuration can be estimated as ~2 milliseconds at this temperature. The mean lifetime of a particular active center is difficult to determine since no accurate ion-pair propagation constant for methyl or ethyl methacrylate is available in toluene. Side reactions of initiators with the monomer produce uncertainties in the determination of the number of active centers generated from a given initiator concentration. Diphenylhexyllithium gives, in toluene, the highest initiator efficiency (~20%) which leads to a $k_p^{\pm}$ of about 4 $M^{-1}s^{-1}$ at -30° [23]. This would lead to an average active center lifetime of ~2 seconds at an initial monomer concentration of 0.2M. t-Butyl methacrylate appears to be much less prone to side reactions with the initiator [24] producing narrow molecular weight distribution polymers and an estimated $k_p^{\pm}$ of ~80 $M^{-1}s^{-1}$ at -30° which would correspond to an average lifetime of ~0.1s under the same conditions. If either is taken as a model for the polymerization of ethyl methacrylate this predicts many cation oscillations even in toluene before an ion-pair adds monomer.

If the living dimer is a close enough model for the ultimate and penultimate units in a growing polymer chain, a more dynamic description than is usually given of the stereoregulation process in toluene would be necessary. This would probably require reaction to occur at a favorable conformation of the chain end which would resemble the static lithium-ester group interactions conventionally used. Since a mobile active center only becomes defined in structure on monomer addition, meso diads will be formed by attack on the pro-meso carbanion face whichever monomer face is presented. The experiments of Fowells et al. [25] using cis-β-d_1 methacrylates show that in fact a definite face is presented in the total absence of added polar impurities. Again lithium interaction with the monomer's ester group was invoked. Successful meso producing encounters in which the opposite monomer face must have been presented were, however, recorded if small amounts of THF were present.

The cis and trans forms of the active centers in vinyl polymerization by contrast only interconvert slowly so that the isomer in the transition state is, as with dienes, important. Information on these systems available so far, indicates that even in THF and for all counter-ions, relaxation to the stable form is likely to be slow on the scale of an active center lifetime [20,26]. It then becomes possible that isomers are only interconvertable on monomer addition. This leads to two site models for stereoregulation which predict non-Bernoullian or even non-Markoffian behavior without invoking penultimate monomer unit effects [26].

References

1. S. Bywater, Polymer J., 12, 549 (1980).
2. W.H. Glaze, and P.C. Jones, Chem. Commun., 1434 (1969).
3. F. Schué, D.J. Worsfold, and S. Bywater, J. Polym. Sci. (Polym. Lett. Ed.), 1, 821 (1969).
4. L. Vancea, and S. Bywater, I.U.P.A.C. Macromolecular Symposium, Preprints Mainz., 1, 140 (1979).
5. W.L. Jenkins, C.F. Tien, and T.E. Hogen-Esch, J. Polym. Sci. (Polym. Lett. Ed.), 16, 501 (1978).
6. M. Morton, and L.J. Fetters, Rubber Chem. and Technol., 48, 359 (1975).
7. G. Fraenkel, A.F. Halasa, V. Mochel, R. Stumpe, and D. Tate, J. Org. Chem., 50, 4563 (1985).
8. S. Bywater, and D.J. Worsfold, J. Organometal. Chem., 159, 229 (1978).
9. T. Clark, C. Rohde, and P.v.R. Schleyer, Organometallics, 2, 1344 (1983).
10. F.W. Stavely et al., Ind. Eng. Chem., 48, 778 (1956).
11. D.J. Worsfold, and S. Bywater, Macromolecules, 11, 582 (1978).
12. A. Garton, and S. Bywater, Macromolecules, 8, 694 (1975).
13. R.T. McDonald, and S. Bywater, Organometallics, 5, 1529 (1986).
14. R.T. McDonald, P.E. Black, and S. Bywater, Unpublished.
15. S. Bywater, Y. Firat, and P.E. Black, J. Polym. Sci. (Polym. Chem. Ed.), 22, 669 (1984).
16. A.F. Halasa, D.N. Schulz, D.P. Tate, and V.D. Mochel, Adv. Organometal. Chem., 18, 55 (1980).
17. D.J. Worsfold, S. Bywater, F. Schué, J. Sledz, and V. Marti-Collet, Makromol. Chem. Rapid Commun., 3, 239 (1982).
18. J.L. Sommerdijk, and E. deBoer in: Ions and Ion-pairs in Organic Reactions, M. Szwarc ed., (Wiley-Interscience, N.Y. 1972), 1, 322.
19. Y. Firat, and S. Bywater, Eur. Polym. J., 18, 265 (1982).
20. L. Vancea, and S. Bywater, Macromolecules, 14, 1776 (1981).
21. V. Warzelhahn, H. Höcker, and G.V. Schulz, Makromol. Chem., 179, 2221 (1978).
22. H. Jeuck, and A.H.E. Müller, Makromol. Chem. Rapid Commun., 3, 121 (1982).
23. S. Bywater in: Comprehensive Chemical Kinetics, C.H. Bamford and C.F.H. Tipper eds, (Elsevier, Amsterdam 1976), 15, Chapter 1.
24. A.H.E. Müller, H. Jeuck, C. Johann, and P. Kilz, Polym. Prepr. Am. Chem. Soc., 27(1), 153 (1986).
25. W. Fowells, C. Schuerch, F.A. Bovey, and F.P. Hood, J. Am. Chem. Soc., 89, 1396 (1967).
26. I.M. Khan, C.C. Meverdon, and T.E. Hogen-Esch, Polym. Prep. A.C.S. Div. Polym. Chem., 24(2), 151 (1983); 27(1), 159 (1985).

MECHANISM OF POLYMERIZATION OF MMA BY GRIGNARD REAGENTS AND PREPARATION OF HIGHLY ISOTACTIC PMMA WITH NARROW MOLECULAR WEIGHT DISTRIBUTION

K. HATADA, K. UTE, K. TANAKA, T. KITAYAMA, AND Y. OKAMOTO
Department of Chemistry, Faculty of Engineering Science, Osaka University, Toyonaka, Osaka 560, Japan

INTRODUCTION

Many papers have been published on the polymerization of methacrylates by organomagnesium compounds since Goode [1] and Nishioka [2] reported the polymerization of methyl methacrylate (MMA) by Grignard reagents in 1960. In several review articles [3-6], the polymerizations of methacrylates by organomagnesium compounds have been summarized and the mechanism of polymerization is discussed in some detail.

However, there still exist some ambiguities regarding the polymerization, particularly on the structure of the polymer, efficiency of initiator, and mechanism of polymerization. The complexity of the factors affecting the polymerization has prevented a clear understanding of the polymerization reaction, and the molecular weight and its distribution of the polymer can not be fully controlled.

Several recent studies have dealt with the effects of temperature and amount of THF or MgX_2 on the tacticity and molecular weight of the resulting polymer [7-9]. Allen and his coworkers [10] studied the mechanism of the polymerization of MMA by butylmagnesium compounds in toluene-tetrahydrofuran (THF) mixtures mainly by a kinetic method.

Structural analysis of the reaction products is another approach to the mechanism of polymerization. Goode and his coworkers [11] studied the initiation and termination processes in the polymerization of MMA by tritium-labeled phenylmagnesium bromide through the analysis of the low molecular weight products of polymerization. Yasuda and his coworkers [12] investigated the elementary steps of the polymerization reaction by butylmagnesium bromides. Bateup and Allen [13] determined the amounts of the unreacted initiator and CH_3OMgBr formed during the polymerizaiton of MMA by butylmagnesium bromides in toluene-THF mixtures.

The present authors have developed a new method for understanding the mechanism of polymerization using totally deuterated monomers [14-19]. When a deuterated monomer is polymerized by an undeuterated initiator, the fragments of the initiator and the terminating reagent incorporated in the polymer or oligomer chain can be easily detected by ^1H NMR spectroscopy without being disturbed by the strong signals of monomeric units. In this work, we studied the polymerization of MMA by butylmagnesium chloride (C_4H_9MgCl) by this technique to clarify the mechanism of polymerization. The results led us to the preparation of highly isotactic PMMA with narrow molecular weight distribution (MWD).

EXPERIMENTAL

MMA and MMA-d_8 were purified in usual manner, dried over calcium dihydride, and then distilled under vacuum just before use. The isotopic purity of MMA-d_8 is as follows; CD_2 99.12%, α-CD_3 99.17%, OCD_3 99.85%.

Grignard reagents were prepared in diethyl ether from the corresponding alkyl halide and magnesium. Concentrations of alkylmagnesium(RMg-), total magnesium(Mg^{2+}) and total bromide(Br^-) were determined by acid-base titration, chelatometric titration with EDTA and precipitation titration (Fajans's method), respectively.

Polymerization was initiated by adding monomer to initiator solution in

Recent Advances in Anionic Polymerization
T.E. Hogen-Esch and J. Smid, Editors

toluene kept at polymerization temperature, and terminated by adding a small amount of methanol containing HCl equivalent to the Mg^{2+} at the polymerization temperature. The polymerization mixture was poured into a large amount of methanol or hexane to precipitate the polymeric products.

NMR spectra were recorded on JEOL JNM-FX100 and -GX400 spectrometers. The number-average molecular weight(Mn) of the polymers was measured on a Hitachi 117 vapor pressure osmometer in toluene at 60°C or in chloroform at 40°C. GPC chromatogram was recorded on a JASCO FLC-A10 GPC chromatograph equipped with a Shodex GPC column A-80M (50 cm × 2) and KF 802.5 (30 cm) with maximum porosity of 5×10^7 and 4×10^4, respectively, using THF or chloroform as an eluent. The GPC chromatogram was calibrated against standard polystyrene samples.

RESULTS

1. Polymerization of MMA with n-C_4H_9MgCl in toluene [20]

Polymerization of MMA-d_8 was carried out with n-C_4H_9MgCl in toluene at -78°C. The results are shown in Table I. ^1H NMR spectra of the polymer

TABLE I. Polymerization of MMA-d_8 with n-C_4H_9MgCl in toluene[a)]

	Yield (%)	Mn[b)]	Mw/Mn[c)]	Tacticity(%) I	H	S
Polymer(MeOH insol.)	71	13,600	3.02	31	18	51
Oligomer(MeOH sol.)	28	1,000	—	—	—	—

a) MMA-d_8 4.9 mmol, C_4H_9MgCl 0.5 mmol, toluene 5 ml, time 72 hr, temp. -78°C.
b) Determined by VPO. c) Determined by GPC.

and oligomer obtained are illustrated in Fig. 1. The signal at 0.8 ppm was assigned to methyl protons of the butyl group incorporated into the chain from the initiator. The triplet at 2.40 ppm is attributed to the carbonyl methylene protons ($-COCH_2-$) of butyl pentadeuteroisopropenyl ketone (BIPK-d_5) unit in the chain [16]. The ketone should be formed through the attack of the initiator on the carbonyl group of MMA-d_8 (Eq. A).

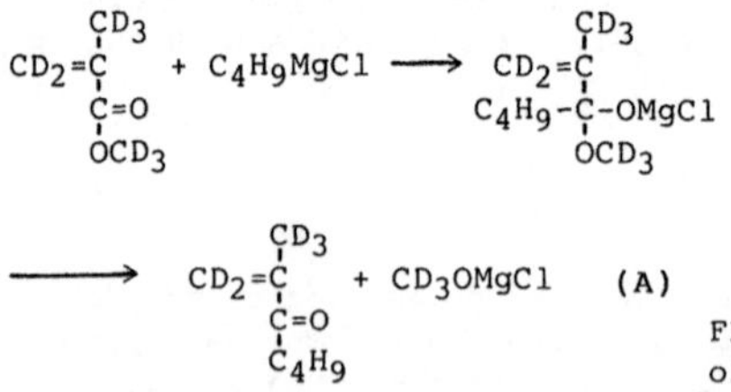

(A)

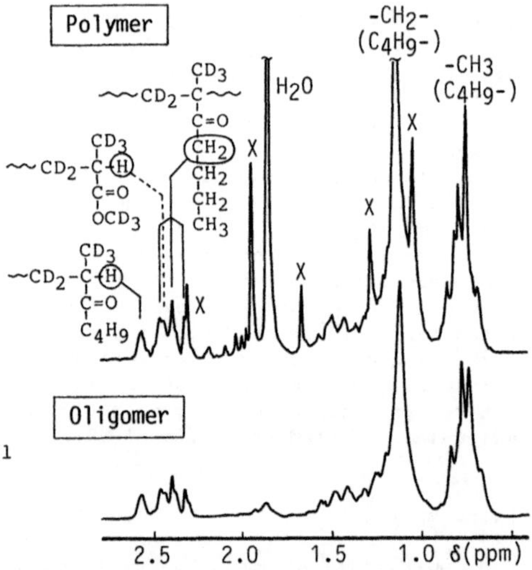

FIG. 1 ^1H NMR spectra of polymer and oligomer of MMA-d_8 prepared by n-C_4H_9MgCl in toluene at -78°C.
The signals "X" are due to the remaining protons in the monomer units.

The presence of BIPK-d_5 unit in the chain was also evidenced by the gas chromatographic analysis of the thermal degradation products of polymer and oligomer.

The singlet at 2.57 ppm was assigned to the methine proton attached to BIPK-d_5 unit at the chain end (BIPK-H). The methine proton attached to MMA-d_8 unit at the chain end (MMA-H) showed the signal at 2.45 ppm which overlapped with the triplet at 2.40 ppm [19]. The signal intensity for MMA-H could be estimated by comparing the spectrum of polymer or oligomer formed in the CD_3OD-terminated polymerization with that of polymer or oligomer formed in the CH_3OH-terminated one. The amount of MMA-H could be determined more precisely by the analysis of the ^1H NMR spectra of the polymer prepared by $CH_3CH_2CH_2CD_2MgCl$. The overlap of signal of the carbonyl methylene protons with that of MMA-H was removed by deuteration at the α-position of the initiator and the amount of MMA-H can be more easily and precisely determined [21].

Table II shows the numbers per chain of C_4H_9- group at the α-end* of

TABLE II. Number of initiator fragment and terminal methine proton per polymer or oligomer molecule of MMA-d_8 produced by n-C_4H_9MgCl in toluene at -78°C

	C_4H_9- (total)	$\boxed{C_4H_9}$-CD_2-C(CD_3)(C=O)(OCD_3)-	-CD_2-C(CD_3)-C(=O)-$\boxed{C_4H_9}$	-CD_2-C(CD_3)($\boxed{H}$)-C(=O)-C_4H_9	-CD_2-C(CD_3)($\boxed{H}$)-C(=O)-OCD_3
Polymer	1.3	1.0	0.3	0.3	0.4
Oligomer	1.4	1.0	0.4	0.4	0.4

Reproduced with permission from Ref. [20]. Copyright 1986, "The Society of Polymer Science, Japan".

the chain, BIPK-d_5 unit, BIPK-H and MMA-H. Both the polymer and oligomer molecules contained one C_4H_9- group at the α-end, and about 0.4 MMA-H at the ω-end*. The polymer and oligomer molecules each contained an equal amount of BIPK-d_5 unit to BIPK-H. This indicates that most of the BIPK unit is located at the ω-end. Mass spectra of the fractionated lower molecular weight oligomers, from dimer to heptamer, evidenced that there are two types of oligomers, (B) and (C), which are ended with MMA and BIPK units, respectively.

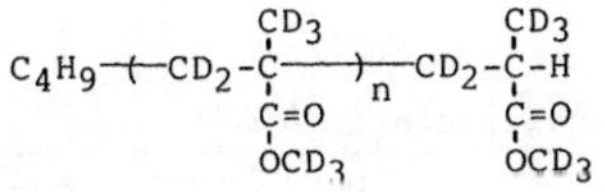
(B)

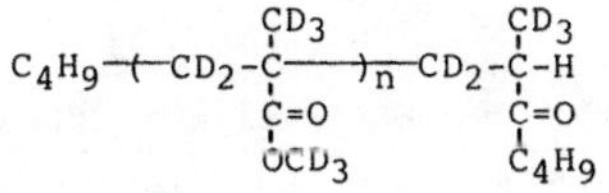
(C)

For example, the trimer fraction is composed of two trimers of different molecular weights. The higher molecular weight trimer showed the EI-mass spectrum which corresponds to the trimer ending with BIPK unit (trimer (C), n=2) as shown in Table III. No fragments which would come from the trimer C_4H_9-(MMA)-(BIPK)-(MMA)-H or C_4H_9-(BIPK)-(MMA)-(MMA)-H were observed in the spectrum. The FD-mass spectrum of the lower molecular weight trimer showed that this is the MMA trimer without a ketone unit in the chain (trimer (B), n=2). Similar results were obtained on the dimer, tetramer, pentamer, hexamer and heptamer.

TABLE III. EI-Mass spectrum of higher molecular weight trimer

M^+/e	Fragment
384	$[C_4H_9\text{-(MMA)-(MMA)-(BIPK)-H}]^+$
257	$[C_4H_9\text{-(MMA)-(MMA)}]^+$
227	$[\text{(MMA)-(BIPK)-H}]^+$
157	$[C_4H_9\text{-(MMA)}]^+$
127	$[\text{(BIPK)-H}]^+$

* IUPAC Commission on Macromolecular Nomenclature recommended (1983) that a nomenclature "α-X-ω-Y-poly(A)" should be used for the polymer "X-$(A)_n$-Y". Following this recommendation, the present authors proposed to use the designations "α-end", to express the beginning of polymer chain and "ω-end" to express the terminating end of the chain [20].

As shown in Table II, the sum total of MMA-H and BIPK-H per polymer or oligomer molecules does not amount to unity, and this suggests the presence of other types of structure at the ω-end of chain. ^{1}H and ^{13}C NMR spectra of the undeuterated oligomers were examined in detail and the characteristic signals due to the following three structures (D, E, F) were observed. The structure (D) formed through intramolecular cyclization, which was proposed as the principal mode of termination in the polymerization of MMA by phenylmagnesium bromide in toluene at 0°C [11]. The structures (E) and (F) were formed by the addition of the propagating chain end onto the carbonyl groups of BIPK and MMA, respectively. The contents of (D), (E) and (F) were 0.05, 0.04 and 0.01 per chain, respectively. Then, the structures of chain ends of about 80% oligomer molecules were clarified. The structures (D), (E) and (F) may also be present in the polymer molecules, but they have not been detected, probably owing to the lower intensities of the NMR signals.

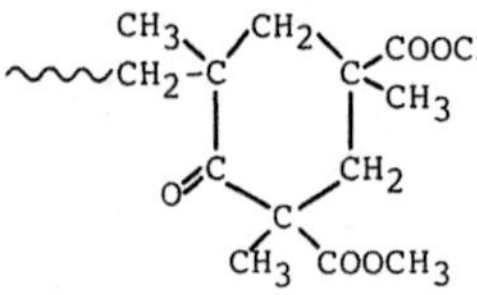
(D)

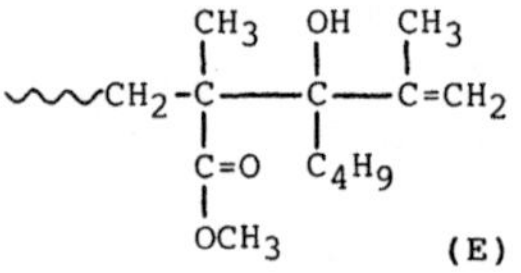
(E)

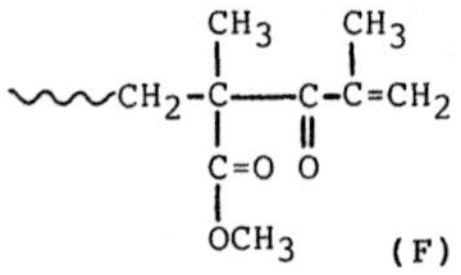
(F)

2. Polymerization of MMA in Toluene with Butylmagnesium Bromides and Preparation of Highly Isotactic PMMA with Narrow MWD [22,23]

Table IV shows the results of polymerizations in toluene at -78°C with n-, iso-, sec- and tert-butylmagnesium bromides. n-C_4H_9MgBr gave syndiotactic or stereoblend PMMA in low yield, and the tacticities depended on the initiator concentration probably due to the coexistence of isotactic and syndiotactic active species. The Mn's of the polymers were much larger than the values expected from the amounts of monomer consumed and initiator used and the molecular weight distributions (MWD's) were very broad. As the alkyl group became bulkier, the isotacticity of the polymer increased greatly and the Mn became closer to the calculated value and the MWD became narrower. t-Butylmagnesium bromide gave highly isotactic PMMA with narrow MWD.

TABLE IV. Polymerization of MMA with RMgBr in toluene at -78°C[a)]

RMgBr		Yield[b)]	Tacticity(%)			Mn[c)]		Mw[c)]/Mn
R-	Mg^{2+}/RMg-	(%)	I	H	S	Obsd.	Calcd.	
n-C_4H_9[d)]	1.0	8.1	11.0	15.3	73.7	12000	1680	13.4
n-C_4H_9	1.0	14.1	20.5	15.2	64.3	7420	760	11.2
iso-C_4H_9	1.1	33.1	92.5	5.4	2.1	5540	2270	2.49
iso-C_4H_9	2.5[e)]	21.7	95.5	4.5	0.0	4410	1140	2.12
sec-C_4H_9	1.2	100	95.5	4.5	0.0	4930	5060	1.29
sec-C_4H_9	2.4[e)]	94.5	96.8	3.2	0.0	4680	4790	1.21
t-C_4H_9	2.2	100	97.4	2.6	0.0	5010	5060	1.18

a) MMA 20 mmol, RMgBr 0.40 mmol, toluene 10 ml, time 72 hr. All the Grignard reagents were prepared in diethyl ether.
b) Hexane-insoluble part. c) Determined by GPC. d) RMgBr 0.10 mmol.
e) Grignard reagents were prepared in the presence of 1,2-dibromoethane ($RBr + Mg + BrCH_2CH_2Br \longrightarrow RMgBr + MgBr_2 + CH_2{=}CH_2$). The solutions of reagents were found to contain excess amount of $MgBr_2$ as indicated above.

Polymerization of MMA was carried out with t-C_4H_9MgBr in toluene at

-78°C at various ratios of monomer to initiator. The results are shown in the Table V. The Mn of the polymer agreed well with the value calculated from the amount of monomer consumed and initiator used. So the Mn of the polymer can be easily controlled by changing the ratio of initial amounts of MMA and initiator. In all the cases, the polymers obtained were highly isotactic and of low polydispersity. The ratio of polymerization was small at -78°C, and was enhanced at -40°C although the MWD became a little broader.

TABLE V. Polymerization of MMA with $t\text{-}C_4H_9MgBr$ in toluene at -78°C[a]

$t\text{-}C_4H_9MgBr$ (mmol)	Time (hr)	Yield (%)	Tacticity(%) I	H	S	Mn[b] Obsd.	Calcd.	Mw[b]/Mn
0.20	24	73	97.2	2.8	0	3510	3700	1.14
0.20	48	100	97.4	2.6	0	5010	5060	1.18
0.10	120	100	98.4	1.6	0	9320	10060	1.19
0.10[c]	145	99	99.2	0.8	0	20400	19900	1.12
0.20[d]	24	100	97.1	2.9	0	5600	5060	1.43
0.40[e]	24	100	2.1	20.0	77.9	14490	2560	3.10

a) MMA 10.0 mmol, toluene 5 ml. b) Determined by GPC.
c) MMA 20.0 mmol, toluene 10 ml. d) Polymerization at -40°C.
e) $(t\text{-}C_4H_9)_2Mg$ was used.

In toluene at -78°C the Mn increased proportionally to the polymer yield. The amount in mmol of polymer molecules was independent of the yield and just corresponded to the amount of $t\text{-}C_4H_9MgBr$ used (Fig. 2). These results indicate that the polymerization has the "living" character which is necessary for the formation of polymer of narrow MWD.

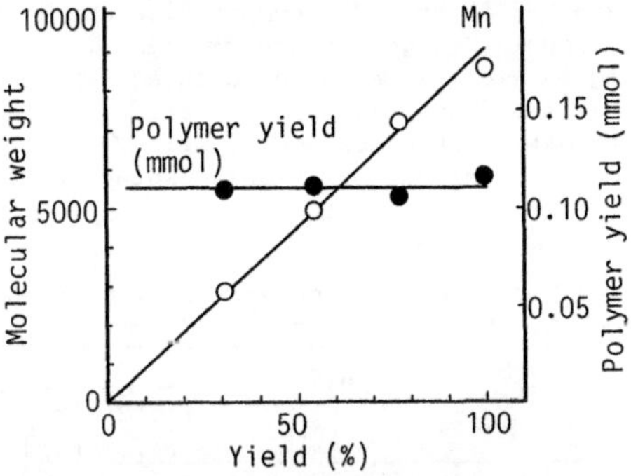

FIG. 2. Polymerization of MMA with $t\text{-}C_4H_9MgBr$ in toluene at -78°C for different polymerization times. MMA 10 mmol, $t\text{-}C_4H_9MgBr$ 0.11 mmol, toluene 5 ml.

The polymer of MMA-d_8 prepared with $t\text{-}C_4H_9MgBr$ at -78°C showed two ^1H NMR signals at 0.81 and 2.46 ppm which were assigned to $t\text{-}C_4H_9$- group at the α-end of the chain and the methine proton attached to the MMA unit at the ω-end, respectively. Measurements of the absolute intensity of these signals and the Mn indicated the polymer to contain one $t\text{-}C_4H_9$- group at the α-end and one MMA-H at the ω-end. No sign of side reaction was observed. The amount of MMA-H was decreased to 0.87, when the polymerization was carried out at -40°C. This may be due to the formation of cyclic ketone at the ω-end, which results in broadening of the MWD.

The polymerization reaction was followed in toluene-d_8 at -78°C by the intensity measurements on the vinylidene methylene proton NMR signals of MMA. The signal of $t\text{-}C_4H_9MgBr$ at 1.58 ppm disappeared upon the addition of MMA, indicating a fast initiation reaction. The apparent rate of polymerization, Rp(mol/l·sec) was described as follows:

$$Rp = 4.3 \times 10^{-4}[t\text{-}C_4H_9MgBr][MMA].$$

Fast initiation and slow propagation as well as living character cause the formation of PMMA with narrow MWD.

Figure 3 shows the NMR signals of carbonyl carbon in PMMA prepared by $t\text{-}C_4H_9MgBr$ in toluene at -78°C. In the spectrum there exist a strong mmmm pentad peak and two weak signals of equal intensity due to mmrm and mmmr pentads, and no other pentad signals are observed. Then, the steric defect in the polymer chain can be depicted as follows:

~~~~DDDDDLLLLL~~~~

This indicates that one racemic enchainment causes the inversion of isotactic propagating species, and that the stereoregulation in this polymerization is chain-end controlled.

Table VI shows the results of block copolymerization of MMA and ethyl methacrylate (EMA) with $t\text{-}C_4H_9MgBr$ in toluene at -78°C. MMA was polymerized first for 24 hr to complete polymerization, and then EMA was added and polymerized for another 48 hr. The composition and molecular weight of the copolymer agreed well with the calculated values, respectively, indicating the formation of block copolymer. The block copolymer was highly isotactic but the MWD was rather broad.

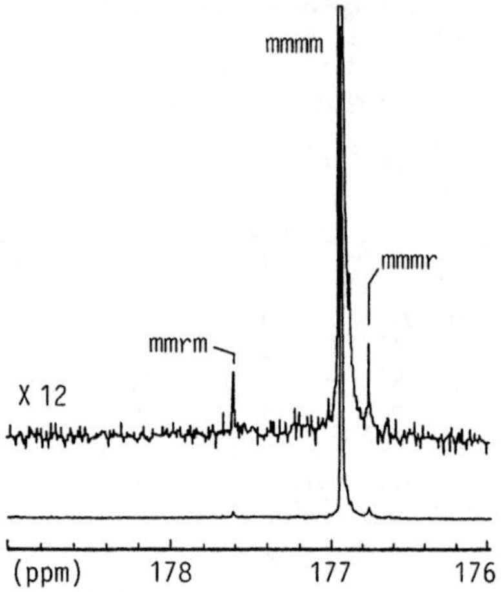

FIG. 3 ^{13}C NMR signal of the carbonyl carbon in the PMMA prepared by $t\text{-}C_4H_9MgBr$ in toluene at -78°C. (Nitrobenzene-d_5, 110°C, 100 MHz) Reproduced with permission from Ref. [23]. Copyright 1986, "The Society of Polymer Science, Japan".

TABLE VI. Block copolymerization of MMA and ethyl methacrylate(EMA) with $t\text{-}C_4H_9MgBr$ in toluene at -78°C[a)]

M_1[b)] (mmol)		M_2[b)] (mmol)		Block copolymer: Yield (%)	$[M_2]/[M_1]$	Tacticity(%) I	H	S	Mn[c)] Obsd.	Calcd.	$\overline{Mw}/\overline{Mn}$[c)]
MMA	5.0	—	0	100	—	97	3	0	1200	1060	1.38
MMA	5.0	EMA	10	100	2.0	95	5	0	3590	3340	1.51

a) $t\text{-}C_4H_9MgBr$ 0.5 mmol, toluene 10 ml.

b) M_1 was polymerized for 24 hr and then M_2 was added to be polymerized further for 48 hr.

c) Determined by GPC.

Reproduced with permission from Ref. [23]. Copyright 1986, "The Society of Polymer Science, Japan".

DISCUSSION

It is well known that in the polymerization of methacrylates by Grignard reagents, all the initiator used cannot be accounted for if each polymer or oligomer molecule is assumed to contain one initiator fragment. In this study, we found that in the polymerization of MMA with $n\text{-}C_4H_9MgCl$ some of the polymer or oligomer molecules contained two initiator fragments, one at the α-end of the chain and the other at the ω-end of the chain as a ketone unit. The sum total of the butyl groups in the polymer and oligomer chains was about 60% of the initiator used (Table VII). About 20% of the initiator remained unreacted during the polymerization and was recovered as
~~~~

butane when the polymerization was terminated. Bateup and Allen [13] observed similar amounts of unreacted C_4H_9MgBr in the polymerization of MMA in toluene-THF mixtures. They suggest that the unreacted C_4H_9Mg- species coordinate with the propagating species affecting the stereoregulation. The C_4H_9- groups in the unreacted BIPK and dibutylisopropenylcarbinol corresponded to 8 and 6% of the C_4H_9MgCl used, respectively. The carbinol was formed through the reaction of BIPK with n-C_4H_9MgCl. Thus most of the initiator fragments are accounted for (Table VII).

TABLE VII. Fate of initiator fragments in the polymerization of MMA in toluene at -78°C for 72 hr (percentage based on C_4H_9MgCl used)

C_4H_9Mg-	(Unreacted)	18
$CH_2=C(CH_3)-COC_4H_9$	(Unreacted)	8
$CH_2=C(CH_3)-C(OH)(C_4H_9)_2$		6
$C_4H_9-CH_2-C(CH_3)(CO_2CH_3)$~~	(Polymer)	7[a]
	(Oligomer)	33[a]
~~$CH_2-C(CH_3)(H)-COC_4H_9$	(Polymer)	2[a]
	(Oligomer)	16[a]
~~$C(OH)(C_4H_9)-C(CH_3)=CH_2$		—
		1
	Total	91%

a) The data refer to the polymerization of MMA-d_8.

Reproduced with permission from Ref. [20]. Copyright 1986, "The Society of Polymer Science, Japan".

From the results mentioned in the previous section, it was concluded that in the polymerization of MMA with n-C_4H_9MgCl in toluene at -78°C the initiator attacks both vinylidene and carbonyl double bonds of MMA in the initiation process to form the initiating species (Eq. G) and BIPK (Eq. A). BIPK is more reactive than MMA as evidenced from the higher Q and e values [17] than those of MMA.

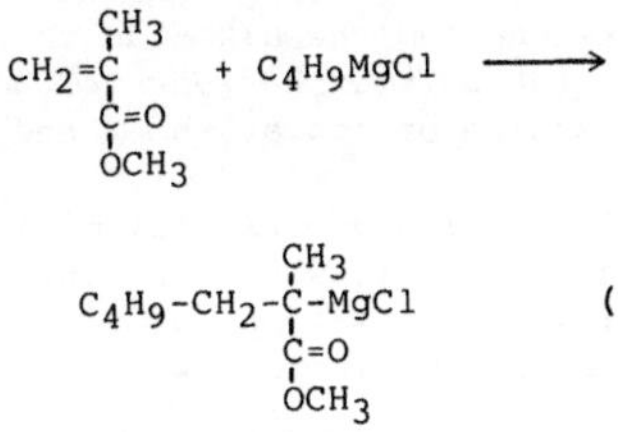

(G)

As a result, a part of the propagating species are attacked by BIPK. The resulting species ending with BIPK unit are less reactive than the species ending with MMA unit and remain unreacted during the polymerization (Eq. H).

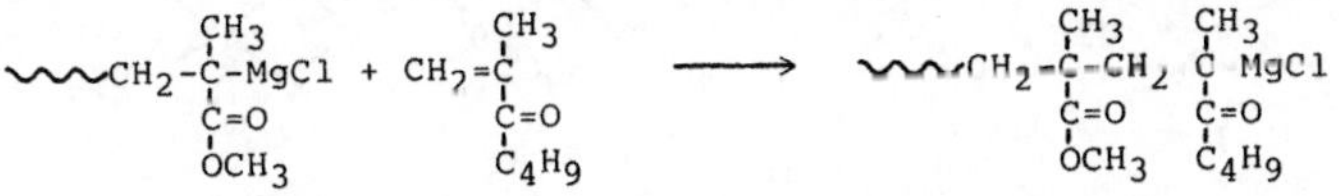

(H)

This is the reason why most of the BIPK units are located at the ω-ends. Thus, participation of BIPK in the polymerization should make the MWD broad.

CH_3OMgCl formed simultaneously with the formation of BIPK (Eq. A) would coordinate with the propagating chain end to produce the species of different activities and stereoregularities. This causes the formation of stereoblend polymer with broad MWD. The coexistence of multiple active species is demonstrated for the polymerizations of methacrylates with organomagnesium compounds [7-9,24,25]. This phenomenon has been ascribed to the coordination of magnesium compounds and ether to active centers [7-9], and stereocomplex formation during the polymerization [24]. In the polymerization of ethyl methacrylate with n-C_4H_9Li in toluene, the propagating species coordinated with the ethoxide forms more syndiotactic polymer than the species free from the alkoxide [18]. This may be the case in the polymerization with n-C_4H_9MgCl.

Coexistence of propagating species has also been ascribed to the Schlenk equilibrium [7-9](Eq. I). Matsuzaki suggested that in the polymerization of MMA by phenylmagnesium bromide active species for isotactic poly-

$$2\ RMgX \rightleftharpoons R_2Mg + MgX_2 \qquad (I)$$

mer were produced from "C_6H_5MgBr" and those for syndiotactic-rich polymer from "$(C_6H_5)_2Mg$" [9]. Isotactic polymerizations have never been observed with dialkylmagnesium initiators [12] with the single exception of dibenzylmagnesium prepared from dibenzylmercury [26]. Thus the Schlenk equilibrium also makes the reaction complex in the polymerization with n-C_4H_9MgCl.

As shown in Table IV, Mg^{2+}/C_4H_9Mg- values for butylmagnesium bromides other than n-C_4H_9MgBr exceeded unity, and the values became larger as the butyl group became bulkier. Excess amount of Mg^{2+} is due to the $MgBr_2$ formed by an Wurtz type coupling reaction of butylmagnesium bromide and butyl bromide during the preparation. t-Butylmagnesium bromide used in this study contained Mg^{2+} 2.2 times as large as t-C_4H_9Mg- group. The excess Mg^{2+} shifts the equilibrium to the side of "t-C_4H_9MgBr" to unify the initiating species. The effect of $MgBr_2$ was confirmed by the fact that the isotacticities of the polymers prepared with iso-C_4H_9MgBr and sec-C_4H_9MgBr were increased and the Mw/Mn ratios were decreased on the addition of excess amount of $MgBr_2$ (Table IV). The ^1H NMR spectrum of t-C_4H_9MgBr solution used showed that the initiator solution did not contain "$(t\text{-}C_4H_9)_2Mg$" which gives syndiotactic PMMA in the polymerization in toluene at -78°C (Table V). Hence, the initiating species for highly isotactic PMMA should be "t-C_4H_9MgBr". The bulky alkyl group prevents the side reaction and the polymerization proceeds in the absence of the ketone and the alkoxide to form highly isotactic PMMA with narrow MWD. So the polymer has one t-C_4H_9- group at the α-end of the chain and one MMA-H methine proton at the ω-end as mentioned in the previous section.

The real structure of initiating species in our isospecific polymerization with t-C_4H_9MgBr is not clear at present, but the following structures, (J), (K), (L) and (M), may be proposed. In this polymerization, efficiency of the initiator is almost 100% and MWD of the polymer is narrow. This means that the initiating sites must be single ones.

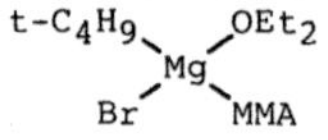

(J)

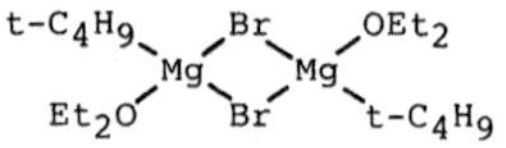

(K)

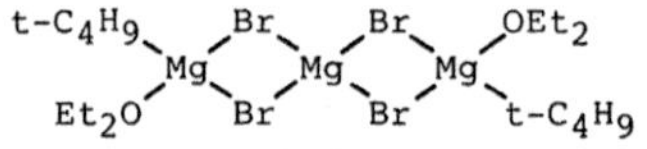

(L)

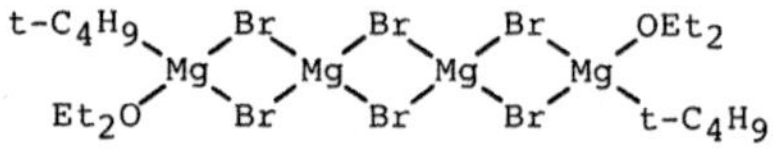

(M)

Allen and Williams [20] have proposed the structure similar to (J) as the initiating species in the polymerization of MMA in toluene with de-etherated t-C_4H_9MgBr prepared in THF on the basis that the isotactic initiating sites should be chiral. However in our polymerization the stereoregulation is chain-end controlled (cf. Fig. 3) and the initiating site is not necessarily chiral. The solubility of ether-solvated $MgBr_2$ in toluene may be low as indicated by Allen and Mair [8]. Moreover, viscosity of the polymerization mixture was decreased greatly when a small amount of methanol was added to quench the living end. Then, the structures (L) and (M) may possibly be considered as the initiating species, particularly (M), in which two t-C_4H_9Mg- groups are located apart enough from each other to act independently as the initiating sites of the same reactivity. If this is the case, the meaning of Schlenk equilibrium should be somewhat changed. The detailed investigation on this problem is now under way.

Several papers have been published on the syntheses of syndiotactic or atactic PMMAs with narrow MWD's. However, there has been no report on the direct preparation of highly isotactic PMMA with narrow MWD, although it could be derived from anionically prepared isotactic poly(triphenylmethyl methacrylate) [27,28]. This work may be the first example of the preparation.

In contrast with the polymerization by t-C_4H_9MgBr, the polymerization by n-C_4H_9MgCl gave methanol-insoluble polymer with broad MWD and also methanol-soluble oligomer (Table I). The oligomer formed rapidly at the initial stage of polymerization and then gradually increased in amount and molecular weight. On the other hand, the amount of polymer first increased slowly and quickly afterwards (Fig. 4). The sigmoidal increase in the amount of polymer is mainly due to the formation of BIPK at the initial stage of polymerization. As described above, BIPK attacks the oligomer anion to form a less reactive anion ending with the ketone unit. This should hardly propagate any more since most of the polymer chains do not contain a BIPK unit in chain. However, there should be a fairly large amount of oligomer containing MMA-H at the ω-end (cf. Table II). This indicates the existence of dormant oligomeric anion which can propagate only very slowly owing to the pseudo-cyclization [29] at the chain end (Eq. N).

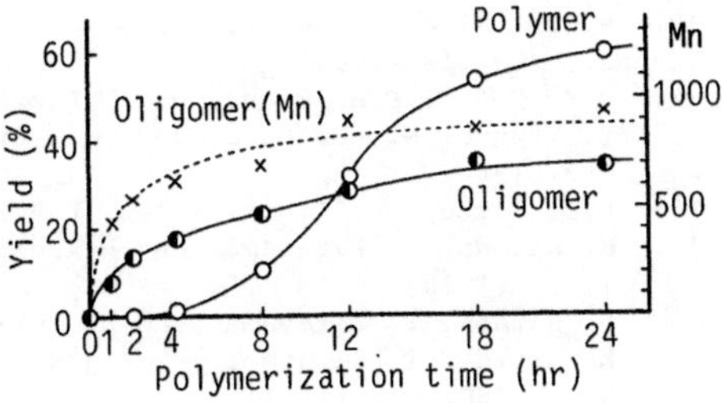

FIG. 4. Time-conversion curve for the polymerization of MMA by n-C_4H_9MgCl in toluene at -78°C. MMA 10 mmol, n-C_4H_9MgCl 1.0 mmol, toluene 10 ml.

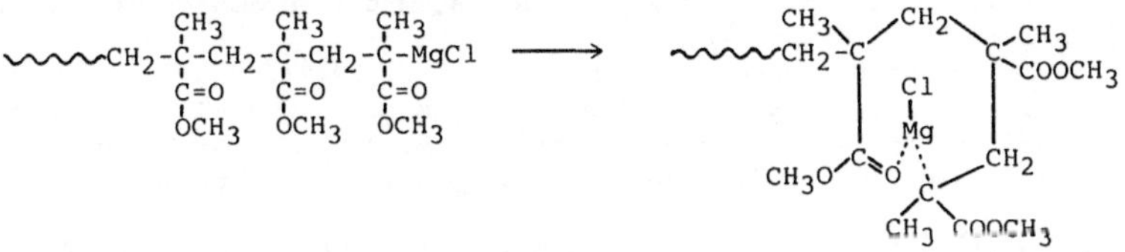

(N)

In the polymerization with t-C_4H_9MgBr in toluene at -78°C the rate of polymerization is very small, however, no side reactions occur and all the initiator molecules are used for the initiation reaction in the early stage of polymerization. Thus, the polymerization proceeds in the absence of the ketone, alkoxide and unreacted t-C_4H_9MgBr. So the propagating chain end in this polymerization might be similar in structure to the structure (N).

ACKNOWLEDGMENTS

The authors wish to thank Mr. Y. Terawaki of Osaka University and Mr. T. Shimamura of Nitto Technical Information Center Co., Ltd. for the measurements of NMR spectra and also Mrs. F. Yano for her clerical assistance in preparing this manuscript.

REFERENCES

1. W.E. Goode, F.H. Owens, R.P. Fellman, W.H. Snyder, and J.E. Moore, J. Polym. Sci., 46, 317 (1960).
2. A. Nishioka, H. Watanabe, K. Abe, and I. Sono, J. Polym. Sci., 48, 241 (1960).
3. P. Pino and U.W. Suter, Polymer, 17, 977 (1976).
4. H. Yuki and K. Hatada, Adv. Polym. Sci., 31, 1 (1979).
5. B.L. Erussalimsky, E.Yu. Melenevskaya, and V.N. Sgonnik, Acta Polymerica, 32, 183 (1981).

6. P.E.M. Allen and D.R.G. Williams, Ind. Eng. Chem. Prod. Res. Dev., 24, 334 (1985).
7. P.E.M. Allen and B.O. Bateup, Eur. Polym. J., 14, 1001 (1978).
8. P.E.M. Allen and C. Mair, Eur. Polym. J., 20, 697 (1984).
9. K. Matsuzaki, H. Tanaka, and T. Kanai, Makromol. Chem., 182, 2905 (1981).
10. P.E.M. Allen, C. Mair, D.R.G. Williams, and E.H. Williams, Eur. Polym. J., 20, 119 (2984).
11. W.E. Goode, F.H. Owens, and W.L.Myers, J. Polym. Sci., 47, 75 (1960).
12. Y. Yasuda, N. Kawabata, and T. Tsuruta, J. Macromol. Sci.-Chem., A1, 669 (1967).
13. B.O. Bateup and P.E.M. Allen, Eur. Polym. J., 13, 761 (1977).
14. K. Hatada, T. Kitayama, and H. Yuki, Makromol. Chem., Rapid Commun., 1, 51 (1980).
15. K. Hatada, T. Kitayama, and H. Yuki, Polym. Bull., 2, 15 (1980).
16. K. Hatada, T. Kitayama, K. Fujikawa, K. Ohta, and H. Yuki, ACS Symp. Ser., 166, 327 (1981).
17. K. Hatada, T. Kitayama, S. Okahata, and H. Yuki, Polym. J., 13, 1045 (1981).
18. K. Hatada, T. Kitayama, S. Okahata, and H. Yuki, Polym. J., 14, 971 (1982).
19. T. Kitayama, K. Ute, and K. Hatada, Polym. J., 16, 925 (1984).
20. K. Ute T. Kitayama, and K. Hatada, Polym. J., 18, 249 (1986).
21. K. Ute and K. Hatada, Polym. J., to be submitted.
22. K. Hatada, K. Ute, K. Tanaka, T. Kitayama, and Y. Okamoto, Polym. J., 17, 977 (1985).
23. K. Hatada, K. Ute, K. Tanaka, T. Kitayama, and Y. Okamoto, Polym. J., 18, No. 12 (1986) in press.
24. T. Miyamoto and H. Inagaki, Polym. J., 1, 46 (1970).
25. K. Hatada, T. Kitayama, H. Sugino, M. Furomoto, and H. Yuki, "Preparation and Properties of Stereoregular Polymers", R. W. Lenz and F. Ciardelli, Ed., D. Reidel Publishing Company, Dordrecht, Holland, 1979, p 185.
26. A. Soum, N. D'Accorso, and M. Fontanille, Makromol. Chem., Rapid Commun., 4, 471 (1983).
27. Y. Okamoto, K. Ohta, K. Hatada, and H. Yuki, ACS Symp. Ser., 166, 353 (1981).
28. A.M. Doherty and T.E. Hogen-Esch, Makromol. Chem., 187, 61 (1986).
29. D.L. Glusker, I. Lysloff, and E. Stiles, J. Polym. Sci., 49, 315 (1961).

KINETICS AND MECHANISMS IN THE ANIONIC POLYMERIZATION OF METHACRYLIC ESTERS

AXEL H. E. MÜLLER
Institut für Physikalische Chemie, Universität Mainz, D-6500 Mainz, Federal Republic of Germany

ABSTRACT
Under appropriate conditions the anionic polymerization of methacrylates proceeds in an ideal manner leading to polymers with narrow molecular weight distributions. From the kinetics of polymerization and from the microstructure of the polymers the existence of different active species is concluded, the most important one being a peripherally solvated contact ion pair. Depending on polymerization conditions two different modes of monomer addition can occur. Homo- and copolymerization experiments show that the influence of the ester group is mainly given by electronic effects. In non-polar solvents the polymerization of *tert.*-butyl methacrylate (in contrast to methyl methacrylate) leads to 100% isotactic polymers with rather low polydispersities.

1. INTRODUCTION

Acrylic monomers are one of the most important classes of vinyl monomers. Especially methacrylates, acrylates and acrylonitrile are of high industrial interest. Moreover, acrylic polymers have some advantages compared to non-polar monomers:
- tacticity can be controlled by the polymerization conditions,
- polymers are easily converted to other interesting polymers, e.g. hydrolysis of poly(alkyl acrylate)s rendering poly(acrylic acid),
- polymers are UV-degradable, e.g. poly(vinyl ketones).

Anionic polymerization of these monomers renders the possibility of preparing polymers with narrow molecular weight distribution (polymer standards), functionalized polymers, blockcopolymers, star polymers, cyclic polymers, networks and other model compounds.
However, in the contrast to non-polar monomers, e. g. styrene, the polymerization mechanism is complicated by the presence of the polar side group, which gives rise to chemical and physical interactions with the living chain end.
The most important potential complications are:

1. Side reactions with the carbonyl group leading to low initiator efficiency or chain termination, e.g:

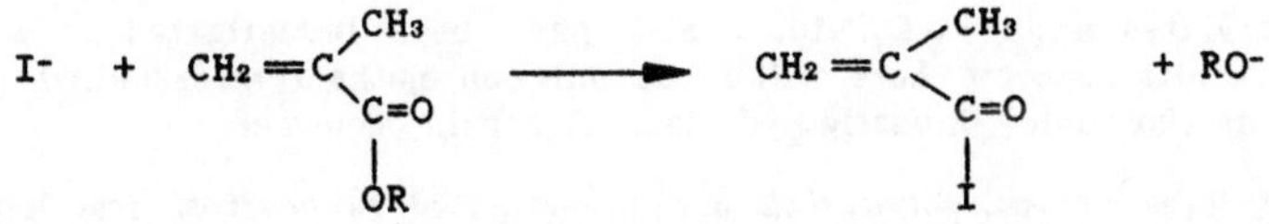

2. Activation of protons in α position to the carbonyl group leading to chain transfer:

Published 1987 by Elsevier Science Publishing Co., Inc.
Recent Advances in Anionic Polymerization
T.E. Hogen-Esch and J. Smid, Editors

3. Due to the bidentate character of the active centres, they may attack the monomer not only by the carbanion (1,2 addition) but also by the enolate oxygen (1,4 addition):

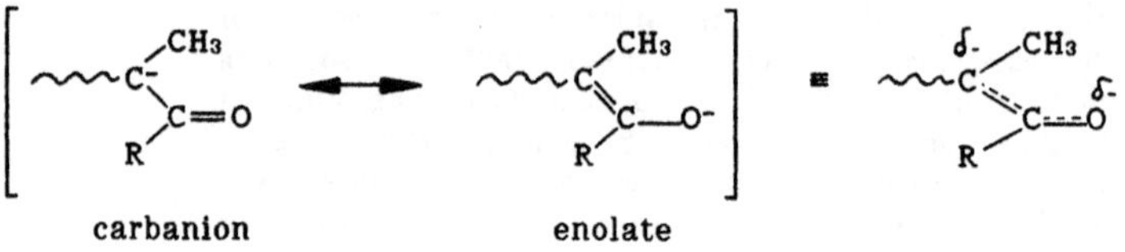

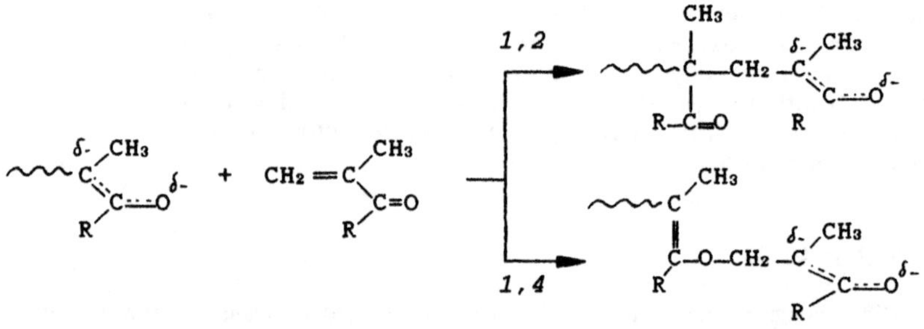

4. The carbonyl group may coordinate with the counterion of the living chain end ("intramolecular solvation") or may lead to association of ion pairs.

5. The chain end may interact with electron donors (e.g. alkoxides).

As a result of these complications, often non-ideal polymerizations have been reported[1], which can be characterized as follows:
- complicated kinetics,
- broad and multimodal molecular weight distributions,
- problems in the preparation of block copolymers or the introduction of functional end groups.

Thus, our primary aim was to find conditions for an "ideal" anionic polymerization of these monomers characterized by simple kinetics and narrow molecular weight distributions. Additionally, it was of interest to determine the factors which influence the kinetics, the amount and nature of side-products, the molecular weight distribution and the microstructure of the polymers. Our second aim was to understand these findings on the basis of mechanisms and the structures of the active centres involved.

Although acrylates and vinyl ketones also have been investigated in recent years, the results reported here will focus only on methacrylates, which have been the most thoroughly investigated class of acrylic monomers.

As the half-lives of the polymerizations investigated range from fractions of seconds to hours, it was necessary to use different types of reactors for the kinetic measurements:
- a flow tube for fast reactions ($0.05 \leq t_{1/2} \leq 2$ sec)[2]

- a stirred tank reactor equipped with a solenoid valve allowing for the withdrawal of samples at intervals of 1 sec or larger for reactions having half-lives ≥ 2 sec.[3]

The advantage of both methods is that for all conversions a polymer sample is obtained which can be characterized for
- molecular weight averages,
- molecular weight distribution,
- activity of chains (after labelling of active chains, e.g. by $^3H^+$),
- microstructure, etc.

2. METHYL METHACRYLATE (MMA)

2.1. Earlier Results:

This monomer has been investigated since 1973. The results up to 1980 were reviewed earlier[4] and will only be summarized shortly here. It was found, given proper conditions, an "ideal" polymerization is achieved. This is characterized by the following:

- instantaneous initiation ($k_i \geq k_p$),

- a linear pseudo first-order decrease of monomer concentration:

$$\ln \frac{[M]_0}{[M]} = k_p \cdot c^* \cdot t \tag{1}$$

  (c^* = concentration of active centres. The constant slope $k_p \cdot c^*$ indicates the absence of chain termination),

- a linear dependence of the number-average degree of polymerization P_n on conversion x_p:

$$P_n = \frac{\text{concentration of polymerized monomers}}{\text{concentration of polymer chains}} = \frac{[M]_0 \cdot x_p}{c_{tot}} \tag{2}$$

  (c_{tot} = total concentration of polymer chains, *including terminated ones*. The constant slope $[M]_0/c_{tot}$ indicates the absence of chain-generating reactions, e. g. transfer),

- narrow molecular weight distributions of the Poisson type ($1.01 \leq M_w/M_n \leq 1.1$ is typical).

The optimal conditions needed for an "ideal" polymerization of MMA are as follows:

- polar solvents: DME > THF » THP,
- low temperatures: T ≤ −75°C for THF,
- large counterions: $Cs^+ > K^+ > Na^+ > Li^+$,
- monofunctional initiators which are reactive enough to give fast initiation but not too reactive towards the carbonyl group of the monomer: diphenylmethyl salts, diphenylhexyl salts, benzyl-oligo-α-methylstyryl salts, and especially metalated esters, like methyl α-lithio isobutyrate, which are models of the active centre[5],
- addition of common-ion salts (e.g. tetraphenylborates) to suppress the dissociation of ion pairs into free ions.

It should be noted that the results reported in the following were obtained by adding excess common-ion salt, except where stated otherwise. Thus, all data given correspond to ion pairs.

2.2. Counterion dependence in THF

More recently, the kinetics of the anionic polymerization of MMA in THF were investigated using different alkali counterions as well as the sodium ion complexed by the bicyclic cryptand 222 (Na^+,222).[6,7] In the latter case, by varying the concentrations of active centres and common-ion salt, it was possible to independently determine the rate constants of the cryptated ion pairs and the free anions. Fig. 1 shows an Arrhenius plot of the rate constants for all counterions investigated. This plot reveals several informations:

- the linearity indicates only one kind of active species,
- except for Na^+, K^+, and Cs^+, a distinct dependence of the rate constants on the counterion is found. This indicates that the active species are contact ion pairs, because a large interionic distance in the contact ion pairs favours the charge separation in the transition state and thus leads to an increase in reaction rate.

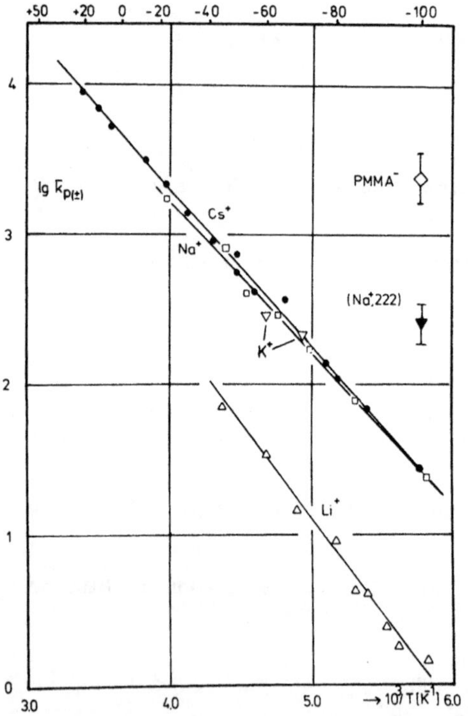

Fig. 1. Arrhenius plot of the propagation rate constants in the anionic polymerization of MMA in THF using different counterions.[7]

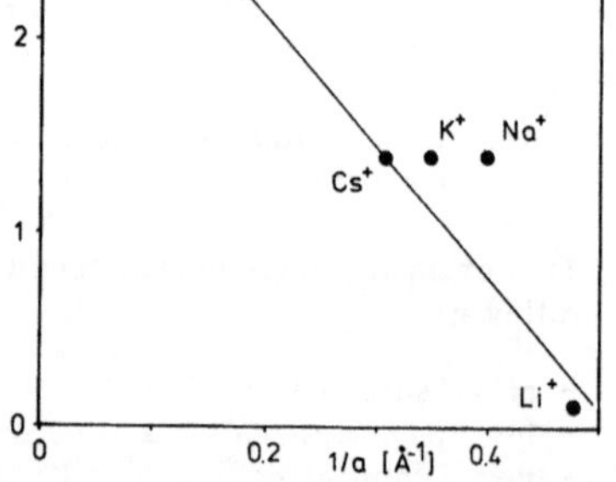

Fig. 2. Dependence of the propagation rate constant on the interionic distance, a, in the anionic polymerization of MMA in THF at −100°C.[7]

This is seen clearly in Fig. 2, which plots log k_p *vs.* the reciprocal interionic distance, *a* (calculated as the sum of the crystal radius of the cation and the electrostatic radius of the anion which is assumed to be 1.5 A). The deviation of Na^+ and K^+ from the straight line indicates that the interionic distances are larger than the values expected from the crystal radii.

These results conform to a model of the active centre described as a contact ion pair peripherally solvated by THF molecules (S denotes solvent):

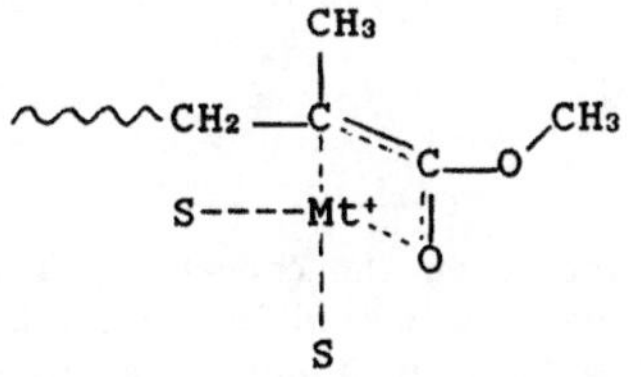

The increased interionic distance of Na^+ and K^+ ion pairs seems to be caused by peripheral solvation spreading the positive charge over a larger area and thus weakening the electrostatic interaction of the cation with the α-carbon and the partially enolated carbonyl oxygen. On the other hand, Li^+ is on the straight line again because of a strong p-π overlap leading to a very tight bond between anion and cation.

It is interesting to note, due to the strong interaction of the counterions with the enolate oxygen, no solvent-separated ion pairs are found. This is in contrast to the anionic polymerization of non-polar monomers in polar solvents.

2.3. Ion pair associates

Although this is suggested by the linearity of the Arrhenius plot, the peripherally solvated contact ion pair is not the only species present in the system. When working at higher initiator concentrations a strong dependence of the rate constant on the initiator concentration is observed[8] as shown in Fig. 3 for Na^+ counterion. Since an excess of common-ion salt was used, this cannot

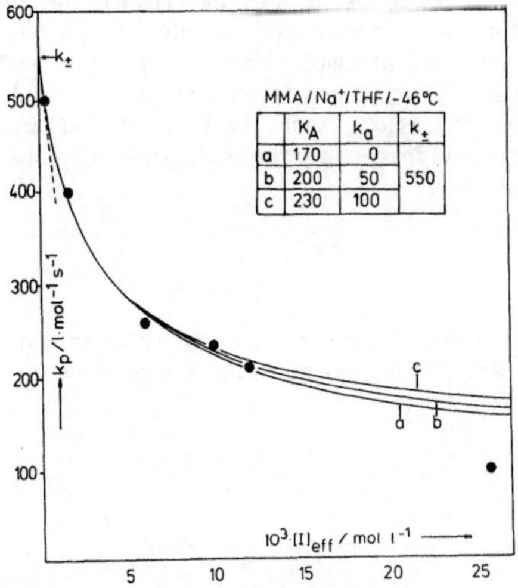

Fig. 3. Dependence of the propagation rate constant on the concentration of active species in the anionic polymerization of MMA in THF at −46°C using Na^+ as the counterion.[8]

be due to the dissociation to free ions. We attribute this effect to the existence of ion pair associates having a very low, if any, reactivity. The kinetic scheme for such a process is given below, assuming dimeric associates only:

The dependence of the rate constant k_p on the concentration of active centres is given by the rate constant of non-associated ion pairs, $k_\pm$, the rate constant of the associates, k_a, and the fraction of non-associated ion pairs α:

$$k_p = \alpha \cdot k_\pm + [(1-\alpha)/2] \cdot k_a = \tfrac{1}{2}k_a + (k_\pm - \tfrac{1}{2}k_a) \cdot \alpha. \qquad (3)$$

$$\alpha = [-1 + (1 + 8 \cdot K_A \cdot c^*)^{1/2}]/4 \cdot K_A \cdot c^*. \qquad (4)$$

For $K_A \cdot c^* \gg 1$

$$\alpha = 1/(2 \cdot K_A \cdot c^*)^{1/2} \qquad (5)$$

and for $K_A \cdot c^* \ll 1$

$$\alpha = 1 - 2 \cdot K_A \cdot c^*. \qquad (6)$$

The lines shown in Fig. 3 result from a fit of these three constants to the experimental data using three different values for k_a. From the data it is not possible to conclude whether or not the associates are active .

When working with bifunctional initiators (e.g. oligo-α-methylstyryl sodium), the chain ends are very close to each other. Even at very low initiator concentrations associates were found in these systems. The degree of association was observed to decrease with increasing degree of polymerization.[9,10]

It is interesting that association effects seem to be of less importance for Li^+ as the counterion. This may be due either to the stronger peripheral solvation of the Li^+ ion or to a stronger tendency to intramolecular coordination with ester groups of the own polymer chain ("intramolecular solvation", *cf.* 2.4). However, for oligomers at very high concentrations there is kinetic[11] as well as direct[12] (vapour pressure osmometry) evidence for the existence of associates.

2.4. "Intramolecular" solvation

Coordination of the counterion with the penultimate or ante-penultimate ester groups leads to an active species different to the peripherally solvated contact ion pair:

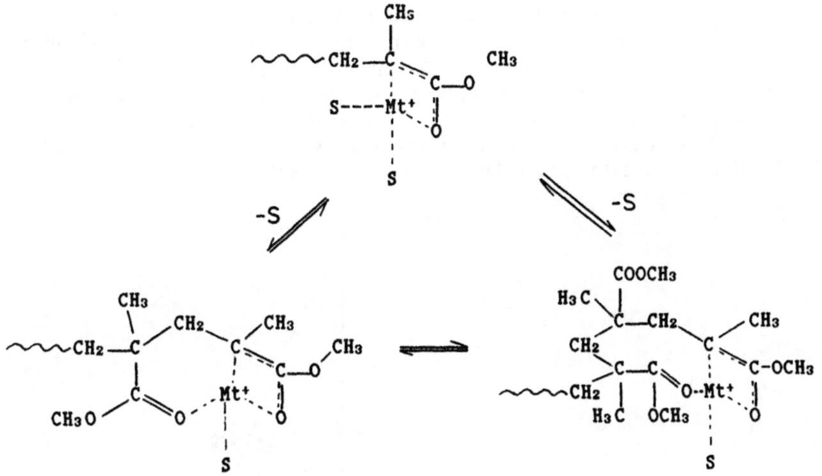

Similar structures were proposed by Fowells et al.[13] on the basis of the stereostructure of partially deuterated poly(acrylates) and methacrylates. There is IR-spectroscopic evidence on the existence of these structures.[14,15,16] However, NMR studies on living oligomers did not give unequivocal results.[17]

The model was successfully applied to the kinetics and stereochemistry of the oligomerization and polymerization of 2-vinyl pyridine in THF.[18,19,20,21] In order to find kinetic evidence in the methacrylate system, we studied the kinetics of the oligomerization of MMA in THF initiated by methyl α-lithio isobutyrate (MIB-Li, the "living monomer", i.e. oligomer of DP=1).[8,11] By using a computer fit procedure, the individual rate constants were determined as a function of the degree of polymerization:

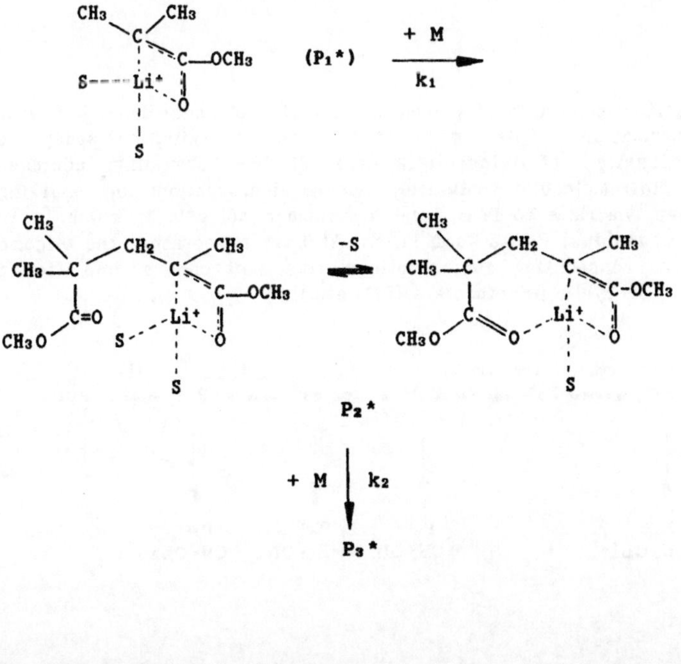

Figs. 4 and 5 show a significant decrease of the rate constant from DP=1 to DP=2 which can be attributed to a lower reactivity of the "intramolecularly" solvated species. The slight increase for higher DP's may correspond to the fact that the additional monomer units increasingly hinder the formation of the intramolecular complex by their motion. Whereas the ratio k_1/k_2 is 3.4 at –46°C, it is >20 at 25°C. This reflects the endothermic nature of "intramolecular" solvation, the ester group being a weaker solvating agent than THF.

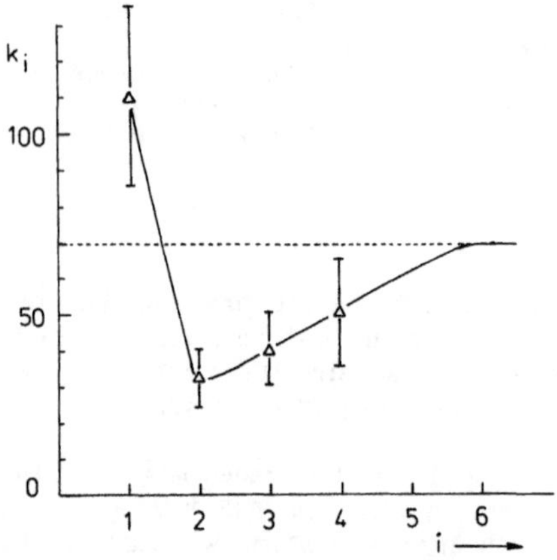

Fig. 4. Dependence of the propagation rate constant on the degree of polymerization, i, in the anionic oligomerization of MMA (THF/Li⁺/–46°C).[8]

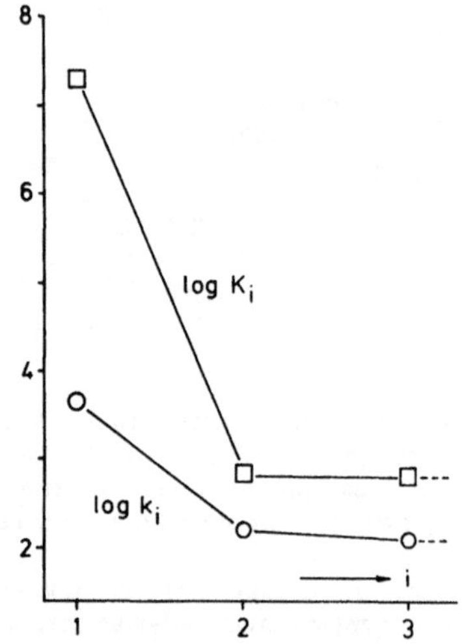

Fig. 5. Dependence of the rate and equilibrium constants of polymerization on the degree of polymerization, i (MMA/THF/Li⁺/+25°C).[11]

The equilibrium constants of polymerization also strongly depend on the degree of polymerization. This was shown in a kinetic study of the "disproportionation" of living oligomers at 25 °C,[22] providing additional evidence for "intramolecular" solvation. During disproportionation, a living oligomer P_i^* depolymerizes to P_{i-1}^* and a monomer molecule M which in turn may be added to residual P_i^* to form P_{i+1}^*. At that temperature the oligomers formed undergo consecutive side-reactions, like autocondensations (formation of linear (P_1^c) or cyclic β-ketoesters (P_3^c etc.)):

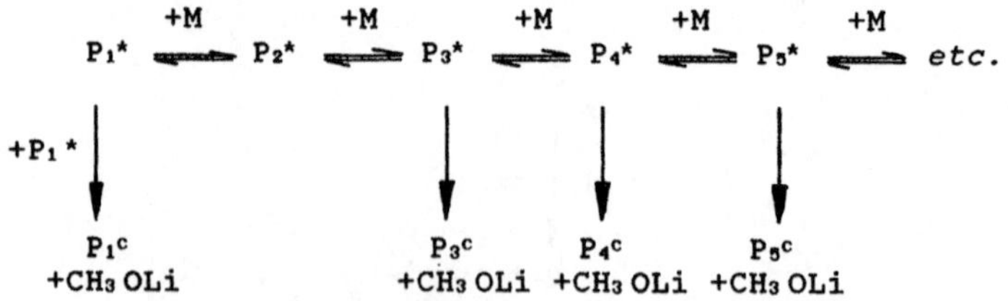

By using computer fit methods, the rate constants of depolymerization (and thus the equilibrium constant of polymerization) can be calculated as a function of the degree of polymerization.[11] Fig. 5 shows that the equilibrium constant decreases by more than four orders of magnitude when going from DP=1 to DP=2, which corresponds to a greatly increased thermodynamic stability of the higher living oligomers as compared to the "living monomer". Although steric effects may contribute to a certain extent, this effect is most reasonably explained by the entropy gain of desolvation in forming "intramolecularly solvated" species.

2.5. Influence of solvent

When working in more polar solvents, (e.g. 1,2-dimethoxyethane; DME), external solvation of the ion pairs presumably increases. Accordingly, the interionic distance in the ion pairs should increase and the degree of inter-ion-pair association should decrease, both leading to higher rate constants. The opposite is expected for less polar solvents, (e.g. tetrahydropyran; THP). Fig. 6 shows that this is true for Na^+ as the counterion (it is also true for Cs^+).[23,24] However, as Fig. 7 shows, this is not true for Li^+ counterion.[25] Here, the rate constants are even smaller in DME, as compared to THF.

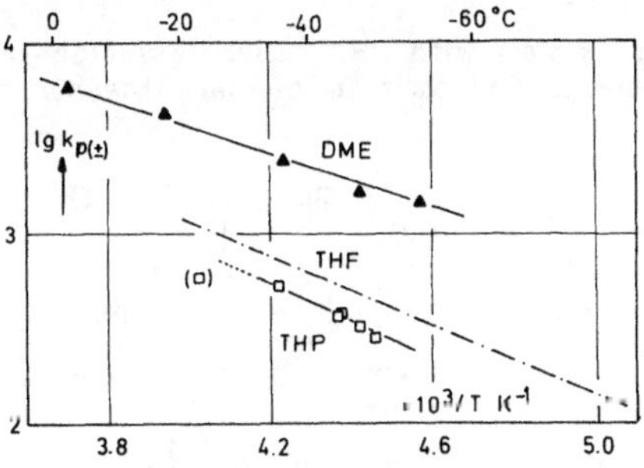

Fig. 6. Arrhenius plot of the propagation rate constants for the anionic polymerization of MMA in various solvents using Na^+ as the counterion.[23,24]

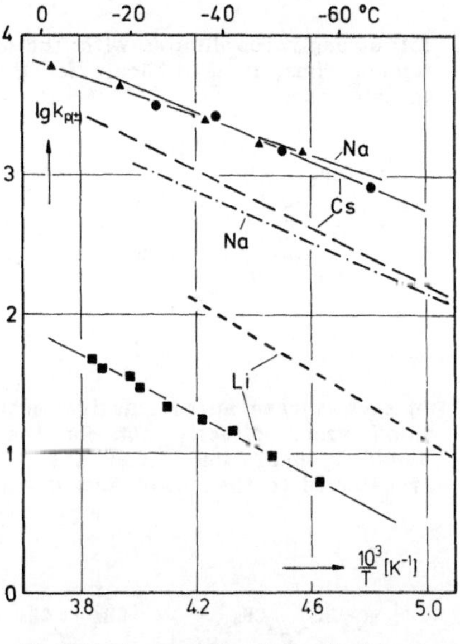

Fig. 7. Arrhenius plot of the propagation rate constants for the anionic polymerization of MMA in THF (---) and DME (——).[23,25]

To understand this effect, it is necessary to have a closer look at the mechanism of monomer addition. Two possible mechanisms are discussed below. In mechanism I, the monomer carbonyl oxygen first coordinates with the counterion displacing a solvent molecule (only one is shown in the scheme). In the transition state the counterion is moved from the active centre to the newly formed one; the counterion is solvated again:

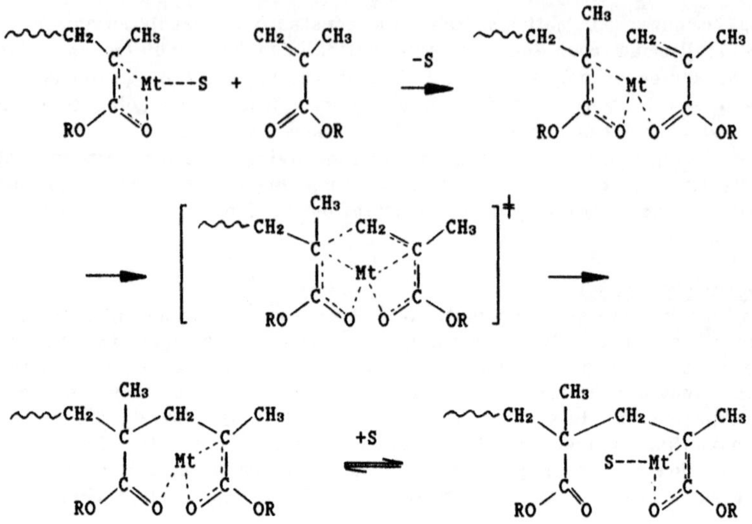

Li^+ strongly coordinates with the anion as well with DME which is a bidentate ligand. Thus, it is difficult for the incoming monomer to displace the solvent in this case:

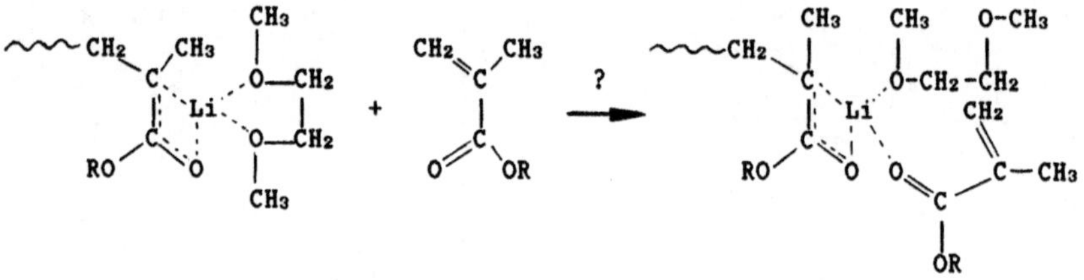

In such a case an alternative mechanism II is more likely. Here, the monomer vinyl group directly attacks the carbanion without coordinating with the shielded counterion. After the transition state the solvated counterion is transferred to the newly formed anion:

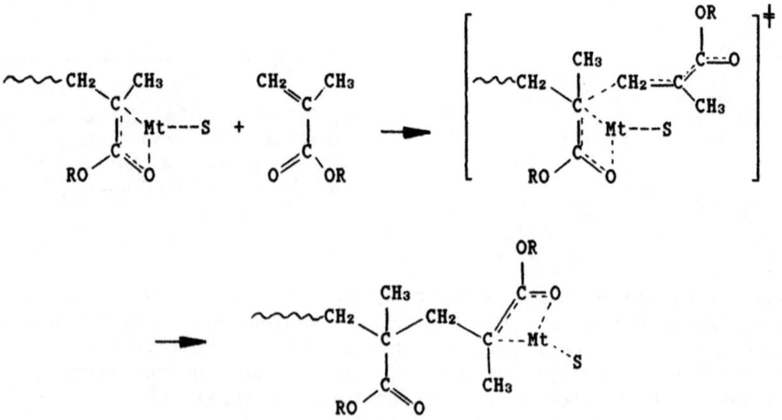

This step is energetically unfavourable. However, mechanism II is the only choice for polymerizations *via* free anions or ion pairs involving large, non-solvated counterions, like cryptated sodium (Na^+,222).

Table 1 shows that for all solvents investigated, Li^+ has the highest activation energies. However, the lower rate constants for Li^+ in DME seem to be due to a higher absolute activation entropy (lower frequency exponent log A), which is in accordance with mechanism II.

TABLE 1: Rate constants and activation parameters in the anionic polymerization of MMA *via* different ion pairs

Solvent		Cs^+	K^+	Na^+	Li^+
DME	E_a /kJ/mol	15.1	-	13.8	21.4
	log A	6.8	-	6.5	6.0
	k_p^{233}	2680	-	2550	16
	Ref.	23	-	23	23
THF	E_a /kJ/mol	19.5	19.3	18.3	24.0
	log A	7.3	7.2	7.0	7.4
	k_p^{233}	860	750	800	100
	Ref.	48	7	31,48	7
THP	E_a /kJ/mol	-	-	21.5	-
	log A	-	-	7.5	-
	k_p^{233}	-	-	480	-
	Ref.	-	-	24[a)]	-

[a)] Due to extensive deactivation, error approx. 30%

2.6. Stereoregulation

Mechanisms I and II should differ significantly in their stereoregulating effect. Fowells et al.[13] proposed that mechanism I should lead to a meso placement. For mechanism II the stereochemistry is not pre-determined. Because of steric reasons, predominantly racemic placements should be found, which is observed in radical polymerization.

Fig. 8 shows the fraction of racemic dyads in the polymers as a function of calculated interionic distance, *a*, and of solvent. The tacticity of polymers prepared *via* the free anion or the cryptated sodium ion is very similar to that of samples polymerized radically at low temperatures.[26,27] Hence, mechanism II must be applicable here. The fact that polymers initiated by Li^+ salts have the same tacticity makes it reasonable to also assume the same mechanism. This means that in the presence of Li^+ there is no coordination of the monomer carbonyl group with the counterion due to the strong peripheral solvation. Only in non-polar solvents like toluene, high meso contents are found[28]. Mechanism I is very likely here, because of the lack of external solvation.

For the larger alkali cations the fraction of racemic dyads decreases with increasing interionic distance as well as with decreasing solvating power of the solvent. It is thus probable that in tose systems both mechanisms I and II are effective to varying extents, the fraction of monomer additions *via* mechanism I increasing with interionic distance and decreasing with polarity of solvent. These assumptions are corroborated by the fact that the statistics of the tactic placements deviate from Bernoullian in the same order (*cf.* persistence ratios ϱ in Table 2):

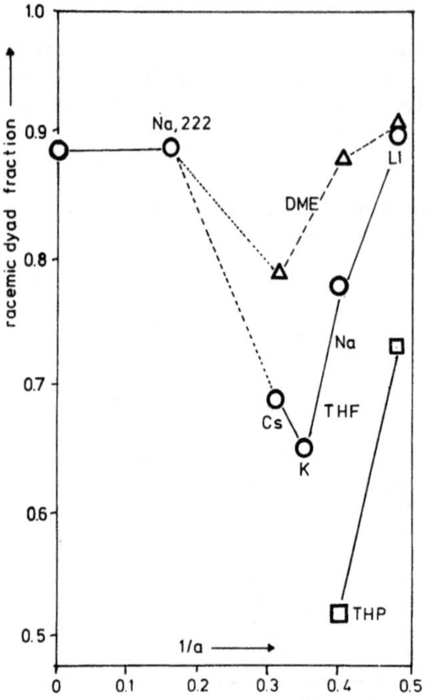

Fig. 8. Fractions of racemic dyads in PMMA prepared in various solvents and using different counterions at ca. -50°C (free anion and cryptated Na^+ at -100°C).

<u>**TABLE 2:**</u> Dependence of the tacticity of PMMA on solvent and counterion of polymerization

Solvent	Counter-ion	Temp [°C]	mm	mr	rr	r	ϱ [a]	Ref.
	Cs^+	-66	.03	.37	.60	.79	.90	23
DME	Na^+	-55	.02	.21	.77	.88	1.00	23
	Li^+	-57	.01	.16	.83	.91	.98	25
	(anion)	-98	.01	.20	.79	.90	.97	29
	$Na,222^+$	-98	.01	.19	.80	.89	1.05	29
THF	Cs^+	-53	.05	.52	.42	.68	.82	30
	K^+	-60	.09	.52	.39	.65	.88	7
	Na^+	-51	.04	.38	.58	.77	.93	31
	Li^+	-45	.01	.22	.77	.88	.96	7
THP	Na^+	-47	.22	.52	.26	.52	.97	24
	Li^+	-35	.06	.32	.62	.78	1.05	32
dioxane	K^+	+13	.14	.56	.30	.57	.87	32
	Li^+	+13	.10	.35	.55	.73	1.12	32
free-radical		-55	.04	.17	.79	.87	1.27	26
		-29	.01	.24	.75	.88	.91	27

[a] persistence ratio $\varrho = m \cdot r/2mr$

3. OTHER METHACRYLATES

3.1. Homopolymerization

Ester groups different from methyl may exert steric and/or electronic effects on the rates of polymerization. Whereas steric effects should always lead to lower rate constants, the electronic effects are difficult to predict. Electron-donating groups, e.g. *tert.*-butyl, decrease the polarization of the monomer and hence its reactivity. On the other hand, they increase the charge density on the carbanion increasing its reactivity. The latter effect may be partially cancelled by ion pairing effects; the increase in charge density in turn may lead to smaller interionic distances which decreases rate constants again (*cf.* 2.2).

Table 3 shows the rate constants and Arrhenius parameters for the polymerization of different methacrylates in THF using Na^+ as the counterion. Taft's σ^* parameter[33] is given as a measure for the electron-withdrawing effect of the substituent.

TABLE 3: Propagation rate constants and Arrhenius parameters in the anionic polymerization of alkyl methacrylates in THF using Na^+ as the counterion.[34, 40]

substituent	σ^*	E_a / kJ/mol	log A	k_p^{198} / l/mol·s	symbol
decyl	-0.12	15.8	6.5	211	✡
methyl[31]	0.00	19.5	7.3	145	□
n-butyl	-0.13	23.6	8.3	128	⬢
ethyl	-0.10	16.5	6.4	100	○
benzyl	+0.23	19.1	7.0	81	●
i-propyl	-0.20	19.5	7.0	72	△
2-ethoxyethyl	-0.17	26.0	8.4	38	▼
t-butyl[35]	-0.32	31.2	8.7	3.0	■
phenyl	+0.60	15.1	4.1	1.4	⬡

Tert.-butyl methacrylate (TBMA) has the highest frequency exponent (i.e. least steric requirements in the transition state). This indicates that steric effects seem to be of minor importance. Moreover, the relatively bulky decyl methacrylate has even higher rate constants than MMA.

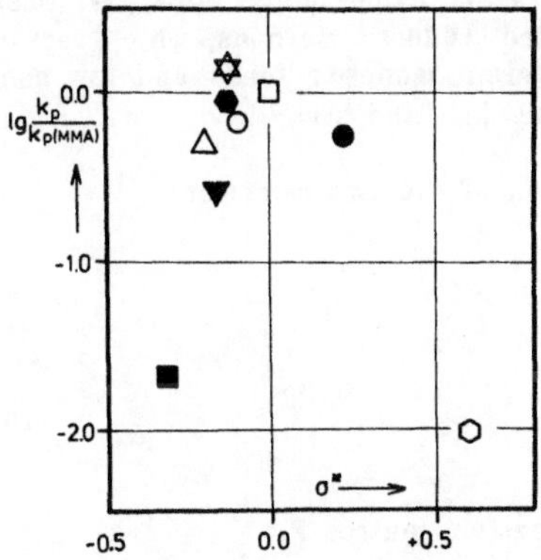

Fig. 9. Dependence of the propagation rate constant of various methacrylates on Taft's σ^* parameter (THF/Na^+/ –75°C).[34] Symbols, cf. Tab. 3.

Fig. 9 shows a Hammett plot[36] of the relative rate constants *vs.* Taft's σ^* parameter. The poor fit indicates that the substituent effect is too complex to be described by one parameter only. In order to gain further information it is necessary to determine separately the relative reactivities of monomers towards a given polymer anion and the relative reactivities of different polymer anions towards a given monomer. This can be done either by direct determination of the rate constants k_{12} and k_{21}[37] or by copolymerization experiments.

3.2. Copolymerization

Assuming the "terminal" model for the kinetics of copolymerization, the following well-known reaction scheme can be adopted:

$$P_1^* + M_1 \xrightarrow{k_{11}} P_1^*$$

$$P_1^* + M_2 \xrightarrow{k_{12}} P_2^*$$

$$P_2^* + M_2 \xrightarrow{k_{22}} P_2^*$$

$$P_2^* + M_1 \xrightarrow{k_{21}} P_1^*.$$

The copolymerization parameters

$$r_1 = k_{11}/k_{12} \quad \text{and} \quad r_2 = k_{22}/k_{21} \tag{7}$$

give information only for the relative reactivities of *monomers* adding to a given living polymer. The relative reactivities of the *active centres* are given by the parameters

$$q_1 = k_{11}/k_{21} \quad \text{and} \quad q_2 = k_{22}/k_{12}, \tag{8}$$

which cannot be calculated from the copolymerization parameters alone. The homopolymerization rate constants are necessary in order to calculate these parameters.

Direct determination of the rate constants k_{12} and k_{21} is not favorable in methacrylate systems. Contrary to aromatic systems, e.g. substituted styrenes, the concentrations of the different active centres cannot easily be measured by spectroscopic methods. Thus, copolymerization experiments were performed. Instead of using the Fineman-Ross[38] or Kelen-Tüdös[39] methods, which require measurements of polymer composition at different monomer feeds and low conversions, a new method was used to determine k_{12} and k_{21}.[40]

The differential equations for the consumption of the two monomers

$$-\frac{d[M_1]}{dt} = (k_{11}\cdot[P_1^*] + k_{21}\cdot[P_2^*])\cdot[M_1] \tag{9}$$

$$-\frac{d[M_2]}{dt} = (k_{22}\cdot[P_2^*] + k_{12}\cdot[P_1^*])\cdot[M_2] \tag{10}$$

and for the change in the concentration of active centres P_1^*

$$\frac{d[P_1^*]}{dt} = k_{21}\cdot[M_1]\cdot c^* - (k_{21}\cdot[M_1] + k_{12}\cdot[M_2])\cdot[P_1^*] \quad (11)$$

were integrated numerically. By varying k_{12} and k_{21}, the calculated time-conversion curves were fitted to the experimental data. Thus, it is possible to investigate the copolymerization kinetics up to high conversions. Fig. 10 shows the result of such a fit procedure.

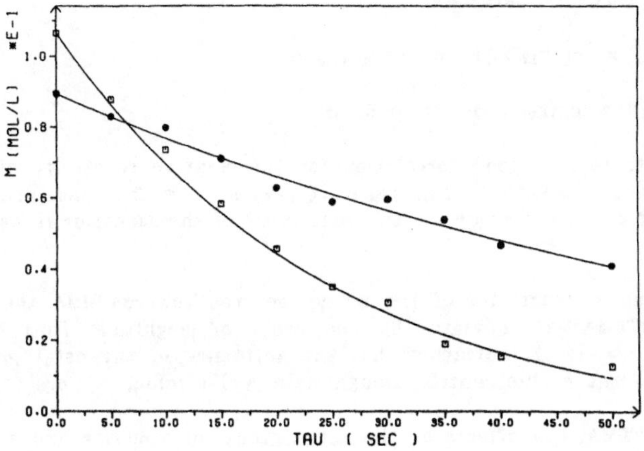

Fig. 10. Fit of the calculated time-conversion function to the experimental data in the anionic copolymerization of MMA (□) and iso-propyl methacrylate (●) in THF at -75°C using Na⁺ as the counterion.[34]

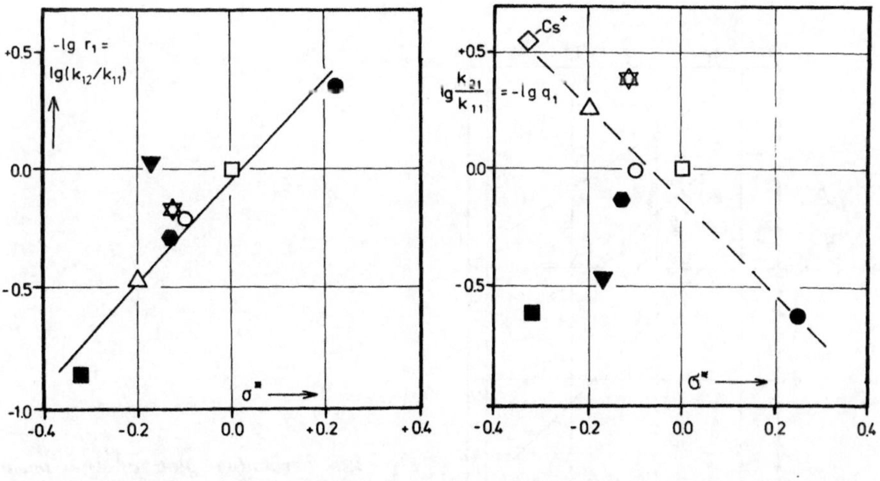

Fig. 11. Dependence of the relative reactivities of (a) different monomers and (b) different polymer anions on Taft's σ parameter in the anionic copolymerization of different methacrylates with MMA in THF at -75°C using Na⁺ as the counterion.*[34] *Symbols, cf. Tab. 3.*

All experiments were performed in THF at −75°C, using Na^+ as the counterion. For most monomers, the "terminal" model is sufficient, i.e. the time conversion curves can be described by four rate constants. Only for *iso*-propyl methacrylate and TBMA the terminal model gives poorer fits probably due to penultimate effects. However, our data are not sufficient to calculate the eight rate constants describing such a model.

Fig. 11 shows the Hammett plots for the relative reactivities of the monomers towards $PMMA^-Na^+$ $(1/r_1)$ and of the living polymers towards MMA $(1/q_1)$ as a function of Taft's σ^* parameter:

$$\log(1/r_1) = \log(k_{12}/k_{11}) = \rho^*_{PMMA\text{-}Na}\cdot\sigma^* \quad (12)$$

$$\log(1/q_1) = \log(k_{21}/k_{11}) = \rho^*_{MMA}\cdot\sigma^* \quad (13)$$

Obviously, there is a good correlation for the relative reactivity of the monomers adding to $PMMA^-Na^+$ ion pairs ($\rho^*_{PMMA\text{-}Na} = 2.2$) indicating that the influence of the ester group on the reactivity of the monomer is predominantly electronic.

For the relative reactivity of the active centres towards MMA the correlation is worse. $PTBMA^-Na^+$ deviates by one order of magnitude from the straight line ($\rho^*_{MMA} = -2.0$) indicating that the influence of the ester group on the reactivity of the active centre, though mainly electronic, is more complex.

Except for TBMA, the effects of the ester group on monomer and active centre roughly cancel each other. This explains the small influence of the ester group on the homopolymerization kinetics.

For the copolymerization of MMA and TBMA the influence of temperature and counterion on the kinetics of homo- and copolymerization was investigated. For Na^+ as the counterion the r values only vary slightly with temperature. An Arrhenius plot of the four rate constants is shown in Fig. 12.

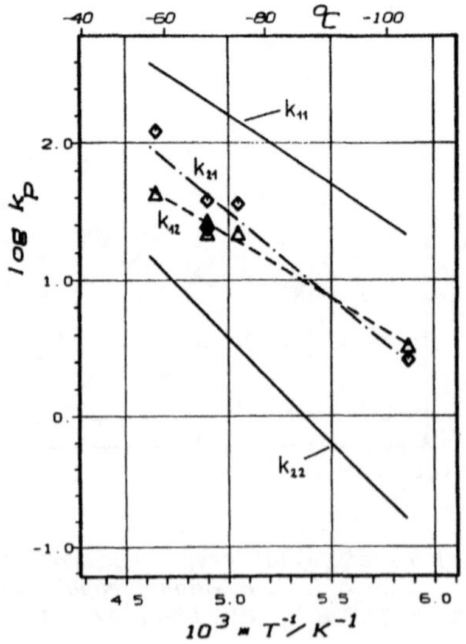

Fig. 12. Arrhenius plot of the homo- and copolymerization rate constants in the anionic copolymerization of MMA (1) and TBMA (2) in THF using Na^+ as the counterion.[34]

At ca. −70°C the copolymerization was investigated using Li^+, Na^+ and Cs^+ as the counterions. The corresponding rate constants and copolymerization parameters are given in Table 4. For Cs^+ as the counterion the "terminal" model can be applied. For Li^+ as the counterion no fit was possible which may be due to penultimate effects. The r values shown are taken from Yuki et al.[41] They were determined using the Fineman–Ross method, which does not take into account possible penultimate effects.

TABLE 4: Rate constants and copolymerization parameters in the anionic copolymerization of MMA(1) and TBMA(2) in THF[34,40]

Counter-ion	Temp [°C]	k_{11}	k_{12}	k_{21} [$l \cdot mol^{-1} \cdot s^{-1}$]	k_{22}	r_1	r_2	$r_1 \cdot r_2$
Cs^+	−67	234	9	850	52	26	0.06	1.6
Na^+	−69	205	25	31	5.4	8.2	0.17	1.4
Li^+	−75	18	-	-	0.6	35	0.43	15 [a)]
						5	0.02	0.1[b)]

[a,b)] r values taken from ref. 41; (a) methanol insoluble fraction, (b) methanol soluble fraction.

The strong counterion effect found is also reflected in the ρ^* values for the addition of different monomers to PMMA ion pairs (cf. eq. 12 and Table 5).

TABLE 5: Hammett ρ^* values for the addition of different methacrylates to PMMA ion pairs in THF at *ca.* −75°C[34,40]

Mt^+	$\rho^*_{PMMA-Mt}$	Comments
Cs^+	4.4	(estimated from two monomers only)
Na^+	2.2	(7 monomers)
Li^+	1.1	(Ref. 41; 8 monomers)

Obviously, selectivity increases with increasing ionic radius. This is unexpected as the reactivity of the ion pairs also increases in this direction. This effect, as well as the unexpectedly low reactivity of the $PTBMA^-Na^+$ ion pairs, may be due to different mechanisms of monomer addition. Further work is necessary to clarify this point.

3.3. Counterion and Solvent Effects in the Polymerization of TBMA

Tert.-butyl methacrylate is a very interesting monomer, not only for mechanistic reasons (*cf.* 3.2), but also because the polymerization is completely undisturbed *even at room temperature.*[35] Thus, it is no problem to prepare polymers or block copolymers of high molecular weight as is shown in Fig. 13. Furthermore, PTBMA is easily hydrolized to poly(methacrylic acid) thus rendering the possibility to prepare narrowly distributed polyelectrolytes or copolymers having electrolyte and non-electrolyte blocks. These polymers have very interesting properties, e.g. as emulsifiers or membrane precursors.

The kinetics of homopolymerization of TBMA have been investigated using Li^+, Na^+ and Cs^+ as counterions in THF[25,35] and using Li^+ in toluene.[42]

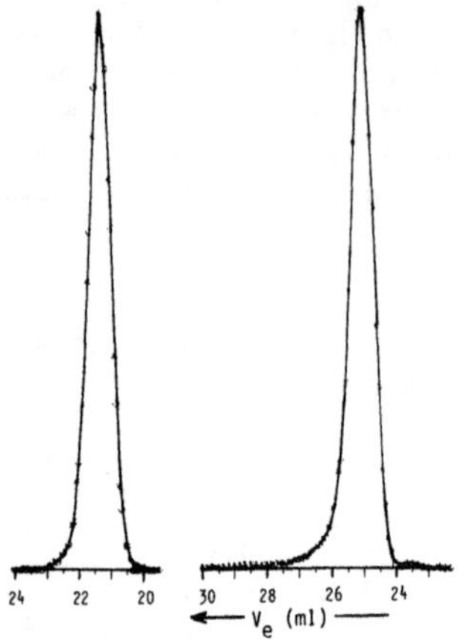

Fig. 13. GPC eluogram of (a) PTBMA and (b) poly(styrene–b–TBMA) prepared in THF using (a) Li^+ and (b) K^+ as counterions.

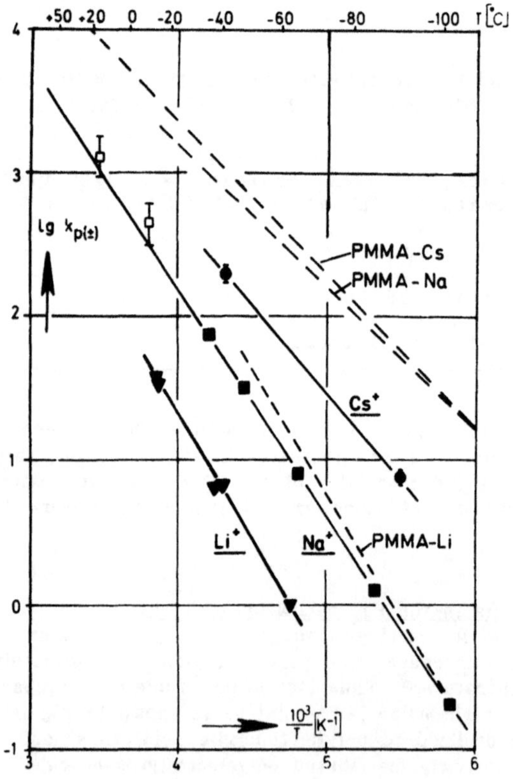

Fig. 14. Arrhenius plot of the propagation rate constants for the polymerization of TBMA in THF using different counterions.[38,42]

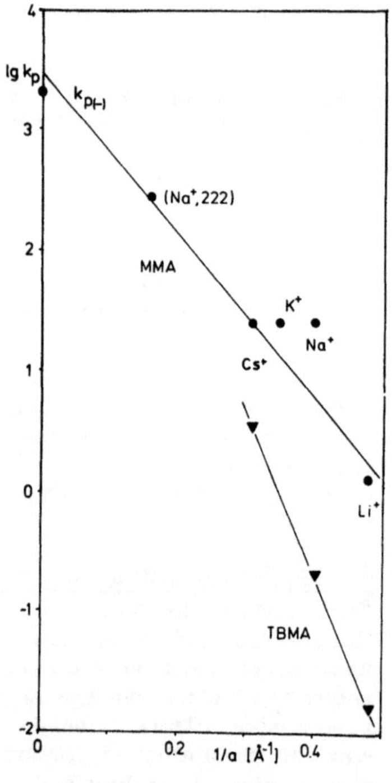

Fig. 15. Dependence of the propagation rate constants on the interionic distance, a, at −100°C.

Fig. 14 shows an Arrhenius plot of the rate constants in THF compared to the data for MMA. The counterion dependence is as expected from the ionic radii, assuming contact ion pairs. It is even clearer than for MMA, as can be seen from Fig. 15, showing the plot of log k_p *vs.* the reciprocal interionic distance *a*. The deviations of Na^+ and K^+ from the straight line were explained by peripheral solvation of these cations (cf. 2.2). The lack of this deviation in the polymerization of TBMA can thus be explained by a weaker effect of peripheral solvation on reactivity, due to the higher negative charge on the α-carbon, leading to a stronger attraction of the counterion.

In the contrast to MMA the polymers are more or less heterotactic, independent of the counterion, resembling the values for PMMA prepared *via* Cs^+ ion pairs (cf. Table 6).

TABLE 6: Tacticities of PMMA and PTBMA polymerized in THF using different counterions.

Monomer	Counter-ion	Temp [°C]	mm	mr	rr	r	ρ[a]	NMR method
	Cs^+	-53	.05	.52	.42	.68	.82	^1H
MMA	Na^+	-51	.04	.38	.58	.77	.93	^1H
	Li^+	-45	.01	.22	.77	.88	.96	^1H
	Cs^+	-42	.04	.51	.45	.70	.87	^{13}C
TBMA	Na^+	-48	.06	.65	.29	.63	.70	^{13}C
	Li^+	-40	.12	.49	.39	.63	.96	^1H[b]

[a] persistence ratio $\rho = m \cdot r/2mr$
[b] transesterified to PMMA for determination of tacticity.

The rather constant values for the tacticity of TBMA irrespective of the counterion seem to indicate that that same mechanism of monomer addition (presumably mechanism II) is prevalent; this contrasts to what is observed in MMA polymerization (*cf.* 2.5/2.6).

The investigation of the kinetics of polymerization in non-polar solvents is of special interest, because only in these media isotactic polymers can be prepared. However, it is known that in non-polar systems, like toluene, the polymerization of MMA is very complex, resulting from severe side reactions and a multiplicity of active species.[1,43] Consequently, very broad, multimodal molecular weight distributions were reported.[1,44] Only the use of Grignard initiators leads to narrow distributions with MMA.[45] On the other hand, these initiators are not favourable for non-polar monomers making the preparation of block copolymers problematic.

Because of the lack of side-reactions during its polymerization in THF, TBMA appears to be a more suitable monomer for non-polar solvents than MMA. In fact, it is possible to prepare isotactic PTBMA of rather low polydispersity in toluene using Li^+ initiators.

In toluene, the kinetics and molecular weight distributions depend strongly on the nature of the initiator anion. When using hydrocarbon initiators, e.g. 1,1-diphenylhexyl lithium (DPH-Li) or 3-methyl-1,1-diphenylpentyl lithium (MDPP-Li), i.e. the adducts of *n*- or *sec*-butyllithium and 1,1-diphenylethylene,

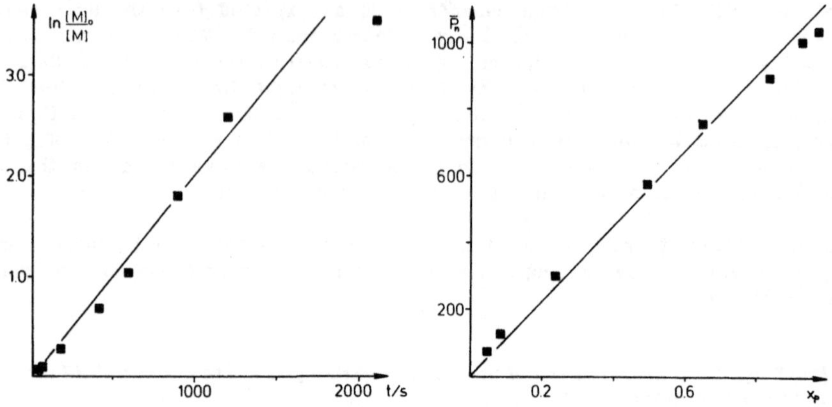

Fig. 16. (a) First-order time-conversion curve and (b) plot of the number-average degree of polymerization vs. conversion in the anionic polymerization of TBMA in toluene at -79°C using DPH-Li as the initiator.[42]

respectively, the kinetics indicate an "ideal" polymerization. Fig. 16 shows the first-order time-conversion plot and the plot of the number-average degree of polymerization P_n *vs.* conversion x_p. Both plots are linear indicating the absence of termination and transfer reactions.

However, as Fig. 17a shows, the molecular weight distributions at lower monomer conversions are bi- or trimodal indicating the coexistence of different active species. At high conversion the peaks merge resulting polymers of rather narrow molecular weight distribution ($M_w/M_n \geq 1.15$; Fig. 17b). Multimodal molecular weight distributions do not appear when the "seeding" technique is used, i.e. addition of few monomer to produce an oligomeric initiator, followed by the addition of a new monomer batch.

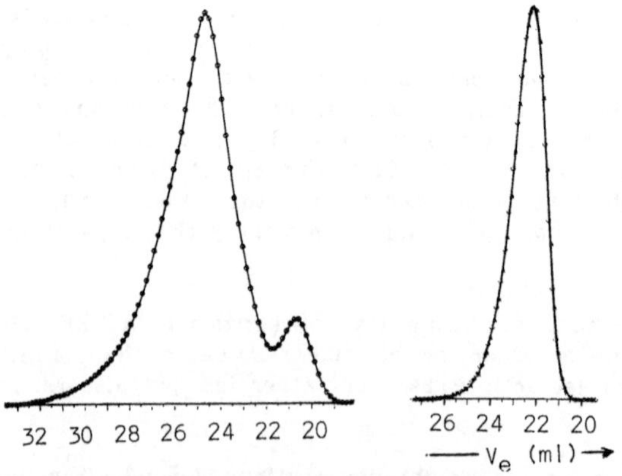

Fig. 17. GPC eluograms of PTBMA samples prepared in toluene using MDPP-Li as the initiator; (a) low, (b) high monomer conversion.[42]

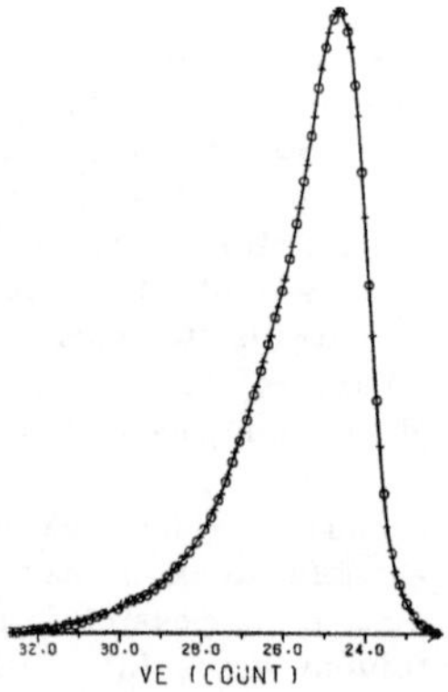

Fig. 18. GPC eluogram of PTBMA prepared in toluene using EIB-Li as the initiator.[42]

When using metalated esters, like ethyl α-lithio isobutyrate (EIB-Li) as the initiator, the molecular weight distributions are unimodal but broader ($M_w/M_n \geq 1.4$; cf. Fig. 18), and the kinetics are non-ideal. Fig. 19 shows that both the first-order time-conversion plot and the plot of P_n *vs.* x_p are non-linear. The curvature is typical for an increase in the number of active centres, i.e. slow initiation. It is known that lithiated esters associate in THF solution; the order of association being dependent on the degree of oligomerization.[12] Thus it appears reasonable to assume that the increase of active centres is due to a decrease of the number of associated centres during polymerization. Slow initiation leads to a tailing in the molecular weight distribution increasing the polydispersity. The rate constants for initiation k_i and for propagation k_p can be calculated separately. The k_p values are identical to those measured using MDPP-Li as the initiator.

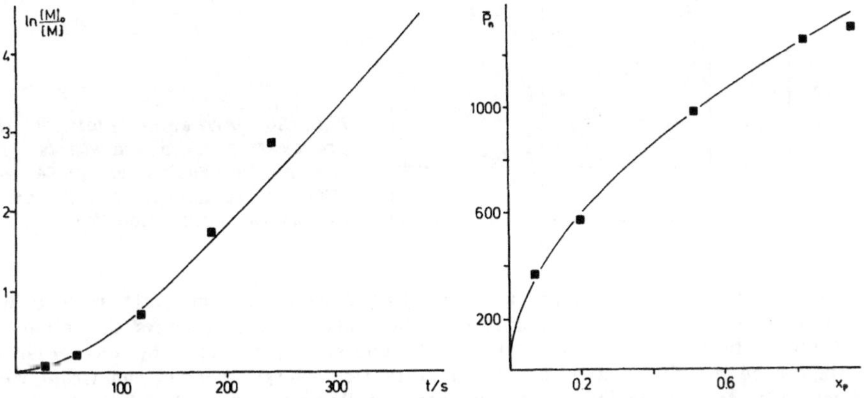

Fig. 19. (a) First-order time-conversion curve and (b) plot of the number-average degree of polymerization vs. conversion in the anionic polymerization of TBMA in toluene at -30°C using EIB-Li as the initiator.[42]

TABLE 7: Rate constants and activation parameters for the anionic polymerization of MMA and TBMA in THF and toluene.

		THF Cs⁺	THF Na⁺	THF Li⁺	Toluene Li⁺
	E_a /kJ/mol	19.5	18.3	24.0	(23)
MMA	log A	7.3	7.0	7.4	(5)
	k_p^{233}	860	800	100	(0.7)
	Ref.	48	31,48	7	44
	E_a /kJ/mol	23.0	31.2	32.8	29.8
TBMA	log A	7.5	8.7	8.2	8.2
	k_p^{233}	180	56	6	45
	Ref.	34	34	42	42

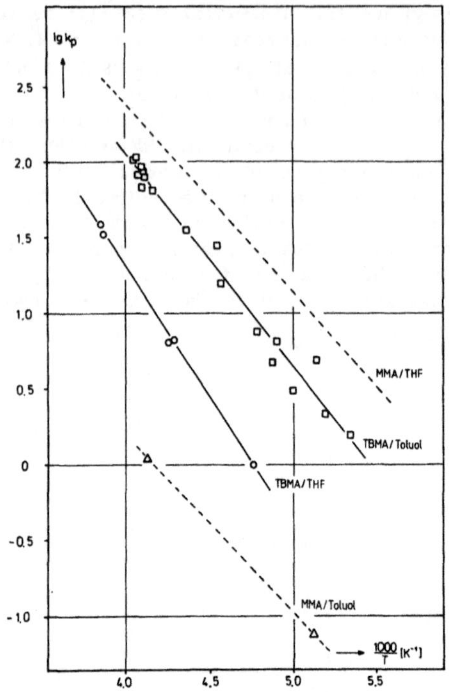

Fig. 20. Arrhenius plot of the propagation rate constants for the polymerization of TBMA and MMA in toluene and THF using Li^+ as the counterion.[42,44]

Fig. 20 shows the Arrhenius plot for the polymerization of TBMA in THF and toluene using Li^+ as the counterion. The activation parameters are given in Table 7. In contrast to MMA, the rate constants are higher by one order of magnitude in toluene due to a lower activation energy. This is consistent with active centres being either non-solvated or "intramolecularly" solvated contact ion pairs:

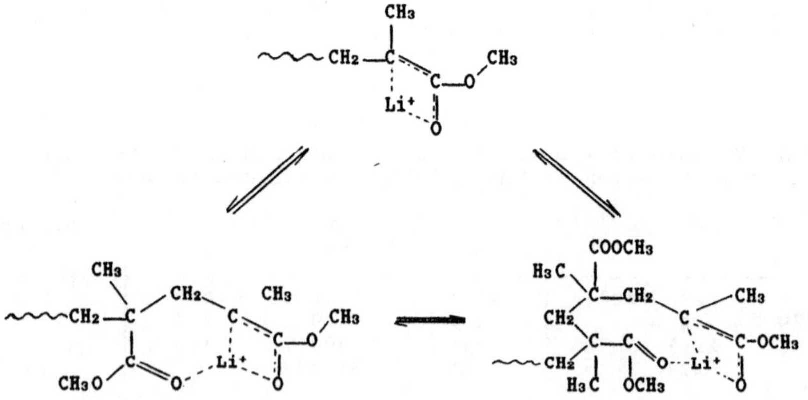

Both species facilitate the coordination of the monomer ester group with the counterion prior to the attack of the vinyl group (mechanism I, *cf.* 2.5/2.6). As a consequence, 100% isotactic polymers are prepared as is shown in Fig. 21.

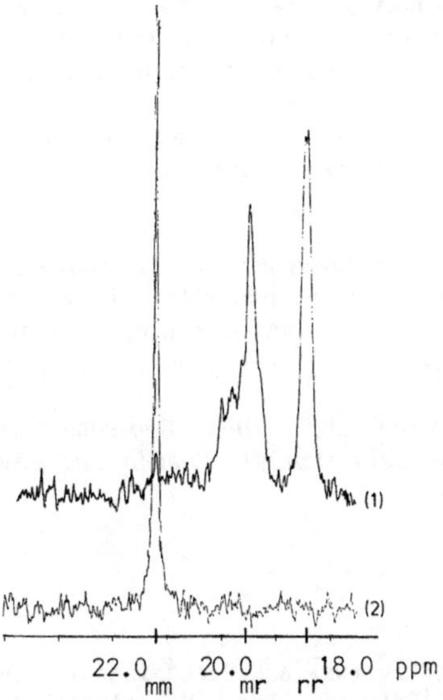

Fig. 21. ^{13}C NMR spectra for PTBMA prepared in THF(1) and toluene (2) using Li^+ as the counterion.[42]

PMMA prepared using the same conditions is only *ca.* 80–90% isotactic depending on monomer concentration and conversion.[28]

The lower rate constants and stereospecificity for MMA in toluene seem to be due to the coexistence of different active species, e.g. associates, which in part add monomer in a *meso* fashion. For TBMA there is no kinetic evidence for polymeric associates. The rate constants do not depend on the concentration of active centres. Lithium *tert.*-butoxide which is known to break associates by complexation with the ion pairs,[46,47] has no influence on kinetics, molecular weight distribution, or tacticity.

4. CONCLUSIONS

Under appropriate conditions (i.e. polar solvents, low temperatures, large counterions) the anionic polymerization of methacrylic esters proceeds in an ideal way leading to polymers with narrow molecular weight distribution. From the linearity and counterion dependence of the Arrhenius plots, it is assumed that in polar solvents the predominant active species is a peripherally solvated contact ion pair. Associated, non-solvated and "intramolecularly" solvated ion pairs may also be present depending on counterion, temperature, and solvent.

Addition of monomer to a peripherally solvated contact ion pair proceeds either by direct attack of the carbanion to the monomer vinyl group or may be preceded by a coordination of the monomer carbonyl group to the counterion and desolvation of a solvent molecule. The two modes of monomer addition are reflected by kinetics and polymer microstructure.

The influence of the structure of the ester group is mainly given by electronic effects. Electron-donating substituents decrease the polarization of the monomer vinyl group and thus its reactivity. In homopolymerizations this effect is more or less cancelled by an increase in negative charge on the carbanion, *tert.*-butyl methacrylate being the major exception. Copolymerization of methacrylates in most cases can be described by using the "terminal" model. Major deviations occur when Li^+ is used as the counterion. The selectivity of a given active centre (i.e. polymeric ion pair) increases with the ionic radius of the counterion.

In non-polar solvents, e.g. toluene, the polymerization of *tert.*-butyl methacrylate (in the contrast to MMA) leads to 100% isotactic polymers with rather narrow molecular weight distribution. Non-solvated contact ion pairs seem to be the major propagating species here.

Acknowledgement: This work is supported by the *Deutsche Forschungsgemeinschaft* within the *Sonderforschungsbereich 41 "Chemie und Physik der Makromoleküle" Mainz/Darmstadt.*

REFERENCES

1) For reviews see: (a) Wiles, D.M., in: "Structure and Mechanisms in Vinyl Polymerization", T. Tsuruta and K. F. O'Driscoll, Eds. (Marcel Dekker, New York 1969), p. 233; (b) Szwarc, M., in: "Ions and Ion Pairs in Organic Reactions, Vol. 2", M. Szwarc, Ed. (Wiley-Interscience, New York 1974), p. 401
2) Löhr, G.; Schmitt, B. J.; Schulz, G. V. *Z. Physik. Chem. (Frankfurt am Main)* **78**, 177 (1972)
3) Warzelhan, V.; Löhr, G.; Höcker, H.; Schulz, G. V. *Makromol. Chem.* **179**, 2211 (1978)
4) Müller, A. H. E. *ACS Symposium. Ser.* **166**, 441 (1981)
5) Lochmann, L., Lim, D. J. Polym. Sci., Polym. Chem. Ed. 12, 2295 (1974)
6) Johann, C., Müller, A. H. E. *Makromol. Chem., Rapid Commun.* **2**, 687 (1981)
7) Jeuck, H.; Müller, A. H. E. *Makromol. Chem., Rapid Commun.* **3**, 121 (1982)
8) Tsvetanov, Ch. B.; Müller, A. H. E.; Schulz, G. V. *Macromolecules* **18**, 863 (1985)
9) Warzelhan, V.; Schulz, G. V. *Makromol. Chem.* **177**, 2185 (1976)
10) Warzelhan, V.; Höcker, H.; Schulz, G. V. *Makromol. Chem.* **181**, 149 (1980)
11) Müller, A. H. E.; Lochmann, L. *Makromol. Chem.* **187**, 1473 (1986)
12) Halaska, V.; Lochmann, L. *Collect. Czech. Chem. Commun.* **38**, 1780 (1973)
13) Fowells, W.; Schuerch. C.; Bovey, F. A.; Hood, F. P. J. Amer. Chem. Soc. **89**, 1369 (1967)
14) Lochmann, L.; Lim, D. *J. Organomet. Chem.* **50**, 9 (1973)
15) Lochmann, L., Trekoval, J. *Makromol. Chem.* **183**, 1361 (1982)
16) Tsvetanov, Ch. B.; Dotscheva, D. T.; Panayotov, I. M. *Makromol. Chem.* **183**, 517 (1982)
17) Vancea, L.; Bywater, S. *Macromolecules* **14**, 1776 (1981)
18) Fisher, M.; Szwarc, M. *Macromolecules* **3**, 23 (1970)
19) Tardi, M.; Sigwalt, P. *Eur. Polym. J.* **8**, 151 (1972)
20) Honnoré, D.; Favier, J. C.; Sigwalt, P.; Fontanille, M. *Eur. Polym. J.* **10**, 425 (1974)
21) Jenkins, W. L.; Tien, C. F.; Hogen-Esch, T. E. *Pure Appl. Chem.* **51**, 139 (1979)
22) Lochmann, L., Trekoval, J.; Müller, A. H. E. *Makromol. Chem.* **185**, 1819 (1984)
23) Kraft, R.; Müller, A. H. E.; Höcker, H.; Schulz, G. V. *Makromol. Chem. Rapid Commun.* **1**, 363 (1980)

24) Gerner, F. J.; Ph. D. Thesis, Mainz 1982
25) Kilz, P.; M. Sc. Thesis, Mainz 1982
26) Reinmöller, M.; Fox, T. G. *Polym. Prepr. (ACS, Div. Polym. Chem.)* **7(2)**, 999 (1966)
27) Müller, M. A.; Stickler, M. *Makromol. Chem., Rapid Comun.* **7**, 575 (1986)
28) Bovey, F. A. "High Resolution NMR of Macromolecules" (Academic Press, New York 1980)
29) Johann, C.; M. Sc. Thesis, Mainz 1981
30) Müller, A. H. E.; Höcker, H.; Schulz, G. V. *Macromolecules* **10**, 1086 (1977)
31) Warzelhan, V.; Höcker, H.; Schulz, G. V. *Makromol. Chem.* **181**, 149 (1980)
32) Pascault, J. P.; Kawak, J.; Golé, J.; Pham, Q. T. *Eur. Polym. J.* **10**, 1107 (1974)
33) Taft, R. W., in: "Rates and Equilibria of Organic Reactions", J.E. Leffler, E. Grunewald, Eds. (Wiley, New York 1963)
34) Jeuck, H., Ph. D. Thesis, Mainz 1985
35) Müller, A. H. E. *Makromol. Chem.* **182**, 2863 (1981)
36) Hammett, L. P. "Physical Organic Chemistry" (McGraw-Hill 1970)
37) Smid, J.; Szwarc, M. *J. Polym. Sci.* **61**, 31 (1962)
38) Fineman, M.; Ross, S. D. *J. Polym. Sci.* **5**, 259 (1950)
39) Kelen, T.; Tüdős, F. *J. Macromol. Sci.* **A9**, 1 (1975)
40) Jeuck, H.; Müller, A. H. E.; in preparation
41) Yuki, H.; Okamoto, Y.; Ohta, K.; Hatada, K.; *J. Polym. Sci., Polym. Chem. Ed.* **13**, 1161 (1975)
42) Kilz, P.; Müller, A. H. E.; in preparation
43) Hatada, K.; Kitayama, T; Fumikawa, K; Ohta, K; Yuki, H. *ACS Symposium Ser.* **166**, 21 (1981)
44) Piejko, K. E.; Ph. D. Thesis, Bayreuth 1982
45) Hatada, K.; Ute, K; Tanaka, K; Kitayama, T.; Okamoto, Y. *Polymer J.* **17**, 977 (1985); *cf.* this book
46) Lochmann, L., Rodova, M.; Trekoval, J. *J. Polym. Sci., Polym. Chem. Ed.* **12**, 2091 (1974)
47) Lochmann, L., Kolarik, J.; Doskocilova, D.; Vozka, S.; Trekoval, J. *J. Polym. Sci., Polym. Chem. Ed.* **17**, 1727 (1979)
48) Kraft, R.; Müller, A. H. E.; Warzelhan, V., Höcker, H.; Schulz, G. V. *Macromolecules* **11**, 1093 (1978)

STEREOELECTIVE ANIONIC POLYMERIZATION OF CHIRAL VINYL MONOMERS VIA INTERCONVERTING ION PAIR EPIMERS

MARK A. BUESE* and THIEO E. HOGEN-ESCH
Department of Chemistry, University of Florida, Gainesville, Florida 32611; *Present Address: General Electric Corporate Research and Development Center, Schenectady, New York 12345

ABSTRACT

The consequences of the possible one and three center controlled stereoelctive polymerizations of chiral monomers are considered. In this treatment, all of the possible diastereomeric ion pair epimers are assumed to be formed initially. Stereoelection is assumed to be absolute and controlled by the chiral centers at the end of the growing chain. The ion pair is considered the last chiral center. The resulting diastereomeric dyads, triads, and tetrads are given for each mechanism. Many of these mechanisms predict that by controlling the optical purity of the monomer feed one may control the stereoregularity of the resulting polymer.

INTRODUCTION

Recently we reported the dimerization of vinyl phenyl sulfoxide initiated by 1-lithio-1-(phenylsulfinyl)-ethane (Scheme 1) [1]. The dimerization occurs with the formation of an s carbon alpha to an R sulfur [2], and a twelve fold stereoelection of S-monomer over R-monomer by 1-lithio-1R-(phenylsulfinyl)ethane. The Li ion of the resulting dimeric carbanions is intramolecularly chelated by the 3-phenylsulfinyl group resulting in the formation of two pairs of ion pair epimers: 1-pro-r-lithio-1S,3R-di(phenylsulfinyl)-3s-butane / 1-pro-s-lithio-1S,3R-di(phenylsulfinyl)-3s-butane and 1-pro-r-lithio-1R,3R di(phenylsulfinyl)-3s-butane / 1-pro-s-lithio-1S,3R-di(phenylsulfinyl)-3s-butane [3].

The interconversion of the epimer pairs appears to be slow relative to dimerization at -78°C. As a result the methylation stereochemistry is dependent on the ratio of monomer to initiator and the time and temperature of storage of the reaction mixture [3]. The methylation appears to occur stereospecifically on the cation side (syn) and thus the stereoselective formation of an r or s carbon upon alkylation reflects the composition of epimers when the electrophile is introduced [3]. For small ratios of monomer to initiator, no interconversion of the epimers occurs after all monomer is consumed. With larger ratios of monomer to initiator the composition of dimers upon methylation suggests that the interconversion of epimers is catalyzed by the monomer (presumably via strong cation coordination) [4]. The distribution of trimers observed for various compositions suggests that monomer addition is stereoelective; however all epimers appear to undergo trimerization. Hence the apparent stereoselectivity of trimerization is also dependant upon the ratio of monomer to initiator [4].

The polymer which results from a racemic mixture of monomer appears to be "atactic". Conversely, the polymer which results from highly optically enriched monomer appears to be very stereoregular [4]. From the oligomerization results one would predict that polymerization occurs via a series of stereospecific and stereoelective vinyl additions. An investigation of various possible processes of this type suggests that the polymerization results for vinyl phenyl sulfoxide would be expected for many cases where the stereoelection is influenced by the chirality of the ion pair [4b].

Recent Advances in Anionic Polymerization
T.E. Hogen-Esch and J. Smid, Editors

Scheme 1

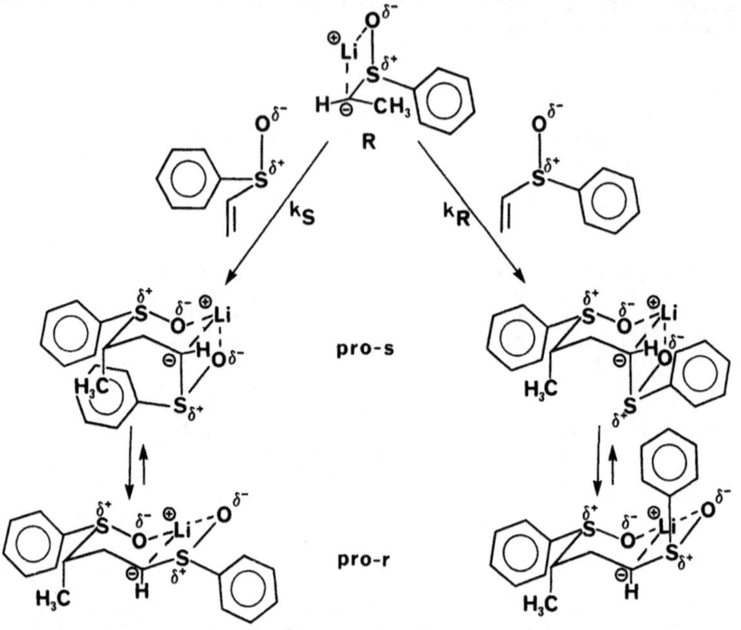

In this paper we shall examine the unique one and three center controlled stereoelective mechanisms where ion pair epimers may interconvert readily. The structural consequences of using racemic and optically pure monomer will be discussed, assuming all dimeric ion pair epimers are formed by the first vinyl addition and the stereoelection of one enantiomer of monomer over the other enantiomer is absolute. Therefore the identity and number of possible stereoisomeric dyads, triads, and tetrads in the polymer are determined by the stereoelection. However, the relative population of the various stereoisomeric sequences depends on the relative population and reactivities of the ion pair epimers. Relative amounts of the various stereoisomeric sequences will not be discussed. This treatment will assume that all epimers are of similar population and reactivity. Therefore the results indicate the maximum number of sequences which could exist throughout the polymer chain. This treatment is general for all polymerizations of monomers which contain a single pendant chiral center, propagate via a chiral active site (ion pair), and occur via a series of stereospecific and stereoelective vinyl additions.

RESULTS AND DISCUSSION

In the polymerization of a chiral vinyl monomer one must consider the six diastereomerically different dyads, the 16 different triads, and the 72 different tetrads which are possible. A modified Fischer projection is used to illustrate the various stereoisomers. As seen below (scheme 2), with poly (vinyl phenyl sulfoxide) as an example, the polymer backbone is in an all trans configuration viewed from the top (view b) with the pendant group (phenylsulfinyl) extending out from the plane of the paper toward the viewer. At the chiral center of the pendant (phenylsulfinyl) group, the lowest priority group (the lone pair of electrons) drops away from the viewer into the plane of the paper and is not drawn in. The other two substituents (phenyl and oxygen) at the pendant chiral center (sulfinyl) reside in a plane parallel to the backbone and are projected toward the viewer. In the modified Fischer projection (view c) only the higher priority of these two groups (oxygen) is drawn. The Fischer projections for the dyads, triads, and tetrads are shown in figures 1-3. The Cahn-Ingold-Prelog notation under the figures (1 and 2) indicates the configuration of the structure. For this notation the pendant group is assigned as the highest priority at the chiral carbon of the polymer backbone. All sequences may be generated in two directions. The first series of notations listed is for the sequence generated with polymer chain to the left of the sequence and the active site (ion pair) immediately to the right, assuming the polymer to be higher in priority than the active site. The second series of notations is for the sequence generated with the polymer to the right (for the projection rotated 180° in the plane of the paper). All of these structures have an enantiomer which is not shown.

Scheme 2

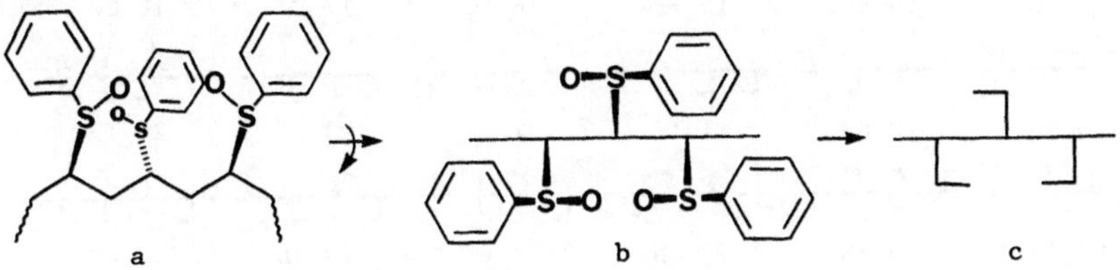

a b c

Figure 1 Diastereomeric Dyads

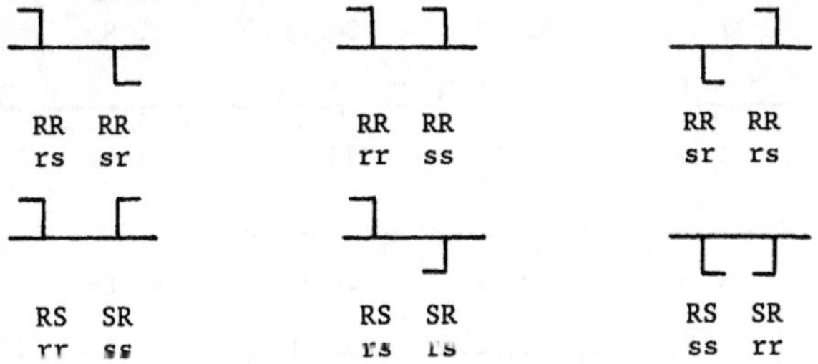

Figure 2 Diastereomeric Triads

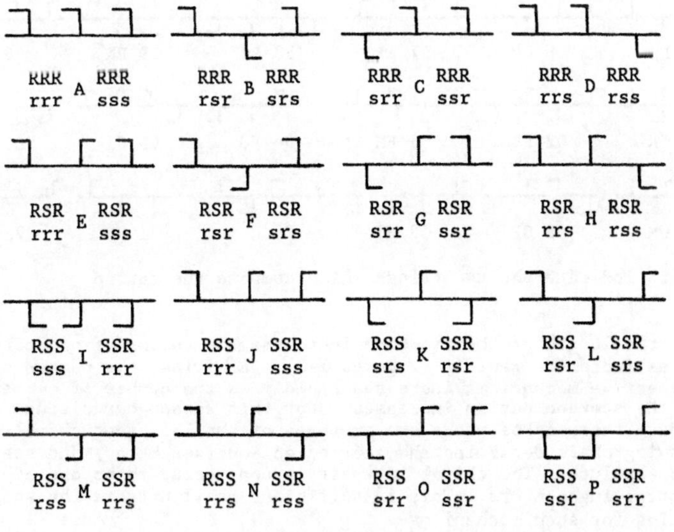

Figure 3 Diastereomeric Tetrads*

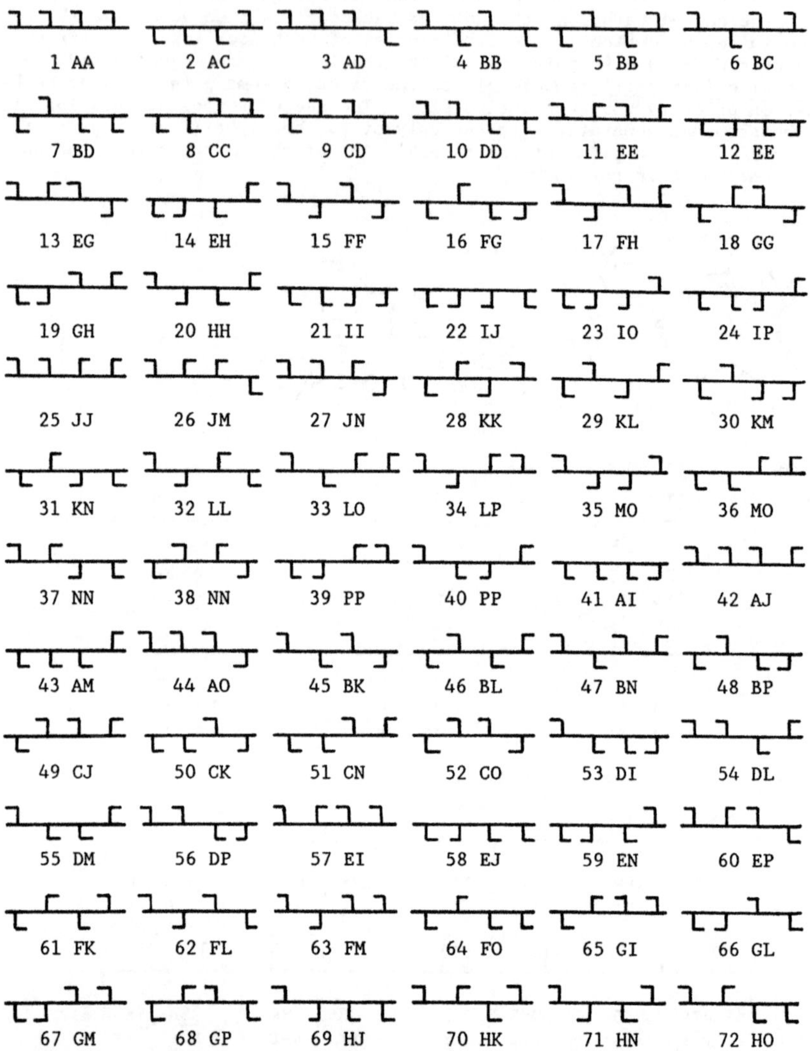

* Letters indicate the two triads which compose the tetrad.

As the length of the sequence increases the number of diastereomeric sequences increases rapidly (Figures 1-3). Likewise, the number of possible stereoelective mechanisms increases rapidly as the number of centers determining the stereoelection increases. For this reason our discussion will be limited to those cases where one or three of the last four chiral centers of the growing chain determine the stereoelection (see below) and stereoelection is absolute. The chiral ion pair is considered to be one of the possible controlling centers. This is sufficient to illustrate the scope of possibilities for such mechanisms.

The different stereoelective polymerizations will be viewed in the following manner (for example see Mechanism 1). The four pairs of epimers are placed at the four corners of a square. Their notation corresponds directly with the dyad of Figure 1 which would be fixed in the polymer chain upon vinyl addition. An arrow is drawn from each chain end to the epimer pair which may result upon stereoelective vinyl addition. The letters adjacent to the arrows are for the triads (Figure 2) which would be fixed in the chain upon subsequent vinyl addition to the two resulting ion pair epimers. Where no letter is given (see Mechanism 5) the epimer pair is not regenerated once it is consumed by vinyl addition, hence the corresponding triad could exist only at the initial end of the chain and generally would be unobservable in an analysis of high polymer. The mechanism is defined immediately before the stereoelection scheme. The last four centers are listed as W,x,Y, and z, where W and Y refer to the penultimate and ultimate pendant groups respectively, and x and z refer to the penultimate and prochiral (ion pair) carbons respectively. The chiral center or centers which control the stereoelection are substituted with the appropriate Cahn-Ingold-Prelog notation, and the enantiomer of monomer which is elected by that chain end is given following the four centers. The tetrads which may form by the mechanism are given below the stereoelective scheme as numbers (Figure 3). The mechanisms are grouped by the similarity of the polymers which would be formed from racemic and optically pure monomer.

ONE CENTER CONTROLLED MECHANISMS

To define stereoelection controlled by one center, only one configuration of the controlling center and the configuration of monomer it elects must be given. There are eight such mechanisms corresponding to a preference for either R or S monomer by each of the four possible controlling centers. In all cases it is assumed that all four pairs of epimers are formed initially with racemic monomer, and that both of the top two pairs of epimers are formed initially with optically pure monomer.

Mechanism 1 $\begin{matrix}RY\\xz\end{matrix}\rightarrow R$

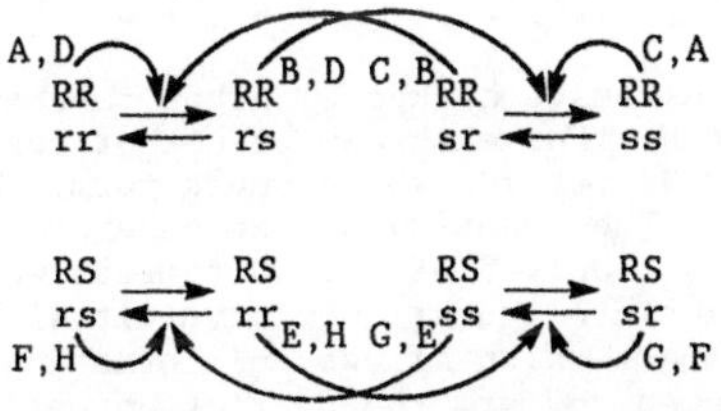

Tetrads:1,2,3,4,5,6,7,8,9,10,11,12,13,14,15,16,17,18,19,20

As seen in Mechanism 1, with stereoelection for like configuration by the penultimate pendant group all six dyads would be observed. However, only eight triads and twenty tetrads would form. The polymer resulting from racemic monomer would actually be a mixture of three stereoisomerically different polymers, each containing three dyads, four triads, and ten tetrads. Two of the polymers would be enantiomeric and would have identical molecular weights. The polymer that is diastereomeric to these polymers could have a different molecular weight and a bimodal molecular weight distribution for the mixture may be observed. The use of optically pure monomer would result in the formation of only one of the three polymers. Its enantiomer and diastereomer could not be formed. Therefore only three dyads, four triads, and ten tetrads would be observed.

Mechanism 2 RY →S
xz

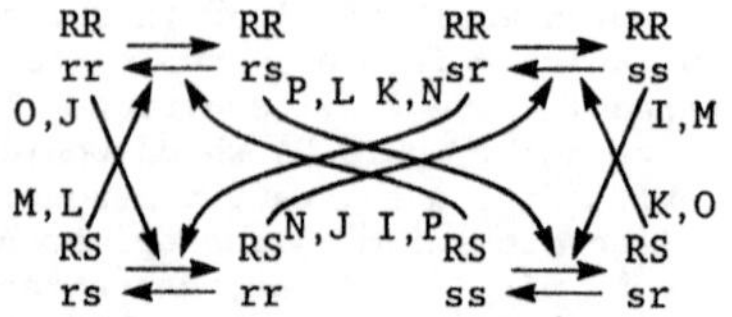

Tetrads:21,22,23,24,25,26,27,28,29,30,31,32,33,34,45,36,37,38,39,40

In contrast to mechanism 1, election of the opposite configuration by the penultimate pendant group (Mechanism 2) results in a single polymer with racemic monomer. In this polymer, two pendant groups of one configuration are followed by two pendant groups of the opposite chirality. The analysis of the dyads may be indistinguishable from the previous mechanism. The eight triads and twenty tetrads formed would be completely different. Because of absolute stereoelection no polymer would result upon the mixing of initiator with optically pure monomer.

Mechanism 3 WY →R
rz

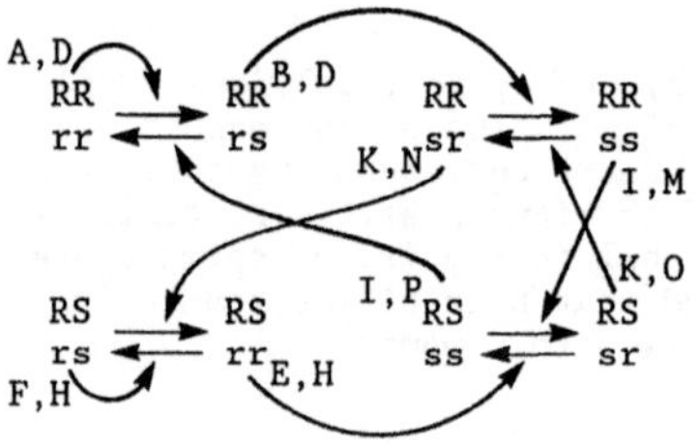

Tetrads:1,3,7,10,14,15,17,20,21, 23,24,28,30,31,35,36,41,45,47, 48,53,55,56,57,59,60,61,70,71,72

Mechanism 4 WY →S
rz

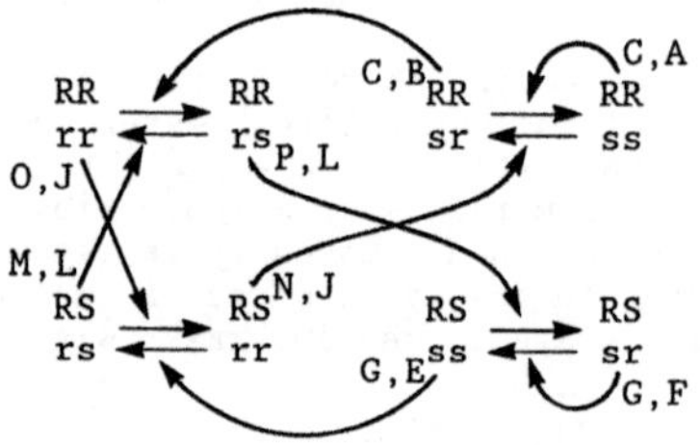

Tetrads:1,2,6,8,13,15,16,18,25, 26,27,32,33,34,35,36,42,46,47, 48,49,51,52,58,59,60,62,66,67,68

Control by the penultimate backbone carbon (Mechanisms 3 and 4) would again yield all six dyads. For election of either configuration of monomer there are 12 triads and 30 tetrads which would form. Eight triads are common to both mechanisms. The identity of the other four triads would distinguish Mechanism 3 from 4. Only 8 of the 30 tetrads would be common to both mechanisms. Only isotactic oligomers would result with optically pure monomer, as the pair of epimers which allows the formation of an isotactic sequence would be consumed to form the pair of epimers which is unreactive toward further vinyl addition.

Mechanism 5 WR →R
xz

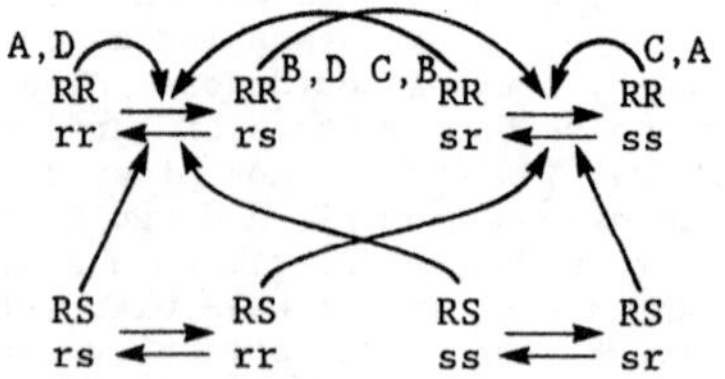

Tetrads:1,2,3,4,5,6,7,8,9,10

Only the lack of optical activity would distinguish the polymer resulting from racemic monomer in Mechanism 5 from the polymer produced in Mechanism 1 with optically pure monomer. Due to the stereoelection of vinyl addition the lower two epimer pairs would be immediately consumed and could not be regenerated. Hence those dyads and the corresponding triads could only exist as the initial sequence of the chain. The polymerization of optically pure monomer would result in the formation of a polymer identical to that resulting from optically active monomer in Mechanism 1.

Mechanism 6 $\underset{xz}{WR} \rightarrow S$

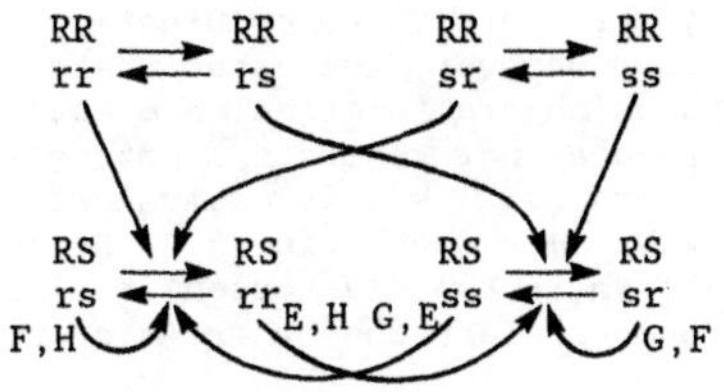

Tetrads:11,12,13,14,15,16,17,18,19,20

Only the three dyads, four triads and ten tetrads with alternating pendant chirality would be formed in the polymer resulting by Mechanism 6. No polymer would form with optically pure monomer.

Mechanism 7 $\underset{xr}{WY} \rightarrow R$

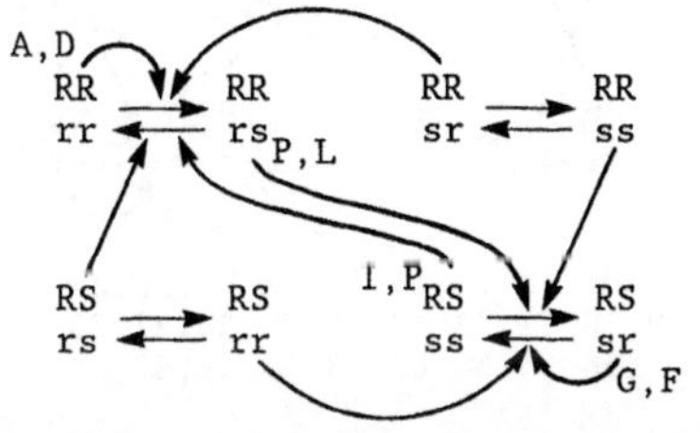

Tetrads:1,3,15,16,24,34,39,40,
41,53,54,56,62,65,66,68

Mechanism 8 $\underset{xr}{WY} \rightarrow S$

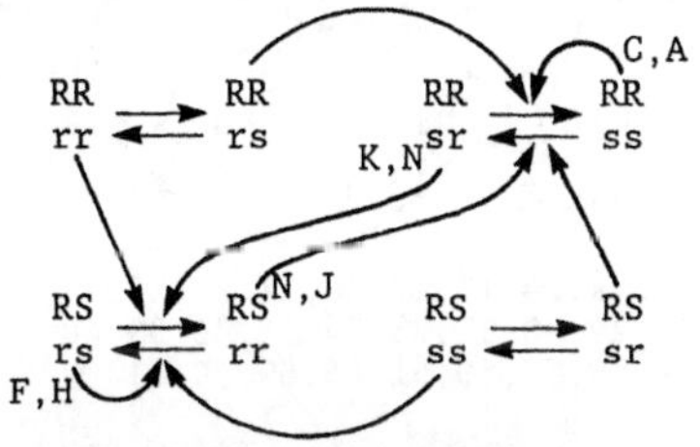

Tetrads:1,2,15,17,27,31,37,38,
42,49,50,51,61,69,70,71

Mechanisms 7 and 8, where the chiral ion pair determines the stereoelection, are the only one center controlled mechanisms which would lead to an atactic polymer from racemic monomer but a stereoregular polymer with optically pure monomer. The two mechanisms of this type differ only in the identity of some of the dyads, triads, and tetrads which result with racemic monomer. In both cases two pairs of epimers are depleted due to the stereoelectivity. Thus only four dyads, seven triads, and sixteen tetrads are formed. With optically pure monomer only one pair of epimers exists after the second vinyl addition. Only one of the epimers would be reactive with the monomer resulting in a purely isotactic polymer.

TWO CENTER CONTROLLED MECHANISMS

To define control by two centers one must define the electivity of the two diastereomeric combinations of centers which control the process. There are no unique two center controlled mechanisms. Any control by a group of two centers is equivalent to control by one of the two centers. For example, the mechanism RxRz → R, RxSz → S is equivalent to WxRz → R (Mechanism 5).

THREE CENTER CONTROLLED MECHANISMS

For a stereoelective mechanism controlled by three centers, one must define the electivity of the four diastereomeric combinations which control the process. There are four ways of grouping three of the last four chiral centers (WxY, Wxz, WYz, and xYz). For each of these groupings there are four diastereomerically different chain ends (ie. RrRz, RrSz, RsRz, and RsSz), and four ways of combining three of these diastereomers. Thus where three of the four diastereomers elect one enantiomer of monomer and the remaining diastereomer elects the other, there are 32 different mechanisms (9-40) which one must consider (4 groupings x 4 combinations x 2 enantiomers). None of these mechanisms is equivalent to a one center controlled mechanism. Additionally there are three combinations of two pairs of diastereomeric chain ends. However, the cases where each pair has two of the three chiral centers which are identical are equivalent to a group of two centers and therefore does not generate a unique mechanism. For example RrRz and RsRz → R, RrSz and RsSz → S is equivalent to RxRz → R, RxSz → S which is equivalent to WxRz → R (Mechanism 5). Hence there are eight unique mechanisms (41-48) where two diastereomers elect one enantiomer of monomer and two diastereomers elect the other (4 groupings x (3-2) combinations x 2 enantiomers).

Mechanism 9

RS/rz → R RS/sz → S RR/rz → S RR/sz → S

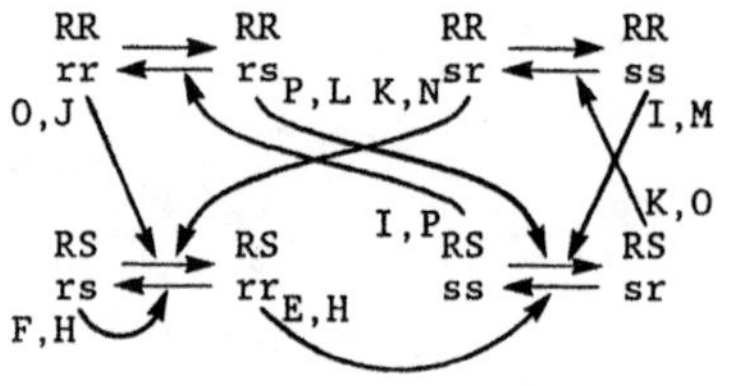

Tetrads:14,15,17,20,21,22,23,24,28,29,30,31,33,34,35,36,39,40,57,58,59,60,61,64,69,70,71,72

Mechanism 10

RS/sz → R RS/rz → S RR/sz → S RR/rz → S

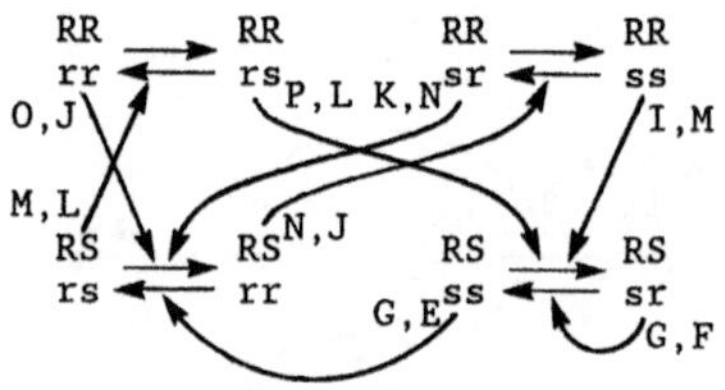

Tetrads:13,15,16,18,22,25,26,27,29,30,31,32,33,34,35,36,37,38,57,58,59,60,62,63,65,66,67,68

Mechanisms 9 and 10 result in all six dyads, 11 triads, and 28 tetrads with racemic monomer. At the triad level they can be distinguished from one another only by the heterotactic triad with alternating pendant chirality. However, at the tetrad level they are very different polymers. No polymer would result from optically pure monomer.

Mechanism 11

RR/rz → R RR/sz → S RS/rz → S RS/sz → S

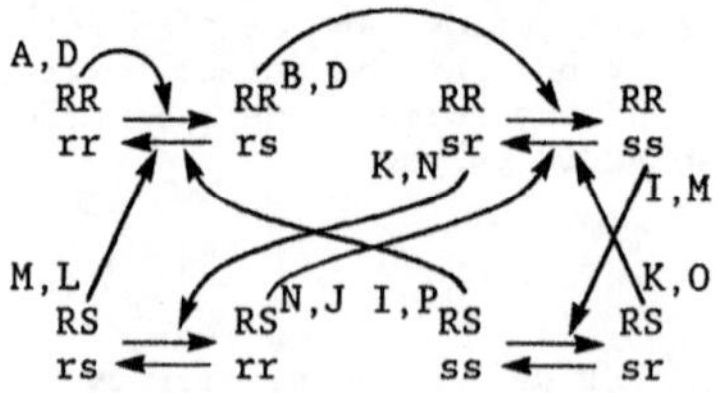

Tetrads:1,3,7,10,21,22,23,24,26,27,28,29,30,31,35,36,37,38,41,43,45,46,47,48,53,54,55,56

Mechanism 12

RR/sz → R RR/rz → S RS/sz → S RS/rz → S

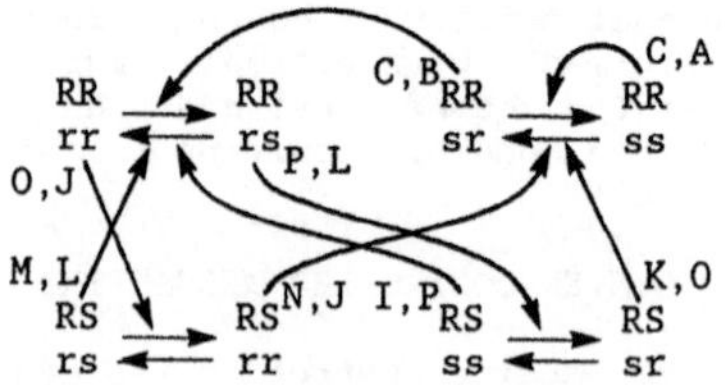

Tetrads:1,2,6,8,22,23,24,25,26,27,29,32,33,34,35,36,39,40,42,44,45,46,47,48,49,50,51,52

As with the Mechanisms 9 and 10, Mechanisms 11 and 12 result in the formation of all six dyads, 11 triads, and 28 tetrads with racemic monomer. At the triad level they can be distinguished from one another only by the heterotactic triad with repeating pendant chirality (C or D). At the tetrad level they are very different polymers. Since they contain tetrads with repeating pendant chirality, they allow some optically pure monomer to be converted to isotactic oligomers before the reactive ion pair is depleted in favor of an unreactive pair of epimers.

Mechanism 13

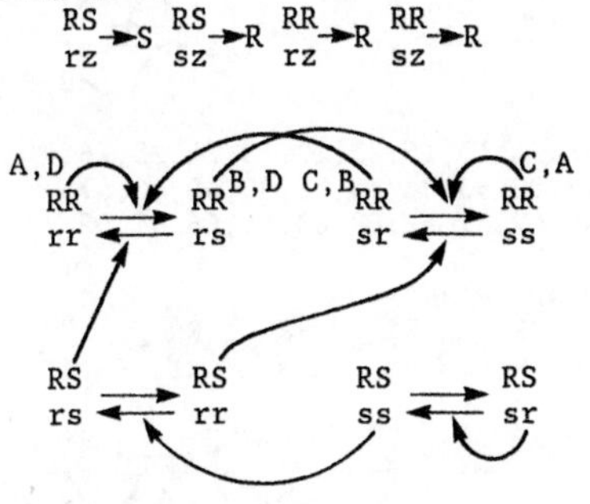

Tetrads:1,2,3,4,5,6,7,8,9,10

Mechanism 14

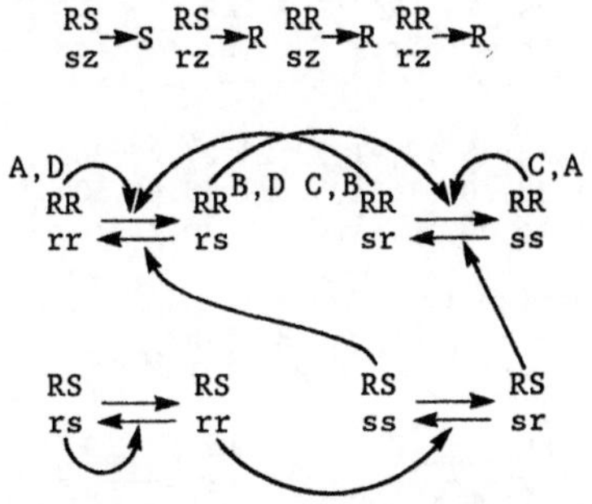

Tetrads:1,2,3,4,5,6,7,8,9,10

Mechanisms 13 and 14 could only be distinguished from each other and from Mechanism 5 by what occurs at the beginning of the chain when racemic monomer is used. A short syndiotactic sequence with alternating pendant chirality may form by these two mechanisms but not by a one center controlled mechanism. The identity of the initial triads formed using racemic monomer would be different for these two mechanisms.

Mechanism 15

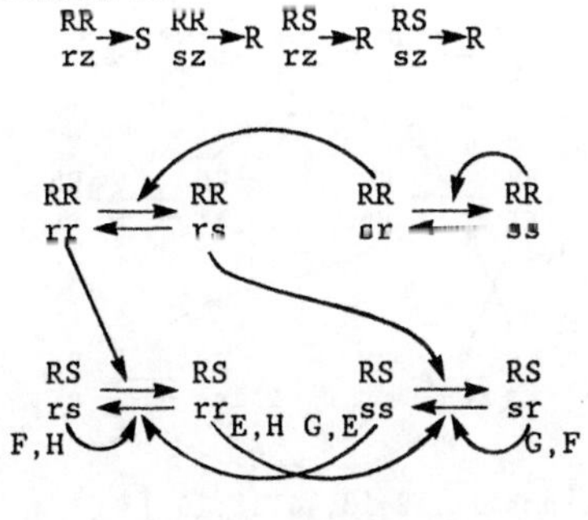

Tetrads:11,12,13,14,15,16,
17,18,19,20

Mechanism 16

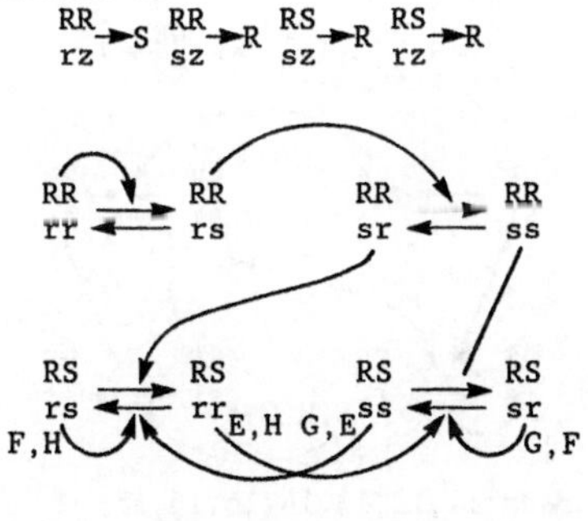

Tetrads:11,12,13,14,15,16,
17,18,19,20

As in the Mechanisms 13 and 14, it may be very difficult to distinguish Mechanism 15 from 16. Again the identity of the initial triads would distinguish these two mechanisms. Unlike Mechanism 6, either of these mechanisms would result in isotactic oligomers with optically pure monomer.

Mechanism 17

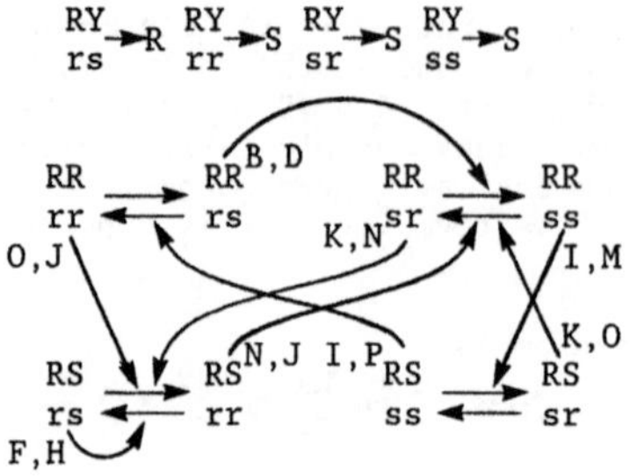

Tetrads:15,17,21,22,23,24,25,26,
27,28,30,31,35,36,37,38,48,50,
51,53,55,56,61,64,69,70,71,72

Mechanism 18

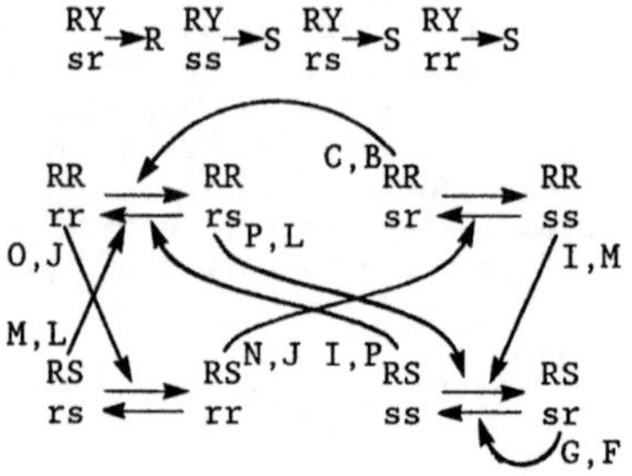

Tetrads:15,16,21,22,23,24,25,26,
27,32,33,34,35,36,39,40,46,47,
48,49,51,52,62,63,65,66,67,68

Mechanism 19

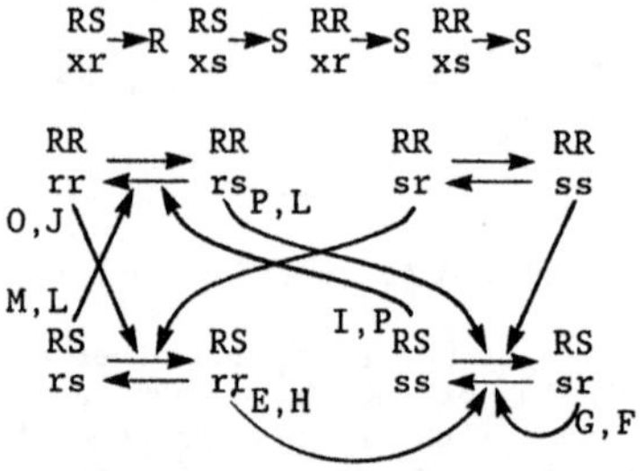

Tetrads:15,16,17,19,22,23,24,
26,32,33,34,35,36,39,40,
57,58,60,62,65,66,68,69

Mechanism 20

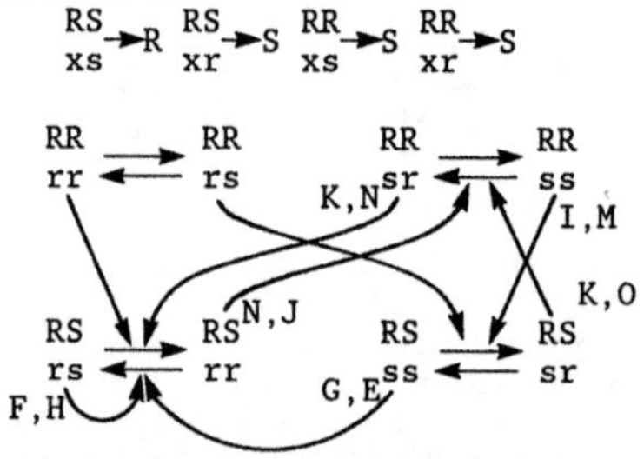

Tetrads:15,16,17,19,22,23,26,
27,28,30,31,35,36,37,38,
57,58,59,61,65,69,70,71

Mechanism 21

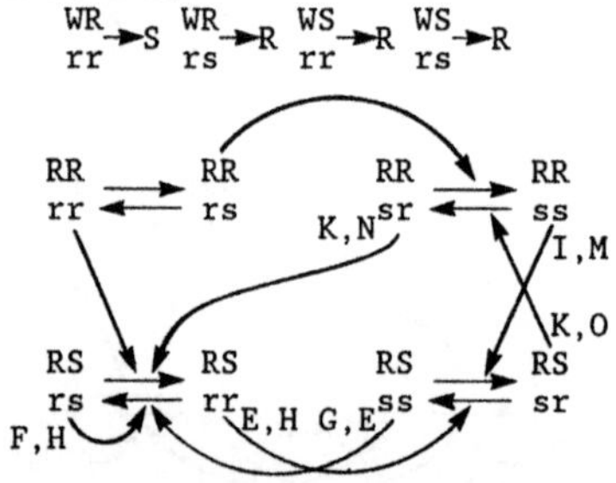

Tetrads:11,12,13,14,15,16,17,
19,20,23,28,30,31,35,36,
57,59,61,65,70,71,72

Mechanism 22

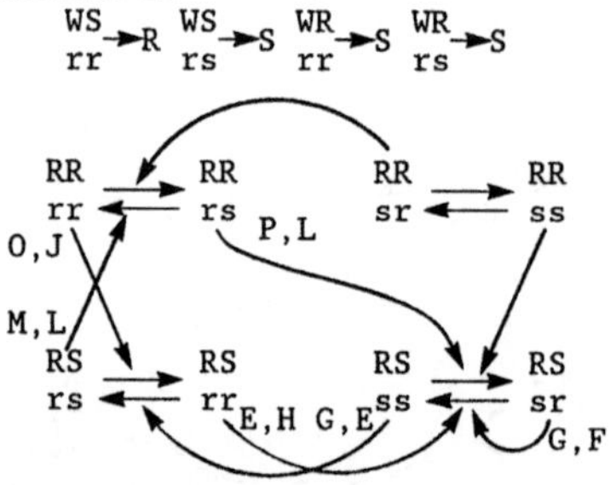

Tetrads:11,12,13,14,15,16,17,
18,19,26,32,33,34,35,36,
58,60,62,66,67,68,69

As seen in Mechanisms 17-22, an influence by the chiral ion pair does not guarantee that a stereoregular polymer would result with optically pure monomer. In these cases no polymer would result with optically pure monomer. For Mechanisms 17 and 18 all six dyads and 11 triads would form with racemic monomer. Only three of the pairs of epimers would participate in the polymerization by the other four mechanisms, resulting in the formation of only five dyads. Mechanisms 19 and 20 would generate ten triads. Mechanisms 21 and 22 would generate only nine triads. Elucidation of dyad structures would be necessary to distinguish between the pairs of mechanisms which generate the same number of triads.

Mechanism 23

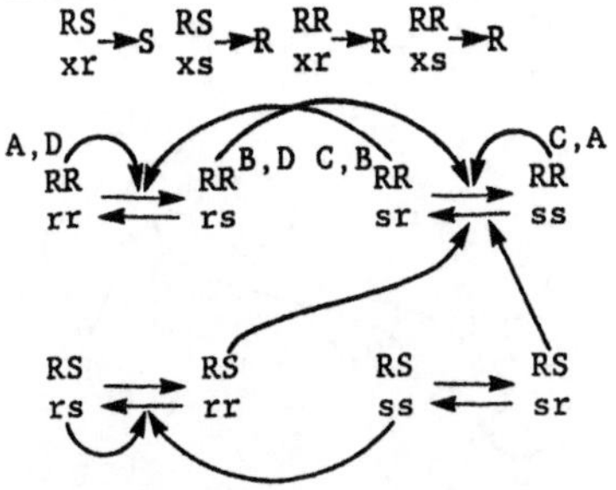

Tetrads:1,2,3,4,5,6,7,8,9,10

Mechanism 24

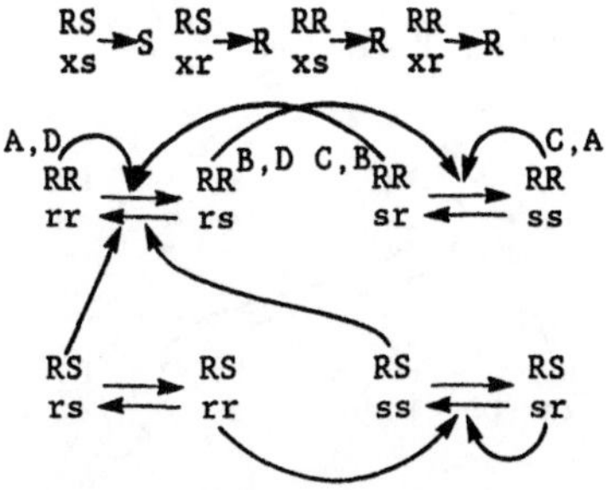

Tetrads:1,2,3,4,5,6,7,8,9,10

Mechanism 25

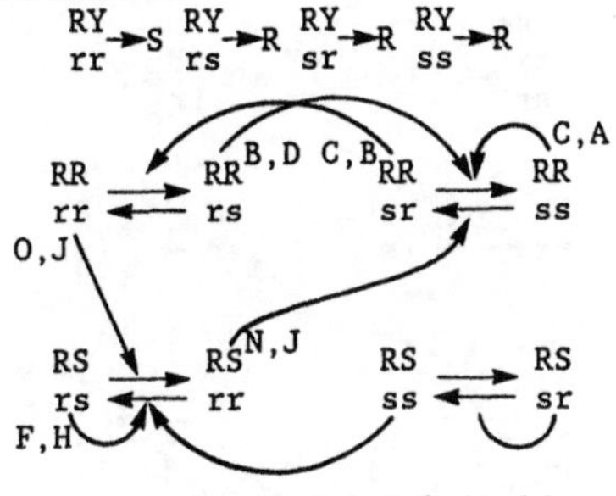

Tetrads:1,2,3,4,5,6,7,8,9,15,17, 25,27,42,47,49,51,52,64,69,71,72

Mechanism 26

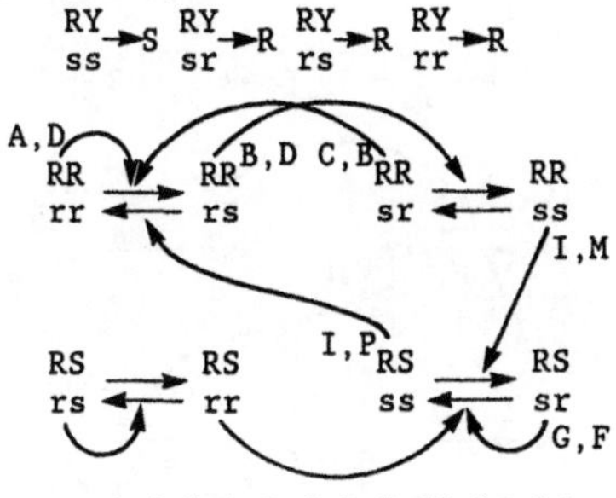

Tetrads:1,2,3,4,5,6,7,9,10,15,16, 21,24,41,48,53,55,56,63,65,67,68

Mechanism 27

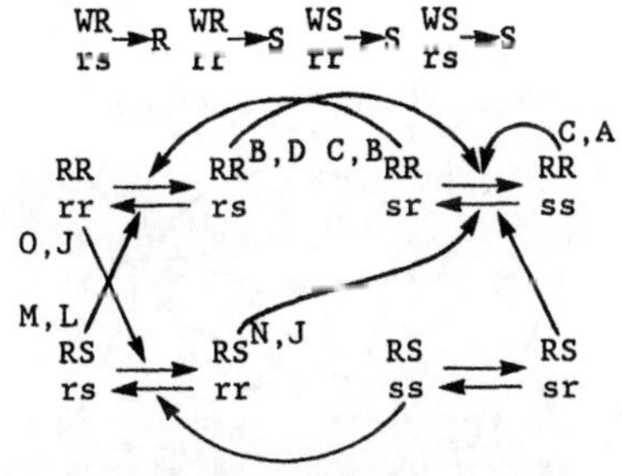

Tetrads:1,2,3,4,5,6,7,8,9,25,26, 27,33,35,36,42,46,47,49,51,52,54

Mechanism 28

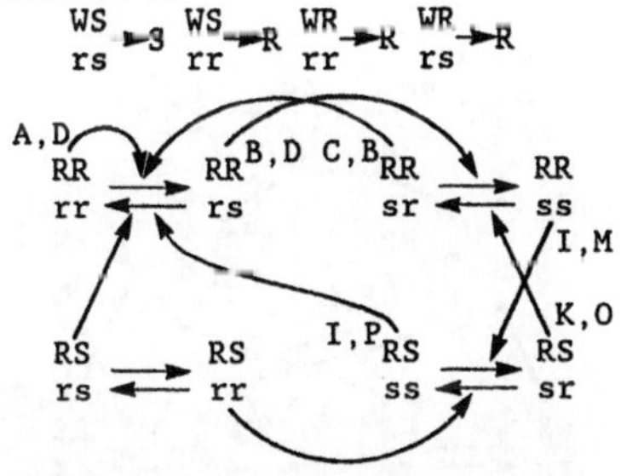

Tetrads:1,2,3,4,5,6,7,9,10,21,23, 24,30,35,36,41,45,48,50,53,55,56

For Mechanisms 23-28, the polymer formed from optically pure monomer would not be stereoregular. With racemic monomer, Mechanisms 23 and 24 would yield only the dyads, triads, and tetrads with like pendant chirality, as in Mechanisms 5,13, and 14. Only the elucidation of the additional triads formed at the beginning of the chain would differ for these five processes. Three pairs of epimers would participate in Mechanisms 25-28. This results in the formation of five diastereomeric dyads and nine diastereomeric triads. Since one of the dyads which would result by Mechanisms 25 and 26 would not be common to both, one may be distinguished from the other at this level. Likewise, Mechanisms 27 and 28 do not share one of the five dyads. The observation of triads with alternating pendant chirality would distinguish Mechanisms 25 and 26 from 27 and 28.

Mechanism 29

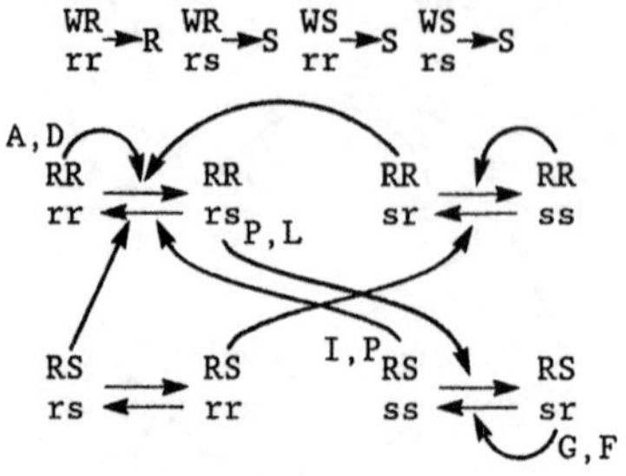

Tetrads:1,3,15,16,24,34,39,40,
41,53,54,56,62,65,66,68

Mechanism 30

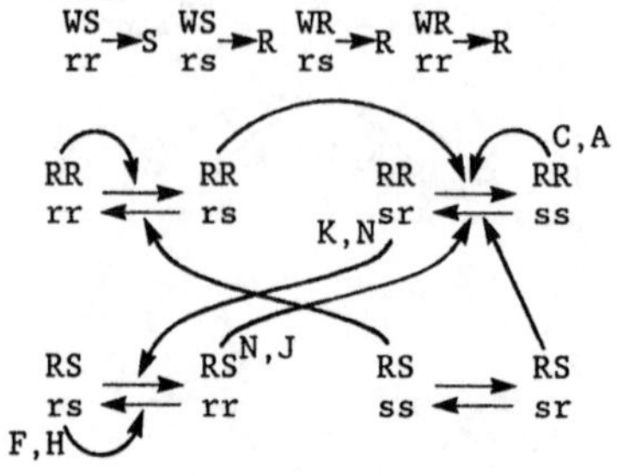

Tetrads:1,2,15,17,27,31,37,38,
42,49,50,51,61,69,70,71

Mechanism 31

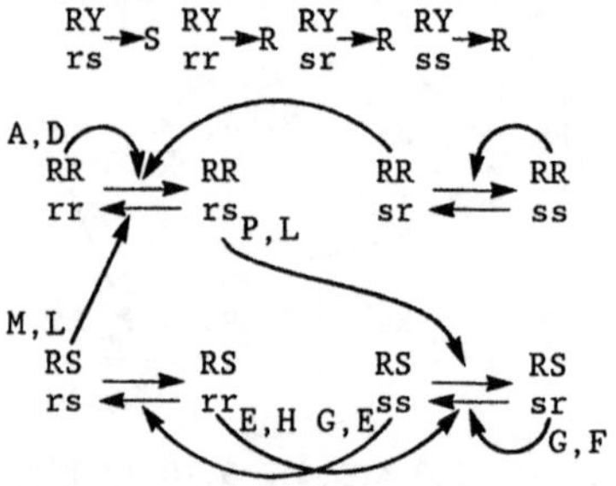

Tetrads:1,3,11,12,13,14,15,16,
17,18,19,32,34,43,54,55,
56,60,62,66,67,68

Mechanism 32

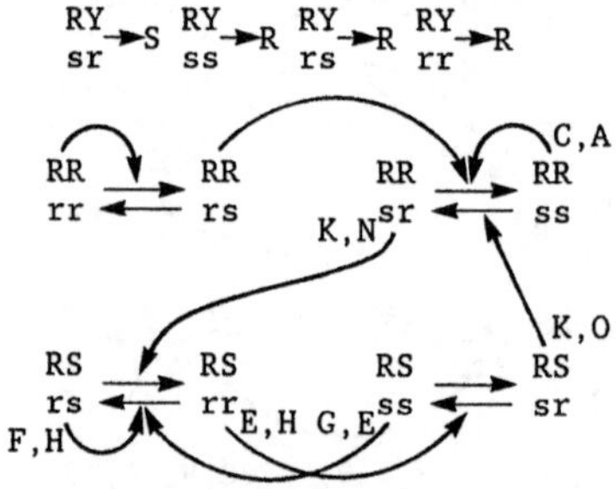

Tetrads:1,2,11,12,13,14,15,16,
17,19,20,28,31,44,50,51,
52,59,61,70,71,72

Mechanism 33

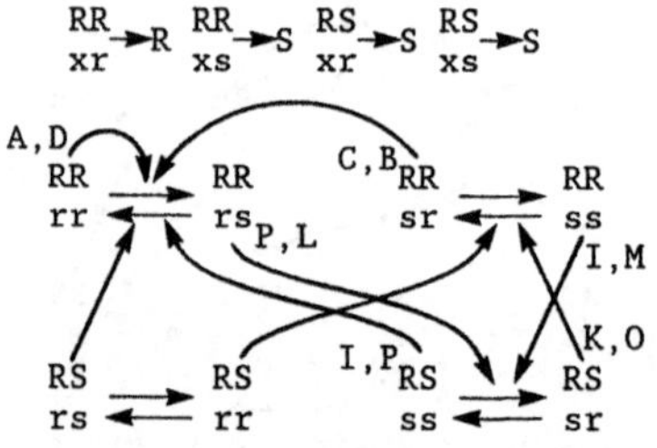

Tetrads:1,2,3,9,21,23,24,29,30,
33,34,35,36,39,40,41,
45,46,48,50,53,54,56

Mechanism 34

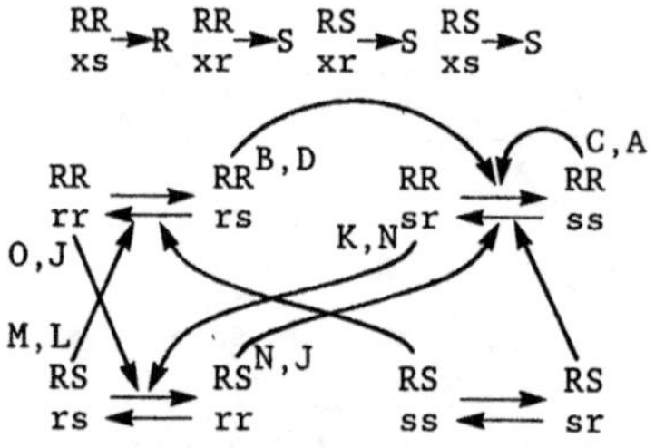

Tetrads:1,2,3,9,25,26,27,29,30,
31,33,35,36,37,38,42,
45,46,47,49,50,51,54

Mechanisms 29-34 would result in purely isotactic polymer when the monomer is optically pure. Mechanisms 29 and 30 would be difficult to discern from Mechanisms 7 and 8 as all four would produce 4 dyads and 7 triads from racemic monomer. Three pairs of epimers would participate in the polymerization of racemic monomer by Mechanisms 31-34, resulting in the formation of five dyads. Mechanisms 31 and 32 could only produce nine triads, while Mechanisms 33 and 34 would produce ten.

Mechanism 35

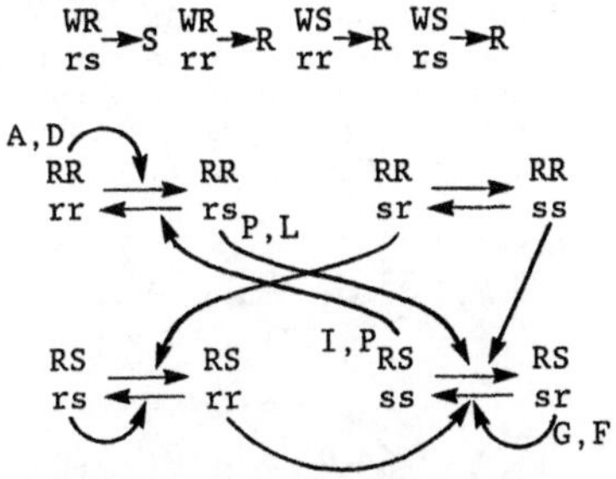

Tetrads:1,3,15,16,24,34,39,40,
41,53,54,56,62,65,66,68

Mechanism 36

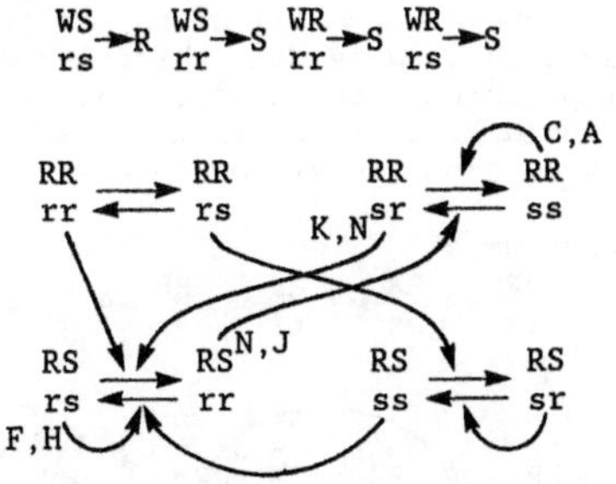

Tetrads:1,2,15,17,27,31,37,38,
42,49,50,51,61,69,70,71

Mechanism 37

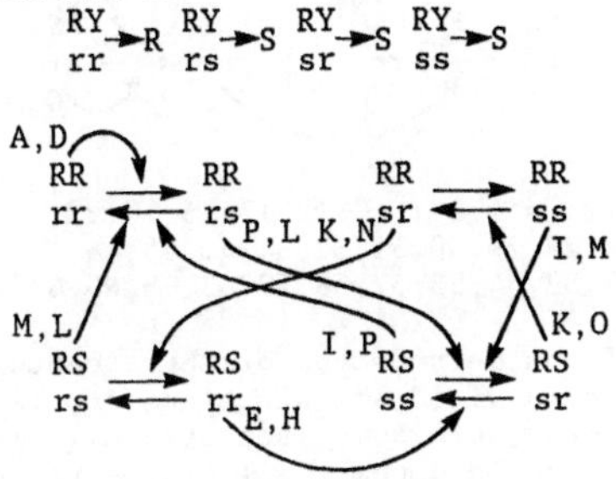

Tetrads:1,3,21,23,24,28,29,30,
31,32,33,34,35,36,39,40,41,43,
53,54,55,56,57,59,60,70,71,72

Mechanism 38

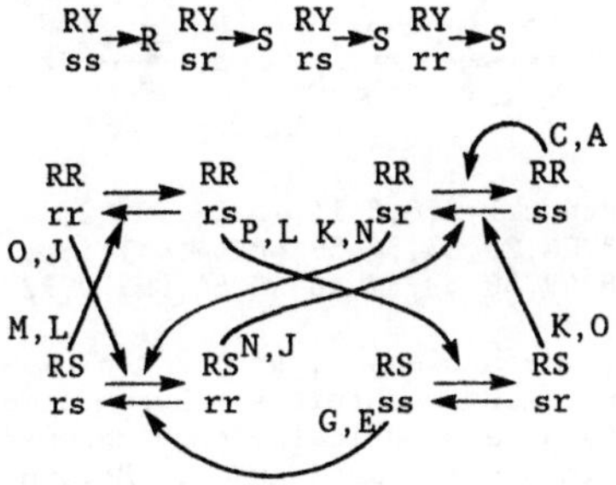

Tetrads:1,2,25,26,27,28,29,30,
31,32,33,34,35,36,37,38,42,44,
49,50,51,52,58,59,60,66,67,68

Mechanisms 35-38 would also result in the formation of isotactic polymer with optically pure monomer. However, as one of the two pairs of epimers which could be formed under these conditions would be unreactive toward monomer, the resulting polymer would be higher in molecular weight than the monomer concentration divided by the initiator concentration even if the system is "living". With racemic monomer, four dyads and seven triads would result for Mechanisms 35 and 36, and five dyads and 11 triads would result for Mechanisms 37 and 38.

Mechanism 39

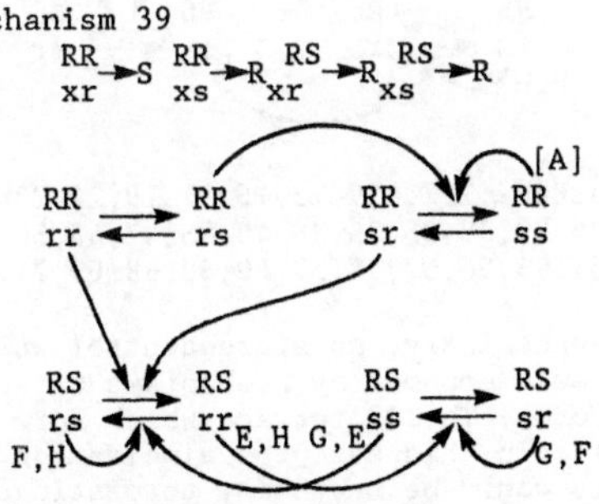

Tetrads:11,12,13,14,15,16,
17,18,19,20

Mechanism 40

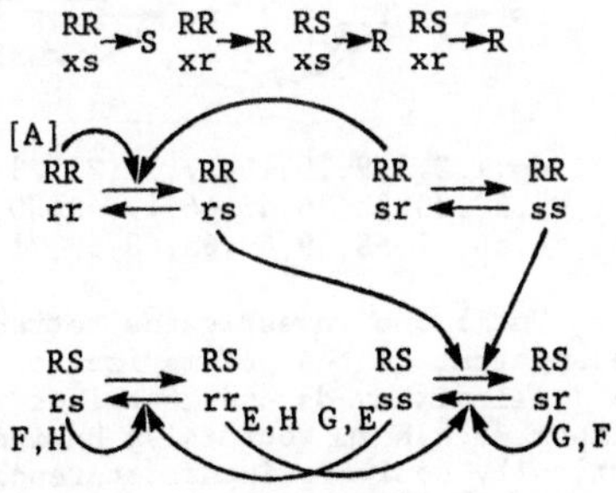

Tetrads:11,12,13,14,15,16,
17,18,19,20

Mechanisms 39 and 40 would generate isotactic polymer with optically pure monomer, and all of the initiator would begin a polymer. An isotactic dyad with like pendant chirality would be difficult to observe in the polymer prepared from racemic monomer, since the ion pair which would generate it would be depleted from the mixture very early in the polymerization. The polymer from racemic monomer would be like that from Mechanism 6.

Mechanism 41

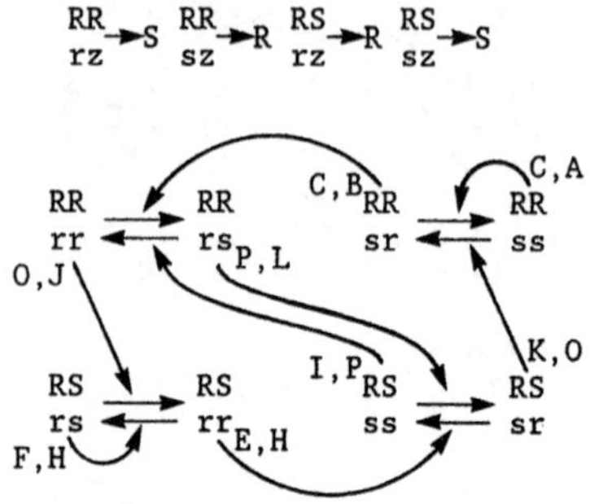

Tetrads:1,2,6,8,14,15,17,20,22,
23,24,29,33,34,39,40,44,45,46,
48,49,50,52,57,58,60,64,69,70,72

Mechanism 42

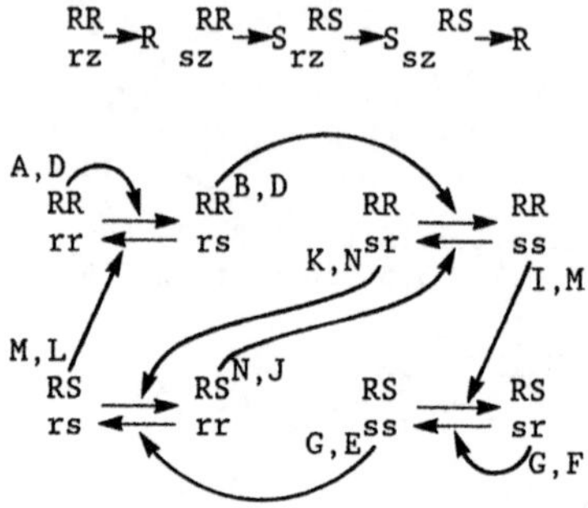

Tetrads:1,3,7,10,13,15,16,18,22,
26,27,29,30,31,37,38,43,45,46,
47,53,54,55,57,58,59,63,65,66,67

Mechanisms 41 and 42 are examples of the second type of three center controlled mechanisms. Racemic monomer would yield all six dyads and 12 triads, and optically pure monomer could only form some isotactic oligomers, as in Mechanisms 3 and 4. However, with racemic monomer all of the various isotactic and syndiotactic triads would result by these two mechanisms.

Mechanism 43

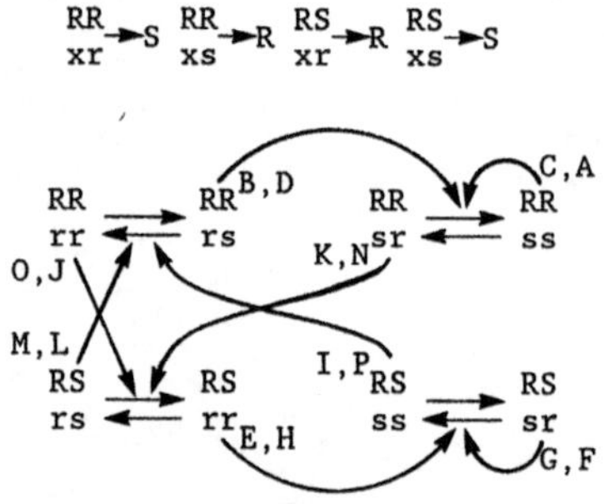

Tetrads:1,2,3,9,15,16,17,19,22,23,
26,29,30,33,35,36,45,46,47,48,50,
51,54,56,57,58,59,60,65,68,69,71

Mechanism 44

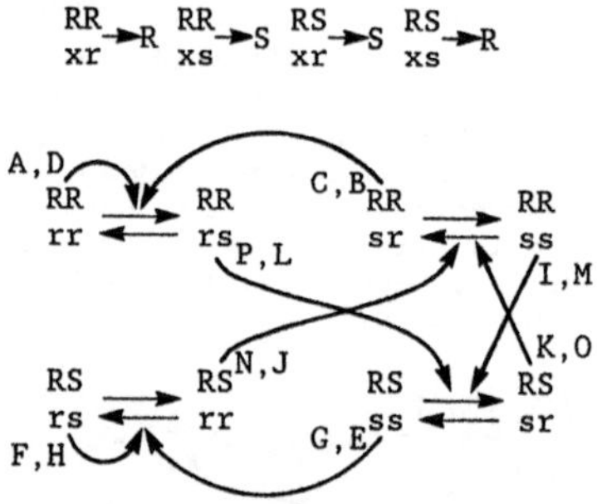

Tetrads:1,2,3,9,15,16,17,19,22,23
26,29,30,33,35,36,45,46,47,48,50,
51,54,56,57,58,59,60,65,68,69,71

Until one investigates tetrad stereochemistry, no stereocontrol would be apparent in the polymerization of racemic monomer by Mechanisms 43 and 44. All six dyads and 16 triads would form. The 32 tetrads which form with either mechanism would also be identical. In this case the elucidation of optically active oligomer stereochemistry would be necessary to distinguish them. The polymer from optically pure monomer would be purely isotactic with either of these stereoelective mechanisms.

Mechanism 45 RY/rr → R RY/ss → R RY/rs → S RY/sr → S

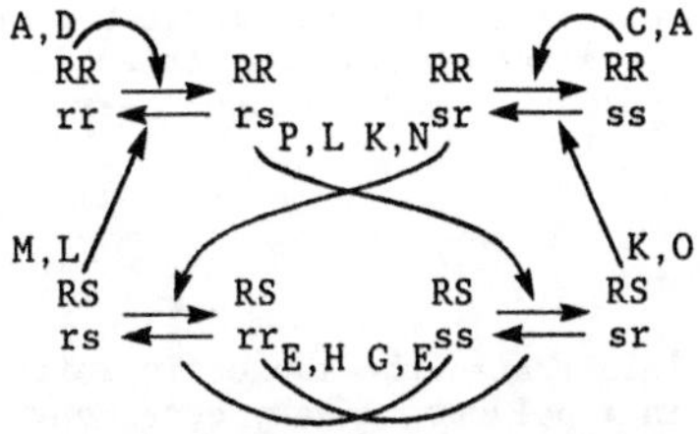

Tetrads:1,2,3,11,12,13,14,28,29,30,31,32,33,34,43, 44,50,51,52,54,56,59,60,66,67,68,70,71,72

The polymer from racemic monomer would contain all six dyads and 12 triads by Mechanism 45. The optically pure polymer would be isotactic and all initiator would generate polymer. However, the isotactic polymer would be generated by two diastereomerically different active centers. Therefore the resulting optically active polymer could be bimodal in its molecular weight distribution.

Mechanism 46 WR/rr → R WS/rs → R WS/rr → S WR/rs → S

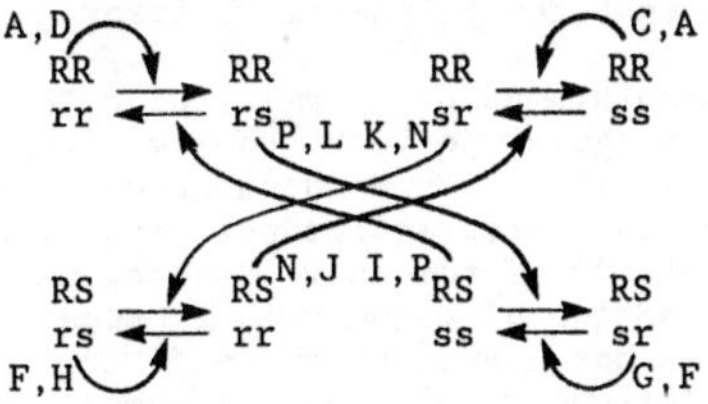

Tetrads:1,3,15,16,24,34,39,40,41,53,54,56,62,65,66,68
1,2,15,17,27,31,37,38,42,49,50,51,61,69,70,71

As in Mechanism 45, optically active polymer by Mechanism 46 would be isotactic and could be bimodal in its molecular weight distribution. Two diastereomerically different polymers would be generated from racemic monomer, one being that which would form by Mechanism 7 and the other being that which would form by Mechanism 8. The two polymers could be of different molecular weight.

Mechanism 47

RY/rr → S RY/rs → R RY/sr → R RY/ss → S

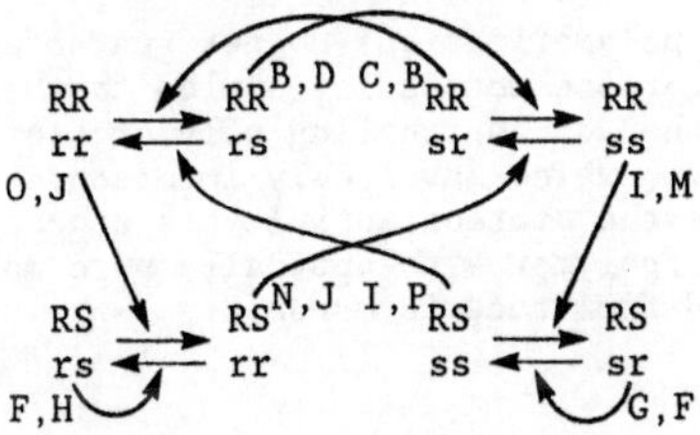

Tetrads:4,5,6,7,15,16,17,21,22,23, 24,25,26,27,47,48,49,51,52,53, 55,56,63,64,65,67,68,69,71,72

Mechanism 48

WR/rs → R WR/rr → S WS/rs → S WS/rr → R

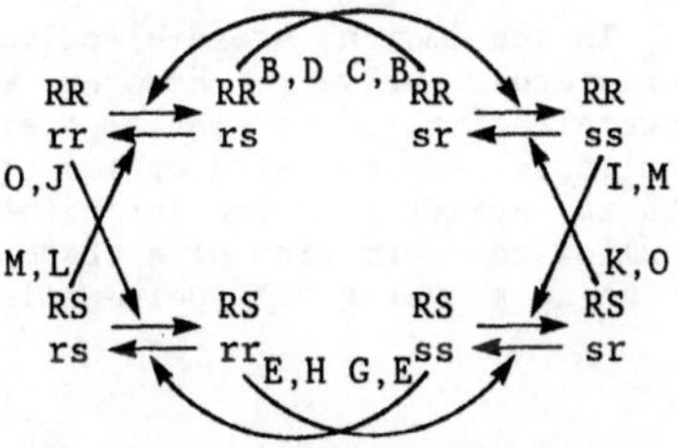

Tetrads:4,5,6,7,11,12,13,14,23,26, 30,33,35,36,45,46,49,50,52,53, 54,55,57,58,65,66,67,69,70,71

Mechanisms 47 and 48 result in the formation of all six dyads and 12 triads. They are unique because the stereoregular polymer generated from optically pure monomer would be syndiotactic. This occurs because the epimer which may react with optically pure monomer converts the active end into the other pair of epimers. Although only one triad is generated with optically pure monomer, there are two dyads and two tetrads formed in equal proportions.

CONCLUSION

The above analysis illustrates the manner in which the determination of stereocontrol and order in a polymer is very dependent upon where one searches for that order. A polymerization with no stereocontrol would yield six diastereomeric dyads, 16 diastereomeric triads and 72 diastereomeric tetrads. Although the above mechanisms are for a series of absolutely stereoelective and stereospecific vinyl addition reactions, when the monomer feed is racemic these conditions are not sufficient to yield an apparently stereoregular polymer. When one searches for order only at the dyad level one may erroneously conclude that no stereocontrol is involved for 20 of these 48 mechanisms as all six diastereomeric different dyads would form. At the triad level only two of these mechanisms would suggest no stereocontrol. However, many of these mechanisms would exclude the possibility of regularity of the carbon backbone, as 28 produce more than eight triads. At the tetrad level, none of these mechanisms could produce even half of the 72 diastereomeric possibilities. At this level the observer would have to be careful not to conclude that the polymer is stereoregular.

Although these mechanisms do not permit the preparation of what is normally considered a stereoregular polymer, the coupling of stereoselective formation of backbone chirality with pendant group stereoelection permits the formation of absolutely stereoregular polymers with optically pure monomer. This would occur via 20 of the above mechanisms. An influence by the ion pair on the stereoelection of monomer is a necessary condition for this to occur. However, it is not a sufficient condition for the preparation of a stereoregular polymer with optically pure monomer. As seen in Mechanisms 23-28, even with its influence an atactic polymer may result.

Although the ability to make stereoregular polymers with optically pure monomer is a very desirable feature of these mechanisms, it is not the most important consequence of these mechanisms, as there are other methods of making stereoregular polymers. The strength of this method is that it allows for the control of the microstructure of a polymer by controlling the feed ratio of the two enantiomers of monomer. Thus by successive addition of racemic and optically pure monomer it is possible to prepare $(AB)_n$ stereo-block polymers very easily if the system is living. It also permits the formation of stereo-gradient polymers, which smoothly vary in their regularity. This could be done by slowly increasing or decreasing the optical purity of the monomer feed, either mechanically or chemically.

In conclusion, stereoelective vinyl polymerizations do not guarantee a stereoregular polymer. However, where stereoselection is coupled to the stereoelection, it becomes a powerful method of controlling polymer stereochemistry. An ion pair ,or its equivalent, which may freely interconvert with its epimer pair and which influences the stereoelectivity is necessary to allow the formation of a stereoregular polymer with optically pure monomer while an "atactic" polymer is formed from racemic monomer.

REFERENCES

1. M.A. Buese and T.E. Hogen-Esch, Macromolecules 17, 118 (1984)
2. The following discussion considers the relative configuration only unless otherwise indicated. Upper case notation refers to the pendant group chirality and lower case refers to the backbone carbon chirality.
3. M.A. Buese and T.E. Hogen-Esch, J. Am. Chem. Soc. 107, 4509 (1985)
4. a) M.A. Buese, PhD Thesis, University of Florida, Dec. 1982
   b) M.A. Buese and T.E. Hogen-Esch, to be published

POLYMERIZATION OF METHYLMETHACRYLATE INITIATED BY COMPLEXED ALKALI ALKOXIDES IN THE HOMOGENEOUS PHASE AND RELATED TACTICITIES

M. VIGUIER*, A. COLLET, F. SCHUÉ**, and B. MULA***
* Laboratoire de Chimie Organique UA CNRS 1097, ** Laboratoire de Chimie Macromoléculaire, *** Laboratoire des Acides minéraux LA CNRS 79. Université des Sciences et Techniques du Languedoc 34060 MONTPELLIER CEDEX FRANCE.

ABSTRACT

The polymerization of methyl methacrylate initiated by alkali alkoxides is greatly promoted in the presence of macrocyclic compounds, cryptands and crown ethers. The solubility in toluene and the reactivity of alkoxides is enhanced by these complexing agents. Polymerizations have been carried out in a non polar solvent in the homogeneous phase. The microtacticities are mainly syndiotactic and consistent with the reactivity of the complexed ion pairs observed in the propagation step.

INTRODUCTION

It is known that alkali metal salts of methanol cannot initiate the polymerization of methylmethacrylate (MMA) in a non polar solvent e.g. toluene [1]. similar behaviour has been observed by Wooding and Higginson [2] using sodium methoxide in dimethyl ether. These results suggest a low reactivity of the methoxides due essentially to their chain association structure. On the other hand, a polar medium solvating strongly cationic species promotes the dissociation of the aggregates and enhances the reactivity of the alkoxide, thus allowing the initiation of MMA polymerization. This has been observed by Goode et al [3] with sodium methoxide in liquid ammonia. More recently Muller [4] has shown that cesium methoxide is able to initiate the polymerization of MMA in THF.

The initiation of MMA polymerization by CH_3O^-/CH_3OH solutions has been studied [5,6] . Oligomers having DP up to six were isolated and characterized. In toluene, alkali metals salts of primary alcohols, cannot initiate the polymerization of MMA [7,8] . To promote the reactivity of primary alkoxides Tomoi et al [8] used alkoxides possessing a methoxy or a dimethyl amino group, such as : $CH_3OCH_2CH_2OM$, $(CH_3)_2NCH_2CH_2OM$ where M = Li, Na, K. Even in these cases, only potassium salts are able to initiate the polymerization of MMA.

On the other hand, in polar solvents, the reactivity of the alkoxides was found to decrease in the order K > Na > Li. Using alkali tertiary alkoxides (Li, Na, K), Lim et al [9,10] were able to polymerize MMA in non polar solvents (benzene, toluene). Trekoval et al [11,12] have shown, in toluene, an activation of lithium tert-butoxide in the presence of piperidine and esters (methyl and butyl isobutyrates). They assume the formation of a transitional activated complex between the tert-alkoxide and piperidine or the ester groups which coexists at equilibrium with aggregated alkoxides.

More recently, the homogeneous polymerization of MMA initiated by potassium salts of methanol and tert-butylalcohol was studied in toluene in the presence of cryptand (222). Complexation of potassium alkoxides by a cryptand greatly increases the reactivity of the alkoxides [13].

Recent Advances in Anionic Polymerization
T.E. Hogen-Esch and J. Smid, Editors

The microstructure of poly(methyl methacrylate) (PMMA) was shown to depend on the structure of the living end of the polymer chain. Modifications of the interactions between the carbanion and the metallic cation, by solvation of the cation, lead to changes in the polymer microstructure.

In non polar solvents, with Li^+ as counter ion, PMMA is mainly isotactic and the temperature has no significant effect on the microstructure [19]. The isotacticity is enchanced by the ability of the cation to be solvated.

In polar solvents, due to the solvation of the cation, polymers are predominantly syndiotactic, Pascault et al [18] and Inoue et al [14] have shown an increase in syndiotacticity at low temperatures. Using mixtures of THF and toluene Inoue et al [14] have shown the dependence of the microstructure according to the THF/toluene ratio. The syndiotacticity increases with the percentage of THF. Schultz et al [15.16.17] observed some variations of tacticity in polar solvents (THF. DME) with Na or Cs as cations. These authors also show that the syndiotacticity increases with the external solvation of the cation. In presence of glyme 4 ($CH_3O-(CH_2CH_2O)_4-CH_3$) or HMPT, the microstructure is independent of the nature of the cation [8.18] .

This paper reports on the homogeneous polymerization of MMA in toluene by potassium methoxide and tert-butoxides of alkali metal (Li. Na. K) in the presence of cryptands (222) and (211) and macrocyclic crown ethers 15C5, 18C6 and DCHE (dicyclohexyl-18-crown 6). The microstructures of PMMA obtained in presence of these complexing agents are discussed.

EXPERIMENTAL

The purification of toluene and MMA as well as the experimental techniques are described elsewhere [20] . The alkali metal salts of the alcohols were prepared by the reaction of the alkali metal with an excess of alcohol in toluene followed by the removal of solvent and alcohol under vacuum at 100°C during at least 15 hours. This gave the alkoxides as snow-white powders. The elementary analysis of the isolated compounds are given in table I.

The small differences observed in the elemental analysis can be attributed to a lack of accuracy.

Cryptand (222) was recrystallized in hexane (mp 69°C) while cryptand (211) was distilled under vacuum. Crown ethers were recrystallized in ether (18C6. mp. 39°C. DCHE mp 50°C) or distilled under vacuum (15C5).

Polymerization procedure

The polymerization (carried out under high vacuum) was initiated by introducing MMA (0.5 to 1 M) in toluene (30 ml) containing the complexing agent and the alkoxide in a given ratio. After a given time, the polymerization mixture was poured into methanol. The precipitated polymer was collected by filtration, washed with methanol and dried in vacuum.

Measurements

The molecular weights were measured by gel permeation chromatography using a Waters Associates 6000 A instrument and are expressed in polystyrene equivalents.

The NMR spectra of the polymer solutions in o-dichlorobenzene were measured at 120°C by a Varian HA 100 or a Varian EM 390 spectrometer. The tacticity was deduced from the proton resonance of the α-methyl group.

TABLE I. Elementary analysis of the alkoxides[a]

Alkoxide compound	C	H	Li	Na	K
$(CH_3)_3COLi$					
calculated %	60.10	11.34	8.67		
found %	57.76	11.04	10.30		
$(CH_3)_3CONa$					
calculated %	49.99	9.44		23.92	
found %	47.01	9.70		24.00	
$(CH_3)_3COK$					
calculated %	42.81	8.09			34.84
found %	42.20	-			31.85
CH_3OK					
calculated %	17.12	4.31			55.75
found %	16.63	4.33			55.19

(Columns C, H, Li, Na, K under "Calculated found %")

a) CNRS, Service Central d'Analyse, Vernaison FRANCE.

RESULTS AND DISCUSSION

Solubility of alkali alkoxides

It has been found that alkali metal alkoxides have a poor solubility in non polar solvents. The results reported in table II show a decreasing solubility in the tert-butoxide series in going from Li to K.

TABLE II. Solubilities of alkali alkoxides in toluene at 20°C[a)]

Alkoxide	Present work	Results from the literature
CH_3OK	insoluble	insoluble
$(CH_3)_3COLi$	$0.4\ M\ l^{-1}$	$0.5\ M\ l^{-1}$ [21]
$(CH_3)_3CONa$	$0.15\ M\ l^{-1}$	
$(CH_3)_3COK$	$< 10^{-5} M\ l^{-1}$	$0.17\ M\ l^{-1}$ [22]

a) Solubilities were determined by acid-base titration

Due to its high polymer chain structure, established by Weiss[23] , potassium methoxide is insoluble in non polar solvents. In the presence of cryptand (222), potassium methoxide is quite soluble in toluene ($1.3 \cdot 10^{-2} M\ l^{-1}$). Using various complexing agents (cryptands or crown ethers) additional significant results have been obtained with alkali tert-butoxides. This increase in solubility can be explained by partial or total rupture of the aggregated structures through complexation of the cation. This increased solubility of alkali alkoxides facilitates the initiation and polymerization of MMA in non polar aprotic solvents. (Methoxide and tert-butoxides are noted respectively MeO and t-BuO).

Polymerization in the presence of MeOK (222) or t-BuOK (222)

Table III and IV show the results of the polymerization initiated with MeOK and t-BuOK in and without the presence of cryptand (222) in toluene at various temperatures between -78 and +25°C. The reactivity of these alkoxides upon complexation with (222) is greatly increased thus allowing initiation by MeOK (222). The yield of the polymer initiated by t-BuOK at -40°C increases from 4 % to 98 % in the presence of (222).

Gel permeation chromatography (GPC) traces shown unimodal molecular weight distribution. Experimental molecular weights $\overline{M}_n$ determined by reference to polystyrene standards are close to or greater than those calculated on the basis of the initial concentration of the reactants. Initiation by t BuOK (222) is not complete and the efficiency E of the initiator measured as $\overline{M}_{n,th}$ x yield/$\overline{M}_{nexp.}$ decreases steadily as the temperature raises from -78 to +20°C.

TABLE III. Results of MeOK/(222) initiated polymerization of MMA in toluene

Temp in °C	Time in h	yield in %	[MMA] mole l^{-1}	[I] mole l^{-1} x 10^3	$\overline{M}_n$ x 10^{-3}	$\overline{M}_w$ x 10^{-3}	$\overline{M}_w/\overline{M}_n$
-78	0.5	100	0.41	0.21	88.2	123.3	1.4
-40	1	95	0.92	3.2	35.0	56.0	1.6
-20	0.5	100	0.69	2.5	52.8	112.5	2.13

[I] = initiator concentration

TABLE IV. Results of t-BuOK/(222) initiated polymerization of MMA in toluene (time of polymerization 0.5 h)

Temp. in °C	[MMA] mol l^{-1}	[I] mol l^{-1} x 10^3	yield in %	$\overline{M}_n$ x 10^{-3}	$\overline{M}_w$ x 10^{-3}	$\overline{M}_w/\overline{M}_n$	E
-78	0.59	3.7	100	19.7	37.0	1.88	0.81
-40	0.99	4	98	37.5	66.4	1.77	0.67
-40[a]	1.24	4	4	48.4	146.6	3.03	-
-20	0.55	2.8	99	34.1	47.0	1.38	0.57
0	0.54	3.1	99	37.9	119.9	3.16	0.46
20	0.96	6.9	87	42.0	102.8	2.45	0.33

a) Polymerization was carried out without cryptand (222)

Polymerization in the presence of t-BuOK complexed by macrocyclic crown ethers

Crown ether compounds like DCHE and 18C6 were used for complexing t-BuOK. In all experiments the molar ratio r = (alkoxide)/(complexing agent) was lower than 0.5. Table V shows the results obtained at different temperatures.

The yields of the polymerization although lower than those obtained with (222), were higher than 80 %. The initiator efficiency of the t-BuOK complexed by crown ethers increases steadily with temperature. The lower polydispersity of the polymers obtained in the presence of crown ethers to those obtained with cryptand (222) can be attributed to the higher reactivity of the cryptated ion pairs. In the experiment at -78°C where t-BuOK was complexed by 18C6, the monomer was added in two steps separated by 30 minutes. The addition of the second monomer fraction does not induce an increase in the molecular weight and the polydispersity stays unchanged. In these conditions it can be concluded that at -78°C the reactivity of the ion pairs complexed by 18C6 is sufficient to allow a total deactivation in 30 minutes.

TABLE V. Results of the polymerization of methyl methacrylate in toluene initiated by t-BuOK complexed by crown ethers (time of polymerization 0.5 h, $0.4 < r < 0.5$).

T°C	[MMA] mol l^{-1}	[initiator] mol.l^{-1} x 10^3	yield in %	$\bar{M}_n$ x 10^{-3}	$\bar{M}_w$ x 10^{-3}	$\bar{M}_w/\bar{M}_n$	E
-78[a)]	0.38	0.89	99	200.2	249.3	1.25	0.21
-40[b)]	0.60	5.4	95	34.5	50.2	1.46	0.21
-20[b)]	0.85	5.2	80	35.5	58.9	1.65	0.36
0[b)]	0.85	6.1	90	26.1	44.8	1.71	0.48
+20[b)]	0.96	6.7	84	21.9	36.4	1.66	0.33

a) Crown ether 18C6 b) Crown ether DCHE

The termination reaction is not intermolecular, in fact no evidence for branching was observed by GPC such as a high molecular weight shoulder. The only possible reactions are an intramolecular reaction or a reaction with the monomer with formation of a methoxide. In these conditions, potassium methoxide complexed by 18C6 does not initate the polymerization of MMA.

Polymerization in the presence of t-BuONa complexed by macrocyclic crown ethers

The polymerization initiated at 0°C was performed to study the effects of the nature of the crown ether and of the ratio "r". Table VI shows the results obtained.

TABLE VI. Results of the polymerization of methyl methacrylate in toluene initiated by t-BuONa complexed by crown ethers (time of polymerization 2h, temperature 0°C).

Crown ether	r	[MMA] mol l^{-1}	[I] mol l^{-1} x 10^3	yield in %	$\bar{M}_n$ x 10^{-3}	$\bar{M}_w$ x 10^{-3}	$\bar{M}_w/\bar{M}_n$	E
-[a)]	-	0.72	1.39	2.5				
18C6	1	0.66	1.37	65	175.1	232.8	1.33	0.18
18C6	0.36	0.67	1.42	77	168.5	230.6	1.37	0.22
15C5	0.50	0.80	1.30	98	101.9	167.3	1.64	0.59

a) Polymerization in the absence of complexing agent

The initiation is incomplete, but the efficiency of the initiator increases with the ability of the crown ether to complex the sodium cation. Table VI shows the enhancement of reactivity due to complexation of the Na^+ cation. Indeed, in the absence of crown ether, the yield of polymerization is extremely low (2.5 %). For a comparable ratio of r, we observe a higher yield for the crown ether 15C5 than for the 18C6, which is due to a better complexing efficiency of the former. On the other hand, for the same crown ether compound 18C6 the yield of the polymerization increases as r decreases.

Polymerization in the presence of t-BuOLi complexed by cryptand (211)

It appears from table VII that the yields are low compared to those obtained with sodium and potassium as counter-ions, but the molecular weights obtained are relatively high. At -40°C, the presence of cryptand (211) is necessary to obtain polymerization [11] . Even in presence of cryptand (211), only a very small amount of t-BuOLi is involved in the initiation.

TABLE VII. Results of t-BuOLi cryptand (211) initiated polymerization of methyl methacrylate in toluene (time of polymerization 2h, (211) excess of 15-20 %)

T°C	[MMA] mol l^{-1}	[initiator] mol l^{-1} x 10^3	yield in %	$\bar{M}_n$ x 10^{-3}	$\bar{M}_w$ x 10^{-3}	$\bar{M}_w/\bar{M}_n$	E
-40[a,b)]	0.94	3.4	7				
-20[b)]	0.39	1.8	4				
-10	0.75	2.4	22	143.4	111.8	1.54	0.05
0[c)]	4.68	20.0	5				
0	0.73	2.4	28	133	260	1.96	0.06
20[c)]	4.68	20.0	17				
28	0.59	2.3	31	72.8	122.3	1.68	0.11

a) Time of polymerization 5 h ; b) Molecular weights not determined due to the presence of microgels

c) Results of Trekoval without cryptand (222) [11]

Microstructures of PMMA initiated by MeOK (222), t-BuOK (222) and t-BuOK (DCHE)

In the presence of cryptand (222) or crown ether DCHE (with a molar ratio (t-BuOK)/(DCHE) lower than 0.5), the polymers are predominantly syndiotactic, excep for the one obtained without a complexing agent (table VIII). As expected, at the same temperature, the tacticities of polymers initiated by MeOK (222) or t-BuOK (222) are very close. The isotactic content increases slightly with an increase in temperature.

Microstructure of PMMA initiated by t-BuONa complexed with crown ethers

The complexation of potassium cation by crown ethers (18C6 or 15C5) increases the syndiotactic content of the polymers. The results reported in table IX show the influence of the ratio "r" on the tacticity. With an excess of complexing agent 18C6 (r = 0.4), the syndiotactic content is enhanced. One can conclude that interactions between complexed cation and living end remain possible in the case of a 1/1 Na^+ - 18C6 complex. The microstructures obtained with Na^+ cation complexed by crown ethers 15C5 or 18C6 (with r < 0.5) are very similar to those obtained with K^+ cation complexed by DCHE or cryptand (222).

TABLE VIII. Tacticity of poly(methyl methacrylate) initated by potassium alkoxides complexed by cryptand (222) or crown ether DCHE[a)]

T°C	I	C(d)	Triads % mm	mr	rr	ρ [b)]
-78	MeOK	222	0.8	25	74.2	0.92
	tBuOK	222	1	24	75	0.94
-40	MeOK	222	2.5	31.5	66	0.95
	tBuOK	222	2	30	68	0.94
	tBuOK	DCHE	2.2	31.8	66	0.93
	tBuOK	c)	39	35	26	1.4
-20	MeOK	222	2.5	36	61.5	0.91
	tBuOK	222	2.3	34.4	63.3	0.91
	tBuOK	DCHE	3.6	32.5	63.9	0.98
0	MeOK	222	3.4	36.7	59.9	0.92
	tBuOK	222	4	41	55	0.90
	tBuOK	DCHE	4.3	36	59.7	0.96
+20	MeOK	222	5.1	38.3	56.6	0.96
	tBuOK	222	6	39.5	54.5	0.97
	tBuOK	DCHE	3.1	36	60.9	0.92

a) The conditions of the polymerization are given in table III, IV and V
b) Persistence ratio $\rho = 2(m)(r)/(mr)$
c) Without complexing agent
d) C = complexing agent
The ratio [I] / [C] lies between 0.9 and 1.0.

TABLE IX. Tacticity of poly(methyl methacrylate) initiated by sodium tert-butoxide complexed by crown ethers.[a)]

C	r	Triads % mm	mr	rr	ρ
b)		22.2	37.8	40	1.29
18C6	1.00	10.2	35.5	54.3	1.13
18C6	0.40	5.3	33.7	61	1.02
15C5	0.50	6.4	36.9	56.7	1.01

a) The conditions of the polymerization are given in table VI
b) Without complexing agent.

Microstructure of PMMA initiated by t-BuOLi (211)

The microstructures obtained with t-BuOLi/(211) (table X) are close to those observed previously with complexed Na^+ and K^+ cations. However the temperature effect on tacticity is reduced. The study of tacticity over a large range of temperatures is limited by the poor reactivity of lithium alkoxide at low temperatures.

TABLE X. Tacticity of poly(methyl methacrylate) initiated by lithium tert-butoxide complexed by cryptand (211)[a)]

T°C	Triads %			ρ
	mr	mr	rr	
-40	4.7	33.5	61.8	1.00
-20	4	34.5	61.5	0.97
-10	4	35	61	0.96
0	4.5	34.5	61	0.99
+28	5.5	36.5	58	0.99

a) The conditions of the polymerization are given in table VII

Influence of complexing agents on the propagation step of MMA polymerization

The propagation mechanism can be studied by examining the dyad and triad information on the polymers obtained. If the propagation is consistent with Bernoullian statistics, the following equations should hold [24, 25] :

$$\text{Dyads : } (m) = Pm,\; (r) = 1 - Pm \quad (1)$$
$$\text{Triads : } (mm) = Pm^2 \quad (2)$$
$$(mr) = 2Pm(1-Pm)$$
$$(rr) = (1-Pm)^2$$

where Pm represents the probability of a meso (isotactic) placement during propagation.

The persistence ratio ρ defined by Coleman and Fox [26] can be calculated from the fraction of dyads and triads, using the relation :

$$\rho = \frac{2(m)\ (r)}{(mr)} \quad (3)$$

If the propagation reaction is controlled by Bernoullian statistics : $\rho = 1$.

When the cation is complexed by cryptand or crown ether with a ratio "r" < 0.5, the values of ρ reported in tables VIII, IX, X are relatively close to 1. From these results it can be concluded that the propagation follows Bernoullian statistics within experimental error. This indicates that there is no effect of center 3 on the stereochemistry of the active center 1 :

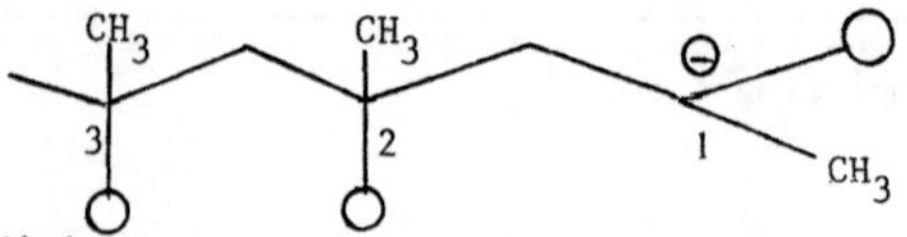

Under this conditions, there is only propagation due to one kind of complexed ion pairs, which has a preference to syndiotactic placement.
Aa shown in figure 1, for a ratio r = 1.0, the propagation does not fit a Bernoullian model.

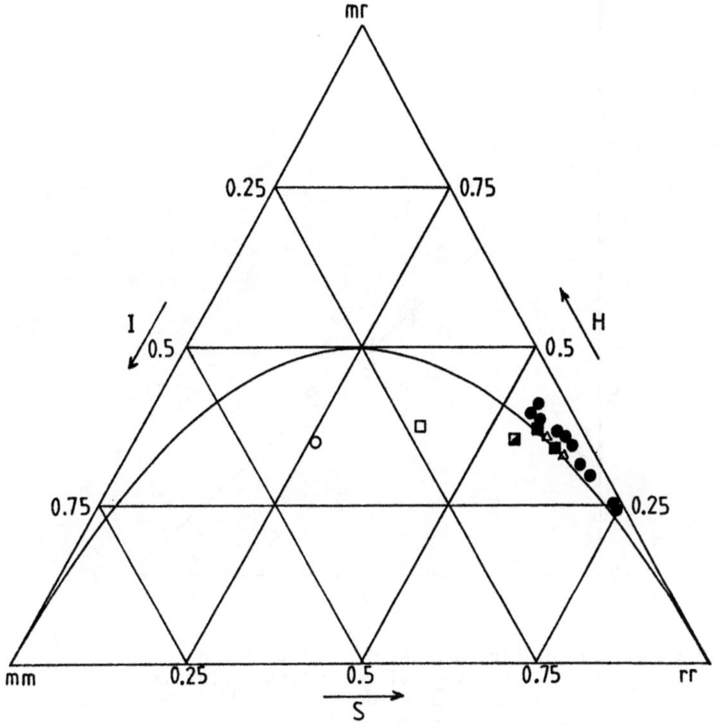

FIG. 1. Tacticity of polymethyl methacrylate represented by a triangular coordinate.
Poly MMA initiated with :
MeOK (222) and tBuOK (222) ●
tBuOK without complexing agent ○
tBuONa : complexed with 15C5 or 18C6 ($r < 0.5$) ■
complexed with 18C6 ($r = 1$) ◪
uncomplexed □
tBuOLi (211) △
Theoretical curve corresponding to $\rho = 1$.

We can, as Furukawa [27] did, use an expression which evaluates the magnitude of the effects in stereospecific polymerization of the ultimate unit as follows :

$$\Delta G^{*} = - RT \, Ln \, (m/r) \qquad (4)$$

Where ΔG^{*} is the difference in activation free energy between the formation of isotactic and syndiotactic placements. Ln(m/r) is plotted against 1/T in figure 2. From this plot, it is shown that ΔG^{*} is linearly dependent on temperature.

ΔS^{*} and ΔG^{*}, being the difference respectively, in the entropy and the enthalpy between isotactic and syndiotactic placements, can be readily calculated from the usual relation :

$$Ln(m/r) = - \frac{\Delta G^{*}}{RT} = \frac{\Delta S^{*}}{R} - \frac{\Delta H^{*}}{RT} \quad (5)$$

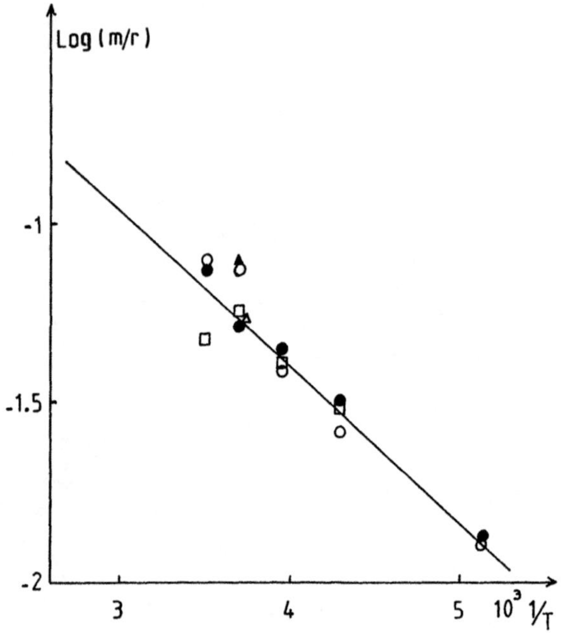

FIG. 2. Plot of Log (m/r) vs. 1/T for poly MMA
Symbols show the polymerization series as follows :
● MeOK (222), ○ tBuOK (222), □ tBuOK DCHE
△ tBuONa 18C6 (r = 0.4)
▲ tBuONa 15C5 (r = 0.5)

The values of 3,63 kJ/mole^{-1} for ΔH^* and 2,94 J/K^{-1}mole^{-1} for ΔS^* were obtained. Thus, in this system, syndiotactic placement appears to be favored by the additional enthalpy of activation required for isotactic placement, the latter being entropically favored.

Very similar results were obtained by Pascault, Kawak, Gole and Pham [18] for the anionic polymerization of MMA and by Elias, Goeldi and Kamar [28] for the radical polymerization.

CONCLUSION

Complexation of alkali cations (Li, Na, K) by monocyclic (crown ethers) or bicyclic compounds (cryptands) increases markedly the solubility of alkoxides and their reactivity. Complexed alkoxides are able to initiate MMA polymerization in the homogeneous phase with high yields. However, some differences in reactivity exist between potassium and lithium alkoxides. From these results it can be concluded that the reactivity of the complexed alkoxides, affecting the different steps of the polymerization are dependent on certain parameters : the nature of alkali cations, the nature of the complexing agents, the ratio r, and the temperature. Reactivities are found to decrease in the order K > Na >> Li. The lower initiation efficiency of the lithium alkoxide confirms the relatively difficult complexation of the lithium cation by the cyclic compounds.

For the potassium alkoxides, the cryptated species show a higher reactivity than crown ether (DCHE or 18C6) complexed species. Bicyclic compounds like cryptand (222) give a better complexation of the cation than monocyclic crown ethers. With a sodium cation, the reactivity of the complexed species decreases in going from 15C5 to 18C6, the cavity size of crown ether 15C5 (0.9 Å) being comparable to the cation radius (0.98 Å). On the other hand, the initiation efficiency is improved when the sodium cation is complexed by at least two crown ether molecules. With r < 0.5 the formation of a sandwich - like structure with two crown ether molecules for one cation can be assumed. This type of complexation leads to a slightly better reactivity. In the case of tBuOLi (211) an increase in temperature promotes the initiation and propagation reactions.

However, at higher temperatures, the termination reactions occur more easily. Even at low temperature (-78°C), living polymethyl methacrylate is unstable.

The microstructures of PMMA initiated by complexed alkoxides are predominantly syndiotactic. The complexation of cation by cryptands leads to structures which are independent of the alkali cation and of the complexing agent. In the case of crown ethers, the microstructure is dependent on the ratio r. When r is equal or less than 0.5, the tacticity is equivalent to those obtained with cryptated cations.

Excepted for Li alkoxides, the isotactic content increases slightly with temperature. The microstructures resemble those obtained in polar solvents ; thus it may be concluded that the reactions of cryptated ions pairs are very similar to those of solvated ions pairs in polar solvents.

The microstructures and the thermodynamic parameters (ΔH^* and ΔS^*) obtained are consistent with the reactivity of complexed ion pairs, observed in the propagation step.

REFERENCES.

1. D.M. Wiles and S. Bywater, Chem. Ind. (London) 1209 (1963).
2. N.S. Wooding and W.C.E. Higginson, J. Chem. Soc., 774 (1952).
3. W.E. Goode, R.C. Fettes and W.H. Snyder, J. Polym. Sci., 42, 367 (1960).
4. A.H.E. Muller, Thesis Mainz Germany (1975).
5. T. Wolker, A. Neumann and M. Baumann, Makromol. Chem., 63, 182 (1963).
6 S. Freireich and A. Zilkha, J. Macromol. Sci. Chem., A6, (7), 1383 (1972).
7. D.M. Wiles and S. Bywater, J. Phys. Chem., 68, 1983 (1964).
8. M. Tomoi, K. Sekiya and H. Kakiuchi, Polymer J. 6, (5), 438 (1974).
9. J. Baca, L. Lochmann, K. Juzl, J. Coupek and D. Lim, J. Polym. Sci., C16, 3865 (1968).
10. J. Trekoval and D. Lim, J. Polym. Sci., C4, 333 (1964).
11. J. Trekoval, J. Polym. Sci., A1, 9, 2575 (1971).
12. A. Kolar, J. Trekoval and P. Vlcek, J. Macromol. Sci. Chem., A12 (5), 757 (1978).
13. M. Viguier, A. Collet and F. Schué, Polym. J. 14, 137 (1982).
14. Y. Inoue, R. Chujo, A. Nishioka, Polymer J. 2, (1) 13 (1971).
15 A.H.E. Muller, H. Hocker, G.V. Schulz, Macromolecules 10, (5) 1086 (1977).
16. V. Warzelhan, H. Hocker, G.V. Schulz, Makromol. Chem. 179, 2221 (1978).

17. R. Kraft, A.H.E. Muller, H. Hocker, G.V. Schulz, Makromol. Chem. Rapid Comm. 1, 363 (1980).
18. J.P. Pascault, J. Kawak, J. Gole, Q.T. Pham, Europ. Polym. J. 10, 1107 (1974).
19. D. Braun, M. Herner, V. Tohnsen, W. Kern, Makromol. Chem. 51, 15 (1962).
20. M. Víguier, M. Abadie, B. Kaempf, F. Schué, Europ. Polym. J. 13, 213 (1977).
21. T.V. Talalaeva, G.V. Tsareva, A.P. Simonov, K.A. Kocheshkov, Bull. Acad. Sci. Ussr, Chem. Sci. 4, 595 (1964).
22. H. Feuer, J.W. Sheperd, C. Savides, J. Am. Chem. Soc., 78, 4364 (1956).
23. E. Weiss, Helv. Chem. Acta., 46, 2051 (1963).
24. F.A. Bovey, Makromol. Chem., 3, 349 (1967).
25. F.A. Bovey, Acc. Chem. Res., 1, 175 (1968).
26. B.D. Coleman, T.G. Fox, J. Polym. Sci. A1, 3183 (1963).
27. J.Furukawa, Kobunshi (High Polymers, Japan) 16, 520 (1967).
28. H.G. Elias, P. Goeldi, S.V.S Kamar, Makromol. Chem. 117, 269 (1968).

ROLE OF E- AND Z-CARBANIONS IN THE STEREOCHEMISTRY OF ANIONIC VINYL POLYMERIZATION

I. M. KHAN and T. E. HOGEN-ESCH*, Department of Chemistry and Center for Macromolecular Science and Engineering, University of Florida, Gainesville, Florida 32611

INTRODUCTION

The occurrence of geometric isomers of type 1 where X and Y may be C, N or O is well documented. [1-8] Recent work in our laboratory has focussed on the occurrence of E and Z isomers of 1 and their role in the stereochemistry of anionic polymerization of vinyl monomers $\bar{C}H_2CHC(Y)X$ where X = 0 or N and Y = 0, N or C.

R = CH_3, Polymer

E-1 Z-1

Scheme I

The reluctance toward interconversion of the two isomers (Scheme I) is quite pronounced. For instance, in the case of alkali salts of 2-ethylpyridine, 2, in THF there is no interconversion on the NMR time scale at temperatures below 100°. [5,7] Similar observations have been made by others. [2-4,6]

The interconversion between the lithio derivatives of E- and Z-2 was shown to have a half life of about 3 hours at 50°. [7] As a result this interconversion is not occurring at -78° during polymerization and the same conclusion is no doubt valid for the enolates of acrylates and other vinyl monomers of the above type. This does not mean, however, that the occurrence of E- and Z isomers is unimportant in the polymerization process. The occurrence of E- and Z isomers has been shown to be of considerable importance in the stereochemistry of electrophilic reactions of carbanions 1 for instance in the alkylation stereochemistry of oxazoline substituted carbanions.[3] In this case the coordination of the counterion with the electrophile as well as with the nitrogen of the carbanion was shown to play a role. It is plausible therefore that this is also the case for the stereochemistry of polymerization for vinyl monomers of the type $CH_2CHC(Y)X$ and we present here a brief account of the statistics of such a polymerization.

RESULTS AND DISCUSSION

Previous work in our laboratory [5,7] showed that the Li salt of 2 generated by reaction of n-BuLi with 2-ethylpyridine in THF at -78° generates predominantly the E isomer in a 16:1 (E:Z) ratio. However, upon addition of 2-vinylpyridine the corresponding E:Z ratio in the dimer anion 3 and the trimer and tetramer anions was found to be close to one. [7] In view of the absence of E-Z interconversion this distribution therefore reflects the mode of monomer presentation at the time of reaction, s-cis and s-trans monomer generating Z- and E carbanions respectively (Scheme 2). The vinyl monomer is thus almost completely unselective with regard to the mode of presentation to the propagating carbanion. Thus, although E and Z isomers do not directly interconvert during polymerization an indirect interconversion is taking place through monomer addition. Since the E- and Z isomers are diastereoisomers they are expected to behave differently with regard to the

Recent Advances in Anionic Polymerization
T.E. Hogen-Esch and J. Smid, Editors

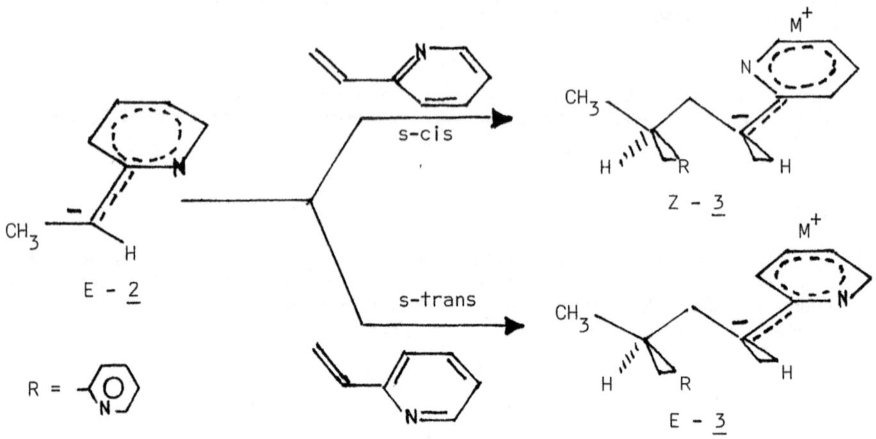

Scheme 2

kinetics and stereochemistry of polymerization and the statistics of such a process will be of interest (Scheme 3). The propagating species may be E or Z and these isomers may be preceded by a meso or racemic dyad. Thus [mE] + [rE] = [E]. If several dyads are specified the one next to the carbanion is indicated last. Thus rmZ specifies a Z carbanion preceded by a dyad meso which is in turn preceded by a racemic dyad. In the addition reactions the carbanion may be either E or Z, the

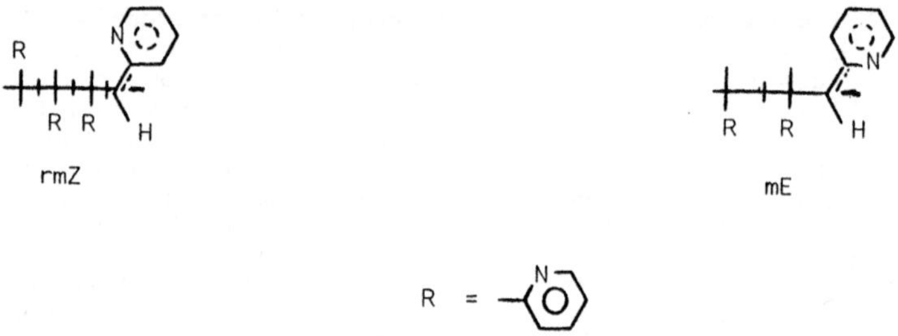

monomer presentation may be s-cis or s-trans and meso or racemic dyads may be formed. As a result, there are eight different processes specified by bimolecular rate constants $k^{R}_{X,Y}$ in which the X specifies the reacting carbanion isomer (E or Z), Y specifies the monomer presentation, s-cis or s-trans (C or T) and R denotes the formation of the new dyad (m or r). A number of simplifying assumptions have been made: a. Only one type of ionic species is present (ion pairs for instance). b. The stereochemistry of vinyl addition is only dependent upon the type of carbanion isomer and the mode of monomer addition, s-cis (C) or s-trans (T) and not upon the center(s) of asymmetry adjacent to the carbanion. Thus both sites propagate according to Bernoullian statistics. For example the mE and rE centers have the same reactivity and react in a stereochemically identical manner. c. The interconversion of E and Z isomers has been shown to be slow on the reaction time scale in the case of 1 and this will be assumed to hold in the general case also. d. Steady state conditions will hold for a particular species. For instance:
d[mE]/dt = d[rmZ]/dt = 0.

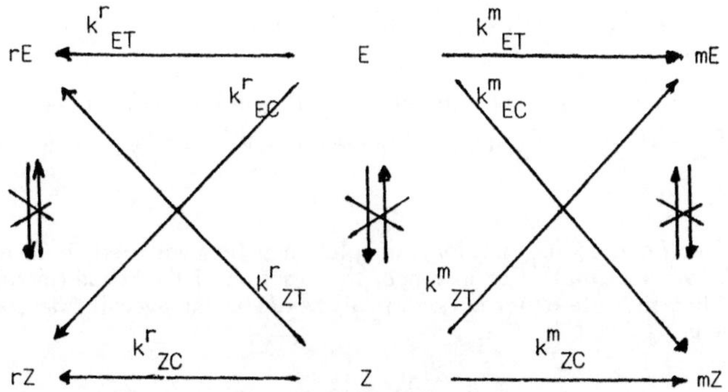

Scheme 3

The steady state concentration of the various species is given by:

$$d[E]/dt = (k^r_{ZT} + k^m_{ZT})[Z][M] - (k^m_{EC} + k^r_{EC})[E][M] = 0$$

$$\text{so that } (E), = [E]/([E] + [Z]) = k_{ZT}/(k_{EC} + k_{ZT}) \qquad (1a)$$

$$\text{and } (Z) = k_{EC}/(k_{EC} + k_{ZT}) \qquad (1b)$$

where [E], [Z] (E) and (Z) are the concentrations and fractions of E and Z isomers respectively, [M] is monomer concentration and $k_{EC} = k^r_{EC} + k^m_{EC}$ and $k_{ZT} = k^r_{ZT} + k^m_{ZT}$. Furthermore:

$$d[D_nmE]/dt = 0 = (k^m_{ET}[D_nE] + k^m_{ZT}[D_nZ] - k_E[D_nmE])\ [M]$$

$$\text{so that: } (D_nmE) = (k^m_{ET}(D_nE) + k^m_{ZT}(D_nZ))/k_E \qquad (2a)$$

where $[D_nmE]$ and (D_nmE) represent the concentration and fraction respectively of the mE sites preceded by a configurational sequence consisting of n stereochemical dyads D_n. (D_nE), (D_nZ) represent the fractions of E and Z sites preceded by n dyads and k_E (the composite rate constant of all E sites) = $k^m_{ET} + k^r_{ET} + k^m_{EC} + k^r_{EC} = k_{ET} + k_{EC}$. The corresponding equations for (D_nmZ) (D_nrE) and (D_nrZ) are:

$$(D_nmZ) = (k^m_{EC}(D_nE) + k^m_{ZC}(D_nZ))/k_Z \qquad (2b)$$

$$(D_nrE) = (k^r_{ET}(D_nE) + k^r_{ZT}(D_nZ))/k_E \qquad (2c)$$

$$(D_nrZ) = (k^r_{EC}(D_nE) + k^r_{ZC}(D_nZ))/k_Z \qquad (2d)$$

where $k_Z = k^m_{ZT} + k^r_{ZT} + k^m_{ZC} + k^r_{ZC} = k_{ZT} + k_{ZC}$.

The configurational composition of the chain itself is calculated from Equations (3).

$$(D_nmm) = (k^m_E(D_nmE) + k^m_Z(D_nmZ))/k_A \qquad (3a)$$

$$(D_nrm) = (k^m_E(D_nrE) + k^m_Z(D_nrZ))/k_A \qquad (3b)$$

$$(D_nmr) = (k^r_E(D_nmE) + k^r_Z(D_nmZ))/k_A \qquad (3c)$$

$$(D_n rr) = (k^r{}_E(D_n rE) + k^r{}_Z(D_n rZ))/k_A \tag{3d}$$

where k_A (= $k_E(E) + k_Z(Z)$ is the apparent overall rate constant for the active site, and where $k^m{}_E = k^m{}_{ET} + k^m{}_{EC}$; $k^r{}_E = k^r{}_{ET} + k^r{}_{EC}$; $k^m{}_Z = k^m{}_{ZT} + k^m{}_{ZC}$ and $k^r{}_Z = k^r{}_{ZT} + k^r{}_{ZC}$.

Any specific n-ad ($D_n$mm for example) may be evaluated in terms of its precursors (D_nmE + D_nmZ). For instance, the fraction of the hexad rmrrm is evaluated by dividing the rate of formation of this hexad by the overall rate constant of polymerization:

$$(rmrrm) = \frac{k^m{}_E\,(rmrrE) + k^m{}_Z(rmrrZ)}{k_A} \tag{4}$$

and the unknowns (rmrrE) and (rmrrZ) may be evaluated in terms of eight rate constants by consecutive substitution of the steady state functions. For example:

$$mm = \frac{k^m{}_E\,(mE) + k^m{}_Z\,(mZ)}{k_A} = \frac{k^m{}_E\;\; k^m{}_{ET}(E) + k^m{}_{ZT}(Z)}{k_A\, k_E} + \frac{k^m{}_Z\;\; k^m{}_{EC}(E) + k^m{}_{ZC}(Z)}{k_A k_Z} \tag{5}$$

Rearrangement results in mm = A(E) + (B(Z) where A and B are listed in Table 2 and (E) and (Z) are given by Equation (1). The dyads, triads and tetrads calculated in this way are listed in Table 1.

The corresponding general expressions for the stereochemical composition of the chain end is given by:

$$(D_n) = (D_n E) + (D_n Z) \tag{6}$$

For instance:

$$(D_n mr) = (D_n mrE) + (D_n mrZ) \tag{6a}$$

where ($D_n$mrE) and ($D_n$mrZ) are evaluated using Equations 2a-2d. For example:(mr⁻) = (mrE) + (mrZ). As a result the stereochemistry of end groups can be readily obtained and compared with that of the chain. In general the stereochemistry of the chain end and the chain are not identical as is also found for Markoff statistics (see below).

Comparison of the Statistics Generated by the E-Z Model with That by Other Models.

It may be shown [7b] from the results of these calculations that the E-Z statistical model does not lead to Bernoullian [9] or first-order Markoff [9] statistics. From an evaluation of the pentad sequences it can be shown that the process does not conform to a second-order Markoff or a Coleman-Fox [10] chain. The E-Z scheme, however, is reducible to Bernoullian or first-order Markoff chains under certain limiting conditions (Table 3).

TABLE I. Dyads, triads and tetrads generated by a two-site E-Z model.

Dyads

$$m = (A + E')(E) + (B + F)(Z)$$

$$r = (C + G)(E) + (D + H)(Z)$$

Triads

$$mm = A(E) + B(Z)$$

$$mr = (E' + G)(E) + (F + H)(Z)$$

$$rr = C(E) + D(Z)$$

Tetrads

$$mmm = \frac{k^m_{ET} \cdot A(E)}{k_E} + \frac{k^m_{ZT} \cdot A(Z)}{k_E} + \frac{k^m_{EC} \cdot B(E)}{k_Z} + \frac{k^m_{ZC} \cdot B(Z)}{k_Z}$$

$$mmr = \frac{1}{k_E} (k^r_{ET} \cdot A(E) + k^r_{ZT} \cdot A(Z) + k^m_{ET} \cdot G(E) + k^m_{ZT} \cdot G(Z))$$

$$+ \frac{1}{k_Z} (k^r_{EC} \cdot B(E) + k^r_{ZC} \cdot B(Z) + k^m_{EC} \cdot H(E) + k^m_{ZC} \cdot H(Z))$$

$$rmr = \frac{k^r_{ET} \cdot G(E)}{k_E} + \frac{k^r_{ZT} \cdot G(Z)}{k_E} + \frac{k^r_{EC} \cdot H(E)}{k_Z} + \frac{k^r_{ZC} \cdot H(Z)}{k_Z}$$

$$mrm = \frac{k^m_{ET} \cdot E'(E)}{k_E} + \frac{k^m_{ZT} \cdot E'(Z)}{k_E} + \frac{k^m_{EC} \cdot F(E)}{k_Z} + \frac{k^m_{ZC} \cdot F(Z)}{k_Z}$$

$$mrr = \frac{1}{k_E} (k^m_{ET} C(E) + k^m_{ZT} C(Z) + k^r_{ET} \cdot E'(E) + k^r_{ZT} \cdot E'(Z))$$

$$+ \frac{1}{k_Z} (k^m_{EC} \cdot D(E) + k^m_{ZC} \cdot D(Z) + k^r_{EC} F(E) + k^r_{ZC} \cdot F(Z))$$

$$rrr = \frac{k^r_{ET} \cdot C(E)}{k_E} + \frac{k^r_{ZT} \cdot C(Z)}{k_E} + \frac{k^r_{EC} \cdot D(E)}{k_Z} + \frac{k^r_{ZC} \cdot D(Z)}{k_Z}$$

The constants A-H are given in Table 2.

TABLE 2. Constants of Table I Expressed as Rate Constants in Scheme 2.

Constant	Expression
A	$\frac{1}{k_E}\left(\frac{k^m_{ET}+k^m_{EC}}{k_A}\right) k^m_{ET} + \frac{1}{k_Z}\left(\frac{k^m_{ZT}+k^m_{ZC}}{k_A}\right) k^m_{EC}$
B	$\frac{1}{k_E}\left(\frac{k^m_{ET}+k^m_{EC}}{k_A}\right) k^m_{ZT} + \frac{1}{k_Z}\left(\frac{k^m_{ZT}+k^m_{ZC}}{k_A}\right) k^m_{ZC}$
C	$\frac{1}{k_E}\left(\frac{k^r_{ET}+k^r_{EC}}{k_A}\right) k^r_{ET} + \frac{1}{k_Z}\left(\frac{k^r_{ZT}+k^r_{ZC}}{k_A}\right) k^r_{EC}$
D	$\frac{1}{k_E}\left(\frac{k^r_{ET}+k^r_{EC}}{k_A}\right) k^r_{ZT} + \frac{1}{k_Z}\left(\frac{k^r_{ZT}+k^r_{ZC}}{k_A}\right) k^r_{ZC}$
E'	$\frac{1}{k_E}\left(\frac{k^m_{ET}+k^m_{EC}}{k_A}\right) k^r_{ET} + \frac{1}{k_Z}\left(\frac{k^m_{ZT}+k^m_{ZC}}{k_A}\right) k^r_{EC}$
F	$\frac{1}{k_E}\left(\frac{k^r_{ET}+k^m_{EC}}{k_A}\right) k^r_{ZT} + \frac{1}{k_Z}\left(\frac{k^m_{ZT}+k^m_{ZC}}{k_A}\right) k^r_{ZC}$
G	$\frac{1}{k_E}\left(\frac{k^r_{ET}+k^r_{EC}}{k_A}\right) k^m_{ET} + \frac{1}{k_Z}\left(\frac{k^r_{ZT}+k^r_{ZC}}{k_A}\right) k^m_{EC}$
H	$\frac{1}{k_E}\left(\frac{k^r_{ET}+k^r_{EC}}{k_A}\right) k^m_{ZT} + \frac{1}{k_Z}\left(\frac{k^r_{ZT}+k^r_{ZC}}{k_A}\right) k^m_{ZC}$

Under conditions 1 and 2 the statistics are Bernoullian because the E and Z sites behave kinetically and stereochemically in an identical or mirror image fashion. Bernoullian statistics are also expected when only E or Z sites are present since these sites are by definition Bernoullian (case 3/4). In cases 5 and 6, E and Z sites generate only m and r dyads respectively or vice versa thereby reducing the E/Z model to first Markoff statistics. This is analogous to a limiting case in Coleman-Fox statistics where exclusively one type of dyad (m or r) results from reaction of each of the two sites. [10] We were unable to locate a set of conditions reducing the E-Z model to second order Markoff statistics.

TABLE 3. Limiting Conditions Reducing the E-Z Scheme to Bernoullian or First-Order Markoff Chains.

No.	Condition (Scheme I)	Statistics
1	$k^m_{ET}=k^m_{ZT}$, $k^m_{EC}=k^m_{ZC}$, $k^r_{ET}=k^r_{ZT}$, $k^r_{EC}=k^r_{ZC}$	Bernoullian[a]
2	$k^m_{ET}=k^m_{ZC}$, $k^m_{EC}=k^m_{ZT}$, $k^r_{ET}=k^r_{ZC}$, $k^r_{EC}=k^r_{ZT}$	Bernoullian[b]
3/4	$k^m_{ZT}=k^r_{ZT}=0$ or $k^m_{EC}=k^r_{EC}=0$	Bernoullian[c]
5	$k^m_{ZT}=k^m_{ZC}=k^r_{ET}=k^r_{EC}=0$	1st Markoff[d]
6	$k^m_{ET}=k^m_{EC}=k^r_{ZT}=k^r_{ZC}=0$	1st Markoff[e]

[a]E and Z sites show identical behavior, [b]S-trans addition to E is identical to S-cis addition to Z and vice versa, [c]Only Z or E sites are present respectively, [d]E sites lead to m-dyads, Z sites lead to r-dyads, [e] E sites lead to r-dyads and Z sites lead to m dyads.

The relationships used to compare the E-Z model with Coleman-Fox statistics are shown in Equations (7)-(10).

$$(mm)-(m)^2 = S \tag{7}$$

$$(rr)-(r)^2 = S \tag{8}$$

$$(mmm)(m)-(mm)^2 = Q \tag{9}$$

$$(rrr)(r)-(rr)^2 = R \tag{10}$$

The constants S, Q and R can be expressed in terms of the eight rate constants. For a Coleman-Fox mechanism all three of the constants take on positive values. [13,14] For the E-Z model, S, Q and R may be positive or negative but will generally differ from zero. Table 4 gives qualitative values for these three parameters for five different types of statistics.

TABLE 4.

	Bernoullian	1st Markoff	2nd Markoff	Coleman-Fox	E/Z
S	0	±	±	+	±
Q	0	0	±	+	±
R	0	0	±	+	±

It is clear that the E-Z mechanism is different from the other mechanisms except the second Markoff chain in that its presence may be more difficult to detect. It should finally be pointed out that the sequence statistics should not vary with monomer concentration in contrast with Coleman-Fox statistics. This dependence may therefore constitute an additional effect by which to distinguish between these two types of statistics.

TABLE 5. Polymers Generated by E/Z Model.

Polymer No[a]	Rate Constants[b]								Main Chain Stereo-chemistry		Chain End Stereo-chemistry		Bernoulli Test	First-Order Markoff Test	Second-Order Markoff Test	Coleman-Fox Test[a]
	k_1	k_2	k_3	k_4	k_5	k_6	k_7	k_8	m	mm	m^-	mm^-	ρ			
1	3	2	3	4	1	6	7	8	0.429	0.184	0.429	0.184	1.000	1.000	1.000	0.000
2	2	0	1	0	0	4	0	6	0.545	0.364	0.613	0.409	1.363	1.000	1.000	6.6×10^{-2}
3	14	0	7	0	0	25	0	6	0.708	0.472	0.697	0.465	0.877	1.000	1.000	-2.9×10^{-2}
4	75	1	25	1	4	5	30	6	0.467	0.320	0.269	0.175	1.254	0.908	1.018	5.04×10^{-2}
5	3	3	7	7	1	1	2	2	0.769	0.592	0.769	0.592	1.000	1.000	1.000	0.000
6	8	1	8	1	9	9	9	3	0.336	0.103	0.362	0.111	0.935	0.989	0.994	-0.97×10^{-2}

[a] $mm - m^2 = S$; [b] $k_1 = k^m_{ET}$, $k_2 = k^m_{ZT}$, $k_3 = k^m_{EC}$, $k_4 = k^m_{ZC}$, $k_5 = k^r_{ET}$, $k_6 = k^r_{ZT}$, $k_7 = k^r_{EC}$, $k_8 = k^r_{ZC}$

Computer Modeling with E/Z Model

In order to obtain a better understanding of the E/Z mechanism, some computer modeling was carried out. Magnitudes for the eight rate constants were entered, and information about the tacticity of the polymers was obtained. These computer generated statistics were subjected to various tests to show consistency with the Bernoullian, Markoffian, or Coleman-Fox chains. Table 5 describes six of such polymers.

Applications

With the advent of high resolution NMR instrumentation, it has been possible to obtain detailed configurational sequence statistics of several polymers. [15-17] Table 6 shows a deviation in the sequence statistics from those expected for a Bernoullian, first-order Markoff and second-order Markoff processes. It has already been discussed that Coleman-Fox mechanism is qualitatively consistent with the E/Z scheme following equations (7)-(10), whereas the E/Z scheme may or may not be consistent with a Coleman-Fox mechanism (Table 4). If the values of S, Q and R are known, the E/Z mechanism can be tested by comparing the predicted pentads and those obtained from NMR spectroscopy. However, extreme care must be taken because the accuracy of the n-ad information obtained from the NMR must be excellent.

In Table 7, the microtacticity of poly(methyl methacrylate) prepared with 9-fluorenyllithium in THF/toluene (4/96) is shown. [18] The tacticity of this polymer has been shown to be inconsistent with Bernoullian, first-order Markoff or second-order Markoff process. [19] It has also been shown by ^{13}C NMR studies that the propagating carbanion of poly(methyl methacrylate) exists as a mixture of two isomers, the exact structure of which was not elucidated. [6,20] However, "living" poly(methyl methacrylate) is capable of existing as E or Z enolate isomers. Therefore, it is of interest to explore the viability of the E/Z model in the interpretation of the sequence statistics of this polymer.

The calculated microtacticity of poly(methyl methacrylate) calculated by the E/Z model and the so-called stereoblend model is compared with the observed tacticity in Table 7. Both show reasonable agreement. The "stereoblend" model is a complicated hybrid Coleman-Fox type model in which a site with Bernoullian statistics coexists with one that propagates according to a first-order Markoff process. Furthermore, the interconversion between the two sites is assumed to be slow compared to the rate of propagation. The stereoblend model is not based upon experimental evidence and would thus appear to be somewhat artificial. Furthermore, the rate of equilibration between various ionic species is often quite fast compared to the rate of polymerization for instance in the case of solvation-desolvation processes. [18] The E/Z model would appear to be more reasonable, being based upon direct observation. The model also shows an equally good agreement with the observed pentad intensities of the polymer.

It is therefore quite plausible that the E/Z model is applicable to the anionic polymerization of methyl methacrylate.

The values of S, Q and R are valuable in ruling out a Coleman-Fox mechanism (Table 4). The poly(2-vinylpyridine) obtained by anionic polymerization with Li^+ as counterion in THF has been shown to be inconsistent with a Bernoulli trial or a first-order Markoff process. [21,22] Furthermore, the value for S for poly(2-vinylpyridine) is a negative quantity (Table 8), and is thus inconsistent with a Coleman-Fox mechanism. The existence of the E and Z geometric isomers in the carbanion intermediate of poly(2-vinylpyridine) was demonstrated in earlier work.

TABLE 6. Sequence Proportions for the E/Z Model.

Triads[a]

$$(mm)-(m)^2 = S$$
$$(mr)-2(m)(r) = -2S$$
$$(rr)-(r)^2 = S$$

Tetrads[b]

$$(mmm)(m)-(mm)^2 = Q$$
$$(mmr)(m)-(mm)(mr)^2 = -2Q$$
$$4(mrm)(r)-(mr)^2 = 4R$$
$$(mrr)(r)-(mr)(rr) = -2R$$
$$4(rmr)(m)-(mr)^2 = 4Q$$
$$(rrr)(r)-(rr)^2 = R$$

Pentads[c]

$$(mmmm)(mm)-(mmm)^2 = Q^2/S$$
$$(mmmr)(mm)-(mmm)(mmr) = -2Q^2/S$$
$$(mmrm)(mr)-2(mmr)(mrm) = 4Q^2/S$$
$$(mmrr)(mr)-(mmr)(mrr) = -4QR/S$$
$$(mrmr)(mr)-4(rmr)(mrm) = -4QR/S$$
$$4(mrrm)(rr)-(mrr)^2 = 4R^2/S$$
$$(mrrr)(rr)-(rrr)(mrr) = -2R^2/S$$
$$4(rmmr)(mm)-(mmr)^2 = 4Q^2/S$$
$$(rmrr)(mr)-2(mrr)(rmr) = 4QR/S$$
$$(rrrr)(rr)-(rrr)^2 = R^2/S$$

[a]Values are zero for Bernoullian statistics; [b] Zero for Bernoullian and 1st-Order Markoff; [c]Zero for Bernoullian, 1st- and 2nd-Order Markoff.

TABLE 7. Microtacticity of Poly(methyl methacrylate).

	Observed[a]	Stereoblend Model	Deviation	E/Z Model	Deviation
Pentads					
mmmm	0.37	0.39	0.02	0.36	0.01
mmmr	0.04	0.01	0.03	0.05	0.01
rmmr	0.02	0.02	0.00	0.01	0.01
mmrm	0.02	0.01	0.01	0.03	0.01
rmrm	0.04	0.04	0.00	0.07	0.03
mmrr	0.06	0.04	0.02	0.04	0.02
rmrr	0.10	0.12	0.02	0.11	0.01
mrrm	0.03	0.02	0.01	0.02	0.01
mrrr	0.08	0.12	0.04	0.10	0.02
rrrr	0.24	0.23	0.01	0.22	0.02
			Σ0.16		Σ0.15

[a] Reference 18, obtained by 25 MHz ^{13}C NMR

Therefore it is quite plausible that the anionic polymerization of 2-vinylpyridine may conform to this mechanism.

TABLE 8. Bernoulli and Coleman-Fox Sequence Statistics TestApplied to Poly(2-vinylpyridine)

T°C	mm	mr	rr	ρ	S^a
-78°	0.44	0.47	0.09	0.93	-1.5×10^{-2}

[a]For Coleman-Fox, this value is positive.

Conclusions

The participation of E and Z- intermediates in anionic vinyl polymerization has been demonstrated in the case of 2-vinylpyridine, MMA and t-butylvinylketone and is probable in many other cases. These two isomers are generally incapable of direct interconversion under polymerization conditions. A statistical model is developed in which E and Z sites are interconverted through monomer addition with each site propagating according to Bernoullian statistics. This model leads to a new type of statistics which is different from other accepted models but may be reduced to Bernoullian or first-order Markoff statistics under certain limiting conditions.

The applicability of this model may extend beyond the area of anionic polymerization. Thus E and Z intermediates should occur in the cationic polymerization of vinyl ethers as a result of charge delocalization onto oxygen. Likewise, these intermediates have been demonstrated in group-transfer polymerization (GTP). [8] Studies on the stereochemistry of GTP of MMA and similar monomers are in progress. Successful strategies for enhanced stereochemical control in the case of new vinyl monomers may include the control of (E)/(Z) ratios to very low or very high values for instance through non-bonded interactions. This may be possible by the selection of monomers where the s-cis or s-trans conformation is strongly preferred. An example of such a monomer is 3-methyl-2-vinylpyridine where the s-cis conformation is strongly preferred. Anionic polymerization of this monomer in THF at -78° in the presence of Li^+ has been shown to lead to highly isotactic polymers. [7b,23]

ACKNOWLEDGEMENTS

We wish to acknowledge the support of the National Science Foundation DMR, Polymers Programs.

REFERENCES

1. H. O. House, Modern Synthetic Reactions, Benjamin, New York, 1965.
2. R. E. Ireland, R. H. Mueller, A. K. Willard, J. Am. Chem. Soc. 98, 2868 (1976).
3. A. I. Meyers, E. S. Snyders, J. J. H. Ackerman, J. Am. Chem. Soc. 100, 8189 (1978): M. A. Hoobler, D. E. Bergbreiter, M. Newcomb, Ibid, 8182 (1978).
4. A. I. Meyers, D. R. Williams and M. Druelinger, J. Am. Chem. Soc. 98, 3032 (1976).
5. T. E. Hogen-Esch and W. L. Jenkins, J. Am. Chem. Soc. 103, 3666 (1981).
6. L. Vancea and S. Bywater, Macromolecules 14, 1321 (1981).

7. a. I. M. Khan and T. E. Hogen-Esch, Makromol. Chem., Rapid Commun. 4, 569 (1983); b. I. M. Khan, Ph.D. Thesis, University of Florida, 1984.
8. W. R. Hertler, Private Communication.
9. F. A. Bovey and L. W. Jelinski, "Chain Structure and Conformation of Macromolecules", Academic Press, 1982, Chapter 3.
10. B. D. Coleman and T. G. Fox, J. Am. Chem. Soc. 85, 1241 (1963).
11. A. Soum and T. E. Hogen-Esch, Macromolecules 18(4), 690 (1985).
12. I. M. Khan and T. E. Hogen-Esch, Polymer Preprints, 26(2), 273 (1985).
13. B. D. Coleman and T. G. Fox, J. Chem. Phys. 38, 1065 (1963).
14. H. L. Frisch, F. A. Bovey and C. L. Heatly, Macromolecules 1, 533 (1968).
15. F. A. Bovey and G. U. D. Tiers, J. Polym. Sci. 44, 173 (1960).
16. L. F. Johnson, F. Heatley and F. A. Bovey, Macromolcules 3, 175 (1970).
17. T. Suzuki, O. Yamada, Y. Watanabe and Y. Takigame, Polym. J. 14, 343 (1982).
18. Y. Inoue, A. Nishioka and R. Chujo, Polym. J. 4, 535 (1971).
19. T. Suzuki and H. F. Harwood, Polymer Preprints 16, 638 (1975).
20. L. Vancea and S. Bywater, Macromolecules 14, 1776 (1982).
21. S. S. Huang, A. H. Soum and T. E. Hogen-Esch, J. Polym. Sci. (Polym. Letters), 21, 559 (1983).
22. A. Soum and T. E. Hogen-Esch, Macromolecules 18(4), 690 (1985).
23. I. M. Khan and T. E. Hogen-Esch, Macromolecules, Submitted.
24. B. C. Bell, Ph.D. Thesis, University of Florida, 1986.

RING-OPENING POLYMERIZATION

PROGRESS IN ANIONIC RING-OPENING POLYMERIZATION

STANISLAW PENCZEK AND STANISLAW SLOMKOWSKI
Polish Academy of Sciences, Centre of Molecular and Macromolecular Studies, 90-362 Lodz, Boczna 5, Poland

INTRODUCTION

The large majority of anionic and cationic ring-opening polymerizations belongs to the nucleophilic substitution reactions. Typical examples are given below:

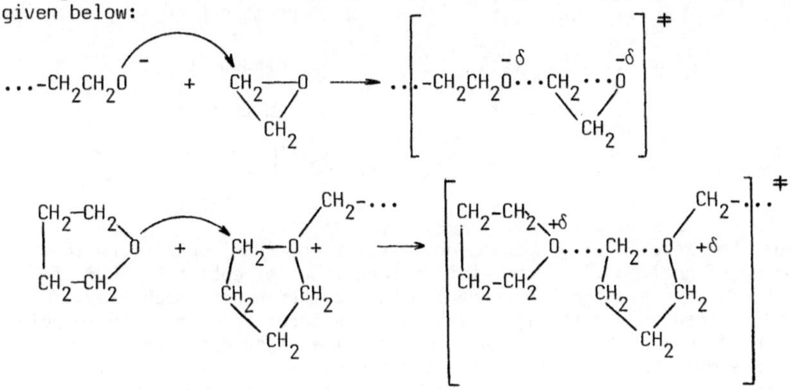

(1)

Some of the pseudoionic processes can be presented in a similar way, e.g. the pseudocationic polymerization proceeding on covalent bonds, such as ...-RX, where X= halogen or ester group (e.g. triflate, fluorosulfonate). The same applies to pseudoanionic processes, involving for example covalent metal alcoholates as the active species, the most prominent being aluminum alcoholates.

Scheme 1 clearly indicates that a common feature of these reactions is the similarity of the transition states. For the general rules governing these reactions it does not really matter whether an active species or a monomer molecule plays the role of a nucleophile or an electrophile. Nevertheless, the division into anionic (comprising the covalent - pseudoanionic) and cationic polymerizations (comprising the covalent - pseudocationic) is useful, although it is for the development of this field more important to find the common features wherever possible, rather than to further separate these processes.

We believe that there is no need to introduce new complications in nomenclature. For instance, it has been proposed to call some cationic polymerizations "electrophilic processes", because the growing species is an electrophile. This leads, however, to the ambiguous internal contradiction that an "electrophilic" reaction proceeds by a nucleophilic mechanism.

The four types of S_N2 reactions are shown below with their corresponding analogues in the ring-opening polymerization:

Reagents	Transition State	Analogy with Ring-Opening Polymerization
I. $RX + Y^-$	$\overset{-\delta}{X}.....R.....\overset{-\delta}{Y}$	 + $^-OCH_2CH_2$-... (anionic)

Recent Advances in Anionic Polymerization
T.E. Hogen-Esch and J. Smid, Editors

Reagents Transition State Analogy with Ring-Opening Polymerization

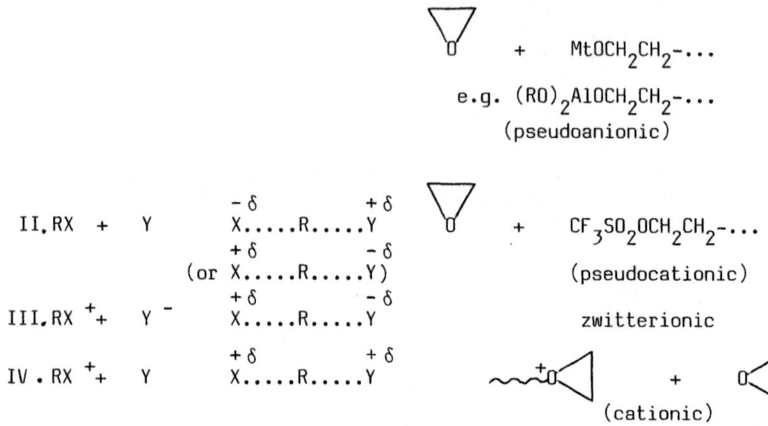

(2)

Thus, the anionic ring-opening polymerization (AROP) belongs to the first group of nucleophilic substitution reactions (exception is AROP of some lactones with propagation according to the tetrahedral mechanism).

AROP differs substantially in its basic features from the anionic polymerization of unsaturated non-polar monomers. The major difference is related to the steric requirements of the S_N2 reactions, namely the nucleophilic species (e.g. anion) should attack the electrophile (monomer) along the bond releasing the leaving group. Thus, in AROP this is, for instance, the C-O or C-S bond in the monomer molecule, whereas stereochemistry of an anion addition to the double bond is entirely different.

On the other hand, several similarities are observed between AROP and the polymerization of polar unsaturated monomers, such as acrylates, methacrylates, or vinylketones [1].

In this paper the progress in anionic ring-opening polymerization is reviewed and whenever possible the results are compared with those in the anionic polymerization of vinyl monomers.

AROP may involve a large variety of chemical structures, due to the ambident reactivity of some cyclic monomers. This is particularly apparent for monomers with unsymmetrically substituted rings but nevertheless able to open in two different ways. Similar phenomena are observed in the polymerization of methacrylates, where carbanions can be in equilibrium with enolates. Thus, the first part of this paper describes methods for determining the structure of the growing species.

The electrical charge is usually more concentrated on the heteroatoms in the AROP than it is in the anionic polymerization of unsaturated compounds. Examples of the charge concentration in models of active centers are given below [2,3]:

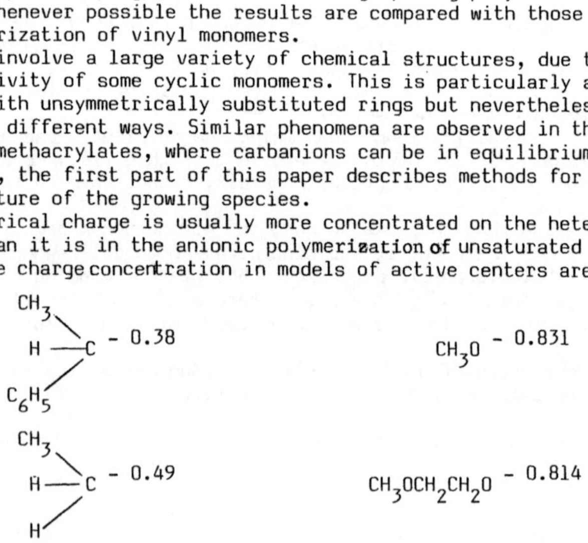

(3)

This high concentration of charges makes dissociation of the ion-pairs more difficult (stronger Coulombic attraction). Therefore, the ion-pairs and their aggregates prevail in AROP.

AROP involves polar monomers with electrondonating heteroatoms, which after polymerization become a part of the polymer chain. This leads to complicated solvation phenomena. Solvation of the fragments of growing species changes with conversion of monomer into polymer, like in the polymerization of ethylene oxide. Thus, counterions in the macroion-pairs can specifically be solvated, internally or externally, with a pertinent kinetic effect (observed also in methyl methacrylate).

The next important feature of the AROP is the formation of macrocyclics. The ring-chain equilibria kinetics and thermodynamics have recently been elaborated on by taking into account two processes, namely end-biting and back-biting of the active centers to any of the polymer units.

Finally, some of the cyclic monomers are composed of rings fitted with exocyclic groups. Ring-opening (reaction of the endocyclic bond) leads to propagation whereas reaction of the exocyclic group leads to the termination of the material chain and transfer of the kinetic chain.

In spite of many complications stemming from these peculiarities of AROP, a number of systems has been found which behave as living ones and allowing well defined polymers to be prepared. These include block and graft copolymers as well as new macromonomers, based, for instance, on AROP of β-propiolactone and giving graft copolymers with crystallizable side chains.

These problems, briefly outlined in this introduction, are being discussed in this paper.

CHEMICAL STRUCTURE OF THE GROWING SPECIES. AMBIDENT REACTIVITY OF MONOMERS AND MULTIPLICITY OF GROWING SPECIES

For some heterocyclic monomers the unique chemical structure of the growing species follows unequivocally from the monomer structure: the bond involving heteroatom is the most susceptible to the chemical change. However isomeric structures have to be taken into account in the case of ambident monomers. For instance, in the polymerization of thietane it was claimed that under certain conditions the carbanions and not the thiolate anions are the growing species [4]. Unsymmetrically substituted monomers can provide active species by α- or β-ring scission, like in the polymerization of substituted α-oxides, e.g.:

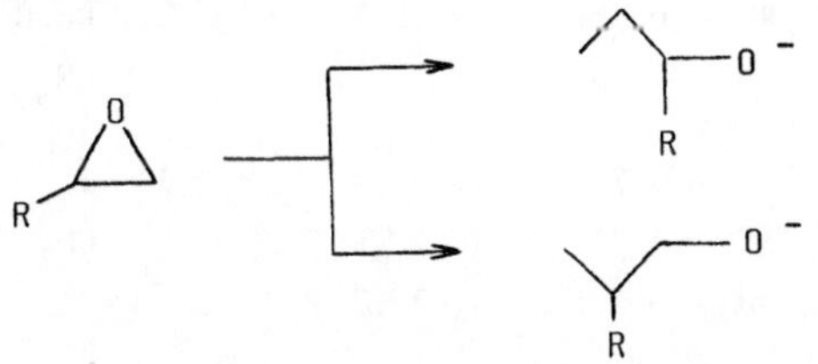

(4)

Lactones and dimers of α-hydroxycarboxylic acids can polymerize by O-acyl or O-alkyl bond scission, giving correspondingly alcoholate or carboxylate anions as the growing species. Transition from one kind of species, formed in the initiation, to the other kind, actually propagating, has also been observed [5].

We have recently developed a method allowing the determination of the chemical structure of active centers, and is based on the anion-capping with $ClP(O)(OC_6H_5)_2$. The structure of the resulting phosphate end-groups is further determined by $^{31}P\{^1H\}$ NMR [6].

$$RX^{-},Mt^{+} \quad + \quad ClP(O)(OC_6H_5)_2 \longrightarrow RXP(O)(OC_6H_5)_2 \quad + \quad MtCl \qquad (5)$$

where X= heteroatom. Chemical shifts of the phosphate end-groups are sensitive to the structure of the macromolecular substituent. Hence, the comparison of chemical shifts of phosphate end-groups with those of model compounds enables determination of the structure of the parent anions. This method is related to the trapping of cations, employing phosphines and was reported earlier from our laboratory [7]:

$$...-R^{+} \quad + \quad PR'_3 \longrightarrow ...-\overset{+}{R}PR'_3 \qquad (6)$$

Some examples of anion capping are given below.

Table 1. Monomers, structures of trapped anions, and parent growing species. Measurements of $^{31}P\{^{1}H\}$ NMR [6]

Monomer	Product of trapping	^{31}P ppm	Growing species
$\overline{CH_2CH_2O}$ (cyclic)	$...-CH_2OP(O)(OC_6H_5)_2$	-11.5	$...-CH_2O^{-}$
$\overline{CH_2CH(C_6H_5)O}$ (cyclic)	$...-CH_2CH(C_6H_5)OP(O)(OC_6H_5)_2$	-13.0	$...-CH_2CH(C_6H_5)O^{-}$
	$...-CH(C_6H_5)CH_2OP(O)(OC_6H_5)_2$	-12.6	$...-CH(C_6H_5)CH_2O^{-}$
$\overline{CH_2CH_2OC(O)}$ (cyclic)	$...-CH_2C(O)OP(O)(OC_6H_5)_2$ ↓ pyrophosphate	-26.0	$...-CH_2CO_2^{-}$
$\overline{(CH_2)_4OC(O)}$ (cyclic)	$...-(CH_2)_4OP(O)(OC_6H_5)_2$	-12.6 *	$...-CH_2O^{-}$
$\overline{(CH_2)_5OC(O)}$ (cyclic)	$...-(CH_2)_5OP(O)(OC_6H_5)_2$	-11.5	$...-CH_2O^{-}$
$\overline{[(CH_3)_2SiO]_3}$ (cyclic)	$...-Si(CH_3)_2OP(O)(OC_6H_5)_2$ ↓ pyrophosphate	-26.0	$...-Si(CH_3)_2O^{-}$
$\overline{CH_2CH(CH_3)S}$ (cyclic)	$...-CH_2CH(CH_3)SP(O)(OC_6H_5)_2$	+18.9	$...-CH_2CH(CH_3)S^{-}$

* Ref. 8

Integration of the signals of capped active centers and of diphenyl chlorophosphate allows also the concentration of active centers to be determined.

Table 2. Determination of concentration of active centers (AC) and initiator (I) (converted according to Eqn (5) into phosphates) by ^{31}P NMR [6]

Initiator	Monomer	$[I]_o$ mol/l	$[I]$ mol/l	$[AC]$ mol/l	$[I] + [AC]$ mol/l
$CH_3O^- K^+$	$\overline{CH_2CH(C_6H_5)O}$	$1.2 \cdot 10^{-1}$	-	$1.4 \cdot 10^{-1}$	$1.4 \cdot 10^{-1}$
$CH_3O^- K^+$	$\overline{(CH_2)_5C(=O)O}$	$2.8 \cdot 10^{-2}$	-	$2.4 \cdot 10^{-2}$	$2.4 \cdot 10^{-2}$
$(CH_3)_3SiO^- K^+$	$\overline{[Si(CH_3)_2O]_3}$	$5.8 \cdot 10^{-2}$	-	$6.0 \cdot 10^{-2}$	$6.0 \cdot 10^{-2}$
$C_6H_5S^- Na^+ \cdot DBC^*$	$\overline{CH_2CH(CH_3)S}$	$1.5 \cdot 10^{-3}$	$6.8 \cdot 10^{-4}$	$8.1 \cdot 10^{-4}$	$1.49 \cdot 10^{-3}$

* DBC - dibenzo-18-crown-6

Thus, capping provides a possibility to determine the isomeric structures during polymerization and to measure their proportions and rates of interconversion. It has to be made sure in every particular system that the rate of capping is sufficently higher than the rate of interconversion.

Another method, elaborated recently, is based on the formation of different structures in the reaction of carboxylate and alcoholate anions with 2,4,6-trinitroanisole [5]. Carboxylate anions react with 2,4,6-trinitroanisole giving picrate anions:

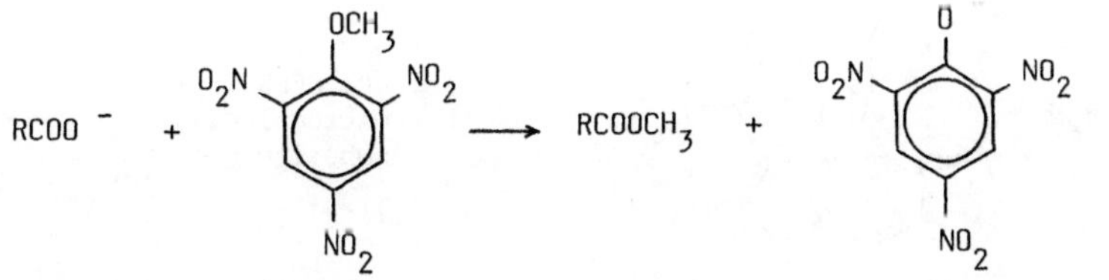

(7)

Alcoholate anions participate in the reversible reaction leading to the formation of a Meisenheimer complex [9]:

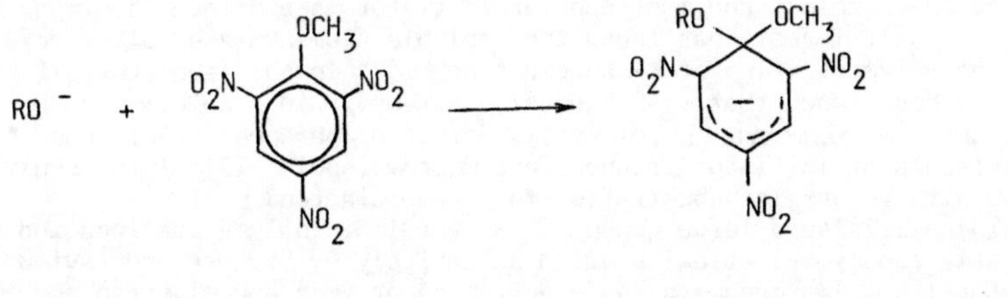

(8)

Picrate anions absorb at $\lambda = 380$ nm, whereas the Meisenheimer complex absorbs at $\lambda = 420$ nm and 495 nm. Thus, differences in the UV spectra of polymerizing mixtures with added 2,4,6-trinitroanisole can be used to discriminate between carboxylate and alcoholate active species. When 2,4,6-trinitroanisole is used in excess the concentration of formed picrate anions, determined from UV spectra, is equal to the concentration of parent carboxylate anions.

Table 3. Concentrations of carboxylate anions and measured concentrations of picric anions after addition of 2,4,6-trinitroanisole. Solvent DMF [5]

[2,4,6-trinitroanisole] mol/l	$[CH_3COO^-K^+DBC]$ mol/l	$[..-CH_2CH_2COO^-K\cdot{}^+DBC]$ mol/l	[Picric anions] mol/l
$1.02\cdot10^{-2}$	$9.38\cdot10^{-4}$	-	$8.70\cdot10^{-4}$
$1.50\cdot10^{-2}$	$2.69\cdot10^{-3}$	-	$2.51\cdot10^{-3}$
$3.08\cdot10^{-2}$	$5.03\cdot10^{-3}$	-	$5.20\cdot10^{-3}$
$8.24\cdot10^{-2}$	$8.32\cdot10^{-3}$	-	$9.20\cdot10^{-3}$
$1.93\cdot10^{-2}$	-	$1.89\cdot10^{-3}$	$1.65\cdot10^{-3}$
$5.76\cdot10^{-2}$	-	$3.96\cdot10^{-3}$	$3.95\cdot10^{-3}$

$...-CH_2CH_2COO^-\ K^+\cdot DBC$ - active centers in the living polymerization of β-propiolactone initiated with $CH_3COO^-\ K^+DBC$.

The described methods resemble the application of Michler's ketone in the determination of living polystyryl anions (Trotman and Szwarc [10]), and even the earlier radical-trapping methods [11].

Recently, Inoue has provided detailed information on the chemical structure of active centers in polymerizations initiated with metalloporphyrines [12,13]. This successful initiating system was characterized directly by ^1H NMR and the structure of the active species was shown to be:

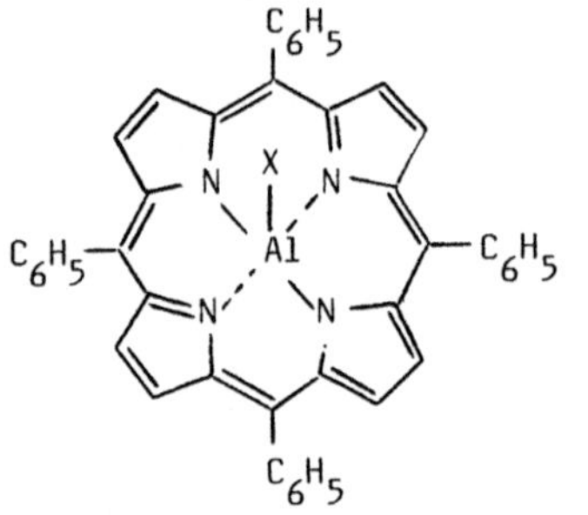

X = OR, OCOR
R = macromolecular substituent (9)

There is an important group of metal alkoxides, particularly aluminum alkoxides, used extensively in the polymerization of cyclic ethers and esters. The simplest and most popular initiator is aluminum isopropoxide, $(i\text{-}C_3H_7O)_3Al$. Teyssié has found that soluble aluminum-zinc alkoxide gives a living polymerization of ε-caprolactone [14]. In our laboratory it has recently been shown that $(i\text{-}C_3H_7O)_3Al$, tetrameric in a solution of CH_2Cl_2, initiates the polymerization of cyclic esters of phosphoric acid and that one molecule of initiator produces one macromolecule [15]. This result agrees with an earlier observation for ε-caprolactone [14].

There is also a large group of zwitter-ionic polymerizations and particularly copolymerizations studied extensively by Saegusa and reviewed recently [16]. Two monomers, an electron donor (M_D) and electron acceptor (M_A) mixed together produce zwitter-ion M_DM_A, giving for many pairs alternating copolymers. The zwitter-ionic structures of active species has also been detected by ^1H NMR. (Ref. 16 and references cited therein).

Although the methathesic ring-opening polymerization of cyclic olefins is out of the scope of the present review, the recent direct NMR observation of the active species, namely tungsten carbenes and tungsten cyclobutane derivatives, complements NMR data of active species in cationic and anionic ring-opening polymerizations. The ^{1}H NMR data, reported by Ivin et al [17], have shown that both species mentioned above can be simultaneously formed:

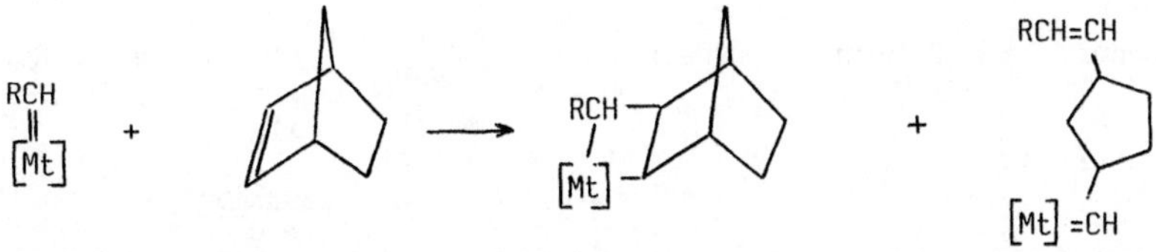

(where Mt = W)

(10)

although the conditions of equilibria involving these species are not yet established.

In all of the systems described above active species are located at the end of the chain; these processes therefore can be classified as "active chain end (ACE)" polymerizations, in contrast to the " activated monomer (AM)" polymerization, where a monomer molecule bears an electrical charge in the actual propagation step. Progress was made in this area when the living polymerization of lactams and the synthesis of monodisperse polyamides by an AM mechanism was disclosed by Sebenda in 1981 [18]. The latter occurs by properly choosing a monomer in which hydrogen atoms, prone to abstraction (and, therefore, responsible for transfer), are substituted by alkyl groups, and the strain of the four-membered ring is high enough to use milder than usual conditions for polymerization. Thus, α-methyl-α-butylbutyrolactam, initiated with its Li salt in the presence of N-pivaloyllactam.

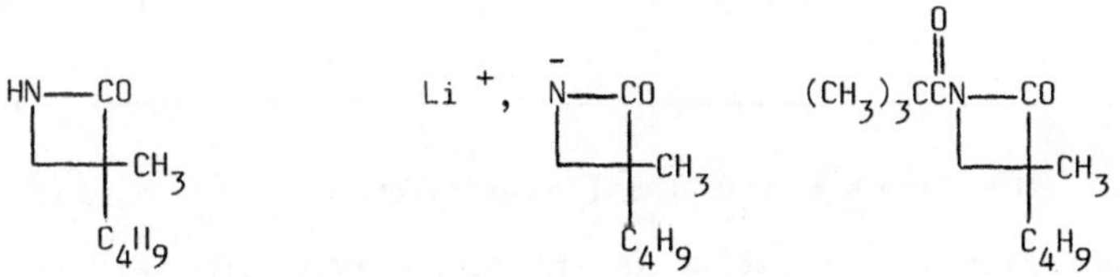

(11)

yields nearly monodisperse polymers in a living process.

There is a number of new initiating systems proposed for AROP, but they usually lead to the already known structures of growing species. It has recently been shown, for instance, that initiation of βPL polymerization with potassium solutions leads to the carboxylate growing species, i.e. the same that operates when CH_3COOK is used as initiator [19]

MACROIONS, MACROION-PAIRS AND THEIR AGGREGATES

There are only a few systems in the AROP for which quantitative data are available, namely, ethylene oxide - alcoholate anions (mostly in THF solvent) [1], β-propiolactone - carboxylate anions (CH_2Cl_2 or DMF) [20], ε-caprolactone - alcoholate anions (THF) [21], propylene sulfide - thiolate anions (mostly THF) [22], and neat cyclosiloxanes - silanolate anions [22]. Dissociation data, summarized in Table 4, are consistent with a high concentration of negative charge on the heteroatom, when compared with the much weaker charge localization on carbanions. For example, the much stronger interaction between the ions in potassium alcoholate ion pairs results in a K_D value which is nearly 10^3 times smaller than for potassium polystyryl salts ($K_D=7\cdot10^{-8}$ mol/l, 25°C, THF [25]).

Table 4. Dissociation constants (K_D) and interionic distances (a) * of some macroion-pairs in the AROP.

Monomer	Solvent	Ion-Pair	K_D mol/l	$\frac{\Delta H_D}{\Delta S_D}$ $\frac{kcal/mol}{e.u.}$	a Å	Ref.
$\overline{CH_2CH_2C(O)O}$	CH_2Cl_2	$..-CH_2CH_2COO^-K^+DBC$	$5.0\cdot10^{-5}$	$\frac{0}{-18}$	2.7	19
- " -	DMF	- " -	$2.5\cdot10^{-4}$	$\frac{-4.8}{-32}$	-	20
$\overline{(CH_2)_5C(O)O}$	THF	$..-C(O)(CH_2)_5O^-K^+$	$4\cdot10^{-10}$	$\frac{-}{-}$	2.8	21
$\overline{CH_2CH_2O}$	- " -	$..-CH_2CH_2O^-K^+$	$1.8\cdot10^{-10}$	$\frac{-6.7}{-68}$	2.7	23
- " -	- " -	$..-CH_2CH_2O^-K^+\cdot[222]$	$3.10\cdot10^{-7}$	$\frac{-}{-}$	3.2	24
$\overline{CH_2CH(CH_3)S}$	- " -	$..-CH_2CH(CH_3)S^-K^+$	$6.0\cdot10^{-10}$	$\frac{-8.2}{-61}$	3.0	25
- " -	- " -	$..-CH_2CH(CH_3)S^-$ $Na^+\cdot[222]$	$4.2\cdot10^{-5}$	$\frac{-0.4}{-20}$	6.9	26

* a : calculated from $K_D=(3000/4\pi Na^3)\exp(-e^2/aDkT)$

By the same token, aggregation is much more important for alcoholate ion-pairs. The size of the counterions also affects the extent of aggregation. For example, in the polymerization of ε-caprolactone [21,28] replacement of Na^+ with the larger K^+, or complexing Li^+ with cryptands in the polymerization of D_3 [22], increases the effective cation radius and completely disrupts the aggregation.

PROPAGATION OF MACROIONS AND MACROION-PAIRS

There are only a few monomers for which the propagation rate constants for ions (k_p^-) and ion-pairs ($k_p^{\pm}$) have been measured, namely: ethylene oxide [22], propylene sulfide [22], β-propiolactone [19,20], and α-methyl-α-propyl-β-propiolactone [29]. In the polymerization of ε-caprolactone with K^+ [21] and of D_3 with Li^+. [211][22] only $k_p^{\pm}$ were determined, because the contribution of ions to the propagation could not be determined due to their low concentration. Some values of the rate constants are given in Table 5. Rate constants for the anionic polymerization of styrene [30] and methyl methacrylate are included for comparison [31,32].

In the polymerization of cyclic siloxanes the major difficulty in measuring individual rate constants in solvents of low polarity is due to the formation of pairs of ion-pairs:

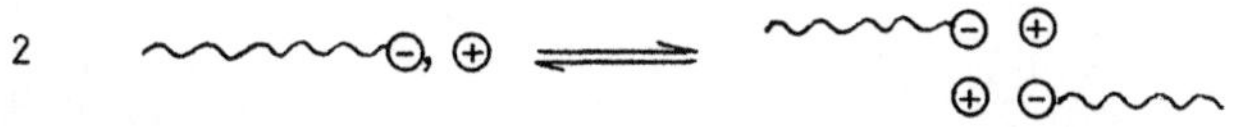

(12)

as well as of higher aggregates [33]. Boileau has found an easy solution to this problem by adding strongly cation-solvating agents (cryptands) in order to disrupt the quadrupoles and higher agglomerates [22]. The value of $k_p^{\pm}$, measured in C_6H_6 solution for D_3 in the presence of Li^+. [211], is listed in Table 5.

Table 5. Propagation rate constants of ions (k_p^-) and ion-pairs ($k_p^{\pm}$) in anionic polymerization

Monomer	Solvent	Temp. °C	Counterion	k_p^- 1/(mol·s)	$k_p^{\pm}$ 1/(mol·s)	Ref.
$\overline{CH_2CH_2C(O)O}$	CH_2Cl_2($[\beta PL]_o$=	25	K^+·DBC	$1.6\cdot10^{-1}$	$7\cdot10^{-4}$	19
- " -	3.0 mol/l)	-20	- " -	$6.2\cdot10^{-4}$	$1.1\cdot10^{-4}$	19
- " -	DMF ($[\beta PL]_o$=	40	- " -	$6\cdot10^{-1}$	$2\cdot10^{-1}$	20
- " -	1.0 mol/l)	20	- " -	$3\cdot10^{-2}$	$4\cdot10^{-2}$	20
$\overline{CH_2C(CH_3)(C_3H_7)C(O)O}$	THF	-20	K^+·[222]	$1.3\cdot10^{-2}$	$1.3\cdot10^{-1}$	29
$\overline{(CH_2)_5C(O)O}$	- " -	20	K^+	-	4.7	21
$\overline{CH_2CH_2O}$	- " -	- " -	- " -	-	$4.8\cdot10^{-2}$	22
- " -	- " -	- " -	Cs^+	-	$1.2\cdot10^{-1}$	22
- " -	- " -	- " -	K^+·[222]	1.67	$2.5\cdot10^{-2}$	22
$\overline{CH_2CH(CH_3)S}$	- " -	-30	Na^+	3.8	$2.5\cdot10^{-3}$	22
- " -	- " -	- " -	Cs^+	-	$2.3\cdot10^{-1}$	22
- " -	- " -	- " -	Bu_4N^+	3.5	1.7	22
- " -	- " -	- " -	Na^+·[221]	3.0	8.9	22
- " -	- " -	- " -	Na^+·[222]	5.6	11.9	22
D_3	benzene	20	Li^+·[211]	-	1.4	22
CH_2=C(CH_3)COOH_3	THF	-75	Cs^+	$6\cdot10^4$	$8\cdot10^1$	31
- " -	- " -	-20	- " -	-	$2.9\cdot10^3$	32
- " -	- " -	- " -	Na^+	-	$2.2\cdot10^3$	32
- " -	DME	- " -	- " -	-	$4.2\cdot10^3$	32
CH_2=CHC_6H_5	THF	25	Li^+	-	$1.5\cdot10^2$	30
- " -	- " -	- " -	Na^+	$6.5\cdot10^4$	$8\cdot10^1$	30
- " -	- " -	- " -	K^+	-	$5\cdot10^1$	30
- " -	- " -	- " -	Rb^+	-	$5\cdot10^1$	30
- " -	- " -	- " -	Cs^+	-	$2\cdot10^1$	30

In Fig. 1 we combined the known data for propagation of propylene sulfide [26] and methyl methacrylate [34], indicating some similarities; $\ln k_p^{\pm}$ is plotted for propylene sulfide as a function of a^{-1} (a is the distance between ions in the ion-pair) and as a function of the (ionic radius)$^{-1}$ for methyl methacrylate. Comparison of the relative reactivities of ions and ion-pairs ($k_p^-/k_p^{\pm}$) shows that in the polymerization of propylene sulfide and β-lactones, in contrast to the reported behavior of vinyl systems [30-32], the reactivity of macroion-pairs can approach or even exceed that of macroions.

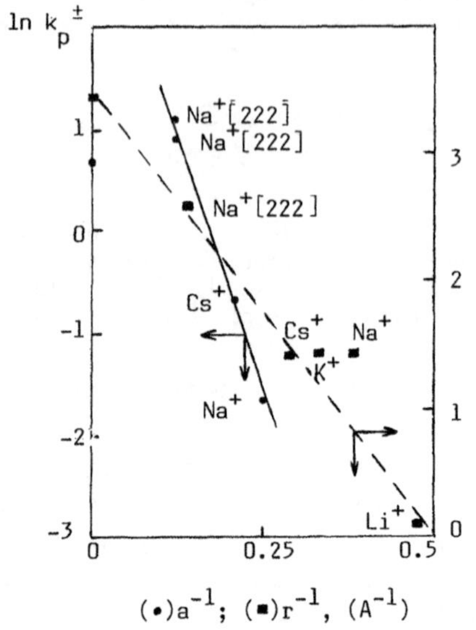

Fig. 1 Comparison of the dependence of $\ln k_p^{\pm}$ in the polymerization of propylene sulfide on a^{-1} (a=distance between ions in ion-pair) [26] and dependence of $\ln k_p^{\pm}$ in the polymerization of methyl methacrylate on r^{-1} (r=ionic radius) [34]

It is worth noting that in the polymerization of ε-caprolactone the rate constant of ...-$CH_2CH_2O^-$ K^+ ion-pairs, are higher than those in the polymerization of ethylene oxide (c.f. Table 5). This manifests the higher reactivity of εCl (active centers are the same, namely alcoholate anions).

Only recently the first data on the kinetics of anionic polymerization by AM mechanism have become available. Frunze et al studied the kinetics of anionic polymerization of ε-caprolactam in the melt, at 150°C [35]. Li, Na, and K salts of ε-caprolactam were used as initiators, and N-acetyl-ε-caprolactam was the activator. From the dissociation constants of ε-caprolactam salts,measured under these conditions,it was concluded that less than 10% of the salt was dissociated into ions ($[salt]_o \cong 10^{-3}$ mol/l). Apparently the initiation is neither instantaneous nor quantitative. Nevertheless, an attempt was made to measure both k_p^- and $k_p^{\pm}$. Their values are not included in Table 5, where only polymerizations proceeding by active chain end are collected. Frunze et al found that $k_p^- > k_p^{\pm}$, and that for a number of salts with different cations k_p^- is the same, whereas $k_p^{\pm}$ increases with decreasing size of the cation. Thus, the actual data measured at 150°C in bulk are $k_p^- = 67$ and $k_p^{\pm} = 1.45$, 0.95, and 0.15 (in l/(mol·s)), for Li^+, Na^+, and K^+ respectively.

SOLVATION

Heterocyclic monomers and their corresponding polymers are capable to solvate cations,due to the nucleophilicity of the heteroatoms. Polymers from unsaturated polar monomers (like poly(methyl methacrylate) or poly(2-vinyl pyridine)) can behave similarly.

These solvation processes are of particular importance in the studies of AROP of such monomers as ethylene oxide. Chain growth produces chains with increasing number of oxyethylene units, changing conditions and ability of solvation as a function of monomer conversion. However, information on solvation of alkali metal cations with polyoxyethylene oligomers was available from the earlier studies of the structure of some pairs (reviewed by Smid [36]), and in the presence of cation solvating additives [37],phase transfer catalysis (c.f. Ref 38 and papers cited therein), and from the studies of anionic polymerization. We do not intend to review these works in this paper.

Kazanskii [39] established more recently by ^{23}Na NMR and conductometry studies, that 6-8 oxygen atoms of poly(ethylene oxide) participate in formation of the solvation shell around each Na^+ cation (nitromethane solvent, temperatures from 25 to 80°C). Panayotov [40] studied interaction of poly(ethylene oxide) with potassium picrate in THF by UV spectroscopy and conductometry. It has been found that in the poly(ethylene oxide) potassium picrate complexes the number of ethylene oxide units coordinated with potassium cation equals 9±1.

According to Kazanskii [39] the solvation of cations with poly(ethylene oxide) is of a cooperative nature:

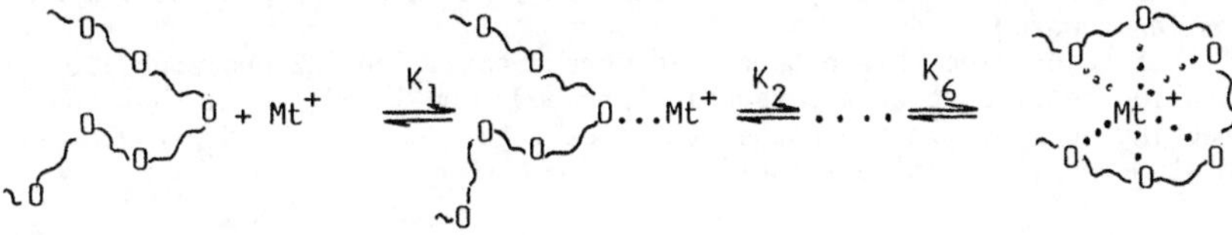

(13)

where $K_1 << K_2 < K_3 ... < K_n$, n is a maximal number of oxygen atoms which can be coordinated with metal cation.

The first equilibrium (K_1) represents a process with a loss of translational degrees of freedom. The coordination with subsequent oxygen atoms is accompanied only with a reduction of vibrational and rotational degrees of the polymer segments. Thus, the loss of entropy is the highest in the first step. Solvation of counterions by the penultimate polymer units was also observed in the polymerization of vinyl monomers with nucleophilic nitrogen or oxygen atoms, e.g. in the polymerization of 2-vinyl pyridine [41] and acrylates [32,42].

Recently Buschmann investigated the solvation of Na^+, K^+, Rb^+, Cs^+, Sr^{2+}, Ba^{2+}, and Ca^{2+} with poly(ethylene oxide) in methanol (solvent is not neutral in these interactions) by using calorimetric titration [43]. It has been found for K^+, Rb^+, and Cs^+ that the number of oxygen atoms of poly(ethylene oxide), contributing to solvation, increases with increase of the polymer chain length and levels off at ca 30 oxygen atoms. However, it does not mean that all 30 oxygen atoms are participating in the solvation shell. Only some of them interact directly at any given moment, and the "interacting" and "noninteracting" donor atoms exchange fast. The average heat of interaction is relatively high; e.g., it is equal to 111 kJ/mol for K^+ and 115 kJ/mol for Cs^+.

When highly polar and easily polarizable monomers are used, specific solvation takes place not only with the polymeric units but also with monomers themselves. For instance, the highly polar β-propiolactone (βPL) solvates the propagating carboxylate anions as indicated by our kinetic data [20]. This is evidenced from the fact that $\Delta H_p(-)$ depends on the starting monomer concentration for the polymerization of βPL in CH_2Cl_2 solvent, and even in the much more polar DMF. In the latter solvent, $\Delta H_p(-)$=21 and 27 kcal/mol for $[\beta PL]_0 = 5\cdot 10^{-1}$ and 1 mol/l, respectively [20]. It is understandable that solvation of the dipolar ion-pairs is much less influenced by changing the initial concentration of βPL. In fact, $k_p^{\pm}$ does not depend on $[\beta PL]_0$ whereas k_p^- does. Thus, the Arrhenius plot for k_p^- and $k_p^{\pm}$ can have an intersection point. For $[\beta PL]_0$ = 1 mol/l in DMF solvent this point is situated within a workable temperature range and was observed at 21°C [20]

A ratio $(k_p^{\pm} / k_p^-) > 1$ has also been reported by Boileau for the polymerization of α-methyl-α-propyl-β-propiolactone (-20°C) [29] and for the polymerization of propylene sulfide (-30°C) [22] (cryptated counterions). The dependence of this ratio on temperature is not known, and it is difficult to conclude whether the given explanation also applies to these systems. Nevertheless, the above described solvation of active centers with polar monomers requires that comparison of the reactivities of ions and ion-pairs takes into account the particular reaction conditions employed (solvent, monomer concentration, temperature).

CHANGES IN THE REACTION MECHANISMS

Sigwalt et al [22,26] observed for the anionic polymerization of propylene sulfide that although $k_p^{\pm}$ is a linear function of 1/a (where a is the interionic distance between ions in the macroion-pairs), k_p^{-} is smaller than $k_p^{\pm}$ for cryptated Li^+, Na^+, and Cs^+ counterions. This puzzling result has been explained by assuming that the additional energy required to pull away the cation from the reacting anion in the ion pairs, is compensated by an activation of monomer by the cation. In other words, a push-pull mechanism is proposed.

Let us examine this proposal in more detail. The S_N2 nucleophilic substitution requires that the reaction proceeds "on-line", i.e., that the attacking nucleophile approaches the substrate along the bond between carbon atom and (in this case) sulfur atom:

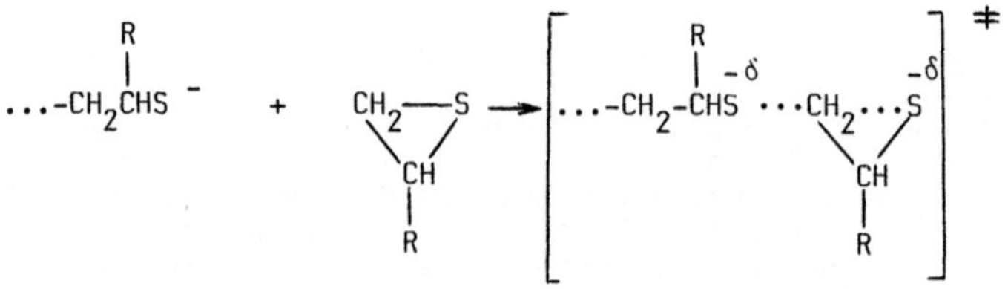

(14)

To preserve the same reaction path for an ion-pair the cation in the transition state would have to be located in a position unfavorable with respect to the nucleophilic sites:

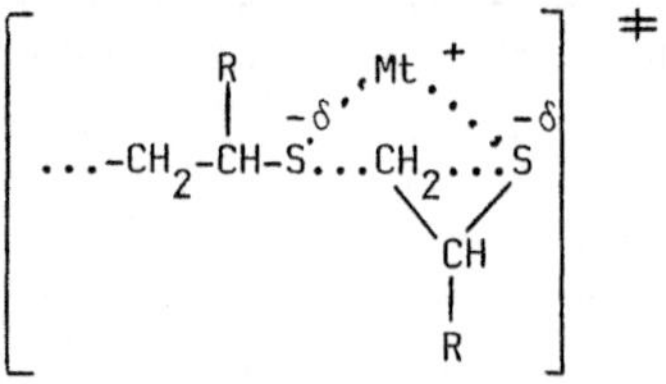

(15)

We would rather expect the positions of atoms in the transition state to be arranged differently, e.g.:

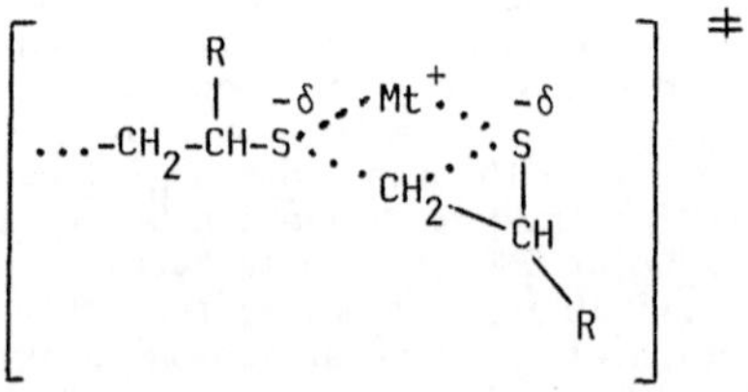

(16)

Thus, it is possible that the change in the mechanism of the reaction (its stereochemical path) is sufficient to cause the reactivity of the ion pair to exceed that of the free ion when Mt^+ approaches a certain size. In that case Coulombic attraction is sufficiently counter-balanced by the stereochemical (entropic) gain, due to the formation of more flexible transition state (Equ (16)).

MACROCYCLIZATION

These processes can be schematically represented in the following way:

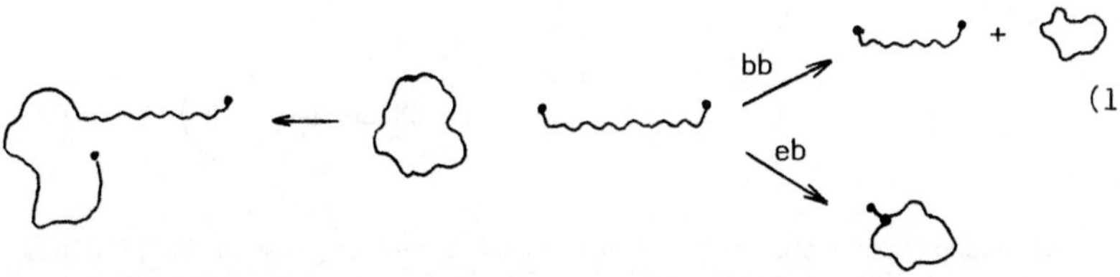

(17)

These interconnected processes are shown separately and as the irreversible ones for the sake of clarity. Thus, to the left hand side propagation proceeds involving cyclic species, i.e., monomer or any higher cyclic oligomer. This process can be considered as the backward reaction for the back-biting (bb).

Recently we analysed polymerizations with macrocyclization by computer simulation [44,45]. When in the scheme above the end-biting (eb) is faster than back-biting (because of higher inherent reactivity or because of a more favorable steric arrangement), then for strained monomers in the early stages of polymerization the concentration of rings should be higher than their equilibrium concentration [44,45]. With increasing chain length of a statistical linear chain this enhancement vanishes because the preferred conformational conditions for chain end encounter gradually disappear. For polymerizations with slow back-biting and slow end-biting the formation of linear polymer may be followed by its partial conversion into macrocyclics after the monomer is completely reacted. Thus, there are polymerizations with kinetic enhancement in macrocyclics as well as in linear polymer units.

Recently we investigated the kinetics of ε-caprolactone polymerization and the formation of cyclic oligomers [28]. From the dependence of $k_p(n)$ (propagation rate constants with participation of macrocycles) on n (number of monomer units in a macrocycle) we concluded that the reactivity of ester groups slightly decreases with increasing ring size. This effect was attributed to "conformational hindrance",resulting from the puckered conformation of rings of medium size. This makes some of the ester groups less accesible. The anionic polymerization of ε-caprolactone is an example of kinetic enhancement in linear polymer.

The observations have lead to synthetic proceedures which result either in predominantly linear of cyclic macromolecules, independently of the proportion and distribution of macrocyclic oligomers,when equilibrium is reached.

ENDO- AND EXOXYCLIC GROUP REACTIVITY

In some heterocyclic monomers there are endo- and exocyclic groups, linked to the site of attack by the same kinds of chemical bonds. The endocyclic bonds are involved in chain growth and the exocyclic ones may participate in chain transfer. The anionic polymerization of cyclic esters of phosphoric acid may be used as an illustration of such a process [46]. Thus, chain transfer is accompanied by a "wrong" monomer addition, e.g., addition of the monomer in a wrong way, giving a cyclic end-group instead of the linear polymer unit (Eq. (18)).

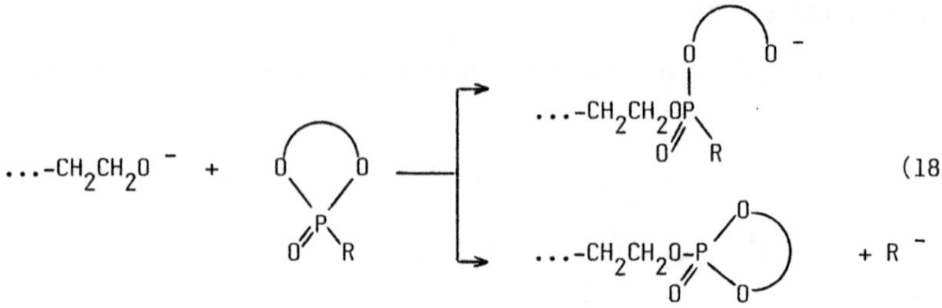

(18)

For example, in the polymerization of 2-methoxy-2-oxo-1,3,2-dioxaphosphorinane ($R=CH_3O-$) initiated with $CH_3CH_2O^- Na^+$ and carried out in bulk at 135°C, the proportion between proper and "wrong" monomer addition is ≈ 4 [47]. Every oligomer molecule contains one cyclic end-group as evidenced from a comparison between the DP_n, determined by vpo, and the proportion of P-atoms involved in these end-groups as measured by ^{31}P NMR.

The proper choice of exocyclic group, for instance R=H, eliminates chain transfer. The same effect was observed when the six-membered phosphoroamidate was polymerized [48]:

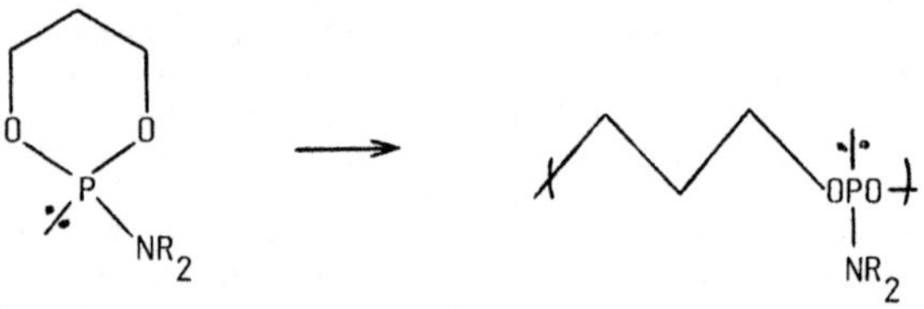

(19)

Propagation involves alcoholate anions and in the propagation step nucleophilic attack on the P atom takes place. This polymerization is a living one when initiated with, for example, $t\text{-}C_4H_9O^- K^+$ [48] in THF solvent.

SOME NOVEL SYNTHETIC APPLICATIONS

These can be classified as follows:

- synthesis of new polymers from some raw materials still available in abundance; these include CO, CO_2, elemental S, SO_2 etc,
- telechelics, macromonomers and derived block and graft copolymers (thermoplastic elastomers, particularly with polar and nonpolar segments, compatibilizing agents, amphiphilic copolymers),
- synthesis of speciality polymers, mostly the development of new applications, like biomedical polymers, polymer electrolytes and conductors, thermo- and photoresists,
- optically active polymers- understanding of the mechanisms of stereoelection and stereoselection, new methods of preparing optically active polymers.

We shall confine ourselves to the discussion of the first three groups, because the major aspects of the optically active polymers, developed mostly in France [49-53] and in Japan [54-57], have been recently discussed by Spassky [58,59] and Tsuruta [60] at this meeting as well as at the 5th IUPAC Symposium on Ring-Opening Polymerization in Blois (June,1986).

Anionic and pseudoanionic copolymerization of CO_2 with heterocyclics originated in Japan [61-63] and has a fairly long history. It led to the preparation and characterization of a number of aliphatic polycarbonates. Several properties of these polymers as thermoplastic materials are discussed in a recently published monograph (Enikolopyan adn Matkovskii), giving more than 400 references for various systems with CO, CO_2, SO_2, etc [64]. "Copolymerization" of CO_2 with cyclic oxides is also used in industrial processes for making cyclic carbonates [65].

Anionic polymerization of cyclic carbonates has been neglected, although these polymers can be of serious importance; for example, as components of solid electrolytes. Höcker et al. recently presented some detailed data on the homopolymerization of the cyclic carbonate: neopentylglycolcarbonate initiated with sec-butyllithium [66]:

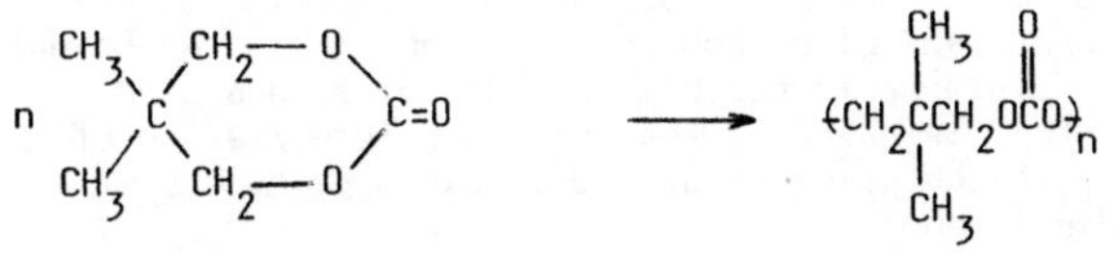

(20)

High molecular weight products were obtained. The ring-chain equilibria shows the $[Mn] = A\cdot n^{-5/2}$ dependence, indicating that this system obeys the Jacobson-Stockmayer theory. There appears to be a certain kinetic enhancement in macrocyclics, but this enhancement, as well as the polymer-monomer equilibrium, have not yet been fully described.

Another group of polymers, based on available raw materials, involves elemental sulfur. The ability of elemental sulfur to copolymerize anionically with cyclic mono- and polysulfides was discovered almost 10 years ago [67]. Since then a number of copolymers were prepared, and the kinetics and thermodynamics of the reactions were elucidated. The major results were already reviewed in 1980 [68]. It has been shown that monomers like ethylene sulfide, propylene sulfide, and isobutylene sulfide copolymerize with S_8 in the presence of simple anionic initiators (e.g. potassium thiophenolate). The cation in these initiators is usually complexed with a crown ether to ensure better solubility in the reaction media and to increase the reactivity. Thus, high molecular weight, elastomeric products were prepared in the direct copolymerization process of elemental sulfur with cyclic monosulfides

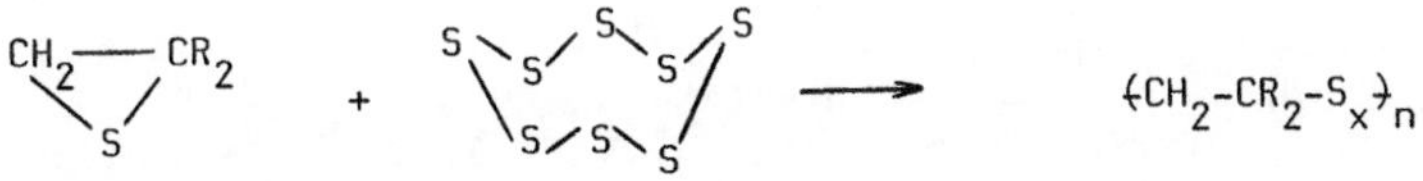

(21)

where R = H,H; H,CH_3; CH_3,CH_3
with x up to 9

A polysulfide bridge of 9 sulfur atoms is a thermodynamic limit because of the high probability to form the eight-membered cyclic sulfur monomer by back-biting of the growing species:

$$\ldots\text{-}CH_2\text{-}CR_2\text{-}\underset{9}{S}\text{-}\underset{8}{S}\text{-}\underset{7}{S}\text{-}\underset{6}{S}\text{-}\underset{5}{S}\text{-}\underset{4}{S}\text{-}\underset{3}{S}\text{-}\underset{2}{S}\text{-}\underset{1}{S}^{-} \underset{slow}{\overset{fast}{\rightleftharpoons}} \ldots\text{-}CH_2\text{-}CR_2\text{-}S^{-} + [S_8] \quad (22)$$

The other cyclic oligomers of sulfur (e.g. $[S_6]$) are much less stable than $[S_8]$, allowing formation of the $-S_9-$ polysulfide bridges.

The high thiophilic reactivity of carbanions has been used by Brossas et al. in their copolymerization of $[S_8]$ with preformed polymeric dianions [69]. The principle of this approach is shown schematically below:

$$Mt^+, {}^-CH_2\sim\sim CH_2^-, Mt^+ + [S_8] \rightarrow Mt^+, {}^-SS_xCH_2\sim\sim CH_2S_yS^-, Mt^+$$

$$Mt^+, {}^-SS_xCH_2\sim\sim CH_2S_yS^-, Mt^+ + Mt^+, {}^-CH_2\sim\sim\sim\sim CH_2^-, Mt^+ \longrightarrow$$

$$\longrightarrow \;\text{-(}CH_2\sim\sim\sim CH_2S_{y-p}CH_2\sim\sim\sim CH_2S_{x-q}\text{)-} \quad + \quad 2\ Mt_2S_r \qquad (23)$$

(where r=p+q+2)

In this way Brossas et al. were able to prepare block copolymers, containing polysulfide bridges and blocks of polystyrene, poly(α-methylstyrene), and polybutadiene.

The demand for new polymeric materials supports interest in telechelics, macromonomers, and graft and block copolymers. Recently, a convenient method for the synthesis of telechelic polysiloxane oligomers has been described [70-72]. This method is based in the anionic polymerization of cyclic siloxanes (e.g. D_4) in the presence of end blockers (e.g. bis(1,3-aminopropyl)tetramethyl disiloxane):

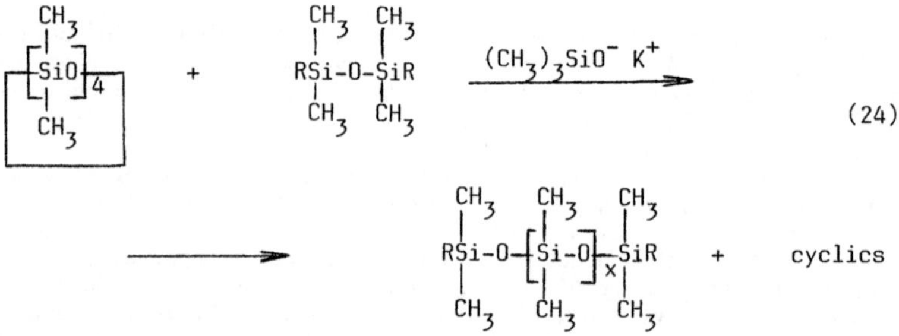

(24)

In an alternative method living silanolate end-groups are converted first into -SiH groups and the resulting polydimethylsiloxane (PDMS) prepolymer, terminated with -SiH, is then reacted with allyl compounds [73]:

$$D_3 \xrightarrow[THF]{nBuLi} nBuPDMSO^-\ Li^+ \xrightarrow{ClHSiMe_2} nBuPDMS\text{-}SiH$$

$$nBuPDMS\text{-}SiH \xrightarrow[Pt/C,\ toluene]{CH_2{=}CHCH_2X} nBuPDMSCH_2CH_2CH_2X \qquad (25)$$

$$X = OH,\ NMe_2$$

In this way polymers with DP_n = 30, 50, and 100 were prepared.

The synthesis of macromonomers has been reviewed recently [74-76], and we only will mention the recent synthesis of a new kind of graft copolymer by means of macromonomers, yielding copolymers with crystallizable side chains [77]. The macromonomers used had acrylic head groups and were obtained by polymerizing β-propiolactone with $CH_2{=}CHCOO^-\ K^+$ DBC as initiator.

Inoue has recently been able to polymerize a number of heterocyclics, mostly cyclic ethers and cyclic esters, using metalloporphyrins as initiators [13,78]. Polymers with a relatively high molecular weight, a narrow molecular weight distribution and a DP_n equal to the ratio of the concentrations of monomer to initiator have been obtained. Block copolymers were prepared with a high efficiency, giving additional evidence of the living character of the polymerization. Although useful in preparing various model macromolecules, the relatively high molecular weight of the initiator makes it difficult to remove it or its components from the polymer formed [78].

There is a continuous search for new polymers for biomedical applications. Particularly, polymers that are water soluble, degradable (either by simple hydrolysis or by enzymatic hydrolysis), and innocuous, or at least nontoxic in their polymeric form as well as in the form of the smaller oligomers formed during hydrolysis are of interest. Metabolites of these oligomers should also be nontoxic. Another requirement is a low immunogenic response. It is not easy to meet all of these requirements simultaneously. Natural polymers change their properties depending on the source, and therefore, synthetic polymers are preferred in a number of applications. Ring-opening polymerization is a method of choice. In this way polymers with backbones comprising of heteroatoms can be prepared. These backbones are more prone to hydrolysis and/or biodestruction.

There are three groups of polymers obtained by anionic ring-opening polymerization which have potential biomedical applications, namely, polyesters of phosphoric acid (poly(alkylene phosphates)), polyesters of carboxylic acids, and poly(α-aminoacids).

The first group has been reviewed by one of us a few times [79-81]. It suffices therefore to describe briefly the mechanism of the anionic polymerization of these monomers and to give some structural information.

Anionic polymerization of cyclic esters of phosphoric acid, phosphorous acid and cyclic phosphoroamidates proceeds by nucleophilic attack of the growing alcoholate anion on the P atom of the corresponding monomer (e.g. for cyclic phosphate):

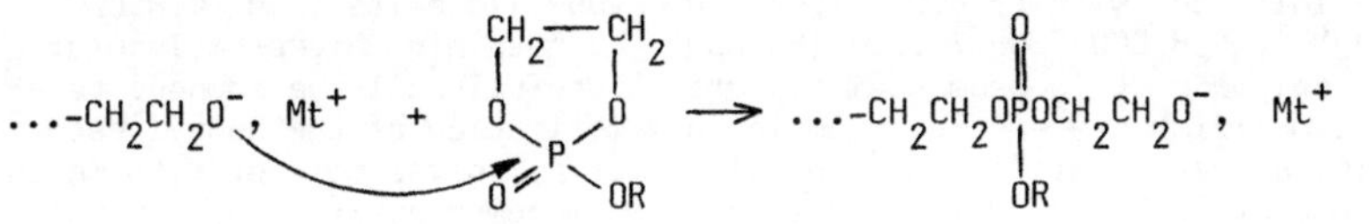

(26)

The most effective initiators,giving the highest polymerization degrees, are Al, Zn, or Mg alkyls or alcoholates. Particularly studied were $(i\text{-}C_3H_7O)_3Al$ [15] and $(i\text{-}C_4H_9)_3Al$ [82]. Polymers with - OP(O)(H)O-, -OP(O)(OR)O-, and -OP(NR_2)O- units in the chain were prepared in this way with $\bar{M}n$ up to 10^5.

The following structures, related to biopolymers, should be mentioned:

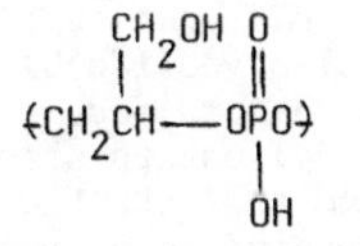

poly(glycerol phosphate)
(model of teichoic acid)

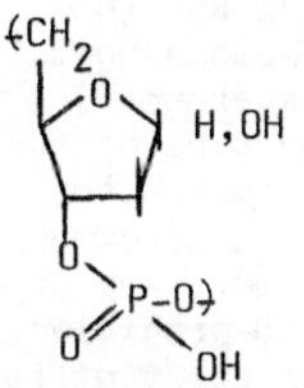

(27)

poly(deoxyribose phosphate)
(NA chain devoid of base)

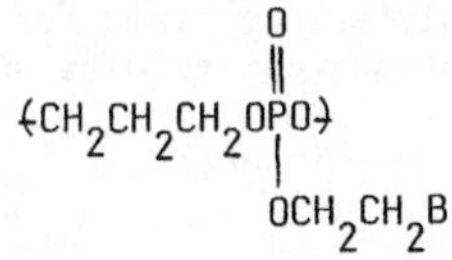

(where B = nucleic acid base (adenine, uracyl))

However, these polymers still have certain irregularities in the chains due to ring-opening in two nonequivalent positions.

The possible use of aliphatic polyesters in medicine as bioresorbable sutures and temporary prosthetic aids drew attention to polymers obtained from substituted β-propiolactones, glycolide, and lactide. Recently, Vert presented a comprehensive review on biomedical applications of polymers derived from lactide and substituted lactones [83]. These polymers are susceptible to slow hydrolysis in vivo to nontoxic hydroxyacids.

The main objectives in this area are:

- the design of synthetic methods enabling control of molecular weight and high purity of polymers,
- to determine the relationship between physical properties and microstructure of these materials,
- modification of their hydrolytic stability (e.g. by copolymerization with other lactones)

The above mentioned polyesters are usually obtained by anionic and pseudoanionic ring-opening polymerization. Substituted β-propiolactones, e.g.,

(28)

$R_1 = C_2H_5$
$R_2 = C_2H_5, C_3H_7, C_4H_9$

(benzyl ester of malolactonate)

can be initiated with triethylamine and carboxylic salts (e.g. $C_6H_5COO^- (C_2H_5)_4N^+$, $C_6H_5COO^- Na^+$ DBC) [84-86]. For obtaining higher molecular weight polymers it is essential to purify carefully all the components of the polymerizing system. For example, polymerization of the benzyl ester of malolactonate, purified only by distillation, gives polymer with Mn in the range of 2000 to 7300, whereas the same monomer purified by two passes through HPLC and two distillations yields polymer with Mn up to 50000 [82].

Glycolide and lactide can be polymerized conveniently in bulk at elevated temperatures (e.g. at 130°C for L-lactide [87]), using stannous octoate as initiator. The detailed mechanism of this polymerization has not yet been clarified. Similar initiators are often used because their residues in polymer are considered to be nontoxic.

Recently, Kricheldorf et al [88] investigated the copolymerization of glycolide and ε-caprolactone initiated with cationic and anionic initiators including $(i\text{-PrO})_3Al$ and $(nBu)_2Sn(OMe)_2$. Copolymers with molar ratios of glycolide/ε-caprolactone units from 1 to 17.5 and with polyglycolide blocks containing on the average from 1.8 to 14 monomer units were obtained.

Physical properties of the related optically active polyesters strongly depend on their microstructure. Prud'homme a.o. [89] found that atactic β-substituted poly- -propiolactones do not crystallize whereas the corresponding isotactic polymers are crystalline. It is worth noting that isotactic disubstituted poly(β-propiolactones) of opposite configurations form stereocomplexes with properties different from their parent polymer. For example, atactic poly(α-methyl-α-ethyl-β- propiolactone) melts at 122°C, isotactic melts at 164°C, and the stereocomplex of R and S polymers melts at 192°C [89]. Such findings are essential in designing polymeric systems for certain end-uses, including the biomedical applications of polyesters discussed above.

Polymerization of the N-carboxy anhydride of α-aminoacids,

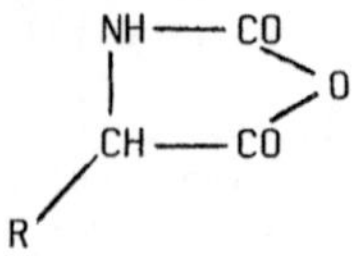

(29)

first applied in 1906 (80 years ago!) by Leuchs in the preparation of polypeptides [90] is still an active field of ring-opening polymerization research. In 1981 alone three papers were published on this topic in Pure & Appl. Chem.: by Sekiguchi [91], Imanishi [92], and Goodman [93] respectively. The AROP, which proceeds to high molecular weight polymers, has been reviewed many times, but the mechanism of propagation is still under debate. The polymerization of NCA has provided a number of important models of biopolymers(polypeptides) and led to the preparation of biomaterials extensively used in the biomedical field. This has been reviewed recently by Walton [94] and by Anderson [95].

Recently Harwood has discussed the existing mechanisms and has critisized the activated monomer mechanism of propagation which is supported by many researchers [96]. There are several mechanisms proposed for the AROP polymerization of NCA. It is agreed, that the mechanism depends on the initiator used and on the polymerization media. Difficulties involved in the studies of these polymerizations are due to the high sensitivity of NCA to traces of H_2O and other common impurities, leading to "self polymerization", and insolubility of polymers in the common solvents used for polymerization. Moreover, the nascent polymer may form helical structure, and this affects the polymerization kinetics.

We shall refrain from discussing the controversies on the polymerization mechanisms. The interested reader may consult a lucid summary of the existing mechanisms, their backgrounds and critiques given by Harwood in the above mentioned paper [96] as well as in a number of review papers by Sekiguchi; e.g. Ref. [97] and papers cited therein. Methods of ions trapping and/or end-capping, coupled with the modern NMR techniques may furnish new data and are likely to make the present discussion, based chiefly on the analysis of kinetic measurements less speculative.

Literature

1. S. Penczek, K.S. Kazanskii, Vysokomol.Soed. A25, 1347 (1983).
2. Yu.Ye. Eizner, B.L. Erusalimsky, Europ. Polym. J. 12, 59 (1976).
3. P.A. Berlin, V.L. Lebedev, A.A. Bagatur'yants, K.S. Kazanskii, Vysokomol, Soed. A22, 1800 (1980).
4. M. Morton, R.F. Kammereck, L.J.Fetters, Br. Polym. J. 3, 120 (1971).
5. A. Hofman, S. Slomkowski, S. Penczek, Makromol. Chem. 185, 91 (1984).
6. S. Sosnowski, A. Duda, S. Slomkowski, S. Penczek, Makromol. Chem., Rapid Commun. 5, 551 (1984).
7. K. Brzezinska, W. Chwialkowska, P. Kubisa, K. Matyjaszewski, S. Penczek, Makromol. Chem. 170, 2491 (1977).
8. A. Hoffman, S. Slomkowski, S. Penczek, Makromol. Chem. submitted.
9. C.A. Fyfe in "The Chemistry of the Hydroxyl Group", Ed. S. Patai, Interscience, London - New York, 1971, vol, 1 p. 95.
10. J. Trotman, M. Szwarc, Makromol. Chem. 37, 39 (1969).
11. P. Sigwalt in "Chimie macromoleculaire", Ed. Champetier, R. Buvet, J. Neel, P. Sigwalt, Hermann. 1970, vol. 1 p.93.
12. T. Aida, S. Inoue, Macromolecules 14, 1162 (1981).
13. S. Inoue, ACS Polymer Prep. 25(1), 225 (1984).
14. A. Hamitou, T. Ouhadi, R. Jerome, Ph. Teyssie, J. Polym. Sci. - Chem. 15, 865 (1977).
15. J. Libiszowski, S. Penczek, in preparation.
16. S. Kobayashi, T. Saegusa in "Alternating Copolymers", Ed. J.M.G. Cowie, Plenum Press, New York, London, 1985, p. 189.
17. J. Kress, J.A. Osborn, R.M.E. Greene, K.J. Ivin, J.J. Rooney, 5th IUPAC Symp. on the Ring-Opening Polymerization, Blois, 1986, Preprints p, 137.
18. J. Sebenda, J. Hauer, Polym. Bulletin 5, 529 (1981).
19. Z. Jedlinski, P. Kurcok, M. Kowalczuk, Macromolecules 18, 2679 (1985).
20. S. Slomkowski, S. Penczek, Macromolecules 13, 229 (1980), S. Slomkowski, Polymer 27, 71 (1986).
21. S. Sosnowski, S. Slomkowski, S. Penczek, J. Macromol. Sci.-Chem. A20, 979 (1983).

22. S. Boileau, ACS Symp. Ser. 166, 283 (1981).
23. A.A. Solovyanov, K.S. Kazanskii, Vysokomol, Soed. A14, 1063 (1972).
24. A. Deffieux, S. Boileau, Polymer 18, 1047 (1977).
25. S. Boileau, P. Sigwalt, Europ. Polym. J. 3, 57 (1967).
26. P. Hemery, S. Boileau, P. Sigwalt, J. Polym. Sci., -Polym. Symp. 52, 189 (1975).
27. D.N. Bhattacharyya, C.L. Lee, J. Smid, M. Szwarc, J. Phys. Chem. 69, 612 (1965).
28. S. Sosnowski, S. Slomkowski, S. Penczek, L. Reibel, Makromol. Chem. 184, 2159 (1983).
29. J. Haggiage, P. Hemery, R.W.Lenz, Polymer 24, 578 (1983).
30. M. Szwarc "Carbanions Living Polymers and Electron Transfer Processes", Wiley, New York, 1968, chapter 8.
31. G. Löhr, G.V. Schultz, Makromol. Chem. 172, 137 (1973).
32. A.H.E. Müller, ACS Symp. Ser. 166, 441 (1981).
33. J. Chojnowski, M. Mazurek, Makromol. Chem. 176, 2999 (1975).
34. A.H.E. Müller, personal communication.
35. T.M. Frunze, V.A. Kotelnikov, V.V. Kurashov, Iu.A. Avakyan, L.B. Danilevskaya, S.P. Davtyan, IUPAC Symp. Macro´83, Bucharest, 1983, Abstracts, Sect. I p. 233.
36. J. Smid in "Ions and Ion Pairs in Organic Reactions", Ed. M. Szwarc, Wiley - Interscience, New York, 1972 p. 86.
37. M. Szwarc, Adv. Polym. Sci. 49, 1 (1983).
38. J.M. Harris, N.H. Hundley, T.G. Shannon, E.C. Struck in "Crown Ethers and PhaseTransfer Catalysis in Polymer Science", Ed. L.J. Mathias, C.E. Carraher, Jr., Plenum Press, New York, London, 1984, p. 371.
39. G.N.Arkhipovitch, S.A. Dubravskii, K.S. Kazanskii, A.N. Shupik, Vysokomol, Soed. A23, 1653 (1981).
40. D.K. Dimov, I.M. Panayotov, V.N. Lazarov, C.B. Tsvetanov, J. Polym. Sci., -Chem. 20, 1389 (1982).
41. C.F. Tien, T.E. Hogen-Esch, J.Am.Chem.Soc. 98, 7109 (1976).
42. W. Fowells, C. Schuerch, F.A. Bovey, F.P. Hood, J.Am.Chem.Soc. 89, 1396 (1967).
43. H.-J. Buschmann, Makromol. Chem. 187, 423 (1986).
44. K. Matyjaszewski, M. Zielinski, P. Kubisa, S. Slomkowski, J. Chojnowski, S. Penczek, Makromol. Chem. 181, 1469 (1980).
45. S. Slomkowski, Makromol. Chem. 186, 2581 (1985).
46. S. Penczek, G. Lapienis, P. Klosinski, Pure and Appl. Chem. 56, 1309 (1984).
47. G. Lapienis, PhD Thesis, Technical University of Lodz, 1975.
48. J. Pretula, K. Kaluzynski, S. Penczek, J. Polym. Sci.,-Chem. 22, 1251 (1984).
49. N. Spassky, ACS Symp. Ser. 59, 191 (1977).
50. Ph. Dumas, N. Spassky, P. Sigwalt, J. Polym. Sci. - Chem. 17 1583 (1979).
51. Ph. Dumas, N. Spassky, P. Sigwalt, J. Polym. Sci. - Chem. 17 1595 (1979).
52. Ph. Dumas, N. Spassky, P. Sigwalt, J. Polym. Sci. - Chem. 17 1605 (1979).
53. M. Sepulchre, A. Khalil, N. Spassky, M. Vert, Makromol, Chem. 180, 131 (1979).
54. T. Tsuruta, S. Inoue, N. Yohida, Y. Yokota, Makromol. Chem. 81, 191 (1965).
55. T, Uryu, H. Shimazu, K. Matsuzaki, J. Polym. Sci., Polym. Lett. 11, 275 (1973).
56. M. Ishimori, T. Hagiwara, T. Tsuruta, Y. Kai, N. Yasuoka, N. Kasai, Bull. Chem. Soc. Japan 49, 1165 (1976).
57. T. Hagiwara, M. Ishimori, T. Tsuruta, Makromol. Chem. 182, 501 (1981).
58. M. Sepulchre, A. Momtaz, N. Spassky, Polym. Prep. 27(1) 173 (1986).
59. C.L. Jun, A. Leborgne, N. Spassky, see ref. 17, p. 43.
60. T. Tsuruta, see ref. 17, p. 29. Full lecture to be published in Makromol. Chem. Symposia.
61. S. Inoue, H. Koinuma, T. Tsuruta, J. Polym. Sci. 81, 287 (1969).

62. S. Inoue, H. Koinuma, T. Tsuruta, Makromol. Chem. 130, 210 (1969).
63. S. Inoue, Chem. Tech. 6, 588 (1976).
64. N.S. Enikolopyan, P.Ye. Matkovskii "Inorganic Gaseous Oxides as Comonomers" (in Russian), Nauka, Moscow, 1985.
65. T. Tsuruta, Hydrocarbon Process. 54, 144 (1975).
66. H. Keul, H. Höcker, see ref. 17, p. 103.
67. S. Penczek, R. Slazak, A. Duda, Nature 273, 738 (1978).
68. S. Penczek, IUPAC Symp. on the Ring-Opening Polymerization of Heterocycles, Kárlový Vary, 1980, Symp. Prep.ML4.
69. J.M. Catala, J.M. Pujol, J. Brossas, see ref. 17 p 129.
70. J.E. McGrath, C.S. Elsbernd, P.M. Sormani, S. Kilic, see ref. 17 p. 9. Full lecture to be published in Makromol. Chem. Symposia.
71. I. Yilgor, E. Yilgor, M. Spinn, J.S. Riffle, R.S. Ward, see ref. 17, p. 91.
72. P.M. Sormani. R.J. Minton. I. Yilgor, P.J. Andolino Brandt, J.S. Riffle, C. Tran, J.E. McGrath, ACS Polym. Prep. 25(1), 227 (1984).
73. Y. Tezuka, H. Kazama, K. Imai, see ref 17, p. 95.
74. P.F. Rempp, E. Franta, Adv. Polym. Sci. 58, 1 (1984).
75. P. Rempp, E. Franta, Polym. Prep. 27(1) 181 (1986).
76. P. Kubisa, K.S. Kazanskii, S. Penczek, Usp. Khim. in press.
77. S. Slomkowski, S. Sosnowski, S. Penczek, G.N. Arkhipovich, K.S. Kazanskii, see ref. 17, p. 99.
78. S. Inoue, T. Aida, see ref. 17, p. 39.
79. S. Penczek, J. Polym. Sci. - Polym. Symp. 67, 149 (1980).
80. S. Penczek "Models of Biopolymers by Ring-Opening Polymerization of Cyclic Phosphorus Containing Compounds" in "Phosphorus Chemistry Directed Towards Biology", Ed. W.J. Stec, Pergamon Press, Oxford - New York, 1980, p. 133.
81. G. Lapienis, S. Penczek "Cyclic Compounds Containing Phosphorus Atoms in the Ring" in "Ring-Opening Polymerization", Ed. K.J. Ivin, T. Saegusa, Elsevier, London - New York, 1984, vol. 2 p. 919.
82. K. Kaluzynski, J. Libiszowski, S. Penczek, Macromolecules 9, 365 (1976).
83. M. Vert, see ref. 17, p. 25. Full lecture to be published in Makromol. Chem. Symposia.
84. R.W. Lenz, E.M. Minter, D.B. Johns, S.Hvilstead, ACS Symp. Ser. 286, 105 (1985).
85. R.W. Lenz, P. Guerin "Functional Polyesters and Polyamides for Medical Applications of Biodegradable Polymers" in "Polymers in Medicine", Ed. C. Chiellini, P. Giusti, Plenum Press, New York - London, 1983, p. 219.
86. P. Guerin, C. Braud, M. Vert, see ref. 17, p. 165.
87. R. Vasanthakumari, A.J. Pennings, Polymer 24, 175 (1983).
88. H.R. Kricheldorf, T. Mang, J.M. Jonte, Macromolecules 17, 2173 (1984).
89. D. Grenier, R.E. Prud'homme, A. Leborgne, N. Spassky, J. Polym. Sci.,-Chem. 19, 1781 (1981).
90. H. Leuchs, Ber. 39, 857 (1906).
91. H. Sekiguchi, Pure & Appl. Chem. 53, 1689 (1981).
92. Y. Imanishi, Pure & Appl. Chem. 53, 715 (1981).
93. M. Goodman, E. Peggion, Pure & Appl. Chem. 53, 699 (1981).
94. A.G. Walton "Polypeptides as Biomedical Materials" in "Biomedical Polymers Polymeric Materials and Pharmaceuticals for Biomedical Use", Ed. E.P. Goldberg, A. Nakajima, Academic Press, New York, 1980, P. 53.
95. J.M. Anderson in "Encyclopedia of Polymer Science and Technology", Second edition, Ed. H.E. Mark, N.G. Gaylord, N.M. Bikales, Interscience, New York, to be published.
96. H.J. Harwood, ACS Symp. Ser. 286,67 (1985).
97. H. Amouyal, B. Coutin, H. Sekiguchi, J. Macromol. Sci. - Chem. A20, 675 (1983).

CHIRAL ANIONIC COORDINATED POLYMERIZATION OF THIIRANES : ENANTIOMERIC RESOLUTION AND DISULFIDE LINKAGE FORMATION

M. SEPULCHRE, A. MOMTAZ AND N. SPASSKY
Laboratoire de Chimie Macromoléculaire, UA 24, Université Pierre et Marie Curie, 4 place Jussieu, 75252 Paris Cédex 05, France

ABSTRACT

The polymerization of monosubstituted racemic thiiranes with zinc (S) 2,2'-binaphtholate as chiral initiator system occurs with high enantioelectivity for methyl and ethyl substituted monomers ($k_S/k_R \sim 15$-20), the magnitude of resolution decreasing however with the bulkiness of substituent group.

For meso type monomers, such as cisdimethyl thiirane or cyclohexene sulfide, optically active polymers were also obtained with the same initiator.

The polymerization reaction is often accompanied by a production of olefins and low molecular weight products containing disulfide linkages.

INTRODUCTION

Ring-opening polymerization of heterocycles containing asymmetric carbon atoms in the ring

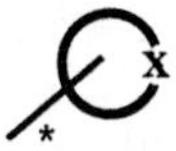

X = O, S, COO

can be achieved using anionic, cationic or anionic coordinated type initiators. The latter are usually obtained by reacting organometallic compounds such as $ZnEt_2$, $CdMe_2$, $AlEt_3$, ... with reagents having an active hydrogen (alcohols, glycols, amines, ...). When the reagent is chiral, optically active products can be obtained. A series of such chiral initiators were prepared and used for the polymerization of substituted oxiranes, thiiranes and β-propiolactones [1,2]. The best results were obtained with initiator system (I) derived from $ZnEt_2$ and R(-) 3,3-dimethyl 1,2 butanediol in 1 : 1 ratio. Recently it was shown that the initiator system (II) resulting from the reaction between $ZnEt_2$ and (S) 2,2'-dihydroxy 1,1'-binaphthyl gives in the polymerization of racemic methylthiirane an enantioelection of almost one order of magnitude higher than that obtained with the best previously known initiator mentioned above. The use of this atropoisomeric system was extended to a series of different monosubstituted and 2,3-disubstituted thiiranes and the results are now reported.

POLYMERIZATION OF MONOSUBSTITUTED THIIRANES

Kinetics equations

When a racemic monomer, an equimolar mixture of R and S enantiomers, is polymerized with a chiral initiator, one of the enantiomers is preferentially incorporated in the polymer chain and an optically active polymer is obtained. At the same time the unreacted monomer is enriched in the enantiomer of the opposite configuration. Thus, this process can be considered as a resolution reaction of a racemic mixture.

Recent Advances in Anionic Polymerization
T.E. Hogen-Esch and J. Smid, Editors

Monomer	Chiral initiator	Polymer		Unreacted monomer
(R + S)	R choice ⟶	(R > S)	+	(S > R)

The consumption equations for each enantiomer are :

$$-\frac{d[R]}{dt} = k_R\,[R]^a \qquad -\frac{d[S]}{dt} = k_S\,[S]^b$$

$$\frac{d[R]}{d[S]} = \frac{k_R}{k_S}\,\frac{[R]^a}{[S]^b} \qquad (1)$$

Assuming that k_R and k_S are constant, equation (1) can be integrated after introduction of the experimental values, i.e. α/α_o-optical purity of unreacted monomer and x-conversion in polymer.

In the case of a first order process, a = b = 1, the following resolution equation is obtained [3,4]:

$$(1 - x)^{r-1} = \frac{1 + \alpha/\alpha_o}{(1 - \alpha/\alpha_o)^r}$$

where $r = k_R/k_S$ is the "stereoelectivity constant" and expresses the preferential choice of the initiator for one of the enantiomers.

Some of the monomers obey a second order process in which a = b = 2 and a different type of equation is obtained [5] :

$$\left(\frac{\alpha}{\alpha_o}\right)^2(1 - x) + \frac{\alpha}{\alpha_o}\left(\frac{\rho_R + 1}{\rho_R - 1}\right) + x = 0$$

where $\rho_R = k_R/k_S$ is the stereoelectivity constant of the second order process.

Thus, depending on the nature of the monomer, the resolution efficiency may be expressed by two kinds of stereoelectivity constants which cannot be directly compared.

It is therefore often more practical to have a direct appreciation of the resolution efficiency by comparing the optical purity of unreacted monomer at a given conversion, particularly at half-reaction $(\alpha/\alpha_o)_{x/2}$.

Polymerization with zinc-(S)-binaphtholate initiator system (II)

Resolution of monomers. With the classical initiator (I) the stereoelectivity constant is equal to 2.3 in the case of the polymerization of racemic methylthiirane, which corresponds to an $(\alpha/\alpha_o)_{x/2}$ of 30 %. Using the atropoisomeric initiator (II) a stereoelectivity constant close to r = 20 was found [6], and the first order stereoelective curve obtained by plotting the optical purity of recovered monomer (α/α_o) versus conversion (x) was not far from the ideal curve corresponding to an exclusive choice of S-enantiomer (Fig. 1). The resolution at half-reaction $(\alpha/\alpha_o)_{x/2}$ approaches 80 % and an almost optically pure R-enantiomer is recovered at 67 % of conversion.

Initiator system (II) was then applied to the polymerization of other substituted thiiranes and the results of resolution are given in Table I. It appears that the resolution efficiency decreases with the bulkiness of the substituent in the order :

$$\text{Me} > \text{Et} >> \text{iPr} > CH_3OCH_2 = \text{tBu}$$

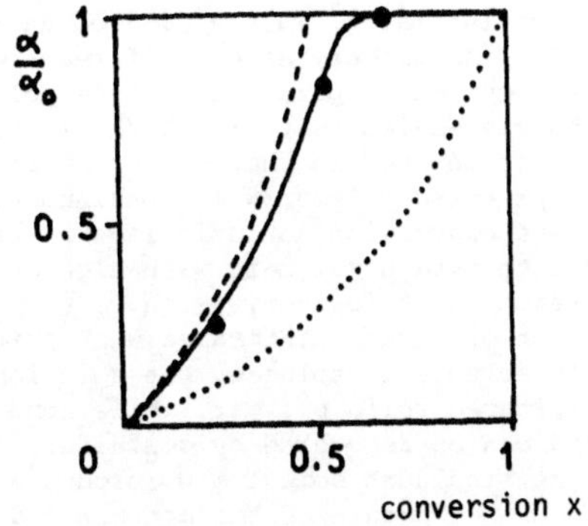

FIG.1. Stereoelective curves calculated for r = ∞ (---) ; r = 20 (——) and r = 2.3 (...). Experimental data obtained with system (II) for methylthiirane (●).

TABLE I. Resolution of racemic thiiranes using chiral initiator systems (I) and (II)

initiator		Me	Et	iPr	CH_3OCH_2	tBu
unreacted monomer	(I)	30	22	45	16	48
$(\alpha/\alpha_o)_{x/2}$ in %	(II)	80	63	31	16	17

This order is almost inverse of that observed with the classical initiator (I) (Table I). Thus, the atropoisomeric initiator (II) is particularly efficient in the cases of Me and Et thiirane, while comparable stereoelections are obtained with systems (I) and (II) for CH_3OCH_2 substituted monomers. System (I) is much better for iPr and t-Bu thiirane. The higher efficiency of system (II) for linear alkyl substituted monomer was substantiated by results obtained with n-butylthiirane. For the latter monomer, the α_o of the pure enantiomer is not yet known. Nevertheless at comparable conversion (∿ 70 %) the optical activity of unreacted monomer is five times higher with system (II) than with system (I) as shown in Table II.

TABLE II. Polymerization of racemic n-butylthiirane with initiator systems (I) and (II)

Initiator	Conversion (%)	α_D^{25} (neat,dm) unreacted monomer
(I)	74	- 8.3°
(II)	68	+ 42.2°

Properties of prepared polymers. In the enantioasymmetric polymerization, the polymer obtained is enriched in one of the enantiomers due to the preferential choice of the initiator. In the case of an ideal process, e.g. exclusive choice of S enantiomer, after half-reaction all the S enantiomer would be consumed and optically pure poly S type polymer should be obtained. In reality the stereoelectivity $r = k_R/k_S$ is not perfect and both of the enantiomers are incorporated in the polymer chain. Therefore the optical activity of the prepared polymer will continuously decrease with conversion and at complete conversion an optically inactive polymer is obtained. Thus, in order to obtain polymers with high optical purity it is necessary to stop the reaction at low conversion.

In the polymerization of racemic thiiranes with chiral initiators the crude polymer is usually soluble in toluene, the solution is poured into methanol giving a precipitated solid polymer, while some l.m.w. polymers remain in suspension and can be recovered by centrifugation. Generally the precipitated solid polymer includes some l.m.w. products. The amount of l.m.w. products depends on the nature of the monomer and of the initiator used.

With initiator system(I) h.m.w. polymers are usually obtained, with a GPC M peak up to 5×10^5 and eventually higher, while only very small amounts of l.m.w. products are present. On the contrary, with initiator system (II) substantial amounts of l.m.w. products ($6\text{-}10 \times 10^3$) are observed (Fig. 2).

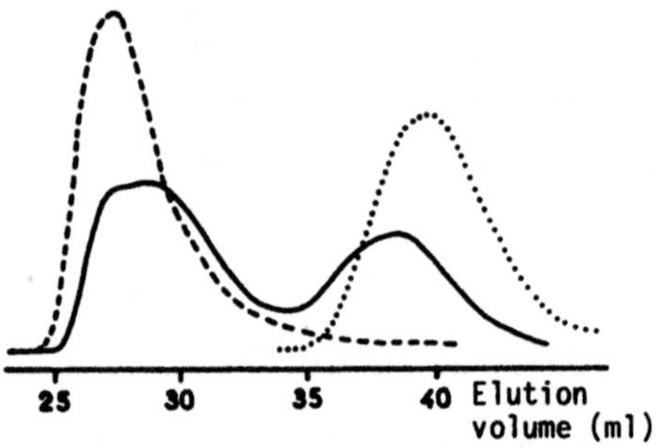

FIG.2. GPC curves of poly(n-butylthiirane)'s prepared with system (I) : (---) and with system (II): h.m.w. product (——), l.m.w. product (...).

The proportion of these products increases with the bulkiness and the degree of substitution on the asymmetric carbon as seen from Table III. In the case of 2-methyl,2-ethyl-thiirane only l.m.w. products are obtained.

TABLE III. Formation of l.m.w. products and disulfide linkages during the polymerization of racemic thiiranes with initiator system (II) at room temperature.

	R_1 , R_2	Me, H	Et, H	iPr, H	secBu, H	nBu, H	Me, Et
	l.m.w. (%)	10	26 13 [a]	43 28 [a]	46	48	100
y	p.s.p.[b]	1.0	1.03	1.11	1.28	1.17	-
y	l.m.w.	n.d.	n.d.	1.46	1.4	1.3	2.05

a) Polymerization at -30°C
b) Precipitated solid polymer

High molecular weight polymers. At low conversions the isolated h.m.w. polymers are of high optical purity, isotactic and usually crystalline. Some of the properties of such polymers obtained from differently substituted thiiranes are given in Table IV.

TABLE IV. Polymerization of thiiranes with initiator (II).

Polymerization in bulk at -30°C.

R_1, R_2	Conv. %	$[\alpha]_D$ (in C_6H_6)	o.p. (α/α_o) in %	m.p. (°C) Init.(II)	m.p. (°C) Other init.
Me, H	24	- 198[a]	∿ 100	63	55
Et, H	45	- 143	73	64	52
iPr, H	9[b]	- 180	66	-	-
nBu, H	28[b,c]	- 160	n.d.	61	45

[a] Fraction soluble in $CHCl_3$, [b] In the presence of tetrahydrothiofene
[c] Polymerization at 25°C.

It is interesting to notice that in the case of poly(methylthiirane) a fraction insoluble in toluene, but soluble in chloroform was isolated with $[\alpha]_D^{25}$ = -198 (C=0.42 $CHCl_3$) m.p. = 63°C which can be considered as an optically pure poly (S) enantiomer. The m.p. of h.m.w. isotactic polymers obtained with initiator system (II) are much higher than those observed before for polymers prepared with any other stereospecific initiator system.

Formation of -S-S-linkages. Simultaneously to the formation of l.m.w. products, a side reaction takes place during the polymerization. The products formed contain an excess of sulfur as compared to the normal monothioether linkage for which y = 1 in the monomer unit.

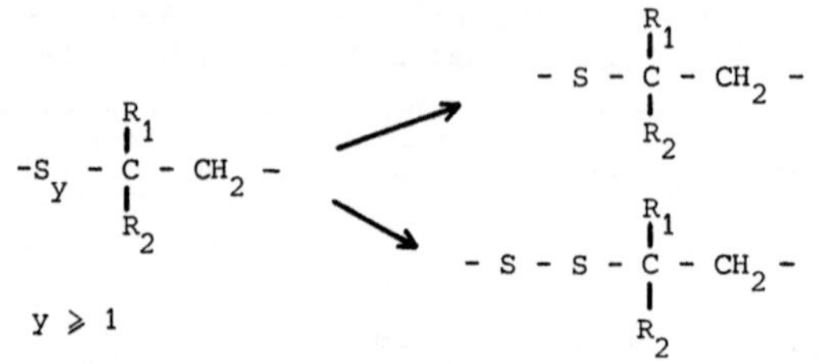

The amount of disulfide linkages varies with the degree of substitution and the bulkiness of substituents on the asymmetric carbon (Table III). While for methylthiirane practically no disulfide linkage was detected, in the case of 2-methyl, 2-ethylthiirane the obtained l.m.w. polymer contains almost exclusively disulfide linkages (y = 2). There are much more disulfide linkages in l.m.w. products than in the corresponding h.m.w. polymers. In fact, it is probable that the latter do not contain any disulfide linkages, and those detected are only due to contamination of h.m.w. polymer by residual l.m.w. products.

The formation of disulfide linkages is accompanied by the evolution of the corresponding olefin which is detected in the recovered monomer.

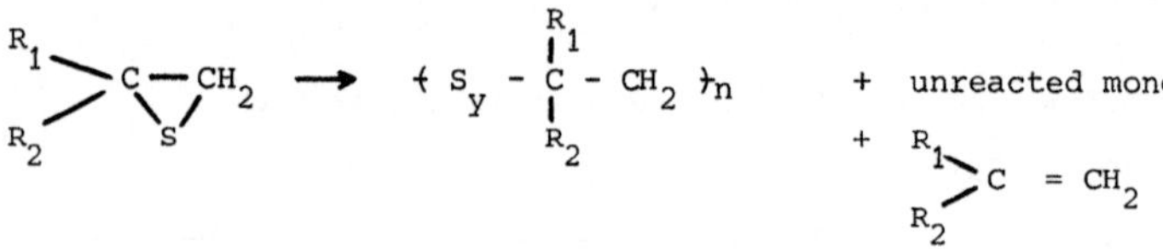

Again the amount of olefin increases with the increase of bulkiness on the asymmetric carbon (Table V).

TABLE V. Presence of olefin in the recovered monomer during the polymerization with Zn/(S)-binaphtholate initiator system (II).

R_1 , R_2	Et, H	iPr, H	nBu, H	tBu, H	Me, Et
Conversion in polymer x (%)	56	50	59	42	60
Olefin in residual monomer (%)	10	34	36	40	92

Identification of -S-S- linkages in the polymers. The average number of sulfur atoms per monomer unit can be calculated from elemental analysis. The presence of -S-S- linkage can be identified by studying the chiroptical properties of the polymers and by looking at the corresponding ^1H and ^{13}C NMR spectra.

Chiroptical properties. The chiroptical properties of products containing -S-S- linkages can be very different to those having the normal thioether linkage. For example, when examining the ORD of poly(n-butylthiirane) products, one finds for the h.m.w. polymer (y = 1.07) $[\alpha]_D^{25} = -119$ (C_6H_6), while $[\alpha]_D^{25} = +162.5$ (C_6H_6) for the l.m.w. (y = 1.30) product. Both are enriched in S enantiomer as concluded from the sign of Cotton effects observed in circular dichroïsm spectra. Cotton effects related to -S-S- linkage are found near 290 nm. They are of opposite sign to those related to normal thioether linkages which appear near 235-238 nm in optically active polythiiranes [7].

NMR spectra. The presence of disulfide linkages substantially modifies NMR spectra. Some typical examples are given.

In the ^{13}C NMR spectrum of poly(n-butylthiirane) containing polysulfide linkages, besides the peaks assigned to normal thioether linkages located at 47.2 and 37.6 ppm for the methine and methylene carbons of the main chain respectively, and at 33.5, 29, 22.6 and 14 ppm for the α,β,γ methylene carbons and the methyl carbon of the side chain as assigned by IVIN [8], several other complex patterns appear, mainly in the methine, methylene and α-methylene regions between 50 and 52 ppm, 44 and 46 ppm and around 36 and 32 ppm.

In the ^1H NMR spectrum (250 MHz, $CDCl_3$) of poly(2-ethyl-2-methylthiirane), prepared with system (I), one observes a triplet centered at 0.97 ppm attributed to the methyl protons of the ethyl group, a singlet at 1.32 ppm attributed to the protons of the methyl group directly linked to the asymmetric carbon atom, a quadruplet centered at 1.62 ppm corresponding to the methylene protons of the ethyl group, and a singlet at 2.68 ppm for the methylene protons of the main chain.

In the spectrum of the polymer prepared with initiator system (II), the main features are the splitting into two peaks at 1.34 and 1.4 ppm of the methyl signal, the broadening of the methylene proton signals of the ethyl group, the disappearance of the main chain methylene signal, and the appearance of a complex pattern between 3.1 and 3.37 ppm with a main peak centered at 3.12 ppm. From a comparison with NMR studies on 2,2-dimethylthiirane-sulfur copolymers, where mono, di, tri and tetrasulfide linkages were shown to appear at 2.85, 3.36, 3.47 and 3.60 ppm [9], it can be concluded that the polymer obtained with system (II) is an almost pure polydisulfide product. The splitting of the methyl signal and the complex pattern between 3.1 and 3.37 ppm is probably due to head-to-head and tail-to-tail irregularities.

Polymerization in the presence of tetrahydrothiofene

When the polymerization using initiator system (II) is run in the presence of small amounts of tetrahydrothiofene (THT), ([THT] / [Monomer] ∿ 1 % in moles) the formation of disulfide linkages is almost stopped and the residual monomer is practically free of olefin as seen in Table VI.

TABLE VI. Polymerization in bulk of substituted thiiranes in the presence of tetrahydrothiofene (THT) using initiator system (II).
[Initiator] / [Monomer] ∿ 2-3 % in moles.

Thiirane	[THT]/[Mon.] moles %	Temp. of polym. (°C)	Conversion in polymer (%)	Olefin in unreact. mon. (%)	$[\alpha]_D^{25}$ polymer (C_6H_6 C=0.4)
iPr	0	- 30	44	31.5	- 96.5
	1.18	- 31	9	0.6	-180
nBu	0	25	67.5	53.5	- 17
	1.4	25	28	2.0	-160

POLYMERIZATION OF 2,3-DISUBSTITUTED THIIRANES

Thiiranes containing two vicinal asymmetric carbons of opposite configuration bearing identical substituents are examples of monomers of meso structure. These thiiranes, when polymerized using chiral initiators, produce optically active polymers due to the preferential attack of one asymmetric carbon atom of a particular configuration involving an inversion of configuration of the latter as demonstrated in the case of oxiranes [10,11].

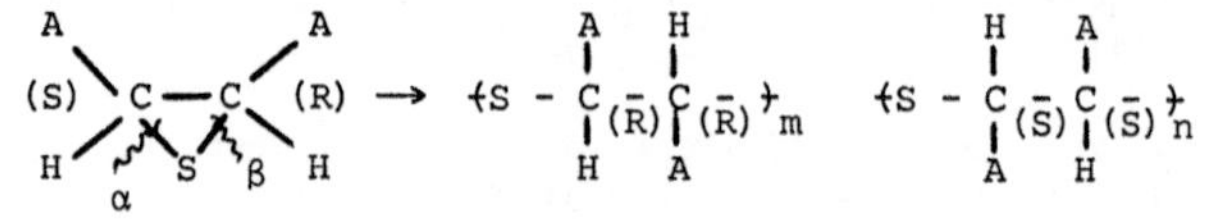

achiral initiator m = n
chiral initiator m ≠ n

Such an enantiogenic reaction corresponds to an asymmetric polymer synthesis.

Some stereochemical and mechanistic aspects of polymerizations using initiator system (I) and few other systems were previously described in the case of cis dimethylthiirane ($A = CH_3$) [12].

It is now reported that in the case of poly(cis dimethylthiirane) and poly(cyclohexene sulfide)tacticity, enantiomeric enrichment, and amount of disulfide linkages are dependent on the nature of initiator, e.g. system (II), the conditions of preparation of initiators and the use of complexation agents.

Poly(cis dimethylthiirane)

Polymerization with different chiral initiator systems. Some results of the polymerization of cis dimethylthiirane with initiator system (I) and (II) are given in Table VII. It appears that the insoluble species (C) of initiator (I) gives an isotactic polymer of high optical activity, while the soluble species (B) leads to an atactic polymer containing -S-S- linkages.

TABLE VII. Enantiogenic polymerization of cisdimethylthiirane using initiator systems (I) and (II)

Initiator system		POLYMER				
Type	Prep. a)	$[\alpha]_D^{25}$ b)	m/n	i % c)	y d)	$\Delta\varepsilon_{237}$ e) sign
	(A)	+ 42	2.9	55	1.1	-
(I)	(B)	+ 2	1	24	1.3	-
	(C)	+100	3.5	56	1.0	-
(II)	(A)	+ 74	3 f)	66	1.55	+
	(D)	-151	11 f)	85	1.05	+

a) Preparation of initiator : (A) The dihydroxy compound reacted with $ZnEt_2$ for 2 h at room temperature in toluene, then dried for 2 h at room temperature. (B) Soluble part of the initiator in toluene. (C) Insoluble part of the initiator. (D) Addition of tetrahydrothiofene simultaneously to the monomer in 1 : 1 volume ratio.
b) In $CHCl_3$ c) Triad content from ^{13}C NMR spectra d) S content in monomer unit e) CD measurement in cyclohexane. f) n/m

With initiator system (II) (A), the obtained polymer has an opposite sign to that expected from the configurational homochiral rules applied to the process [2]. Obviously the high amounts of -S-S- linkages are responsible for the inversion of the sign of rotatory power. Initiator (II) could not be fractionated in soluble and insoluble species. The formation of disulfide linkages was, however, almost completely suppressed by working in the presence of tetrahydrothiofene (D). In this case an isotactic levorotatory polymer with a high optical activity and a high melting point (130°C) was obtained.

Identification of -S-S- linkages.

CD spectra. CD measurements show that both polymers prepared with initiator system (II) have the same positive Cotton effect at 237 nm, i.e. the predominance of the same configurational unit in the chain. The inversion of sign observed in ORD curves is due to the presence of chromophoric groups related to -S-S- linkages which perturb the optical activity [12]. Two negative Cotton effects located at 283 and 318 nm are observed in the visible part of the spectra. They disappear for the polymer prepared in the presence of tetrahydrothiofene (Fig. 3).

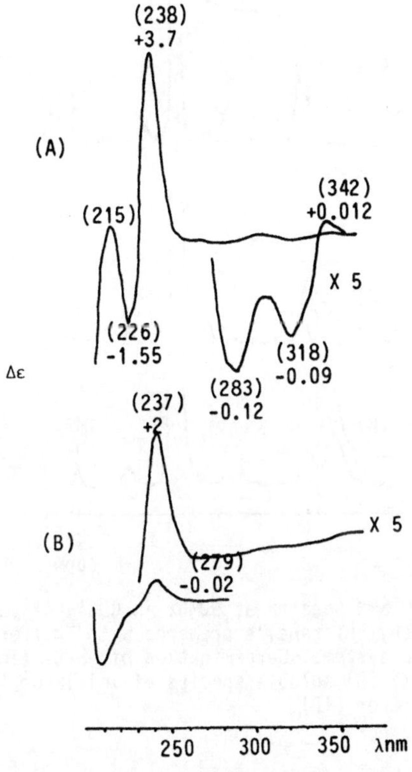

FIG.3. CD spectra in cyclohexane of poly(cis-dimethylthiirane)'s prepared with initiator (II). (A) polymer prepared in absence of THT [α] = +74 ; (B) polymer prepared in presence of THT [α] = -150.

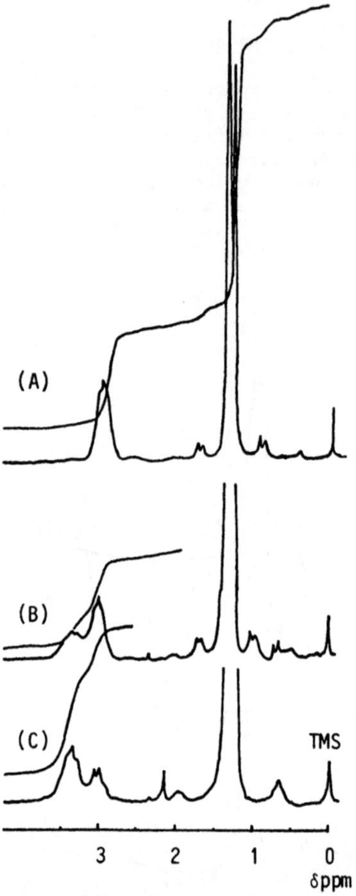

FIG.4. ^{1}H NMR spectra at 90MHz in $CDCl_3$ of poly (cisdimethylthiirane)'s prepared with different initiator systems. Determination of -S-S- linkages. (A) tBuOK, (B) soluble species of initiator (I), (C) initiator (II).

NMR spectra. The presence of disulfide linkages can be qualitatively and quantitatively determined by ^{1}H and ^{13}C NMR. In the ^{13}C NMR spectrum of a polymer prepared with initiator (II) several large patterns assigned to the presence of -S-S- linkages appear very clearly. For the methine carbon atom they are located between 47.0 and 51.0 ppm downfield of the normal signals due to monosulfide linkage. On the contrary, in the methyl carbon atom region, the peaks due to -S-S- linkages appear upfield to the normal signals [12].

^{1}H NMR spectra of poly(cis dimethylthiirane)'s were run at 90MHz in $CDCl_3$ solution. Two patterns are observed, one around 1.33 ppm, which is assigned to CH_3 protons, while the CH protons are found around 3.0 ppm. In the polymers containing an excess of sulfur, the central position of these peaks is slightly shifted downfield (0.02 ppm). The most interesting feature is the appearance of signals at 3.33 ppm due to -S-S- linkages. They are well separated from the signals due to monosulfide linkages (3.0 ppm) and can be easily integrated (Fig. 4).

The quantitative evaluation of different amounts of disulfide linkage in various polymers is given in Table VIII. The results due to ^{1}H NMR analysis are in good agreement with those calculated from the elementary analysis of sulfur. The examination of ^{1}H NMR spectra also allows one to distinguish the type of polysulfide linkage involved in the polymer (in the present case only disulfide linkages were found) and evaluate the tacticity. These results will be described later in a separate paper.

TABLE VIII. Disulfide linkage formation in the polymerization of cis DMT with different initiators.

$\left(S - CH(CH_3) - CH(CH_3) \right)_x$ $\left(S - S - CH(CH_3) - CH(CH_3) \right)_z$

Init.	x	z	% S a)	
			^{1}H NMR	elem. analy.
tBuOK	1.0	0	36.4	36.3
(I) b)	0.67	0.33	43.3	42.9
(II)	0.35	0.65	48.6	46.9

a) For x = 1, z = 0 % S calc. = 36.4
b) Soluble species : see (B) in Table VII.

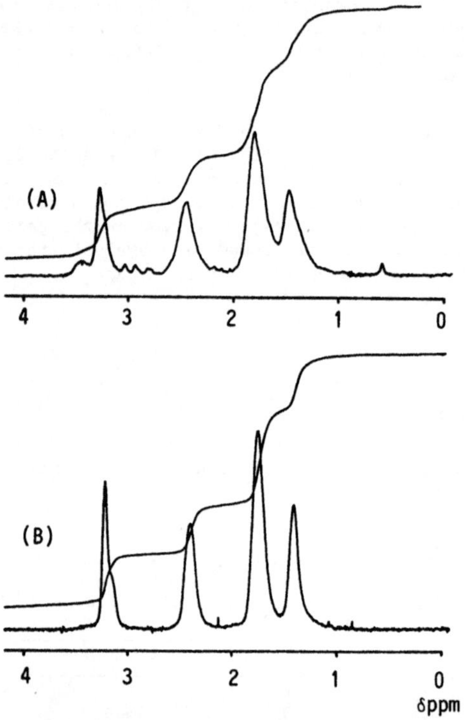

FIG.5. ^1H NMR spectra at 250MHz in C_6D_6 of poly(cyclohexene sulfide)'s prepared with different initiator systems. Determination of -S-S- linkages.
(A) initiator (II), (B) initiator (I).

Poly(cyclohexene sulfide)

Polymerization with different chiral initiators. Cyclohexene sulfide polymerizes readily with chiral initiator (I) leading to optically active polymers e.g. $[\alpha]_D^{25} = 42$ (C = 0.4 $CHCl_3$) . CD measurements in cyclohexane show the presence in dextrorotatory polymers of a negative Cotton effect located at 235 nm which, as in the case of poly(cisdimethylthiirane)'s, indicates a predominance of RR type configurational units in the chain. The polymer prepared with initiator system (II) contains an excess of sulfur but shows no optical activity at the D line. However the ORD curve reveals at 365 nm an optical activity $[\alpha]_{365}^{25} = -30$ (C = 0.45 C_6H_6). No comparative CD measurements could be performed due to the insolubility of this polymer in cyclohexane. The possibility of the participation of binaphthol end-groups to optical activity is not excluded. According to GPC measurements this polymer has indeed a much lower molecular weight ($M_n \sim 4\,500$) than the one prepared with initiator system (I) ($M_n \sim 60\,000$). Experiments in the presence of tetrahydrothiofene were not yet attempted.

Identification of -S-S- linkages by NMR. The polymer prepared with initiator (II) is insoluble in $CDCl_3$ and therefore the spectra were run in C_6D_6 solution.

The ^{13}C NMR spectrum is rather complex, many supplementary peaks are present. They are found downfield and upfield to the pattern of normal C_1 methine peaks located around 49 ppm, while for the carbons C_2 and C_3 of the cyclohexane ring, located respectively around 31.5 and 23.7 ppm, the patterns due to -S-S- linkages appear upfield to the main peaks.

The ^1H NMR spectrum (Fig. 5) of poly(cyclohexene sulfide) with normal thioether linkages is composed of four main peaks located at 3.21, 2.39, 1.74 and 1.40 ppm with respective intensities of 1 : 2 : 1 : 1. The peak at 3.21 ppm assigned to methine protons of the chain is stereosensitive to tacticity. In the polymer prepared with initiator (II), there is some broadening of all peaks and an appearance of new signals due to disulfide linkages. From integration of the peak located at 3.38 ppm downfield to the methine peak it is possible to determine the amount of disulfide linkage. The determined value (y = 1.18) is very close to the experimental value obtained from elemental analysis of sulfur (y = 1.16).

It is also worthnoting that in the case of poly(cyclohexene sulfide)'s prepared with initiator (I), the ^{13}C NMR spectra are strongly temperature and solvent dependent. Moreover, for polymers with normal thioether linkages the ORD curves are of opposite sign, i.e. negative in dioxane and positive in cyclohexane, chloroform, THF and benzene. These features are presently under investigation.

DISCUSSION AND CONCLUSION

From the results presented above some general conclusions can be drawn concerning the polymerization of resolvable and unresolvable monomers using chiral initiator systems, and more particularly initiator (II).

With this initiator the highest stereoelectivity, up to now, is observed in the case of substituted thiiranes. However, this almost ideal resolution works only for thiiranes which bear not very bulky substituents, e.g. methyl or ethyl groups, and, contrary to the results observed with initiator system (I), the stereoelectivity decreases with the increase of the bulkiness of the substituent group. For very bulky substituents, e.g. t-butyl group, this resolution efficiency is very low. It is also shown that chiral initiator systems are able to distinguish a particular chirality in meso type monomers, such as cisdimethylthiirane. Therefore, it is possible to prepare, particularly with initiator (II) under certain conditions, polymers with high enantiomeric enrichment from unresolvable monomers. Such an enantiogenic process corresponds to a polymer asymmetric synthesis.

However, the polymerization with system (II) is always characterized by some particular features.

i) In conjunction with h.m.w. polymer, l.m.w. products are formed. The amount of the latter increases with the bulkiness of the thiirane substituent group and, for example in the case of disubstituted thiiranes, e.g., 2-ethyl-2-methylthiirane, only l.m.w. are obtained.

ii) The l.m.w. products contain an excess of sulfur as disulfide linkages.

iii) The polymerization reaction is accompanied by the production of olefin which can be detected in the residual monomer.

Similar desolefination reactions with production of polysulfide linkages were already observed when using some anionic type initiators like zinc thiolate [13] and particularly complex lithium initiators such as nBuLi.LiOR and RSLi.xLiOR [14].

A mechanism similar to that proposed by ALIEV et al. [14], for polymerization with coordination type initiators involving the attack of the sulfur atom in the monomer by the thiolate growing chain with subsequent olefin elimination, can be proposed.

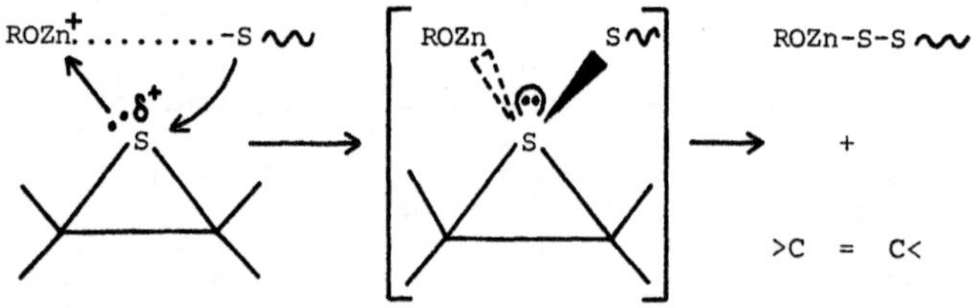

Another mechanism, with a 5-membered ring cyclic transition state involving the coordination of sulfur atoms in the polymer chain on the metallic atom and elimination of the olefin as before, is also possible :

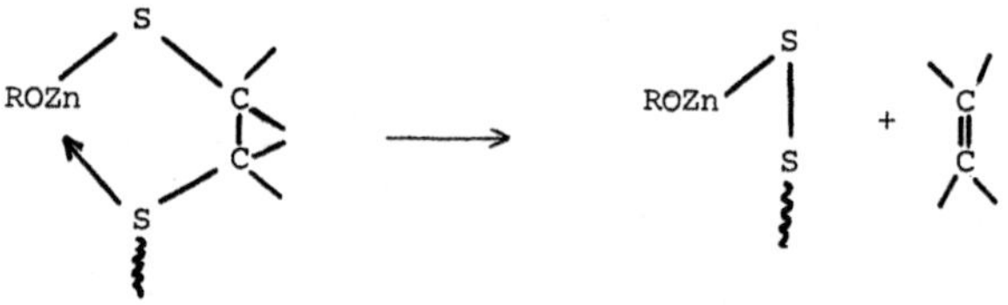

It was shown that in the case of initiator system (I) the soluble species are responsible for the formation of disulfide linkages. In the case of initiator system (II), it also seems that two types of species are present. Type A species produce h.m.w. polymer, type B species seem responsible of the formation of l.m.w. products and -S-S- linkages. GPC studies as a function of conversion indicate that the formation of l.m.w. products is a slower process than that of h.m.w., and eventually a scrambling occurs in the Mw distribution curves.

The role of tetrahydrothiofene consists in the inhibition of the activity of type B species, probably through preferential complexation of THT on the site. It was additionally checked that initiator system (II) is unable to polymerize tetrahydrofuran. This probably rules out a cationic type polymerization process in which addition of THT formed a stable sulfonium salt unable to reinitiate polymerization [15].

Finally, it must be pointed out that while chiral initiator (I) is a very versatile initiator, able to produce stereoelective polymerization of thiiranes, oxiranes and β-lactones, initiator system (II) is much more restricted. Although it is excellent for a few thiiranes, it polymerizes with great difficulty methyl oxirane, giving no appreciable stereoelection, and is not very efficient in the case of β-lactones.

REFERENCES

1. N. Spassky, A. Leborgne, M. Sepulchre, Pure and Appl. Chem., 53, 1735 (1981)
2. N. Spassky, A. Leborgne, A. Momtaz, M. Sepulchre, J. Polym. Sci., Polym. Chem. Ed., 18, 3089 (1980)
3. J.Furukawa,S. Akutsu, T. Saegusa, Makromol. Chem. 81, 100 (1965)
4. M. Sepulchre, N. Spassky, P. Sigwalt, Macromolecules 5, 92, (1972)
5. Ph. Dumas, N. Spassky, P. Sigwalt, J. Polym. Sci., Polym. Chem. Ed. 17, 1583 (1979)
6. M. Sepulchre, N. Spassky, C. Mark, V. Schurig, Makromol. Chem., Rapid Commun. 2, 261 (1981)
7. N. Spassky, P. Dumas, M. Sepulchre, P. Sigwalt, J. Polym. Sci., Symposium n° 52, 327 (1975)
8. K.J. Ivin, E.D. Lillie, I.H. Petersen, Makromol. Chem., 168, 217 (1973)
9. A. Duda, S. Penczek, Makromol. Chem. 181, 995 (1980)
10. E.J. Vandenberg, J. Polym. Sci., A1, 7, 525 (1969)
11. M. Yokoyama, H. Ochi, H. Tadokoro, C.C. Price, Macromolecules 5, 690 (1972)
12. N. Spassky, A. Momtaz, P. Sigwalt, in Coordination Polymerization Ed. by C.C. Price and E.J. Vandenberg, vol. 19 (Plenum Press) (1983), p.111
13. W. Cooper, Br. Polym. J. 3, 28 (1971)
14. A.D. Aliev, B.A. Krentzel, S.L. Alieva, Eur. Polym. J. 19, 71 (1983)
15. E.J. Goethals, Advances in Polym. Sci. vol. 23, (Springer Verlag) (1977) p. 103

SOME MECHANISTIC ASPECTS OF THE ANIONIC BLOCK COPOLYMERIZATION OF CAPROLACTAM AND POLYETHER DIOLS

M. K. AKKAPEDDI, G. J. DEGE, T. D. GALLAGHER AND M. S. WALSH
Polymer Laboratory, Corp. R&D, Allied-Signal, Morristown, NJ 07960

INTRODUCTION

The anionic polymerization of caprolactam was first reported by Joyce and Ritter in 1941 [1]. The polymerization catalyzed by metal caprolactamate is characterized by a high rate of conversion but with an induction period associated with the initiation step viz, the formation of the key acyllactam growth center. The mechanism (see Scheme I) accepted to date was first proposed by Hall [2] Mottus [3] and Sebenda [4] independently. The polymerization can be activated by the addition of suitable cocatalysts such as acyllactams [3,4] and carbamoyl lactams [5] which essentially eliminate the induction period.

Scheme I. MECHANISM OF ANIONIC POLYMERIZATION OF CAPROLACTAM

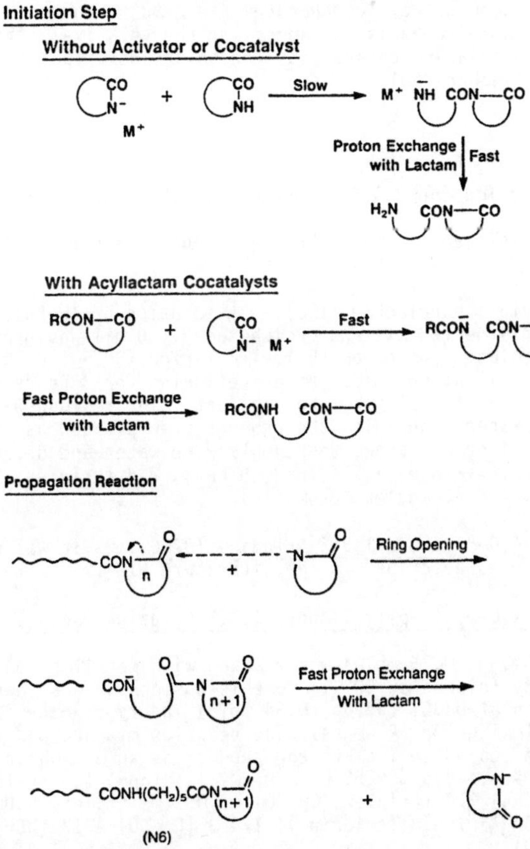

(N6)

Recent Advances in Anionic Polymerization
T.E. Hogen-Esch and J. Smid, Editors

Other cocatalysts which have been studied to a limited extent include certain sym-triazine compounds such as cyanuric chloride and its derivatives [6] which are capable of forming the active, caprolactam substituted triazine intermediates (Scheme II). The activating effect of triazine ring on the lactam is due to the strong electron withdrawing character of the ring nitrogens.

The fast rate of activated anionic polymerization permits the use of low reaction temperatures, below the melting point of nylon. This fact, coupled with the good mechanical properties of nylon 6 resulted in important industrial applications such as casting [7] and reaction injection molding (RIM)[8]. In addition, the process is also suitable for incorporation of elastomeric soft segments in the nylon chain via anionic block copolymerization. Use of hydroxy and amine terminated polyethers in conjunction with di and multifunctional cocatalysts for the block copolymerization of caprolactam has been reported in the recent patent literature [9,10] including that of the present author [11] with the primary objective of improving the impact strength of nylon 6.

In principle, the block copolymerization requires the attachment of cocatalyst moiety to polyether chain ends so that nylon 6 chain initiation can take place. The reaction between the difunctional cocatalyst and polyether diol leading to the prepolymer, may be carried out separately in the absence of lactam or may be generated "in-situ" (Scheme IV). In this paper we present some mechanistic aspects of the relative efficiency of bisacyllactams vs triazine cocatalysts for this block copolymerization reaction with polyether diols.

Experimental

Cocatalysts: -

Cyanuric chloride and 2,4,6-triphenoxy-1,3,5-triazine were obtained from Aldrich.

Terephthaloylbiscaprolactam (TBC) was prepared according to the literature [10].

Terephthaloylbislaurolactam (TBL): - A solution of laurolactam (59.3g, 0.3 mole) in pyridine (70 ml) and cyclohexane (100 ml) was gradually treated with a solution of terephthaloyl chloride (30.5g, 0.15 mole) in cyclohexane (100 ml) at 60-80°C. After refluxing for 3 hours the solution was concentrated by stripping off the solvent. The residue was treated with excess ice water upon which the product precipitated as creamy-white solid. It was filtered, washed thoroughly with water and dried (91%). mp 114-116°C; ^{1}H NMR: - δ(ppm) 1.4 (broad, 9 CH_2), 2.4 (triplet, CH_2CO) 4.0 (triplet, CH_2N), 7.65 (singlet, aromatic).

2-Dimethylamino-4,6-dichloro-1,3,5-triazine (DADCT): - It was prepared from cyanuric chloride as described in the literature [12].

2-Dimethylamino-4,6-bis(α-pyrrolidonyl)1,3,5-triazine (BpT)

2-Pyrrolidone (64.3g, 0.75 mole) was reacted with metallic sodium (15.7g, 0.68 mole) in dry toluene at 90°C. To this suspension was then added gradually a solution of DADCT (65.6g, 0.34 mole) in dry dixoane (260ml) followed by refluxing for 12 hours. The reaction mixture was filtered hot and the filtrate cooled to isolate the product as an off-white crystalline solid (58.2g, 60% yield); mp 255°C. ^{1}H NMR - δ (ppm) 4.0 (triplet, CH_2N); 3.2 (singlet, NCH_3) 2.6 (triplet, CH_2CO); 2.05 (multiplet, $-CH_2$) ^{13}C NMR - δ (ppm) 36.1 (NCH_3), 165.8 (triazine); 174.3 (C = O) 47.3 (NCH_2); 17.7 (CH_2); 34 (CH_2CO)

2,4,6 - Tris(α-pyrrolidonyl) 1,3,5 - triazine (TPT): - It was prepared from cyanuric chloride and sodium pyrrolidone in 30% yield as described above. mp 300°d. ^1H NMR: - δ (ppm) 4.2 (triplet, CH_2N); 2.6 (multiplet, CH_2CO); 2.1 (multiplet, CH_2)

2,4,6 - Tris(Imidazolyl) 1,3,5 - triazine (TImT) To a solution of imidazole (6.6 moles, 449.3g) in THF (1200ml) was added slowly a solution of cyanuric chloride (184g, 1.0 mole) in THF (675ml) and then refluxed for 6 hours. The contents were diluted with water (4-5 liters) and the product filtered off. Yield: - 94%; mp 255°C; ^1HNMR: - δ (ppm) - 7.4, 8.2, 8.8 (imidazole ring protons)

Catalyst: - Bromomagnesium caprolactam (CMB) solutions (0.2 to 0.4 M) in caprolactam were prepared by adding ethylmagnesium bromide (3.0 M in ether) to anhydrous, molten caprolactam at 100°C under nitrogen, while distilling off simultaneously the volatiles.

Polyether Diols: - Polyoxypropylene glycols, PPG 4025 (disecondary, Mn = 4,000) Niax E351 (etylene oxide capped, diprimary diol, Mn = 2,800) were obtained from Union Carbide. Poly G5528 (diprimary, Mn = 4,000) was obtained from Olin.

Polyether Prepolymers: - Polyether diol Poly G5528 (60g, 0.015 mole) BpT (8.7g, 0.03 mole), sodium methoxide (0.035g, 6×10^{-4} mole) and DMF (10 ml) were mixed and heated at 130°C for 2 hours, followed by vacuum stripping. The completion of reaction was monitored by IR (OH bond) and G.P.C. The final prepolymer typically showed three overlapping broad peaks of higher molecular weight compared to the starting polyol, suggesting end capped product with statistical distribution of chain extended by-products. Prepolymer from Niax E351 and TBC was similarly prepared.

α,ω-Dimethoxy Poly(Oxypropylene): - PPG 4025 (100g, 0.025 mole) was treated with metallic sodium (1.1g, 0.05 mole) at 100°C under nitrogen. After cooling to room temperature, methyl iodide (14.2g, 0.1 mole) was added and the contents were left stirring overnight. The excess methyl iodide was stripped off under vacuum. I.R. showed complete etherification. The product was used without further purification.

Polymerizations: - Reservoirs containing freshly prepared solutions of known concentrations of catalyst ("A") and cocatalyst with or without polyether diol/prepolymer ("B"), were stored under nitrogen at 160°C. For rate assay, a glass tube (25 mm O.D.) immersed in an oil bath maintained at 160°C was charged with a known amount of the catalyst solution ("A") under nitrogen using a preheated syringe. The tube contained a thermocouple positioned at the center and attached to a sensitive, temperature recorder (Fig. 1).

The tube containing "A" solution at 160°C was then charged with a known amount of the cocatalyst ("B") solution transferred from the reservoir at 160°C with a heated syringe. The contents were thoroughly mixed by partially pushing up and down the syringe piston quickly a few times. Although the mixing was not strictly adiabatic, care was taken to minimize the heat loss during transferring of "B" solution. The time-temperature profile of this reaction mixture was recorded as soon as the A & B solutions were mixed. The instantaneous (initial) temperature of the reaction mixture was generally no less than 150°C. The time to reach the maximum temperature rise ($t_{\Delta Tmax}$) was recorded and was fairly reproducible for a given system. In some cases addition of A & B solutions was reversed. Similar measurements could also be made using a laboratory Reaction Injection Molding apparatus consisting of a heated, twin-barrel, metal

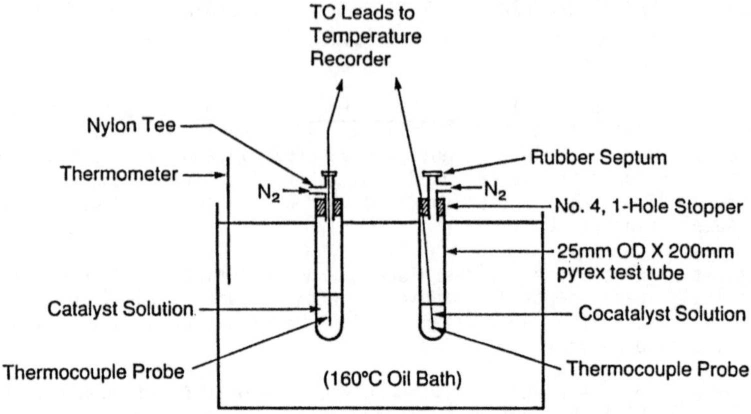

Fig. 1. MEASUREMENT OF TEMPERATURE RISE IN THE ANIONIC POLYMERIZATION OF CAPROLACTAM IN GLASS TUBES

syringe equipment in which the two piston rods are connected to a common discharge handle. The mixing head consisted of a Swagelok "T" fitting. For Reaction Injection Molding (RIM) of 6" x 6" x 1/8" nylon plaques an in-house modified version of laboratory RIM machine with hydraulically driven shaft or a commercial, circulating RIM machine (Battenfeld) were employed. The details of all these apparatus will be reported elsewhere. In principle, "A" and "B" solutions from heated reservoirs (160°C) were pumped at equal volume ratio via a mixing head into a preheated, (160°C) plaque mold (Fig. 2). After the normal curing time of about 2 to 3 min, the nylon plaques were removed and allowed to cool. Under these conditions >95% conversion was achieved as evidenced by methanol extraction.

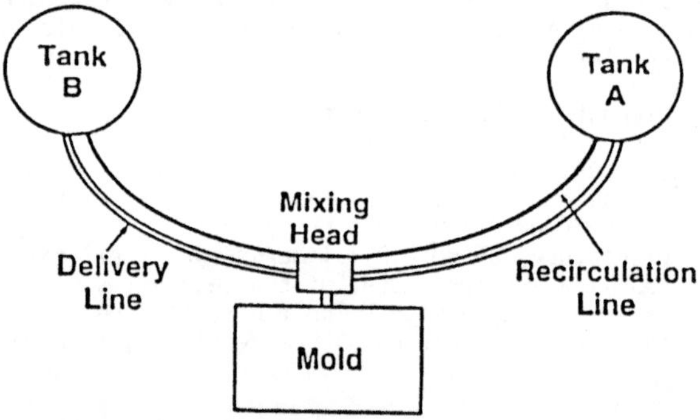

TANK A: - CATALYST SOLUTION IN MOLTEN CAPROLACTAM

TANK B: - COCATALYST, POLYETHER DIOL OR PREPOLYMER IN MOLTEN CAPROLACTAM

Fig. 2. SCHEMATIC DIAGRAM OF NYLON 6 REACTION INJECTION MOLDING (RIM)

Alcoholysis Reactions: - Methanol or 2-methoxyethanol containing 1 mole % of the corresponding sodium alkoxide, was refluxed with BpT or TBC for 3 hours. The reaction products were directly analyzed by L.C. and FAB mass spectrometry.

Results and Discussion

Triazine Cocatalysts

Several triazine derivatives were prepared from the inexpensive cyanuric chloride via nucleophilic displacement of the chlorides (Scheme II). Among them, the imidazole substituted derivative, TImT and the pyrrolidone substituted derivatives, BpT and TpT are new and could be prepared in reasonable yields and purity.

Scheme II . SYNTHESIS OF TRIAZINE COCATALYSTS

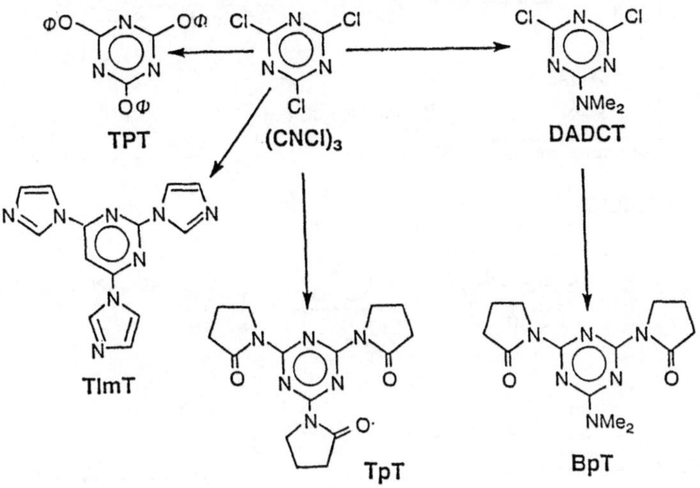

The synthesis of the desired caprolactam substituted triazines could not be achieved in acceptable purity. Use of mild organic and inorganic bases for the reaction between chlorotriazines and anhydrous caprolactam resulted in incomplete substitution and formation of hydroxy triazines from as yet unexplained side reactions. Use of strongly basic sodium or lithium caprolactamate gave oligomeric ring-opened by-products.

Comparative Evaluation of Homopolymerization Rates

All the triazine derivatives on hand were evaluated for their efficiency in initiating the anionic homopolymerization of caprolactam using caprolactam magnesium bromide (CMB) as catalyst. We have earlier found [13] that among all the metal caprolactamate catalysts, CMB was the most effective for triazine cocatalyst systems as was also the case for acyllactam cocatalysts [14]. For the sake of comparison, the efficiency of acyllactam cocatalysts were also evaluated under the same conditions. The comparison was based on the measurement of the temperature rise due to the exothermic polymerization and the time needed to reach maximum temperature ($t_{\Delta Tmax}$) under nearly adiabatic conditions (Fig. 1). According to Wittmer & Gerens [15] the extent of reaction is directly proportional to the adiabatic temperature rise assuming equal heat capacities for monomer and polymer. Although we have not attempted to obtain exact kinetic data, measurement of $t_{\Delta Tmax}$ gave a relative estimate of the over-all polymerization rates.

Triazine derivatives with chloro, phenoxy or imidazole leaving groups act as precursors for the formation of the primary initiating species, viz., the caprolactam substituted triazine. The rate at which these functional groups are displaced by lactam anion via a nucleophilic aromatic substitution involving an addition-elimination mechanism, determines the initiation rate and the over all polymerization rate (Scheme III).

Scheme III. **Initiation Mechanism for the Triazine Cocatalyst System**

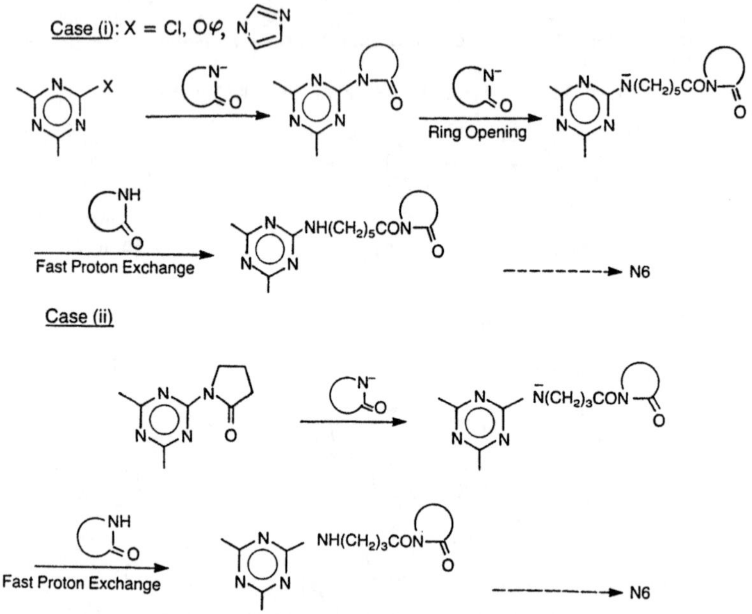

Cyanuric chloride, 2-dimethylamino-4,6-dichloro-1,3,5-triazine (DADCT) and 2,4,6-triphenoxytriazine (TPT) showed slower cure times ($t_{\Delta Tmax}$) (Table 1) associated with slower initiation rates. In addition these cocatalysts consume an equivalent of catalyst (caprolactam anion) for the displacement of neutral chloride or weakly basic phenoxide anions, hence a catalyst/cocatalyst ratio of 2:1 equivalents was needed. On the other hand, trisimidazolyl triazine (TImT) showed good polymerization efficiency even at a catalyst/cocatalyst ratio of 1:1 equivalents in spite of the fact that the reported pk_a of 14.5 for imidazole suggests that it is more acidic than caprolactam [16]. Sekiguchi and coworkers [17,18] determined the pK_a of caprolactam to be 31.3 using 1,2 dimethoxyethane (DME) as solvent for the potentiometric titration method. The pK_a (in DME) values for water, methanol and t-butanol were found to be 17, 24 and 26 respectively, while the corresponding pK_a's based on water scale are known to be 14, 15.5 and 19 respectively. Our results indicate that the displaced imidazole anion is basic enough to regenerate the catalyst via proton exchange with caprolactam and hence pK_a of imidazole is not significantly far from caprolactam.

The new pyrrolidone substituted derivatives (BpT and TpT) were the most effective among the triazine cocatalysts approaching the activity of the acyllactam cocatalysts (Table 1). This is attributed to the facile manner in which the activated pyrrolidone can be opened and is not displaced by the caprolactam anion. Intuitively, one would expect the hypothetical bis and tris caprolactam substituted triazines to be more effective than the pyrrolidone substituted triazines because of the greater tendency for ring opening for caprolactams. In fact, the high reactivity of the caprolactam moieties attached to the triazine, precluded their synthesis and isolation

Table 1. Comparative Rates of Anionic Homopolymerization of Caprolactam - Triazines vs Acyllactam Cocatalysts and Bromomagnesium Caprolactam as Catalyst(a)

Cocatalyst	$t_{\Delta Tmax}$ (min)(b)
N-Acetyl Caprolactam (NACL)	0.9
Terephthaloyl biscaprolactam (TBC)	0.7
Terephthaloyl bislaurolactam (TBL)	0.8
Cyanuric Chloride ($(CNCl)_3$)	3.3(c)
2-Dimethylamino 4,6 - Dichloro - 1,3,5 - Triazine (DADCT)	2.0(c)
2,4,6 Triphenoxy-1,3,5 - Triazine (TPT)	1.7(c)
2,4,6-Tris(Imidazolyl)-1,3,5-Triazine (TImT)	2.5
2-Dimethylamino-4,6-bis (α-Pyrrolidonyl)- 1,3,5 - Triazine (BpT)	1.0
2,4,6-Tris(α-Pyrrolidonyl) 1,3,5-Triazine (TpT)	0.9

(a) Tube polymerization assay - adiabatic temperature rise method; T_0 = 150°C; T_{bath} = 160°C

(b) $t_{\Delta Tmax}$: Time to reach maximum temperature (conversion) at [catalyst] = [cocatalyst] = 1 equivalent/100 equivalents of caprolactam

(c) One equivalent of excess catalyst used.

in pure form. Although the cocatalyst activity of bis or tris lactam triazine still remains unknown among all the co-catalysts studied thus far, terephthaloyl biscaprolactam (TBC) appears to be the most effective.

Thus from Table 1, it is quite apparent that, in spite of the common propagation step, the various cocatalysts differ in their rates of initiation and can therefore have significant effect on the over all rates of polymerization.

Block Copolymerization With Polyether Diols (Polyols) and Prepolymers: - Block copolymerizations of caprolactam with polyether diols may be carried out "in-situ" or via the prepolymer technique. Experimentally, this involves mixing a solution of catalyst (metal caprolactamate) in molten lactam, designated as "A" solution, with a solution of cocatalyst, polyether diol or prepolymer in molten lactam, designated as "B" solution. A two-component, Reaction Injection Molding (RIM) apparatus was used for fast, simultaneous injection of A and B solutions into a mold cavity, via a mixing chamber (Fig. 2). In both processes, the key intermediate is a polyether "prepolymer" with its chain end carrying the cocatalyst moieties viz., acyl caprolactam or triazine derivatives with reactive substituents. This prepolymer is really the reaction product of the polyether diols with the corresponding difunctional cocatalyst e.g. terephthaloylbiscaprolactam (TBC) or 2-dimethyl amino-4,6-bis(α-pyrrolidonyl)triazine (BPT) (Scheme IV).

Scheme IV. Anionic Block Copolymerization of Lactams With Polyether Diols

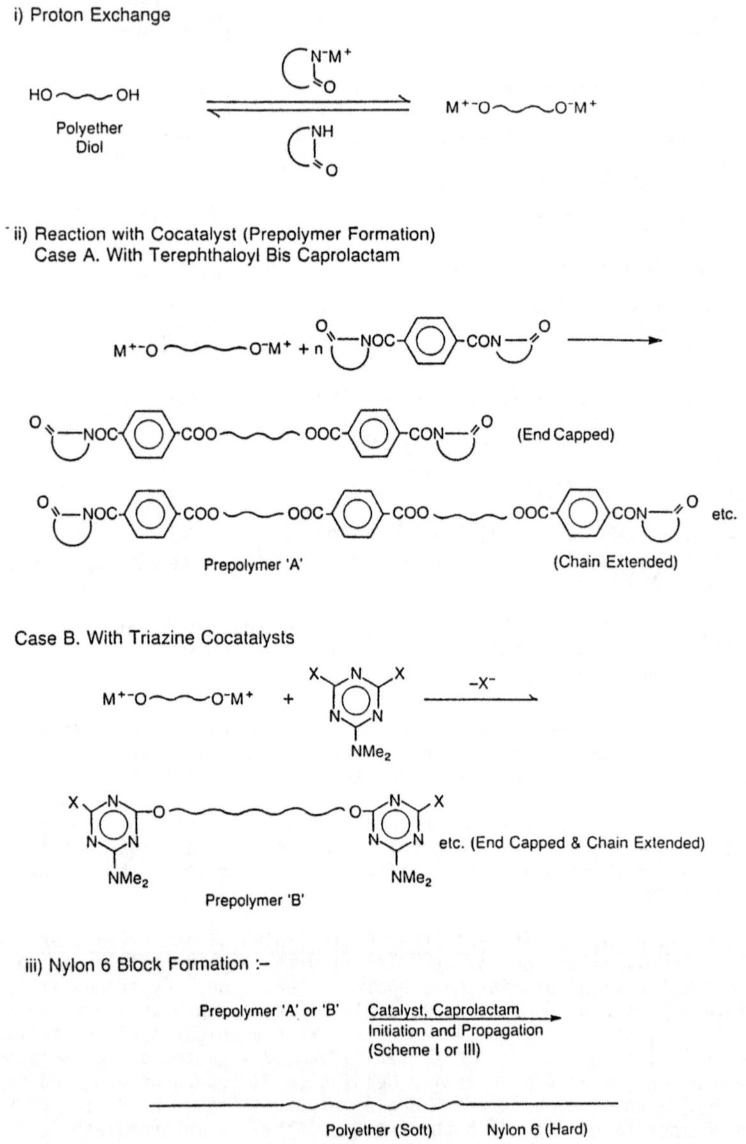

The prepolymer may be prepared in the absence of caprolactam with sodium methoxide as catalyst when the product is obtained as a mixture of the end-capped and chain extended polyether species with cocatalyst moieties attached to the chain ends as indicated by GPC (Fig. 3).

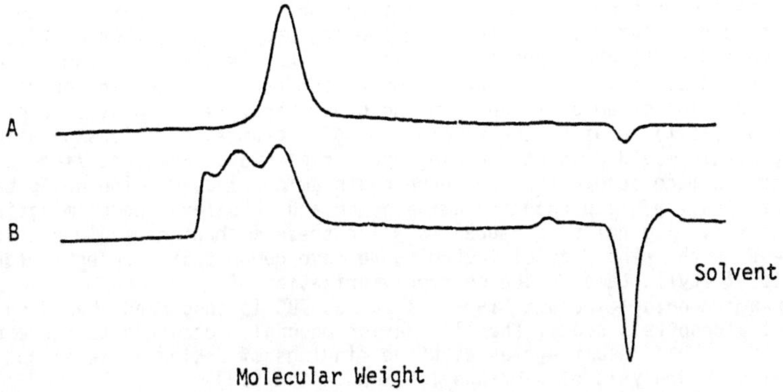

Figure 3. GPC of (A) Poly G5528, Polyetherdiol (Mn = 4,000)

(B) Poly G5528, BpT (1:2) Prepolymer

Subsequent polymerization of caprolactam with this prepolymer gives a block copolymer of ABA structure with the middle block consisting of the "soft" polyether segment. The same prepolymer may also be generated "in-situ" by anionic block-copolymerization of caprolactam with polyether diols using the same difunctional cocatalysts at proper stoichiometry (Scheme IV). Normally an equivalent excess of cocatalyst is provided to react with the hydroxy groups. In the in-situ process, the first step is the reaction of lactam anion with the more acidic polyol by fast proton exchange forming the alkoxide ions. This assumption is based on the known [17] higher pK_a value for caprolactam (31.3) compared to alcohols (17 to 26) in 1,2 dimethoxy ethane (DME). The equilibrium constant for the proton exchange (Ke) may be expressed as $pK_e = pK_a$ (polyol) - pK_a (caprolactam). Nucleophilic reaction of this alkoxide anion of polyol with the cocatalyst results in the prepolymer with terminal cocatalyst moieties (Scheme IV). Depending upon the cocatalyst structure the prepolymer formation can occur either via displacement or ring opening pathways (Scheme V).

Scheme V <u>Mechanism of Prepolymer Formation</u>

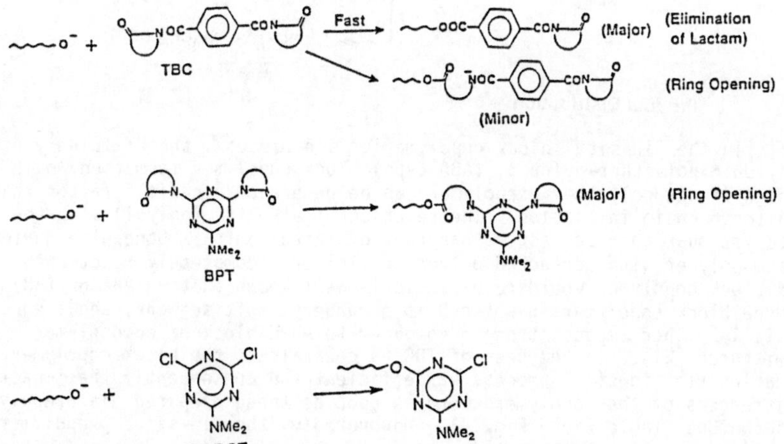

Model alcoholysis reactions indicate that with acyllactams such as TBC, the displacement mechanism is predominant while with the pyrrolidone substituted triazine, BpT, the ring opening predominates (Scheme VI). In the case of TBC, nucleophilic reaction of alkoxide at the lactam carbonyl via the usual addition-elimination mechanism requires the ring opening with the formation of amide anion. On the other hand, reaction at exocyclic carbonyl(acyl) requires the elimination of lactam anion. Consideration of ring strain would predict the ring opening pathway. However, since caprolactam is more acidic than the open chain amide, lactam anion would be a better leaving group than the amide anion and thus favoring elimination mechanism. The relative degree to which these mechanisms would operate may depend on the ring size of lactam as we have demonstrated recently for aminolysis of acyllactams in the self polymerization of N-(p-aminobenzoyl)lactams [19]. As far as TBC is concerned, based on the model alcoholysis study, the elimination mechanism occurred to the extent of >70%. This result agrees with the findings of Sekiguchi et al on the alkaline hydrolysis of N-hexanoyl hexanolactam [20].

In the case of BpT, nucleophilic reaction of alkoxide favors ring opening mechanism due to the extensive delocalization of the nitranion by the triazine ring.

Scheme VI . MODEL ALCOHOLYSIS REACTIONS AND THE RELATIVE PRODUCT DISTRIBUTION DETERMINED BY L.C. AND FAB MASS SPECTRAL ANALYSIS

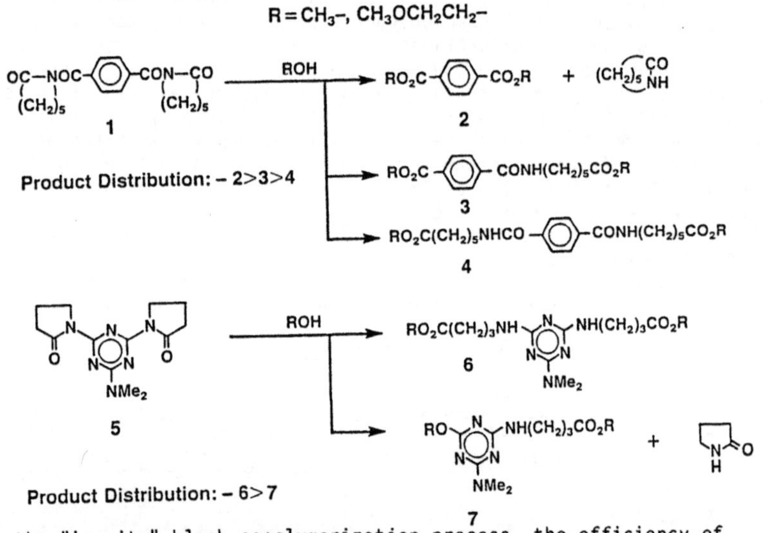

In the "in-situ" block copolymerization process, the efficiency of nylon 6-polyether-nylon 6, (ABA type) block copolymer formation would be high if the polyether prepolymer can be generated "in-situ" faster than nylon 6 chain initiation. The reaction of alkoxide (polyol) with the cocatalyst must then be faster than that of lactam anion. Otherwise, nylon homopolymer with unreacted polyether diol or incompletely reacted AB diblock copolymer would result. It is well known that an ABA or $(AB)_n$ A type block copolymer in which B is a rubbery, soft segment, shows substantially higher impact strength compared to AB diblock or homopolymer mixtures [21]. In the case of TBC as cocatalyst, the block copolymer formation via "in-situ" process was efficient and consequently the impact strengths of the copolymers were as good as those prepared via prepolymer technique (Table 2 and Fig. 4). Undoubtedly, the "in-situ" prepolymer formation was quite fast in this case.

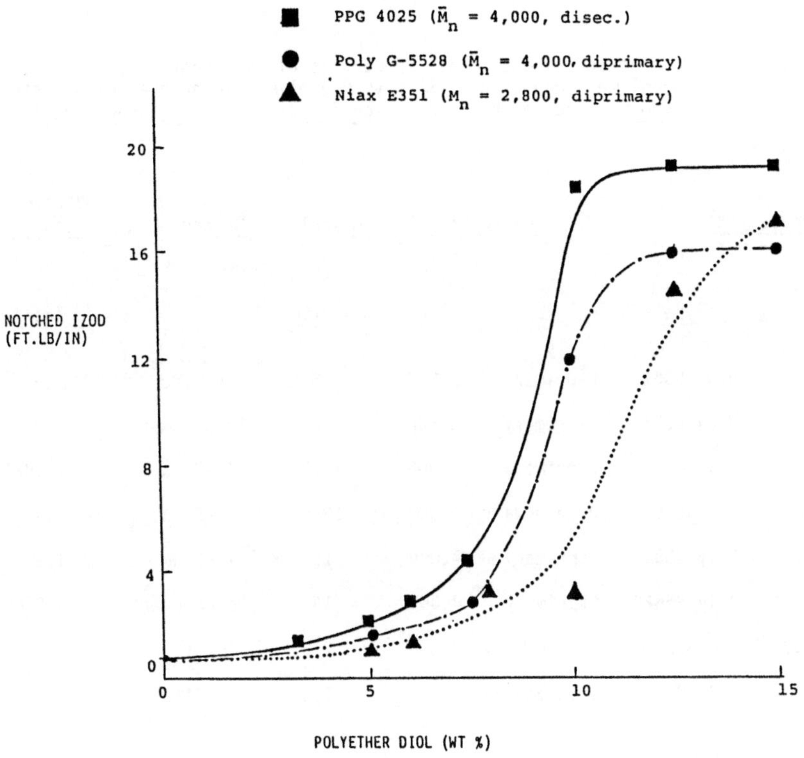

Fig. 4. IMPACT STRENGTHS OF NYLON 6 BLOCK COPOLYMERS MADE VIA "IN-SITU" COPOLYMERIZATION WITH POLYETHER DIOLS (COCATALYST: - TBC; CATALYST: - CMB)

For BpT, the in-situ pre-polymer formation was not so efficient and resulted in a product with much lower impact strength compared to the copolymer prepared from the corresponding prepolymer (Table 2). The product obtained with BpT and polyether diol via the "in-situ" copolymerization method showed no extractable polyether and hence is presumed to be a mixture of AB diblock and homopolymer. When we used α,ω-poly(oxypropylene) in our anionic polymerization process, we obtained a nylon product with very low impact strength (notched izod = 0.7 ft.lbs/in), containing the unreacted polyether as a separated phase which can be extracted out.

The difference between TBC and BpT in their relative efficiency of "in-situ" prepolymer formation, may be visualized as follows:

- Both alkoxide anion and the caprolactam anion existing in equilibrium with each other can react with TBC either at exocyclic acyl carbonyl (elimination mechanism) or at the endocyclic, lactam carbonyl (ring opening mechanism) (Scheme V). However, with lactam anion, only the latter process is effective in initiating nylon 6 polymerization while the former is simply a lactam exchange process. Prepolymer, on the other hand, can be formed by both mechanisms and as alcoholysis already indicated the elimination pathway is faster. The over-all rate of the prepolymer reaction seems to be faster than nylon 6 initiation based on this difference alone.

Table 2. Comparison of the Impact Properties of Polyether - Nylon 6 Block Copolymers Made Via In-Situ vs Prepolymer Techniques of Anionic Copolymerization Under Reaction Injection Molding Conditions[1]

Cocatalyst[2]	Type	Polyoxypropylene Glycol	Mn	Wt%	Method	Notched Izod (ft lbs/in)
TBC	Niax E351	Diprimary	2,800	15	In-Situ	16
TBC	Niax E351	Diprimary	2,800	16	Prepolymer	9
TBC	Poly G5528	Diprimary	4,000	12.5	In-Situ	15(NB)
TBC	PPG 4025	Disecondary	4,000	12.5	In-Situ	19
BpT	Poly G5528	Diprimary	4,000	15	In-Situ	1.9
BpT	Poly G5528	Diprimary	4,000	15	Prepolymer	>10(NB)
DADCT[3]	PPG 4025	Disecondary	4,000	7.5	In-Situ	>15(NB)
DADCT[4]	PPG 4025	Disecondary	4,000	7.5	In-Situ	3

1 Polymerization time: - 2 to 3 min., two component RIM apparatus. Polymerization temperature: - 130° to 160°C.

2 [Cocatalyst] - [hydroxyl] = [catalyst] = 1 equivalent/100 equivalents of caprolactam.

3 Fresh solution in neat polyether diol.

4 Solution in caprolactam/polyether diol, aged at 110°C for 2 hrs.

- In the case of BpT, although displacement mechanism (nucleophilic addition-elimination across triazine ring) is possible, it appears that the alternate ring opening mechanism is preferred as suggested by model alcoholysis study. Hence both alkoxide and lactam anion compete for BpT via the ring opening mechanism. The rates at which these reactions can occur may differ and in this case homopolymerization was favored over prepolymer reaction. This appears to be the only distinguishing feature from TBC.

With 2-dimethylamino-4,6 dichloro-1,3,5 triazine (DADCT) as cocatalyst, high impact strength was achieved with as low as 7.5 wt% polyether diol (Table 3). Obviously, the reaction of alkoxide with the chlorotriazine was competitive enough compared to that of lactam anion so that prepolymer could be generated efficiently in-situ. The morphologies of all the block copolymers will be discussed elsewhere.

It was found that DADCT was effective only when freshly prepared solution in neat polyether diol was mixed directly with the catalyst solution.

TABLE 3. IMPACT PROPERTIES OF NYLON 6 - POLYETHER BLOCK COPOLYMERS PREPARED VIA IN-SITU, ANIONIC BLOCK COPOLYMERIZATION OF CAPROLACTAM AND POLYETHER DIOL UISNG 2-DIMETHYL AMINO 4,6-DICHLORO 1,3,5 TRIAZINE AS COCATALYST

POLYETHER DIOL (PPG 4025)(wt %)	NOTCHED IZOD (ft.lbs/in)	
	23°C	-29°C
0	0.7	0.6
3	3	1.6
5	4.1	2.6
7.5	15	4.2-5
9	NB	3.9-5.4

1. Catalyst: - Caprolactam magnesium bromide
2. Monomer/cocatalyst/catalyst = 100:(1+x):(2+x) equivalent ratio where x = hydroxyl equivalents calculated for each of the above formulations.
3. Mold temperature = 160°C
4. Freshly prepared cocatalyst solution in neat polyether diol was added to catalyst solution in molten caprolactam.
5. N.B. = No break

Unlike cyanuric chloride, DADCT was fairly stable to heat-aging in polyether diol (Fig. 5). Cyanuric chloride, on the other hand, probably reacts with hydroxyl groups via an SNi mechanism as reported in the literature [22] with low molecular weight alcohols.

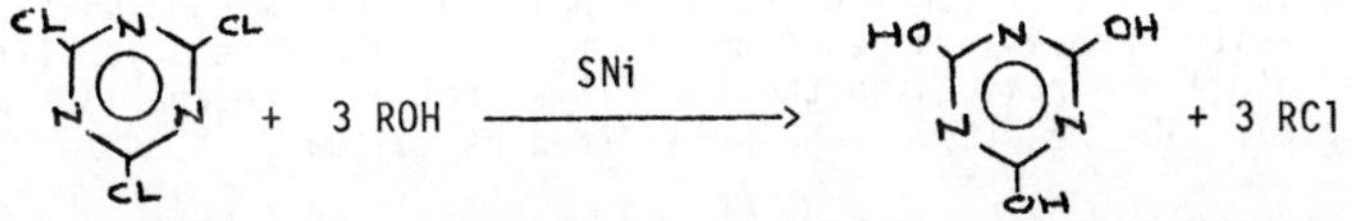

All chlorotriazines were found to deteriorate in their cocatalyst activity as well as polyether coupling efficiency upon prolonged heating in caprolactam due to complex side reactions. Dicaprolactim ether and the corresponding hydroxytriazines have been identified by mass spectral analysis as the products of these degradative reactions (Scheme VII). In the reaction between caprolactam both N-substitution, O-substitution can take place analogous the known N- or O-alkylation of lactams [23]. In our proposed mechanism nucleophilic addition of lactam to the lactim ether double bond (proton assisted) will result in the formation of dicaprolactim ether followed by elimination of hydroxy triazine. The high acidity of hydroxy triazine (or the high stability of the oxyanion) makes it a good leaving group. Dicaprolactim ether is a monofunctional cocatalyst and hence can only be effective for nylon 6 homopolymerization but not for block copolymerization with polyether diol.

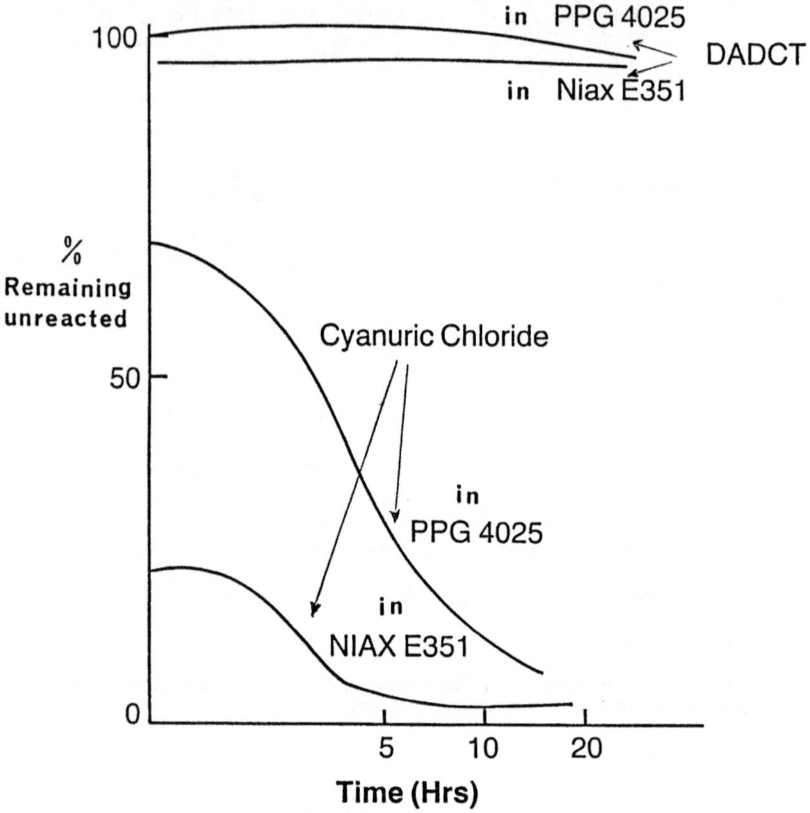

Fig. 5. THERMAL STABILITY OF CHLOROTRIAZINE SOLUTIONS IN NEAT POLYETHER DIOLS (2:1 MOLE RATIO, 100°C)

In conclusion, the efficiency of nylon 6-polyether block copolymerization in an "in-situ" process, depends on how fast the alkoxide anion (from the polyether diol) can react with the cocatalyst relative to the lactam anion (catalyst). The results may be summarized as follows:

(i) With terephthaloylbiscaprolactam as cocatalyst, polyether diol may be used directly for copolymerization with caprolactam since the prepolymer reaction is fast and ABA type block copolymer is formed efficiently as indicated by high impact strengths.

(ii) With 2-dimethylamino-4,6-bis(α-pyrrolidonyl)1,3,5-triazine (BpT) as cocatalyst, in-situ prepolymer formation was inefficient due to slow nature of the reaction of the alkoxide with the cocatalyst.

(iii) With 2-dimethylamino 4,6-dichloro 1,3,5 triazine (DADCT), as cocatalyst, "in-situ" prepolymer formation can be efficient, in freshly prepared solutions. Although the over-all rate is slow, the resulting block copolymers exhibit high impact strengths.

Scheme VII . POSSIBLE DEGRADATIVE REACTIONS OF 2-DIMETHYLAMINO 4,6 DICHLORO 1,3,5 TRIAZINE IN MOLTEN CAPROLACTAM AND THEIR EFFECT ON NYLON BLOCK COPOLYMERIZATION PROCESS

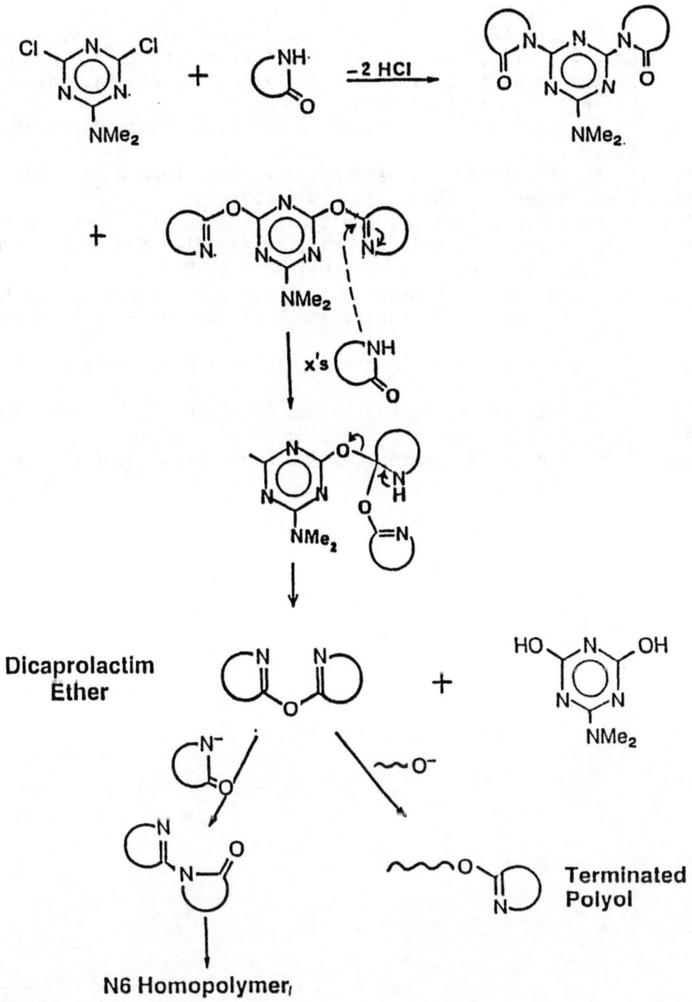

REFERENCES

1. R. M. Joyce and D. M. Ritter, US Pat. 2,251,519 (1941).
2. H. K. Hall, Jr., J. Am. Chem. Soc. 80, 6404 (1958).
3. E. H. Mottus, R. M. Hedrick and J. M. Butter, US Pat. 3,017,391 (1962).
4. J. Sebenda and J. Kralicek, Collect. Czech. Chem. Commun. 23, 766 (1958).
5. J. Stehlicek, K. Gehrke and J. Sebenda, Collect. CZech. Chem. Commun. 32, 370 (1967).
6. E. W. Pietrusza, R. Pinter and J. R. Pederson, US Pat. 3,448,085 (1969).
7. G. Carlyon, "Monomer Casting", chapter 13, p. 457 in "Nylon Plastics", M. Kohan (Ed.), John Wiley & Sons (1973).

8. G. Ferber, Jr., "New Materials for Reaction Injection Molding", presented at SPE, NATEC, Detroit, MI, November 6-8, 1979.
9. J. Lincoln, Brit. Pat. 1,150,725 (1969).
10. R. M. Hedrick and J. D. Gabbert, US Pat. 4,031,164 (1977).
11. M. K. Akkapeddi, US Pat. 4,487,290 (1984).
12. J. T. Thurston, et al, J. Am. Chem. Soc. 73, 2381.
13. M. K. Akkapeddi and G. J. Dege, "Effect of Catalyst Structure on the Relative Efficiency of Anionic Polymerization of Caprolactam" presented in part at Symposium on Nylon RIM, ACS National Meeting, Philadelphia (August 1984).
14. R. Z. Greenley, J. C. Stauffer and J. E. Kurz, Macromolecules 2, 561 (1969).
15. V. P. Wittmer and H. Gerrens, Makromol. Chemie 89, (1951), 27 (1965).
16. D. J. Brown, J. Chem. Soc. 1974 (1958).
17. H. Sekiguchi, "Lactams and Cyclic Imides" in "Ring Opening Polymerization", Vol. 2, chap. 12, p. 814, K. J. Ivin and T. Saegusa, Elsevier Applied Science Publishers (1984).
18. H. Sekiguchi, G. Froyer; J. Polymer Sci. Symp. 52, 157 (1975).
19. M. K. Akkapeddi, A. Housny and M. S. Walsh, Polymer Preprints 27, 120 (1986).
20. H. Sekiguchi and B. Coutin, J. Polymer Sci., Polymer Chem., 11, 160 (1973).
21. J. McGrath and A. Noshay "Block Copolymers", Academic Press (1977).
22. S. R. Sandler, J. Org. Chem. 35, 3967 (1970).
23. R. Benson and T. Cairns, J. Am. Chem. Soc. 70, 2115 (1948).

ANIONIC POLYMERIZATION OF SILOXANES NON-TRIVIAL MODEL SYSTEM FOR STUDIES OF COMMON PHENOMENA

MIECZYSŁAW MAZUREK*
3M Company, Saint Paul, MN 55144

I. INTRODUCTION

Polysiloxanes are among the polymers most widely used in various areas of technology. Their surface activity, low and high temperature performance, low thermal expansion coefficient, bioinertness and most of all - the availability of a broad spectrum of low and high molecular weight homologous series with a number of different functionalities - account for ever growing interest in this class of materials. A long list of existing and potential applications of silicones, as they are traditionally called, can be found in numerous technical articles and encyclopedias.

Early academic interest in siloxanes was primarily concerned with their surface activity [1]. The equilibrium mixture obtained by ionic equilibration of the cyclic siloxanes appeared to be the most convenient system for thermodynamic studies of equilibrium ring opening polymerization systems [2]. Conformational studies of the polymeric systems were performed, and the Jacobson and Stockmayer theory of cyclization was experimentally verified by using siloxanes as models. The effect of substituents on the ring chain equilibria was studied in great detail as a spectrum of cyclic siloxanes with various silicon atom substituents can be easily prepared.

Basic mechanistic studies of the polymerization reactions of siloxanes have attracted less attention, although both anionic and cationic polymerization constitute rather interesting systems. The polymerization of cyclic siloxanes was recently reviewed in a comprehensive article by P.V.Wright [3].

In that article the emphasis was put on the thermodynamics of the system, although mechanistic studies of both the cationic and anionic polymerization were also extensively discussed. With regard to the mechanistic studies the contributions from the group of Chojnowski (of which I was formerly privileged to be a member) have been broadly quoted. This makes the present paper, intended primarily as review of the activities of that group in the field of anionic polymerization of siloxanes, somewhat repetitive. However, I will try to justify this repetition by attempting to demonstrate that the anionic polymerization of siloxanes represents a particularly suitable model system for studies of many phenomena common to the anionic ring opening polymerization of various heterocycles.

Cyclic siloxanes (often designated D_n, where D denotes the [$-SiMe_2O-$] unit) undergo anionic polymerization in the presence of strong nucleophile to give a mixture of linear polymer with a broad spectrum of cyclic species.

$$D_4 \longrightarrow \text{polymer} + D_n$$

KOH and other strong bases [4] are used for the synthesis of silicone gums on a commercial scale. In early studies, it was established that

* Previous address: Polish Academy of Sciences, Center of Molecular and Macromolecular Studies, 90-362 Łódź, Poland

Recent Advances in Anionic Polymerization
T.E. Hogen-Esch and J. Smid, Editors

the reaction, when run in non-polar media (siloxane bulk), obeys first order kinetics with respect to the monomer and the reaction order with respect to the initiator is 1/2 [5]. This result was interpreted in terms of the dissociation of the silanolate group into free ions, which were believed to constitute the actual active centers of polymerization. This interpretation was carried throughout the literature despite serious contradictions. It was not until the pioneering work of Szwarc et al. [6] and Schultz et al. [7] on mechanisms of anionic polymerization brought into polymer chemistry Winstein's concept of ion pairs and their aggregates that the mechanistic studies of ionic polymerization of siloxanes received new impetus.

From a certain point of view the anionic polymerization of siloxanes can be considered as a rather simple system. This simplicity derives from the fact that when silanolate is used as an initiator, one general reaction scheme can be written for all of the component reactions of this equilibrium ring opening polymerization.

$$\equiv Si^{*}OMt \;+ \equiv SiOSi\equiv \;\rightleftharpoons\; \equiv SiOMt \;+ \equiv Si^{*}O\;Si\equiv$$

Initiation, propagation, intramolecular chain transfer (back biting) or intermolecular chain transfer, as well as the cross-over reaction of copolymerization all comprise the same reaction of the nucleophilic attack of silanolate on the silicon atom in a siloxane bond, leading to Si-O bond splitting and to the formation of another Si-O bond and a new silanolate group. This suggests that one model reaction for studies can be chosen.

The initiation reaction is unique as it permits direct studies of "living" species in the polymerization system. Earlier works dealt almost exclusively with the propagation step of polymerization; thus, problems related to the mechanism of siloxane bond opening could be addressed only indirectly. By studying the kinetics of the propagation reaction, only the "external order " of the reaction with respect to the active groups could be determined.

The apparent simplicity of the anionic polymerization of siloxanes might be misleading. All of the component reactions of polymerization are complicated by a number of phenomena, including the following, which are the subject of this paper:

- aggregation of silanolate groups in the polymerization medium
- assistance of the siloxane chain in siloxane bond cleavage
- differences in reactivities of various silanolate groups in the system

It should be recalled that in a typical polydimethylsiloxane equilibration reaction silanolate groups can undergo reaction with both cyclic monomer as well as side reactions with the siloxane chain of the product. These side reactions can be either with siloxane bond along the chain, or, much faster, with siloxane bonds in the vicinity of the siloxanolate group [8]. At this point, the reaction does loose some of its simplistic appearance and this is why the model system had to be carefully chosen for the initial mechanistic studies of anionic polymerization to eliminate the effect of the side reactions.

The approach to mechanistic studies presented in this paper, often utilized by Chojnowski and coworkers, is characteristic of a certain school of research which does not necessarily attempt to determine absolute values of, e.g., absolute rate constants, but rather first seeks a more phenomenological approach to complicated systems. Attempts to devise more and more adequate models to fit the experimental results would be characteristic of such an approach. Quantitative or semiquantitative data, if possible to obtain, would then be sought only as a last step.

II. AGGREGATION - KINETIC EFFECTS

II. 1. Ideal systems of aggregation

The interpretation of fractional order with respect to initiator in the anionic polymerization of siloxanes in terms of the equilibrium between active groups - ion pairs of the silanolate groups, and the dormant centers - aggregates of the ion pairs, can be attributed to Laita [9], who supposedly developed the idea by analogy to his earlier work with Szwarc on the kinetics of the initiation of 1,1-diphenylethylene with BuLi [10]. The mechanism of anionic polymerization involving the association of silanolate groups into double complexes has been postulated by Juželevskii [11].

$$2\ \equiv SiOMt \rightleftharpoons (\equiv SiOMt)_2$$

To study aggregation phenomena a system was needed where the reaction of silanolate groups with siloxane bonds could be observed with no interference from side reactions of redistribution, as mentioned previously.

The reaction of triorganosilanolates with 2,2,5,5-tetramethyl-1-oxa-disilacyclopentane (ODSCP) provided a good model [12]. The reaction of the phenyl substituted silanolate with the siloxane bond of the monomer was monitored by UV. Redistribution reactions could be prevented because the siloxane groups are isolated by ethylene units, thus having extremely low reactivity with silanolates.

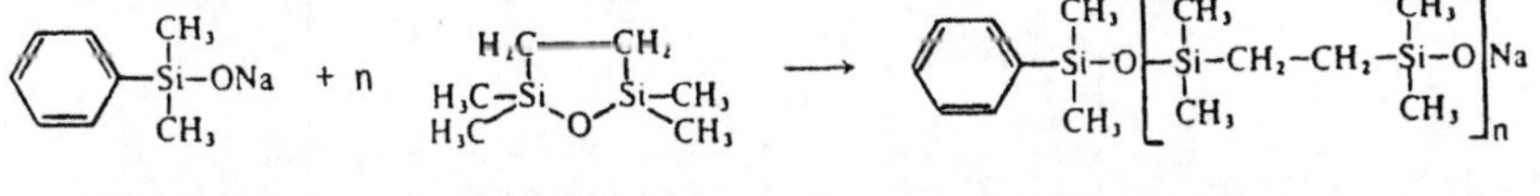

The kinetic investigations were carried out in a heptane/dioxane (95:5) mixture with a large excess of the monomer, to render pseudo-order conditions. The reaction always followed a simple first-order kinetic curve practically throughout the entire process. The external order of the reaction in silanolate was found to be close to 1/4.

The rate of the propagation step in the polymerization of the monomer was measured dilatometrically in the analogous system, by using ω-sodium polysilethylenesiloxane as initiator. The reaction followed first order kinetics, with a pronounced induction period characteristic of slower initiation than propagation. From linear portions of the graph, the kinetic order with respect to the living ends was found to approach a value of 1/4.

To provide interpretation of these results kinetic analysis was performed, starting from the simplest case of the double aggregates formed in the reaction medium.

Aggregates AA, BB, AB and BA must be considered. The initiation reaction follows the scheme:

$$A + A \underset{}{\overset{K_0}{\rightleftharpoons}} (AA), \quad K_0 \gg 1$$

$$A + B \underset{}{\overset{2K_1}{\rightleftharpoons}} (AB), \quad K_1 \gg 1$$

$$B + B \underset{}{\overset{K_2}{\rightleftharpoons}} (BB), \quad K_2 \gg 1$$

$$A + M \xrightarrow{k} B$$

Simple mathematical transformations lead to the expression for the rate of the initiation reaction (a and a-x - initial and actual concentrations of initiator, respectively)

$$R = \frac{k[M]}{(2K_0)^{1/2}} \, \frac{a-x}{a^{1/2}}, \quad \text{if } K = 1$$

for the case when the change in free enthalpies ΔG of the complex formation resulting from the change from A to B in both AA and AB have the same value. This corresponds to the case when $K = K_0K_2/K_1^2 = 1$. The reaction should follow first order kinetics with respect to the initiator, while the external order is 1/2.

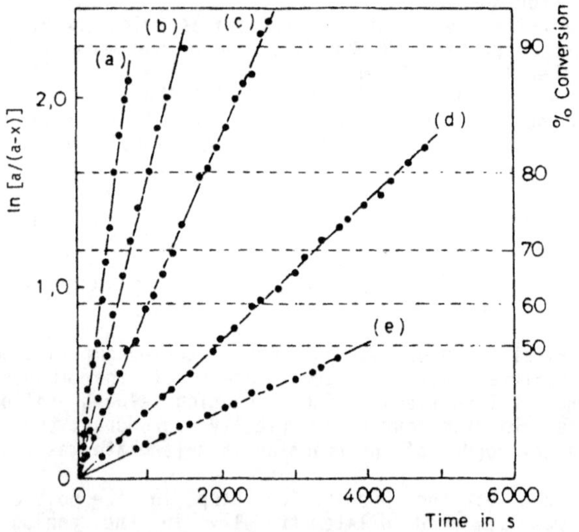

Figure 1. First order kinetic plots for the initiation of the polymerization of ODSCP with sodium phenyldimethylsilanolate at 30°C in heptane/dioxane (95:5v/v); [ODSCP] = 1.0 mol·dm^{-3}*. Initial concentration of silanolate (x10*^{3}*): (a) = 1.0, (b) = 2.3, (c) = 12.0, (d) = 21.0 and (e) = 50.9 mol·dm*$^{-3}$*. (Ref [12], by courtesy of the publishers, Hüthig & Wepf Verlag)*

The analysis, extended for a mechanism including the formation of aggregates A and B of multiplicity n, leads to the following scheme:

$$nA \overset{\binom{n}{0}K_0}{\rightleftharpoons} (A_n)$$

$$(n-1)A + B \overset{\binom{n}{1}K_1}{\rightleftharpoons} (A_{n-1}B)$$

....

$$(n-k)A + kB \overset{\binom{n}{k}K_k}{\rightleftharpoons} (A_{n-k}B_k)$$

....

$$A + (n-1)B \overset{\binom{n}{n-1}K_{n-1}}{\rightleftharpoons} (AB_{n-1})$$

$$nB \overset{\binom{n}{n}K_n}{\rightleftharpoons} (B_n)$$

$$A + M \xrightarrow{k} B$$

In the case where the constant free enthalpy change requirement is satisfied in the series of $(A_{n-k}B_k)$ aggregates, the final equation of the rate of initiation is the following:

$$\frac{-d(a-x)}{dt} = \frac{k}{(nK_0)^{1/n}}[M]\frac{(a-x)}{a^{(n-1)/n}}$$

The kinetic data presented fit the equation indicating that aggregates with average multiplicity of 4 seem to be formed in the polymerization system. The general kinetic law for propagation is

$$\frac{-d[M]}{dt} = k^p_{app}[LE]^{1/n}[M]$$

where [LE]-concentration of living ends.

The apparent rate constants of initiation k_{app} and propagation k_{app} are:

$$k^i_{app} = \frac{k_i}{(nK_0)^{1/n}} \qquad k^p_{app} = \frac{k_p}{(nK_n)^{1/n}}$$

where k_i and k_p are the absolute initiation and propagation rate constants and K_0 and K_n are the equilibrium constants for the aggregation of initiator and living ends, respectively. Despite their complexity, these parameters provide the basis for a rather detailed mechanistic analysis [12].

The requirement of $\Delta\Delta G_{A\rightarrow B}$ = const might be satisfied if the aggregates have structure, e.g., such as shown below, where the displacement of initiator with polymer in A or B does not change the structure of the ring or cage skeleton because the bonding part of the aggregate is not involved in the displacement reaction. Thus, analogy to the free energy relationship can be expected.

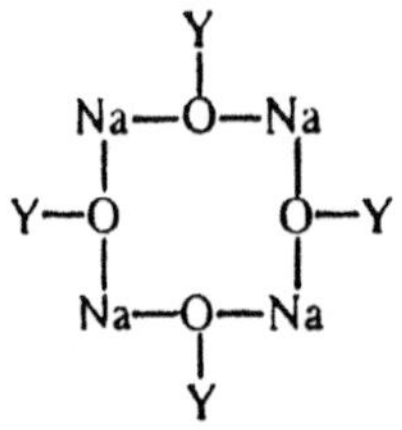

(A)

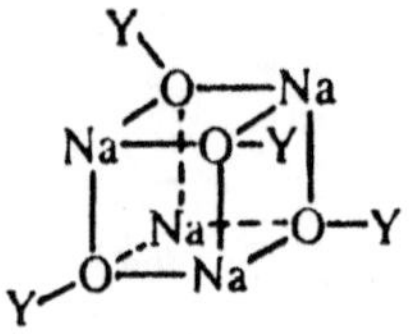

(B)

The quite adequate description of the model polymerization system provided the starting point for mechanistic studies of more complicated systems - polydimethylsiloxanes.

II. 2. Non-Ideal Systems of Aggregation

This simple situation by no means holds in typical polydimethylsiloxane systems. Not only do siloxane molecules easily undergo redistribution reactions, e.g., back biting depolymerization and chain transfer (which can be classified as non-specific redistribution reactions), but, also, very fast reaction of the silanolate with the siloxane unit in the proximity of the living end of the polymer takes place. The latter reaction, observed by Selin [13] and studied in detail by Frye et al. [8] and Fessler et al. [14] can be used for synthetic purposes, as in a non-polar medium organolithium compound reacts much faster with siloxanolates than with cyclic siloxanes to form silanolate, leaving unreacted siloxane, if used in excess.

$$BuLi + D_3 \longrightarrow BuMe_2SiOLi + 2/3\ D_3$$

This reaction, most unexpected from the point of view of electronic considerations, suggested that aggregation is involved in the reaction pattern, as the reaction might be treated as taking place within the complex. This would account for its extremely fast rate even at low concentration of the silanolate groups in the medium. However, an alternative mechanism was recently proposed by Chojnowski et al. [15] suggesting that the redistribution could be interpreted in terms of the elimination-addition sequence.

Considering the above-mentioned complications of the mechanism one would not expect any simple kinetic path of the reaction of silanolates with dimethylsiloxanes. Moreover, even when experimental conditions were found which practically allowed for elimination of redistribution reactions, the complicated pattern of the kinetics of the initiation reaction persisted. This clearly pointed out that the aggregation phenomena in this case are more complicated.

Development of a more general approach to the aggregation phenomena was needed to describe the system, and it seemed to be even more justified, because the processes involving cross-aggregation include some of the component reactions of many important polyreaction systems, such as anionic polymerization of diene and vinyl monomers initiated with alkyl lithium reagents [16] and anionic polymerization of oxiranes [17] and some other heterocycles [18].

II. 2.a. Non-ideal binary complexes

When for some reason the linear free enthalpy relationship is not fulfilled in the series of aggregates

$$\Delta \Delta G_{A \rightarrow B} \neq \text{const}$$

a non-ideal system would be formed. The simplest case of binary complexes will be discussed first. As was shown elsewhere [19], there is no analytical solution of the kinetic equations in such a case. However, a relative rate of the reaction as a function of f (fraction of unreacted substrate) can be simulated for a given value of the equilibrium constant K. The parameter K can be considered as a measure of the deviation of the system from the ideal case and its tendency for cross-aggregation. The simulated curves are shown in Figure 2.

The two limiting curves and the ideal case are of particular interest. When K=1 (ideal system) the relative rate of the reaction is a linear function of f. When simple complexes are much stronger than the cross-aggregates (K= ∞) reaction should follow 1/2 order internal kinetics, while the formation of very strong cross-aggregates (K=0) practically stops the reaction at 50% conversion of the substrate.

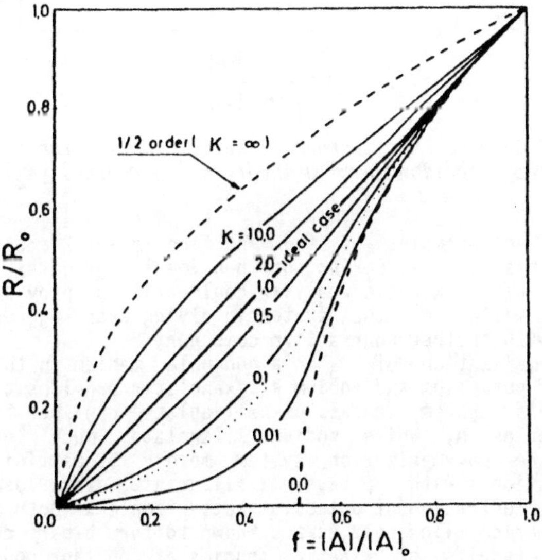

Figure 2. Variation of the relative rate (in relation to the initial rate) with the change of the substrate fraction f= [A]/([A]+[B]) = [A]/[A]$_0$ for the reaction involving a series of binary complexes (A_2), (AB), (B_2), [19].

The integrated first order kinetic plot with normalized time scale (Figure 3) predicts very characteristic shapes of the kinetic curves, unequivocally defined by the value of the equilibrium constant K, with the initial and final portions of the curves being linear. The normalization is done by multiplying the time scale by the initial first order rate constant observed, $k_{obs} = k[M]/(2K_0[A]_0)^{1/2}$.

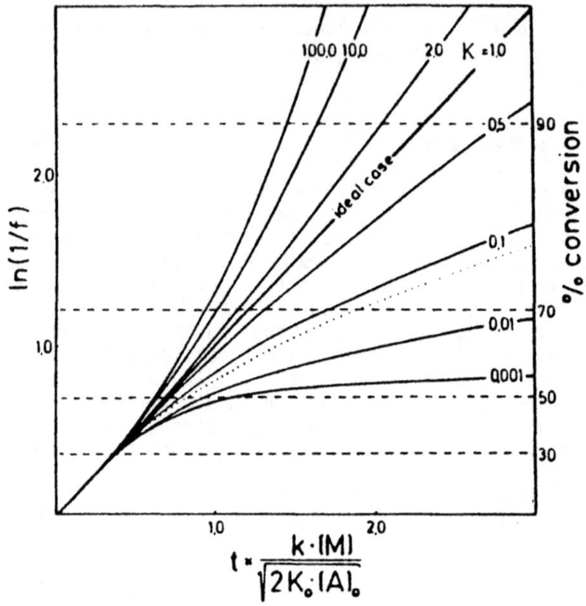

Figure 3. Simulated first-order kinetic curves with normalized time scale for the reaction A ⟶ B involving aggregates (A_2), (AB), (B_2), [19].

Although no examples of the reaction A ⟶ B could be found to illustrate this simple series of non-ideal aggregates, experimental verification of the kinetic analysis could still be provided, since the analysis is valid for the system involving cross-aggregation of the substrate A with another nonreactive component B.

The polymerization of D_3 in a non-polar medium in the presence of a mixture of potassium and sodium siloxanolates provides a good example [19]. In this system, potassium silanolate can be treated as the reactive species A, while sodium silanolate, much less reactive, influences the polymerization system merely by participation in the cross-aggregation with potassium silanolate, thus decreasing the effective concentration of potassium active centers. Both potassium and sodium polymeric siloxanolates are known to form binary complexes. The experimental results of kinetic studies at constant concentration of [~$iOK + ~$iONa], and at variable ratio of the potassium to the total amount of counterion, are represented by circles in Figure 4. The experimental results fit well to the curve simulated for $K = 5x10^{-2}$. The data points to a large tendency toward cross-aggregation between sodium and potassium polysiloxanolates, which accounts for the known phenomenon of polyelectrolytic inhibition of anionic polymerization of siloxanes. Good fitting of the experimental data to the very characteristic, simulated curve supports the correctness of the analysis.

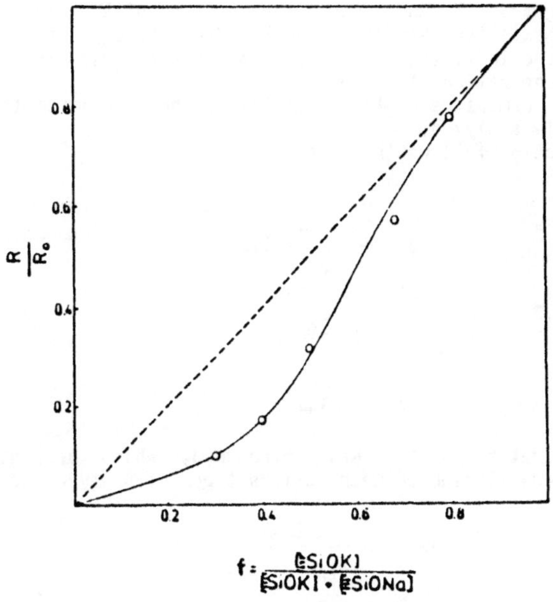

Figure 4. The relative rate of the polymerization of D_3 using potassium silanolate (A) catalyst in the presence of sodium silanolate at variable fractions of potassium silanolate, [19].

II. <u>2.b. Complexes with optional, constant multiplicity</u>

This approach may be extended to reactions involving a simple series of aggregates with a multiplicity, n, greater than 2. The kinetic equation includes n-1 equilibrium constants determining the shape of the kinetic curve in normalized first order kinetic plots [19].

The main feature of the approach presented here is that a knowledge of the values of the constants of aggregate formation is not necessary to describe the kinetics of reaction when cross-aggregation occurs. Instead, the parameters used are related to constants of the equilibria between aggregates. They allow for the calculation of the relative concentrations of complexes as functions of the conversion of the reagent A, as well as for simulation of the kinetic curves in the normalized first-order plot, analogous to the one shown in Figure 3.

II. <u>2.c. Complexes with variable multiplicity</u>

A yet more complicated system was found to occur in the initiation of siloxane polymerization with potassium silanolate in a non-polar medium. Analysis of the dependence of the initial rates of the reaction, as found in the first order kinetic plot, on the initial concentration of silanolate suggested that both double and triple aggregates might be in equilibrium with the reactive silanolate initiator, which tends to the binary complexes on dilution. These complexes might be in equilibrium with primarily double cross-aggregates AB and double aggregates of the living ends of the polymer. The latter was determined based on the dependence of the final rates of the initiation reaction found in the first order plots on the initial concentration of silanolate.

It was demonstrated [19] that this rather sophisticated system can be rigorously analyzed.

The reaction follows the scheme:

$$2A \underset{}{\overset{K_0}{\rightleftharpoons}} A_2 \overset{K_0'}{\rightleftharpoons} 2/3\ A_3$$

$$A + B \overset{K_1}{\rightleftharpoons} AB$$

$$2B \overset{K_2}{\rightleftharpoons} B_2$$

$$A + M \overset{k}{\rightarrow} B$$

A parameter alpha was introduced, which accounted for the additional equilibrium between various types of A_n type aggregates:

$$\alpha = \frac{3[A_3]}{3[A_3]+2[A_2]} \neq 1$$

Assuming a pair of values for the equilibrium constants K (as in eq. for binary aggregates) and K_0' to allow for the equilibrium $A_2 \rightleftharpoons A_3$, it is possible to simulate by numerical calculations the kinetic curves for a chosen initial concentration of the initiator.

Since no details of the analysis can be provided here because of the limitations of space, a reference is made to the original paper [19] with the conclusion that the appropriate nomograms, based on the numerical calculations, allowed the equilibrium constants to be determined unequivocally.

For the reaction of siloxanes with potassium silanolate in non-polar medium $K = K_0K_2/K_1^2 = 0.07$, which suggests a strong tendency toward formation of mixed aggregates AB; $K_0' = 2000$ mol/dm^{-3}.

It was also demonstrated that the kinetic system involving A_2, A_3, AB and B_2 complexes can be clearly distinguished from the A_2, A_4, AB, B_2 system. It should be noted that cryoscopic studies of the state of aggregation of $PhMe_2SiOK$ in dioxane showed an average multiplicity of aggregation close to 4, so one could assume the formation of higher aggregates in the kinetic analysis.

In the studies of the initiation of the reaction of $PhMe_2SiOK$ with siloxanes, a remarkable aggreement was found between the kinetic curves simulated in the normalized first order plot and the experimental data (Figure 5). Reactions of MM and MDM with silanolate obey first order kinetics while all D_n and MD_nM for n>5 are represented by the same normalized curve, in spite of large variation of the specific reaction rate. The analysis actually provided a rationale for using initial apparent rate constants of the reaction of silanolates with siloxanes as a measure of siloxane reactivity. A postulated mechanism which may account for the large differences in the rates of the reactions of the series of linear and cyclic siloxanes will be dealt with in the next section, along with a discussion of some of the reasons for non-ideality of the aggregation systems.

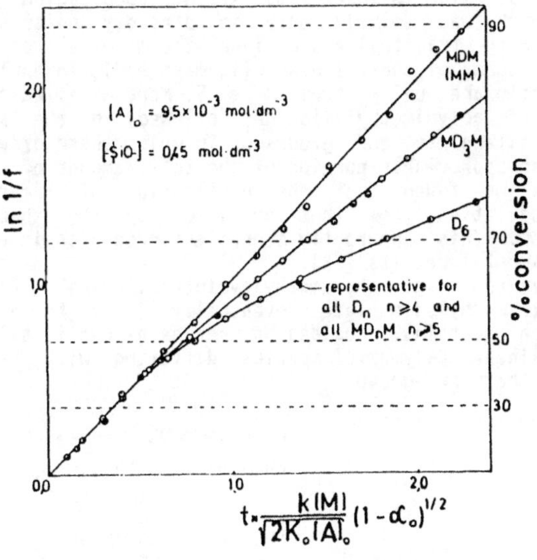

Figure 5. First order kinetic plot with normalized time scale for the reaction of $PhMe_2SiOK$ with linear (MD_nM) and cyclic (D_n) siloxanes, [19].

This "reader's digest" form of presentation of rather elaborate mathematical analysis, combined with tedious kinetic experiments, is intentionally chosen to demonstrate that it is possible to systematically study aggregation phenomena in the polymerization systems by using kinetic methods.

It is well-established that in many systems aggregation cannot be avoided, as in the polymerization of vinyl monomers or dienes, since low polarity of the medium is critical in determining the tacticity of the polymers. On the other hand, attempts to simplify the systems, by using strongly solvating additives (crown ethers, cryptands) to break the aggregates, can raise more questions concerning the mechanism of a given reaction in these new media, and no simple answers can be sought as to the mechanism of the elementary reactions involved. The kinetic studies of the polymerization of siloxanes on cryptated lithium silanolates by Boileau et al. [20] serve as an example.

II. 3. Thermodynamic Enhancement of Short Living Oligomers

It has long been known that there is relatively little difference in viscosity between living polydimethylsiloxane polymers (containing silanolate groups) and those which have been neutralized. Experiments were performed in a vacuum-sealed viscometer [21] using potassium polydimethylsiloxanolate above the entanglement molecular weight to show that on "killing", the viscosity of the system decreases much less than expected based upon the known dependence of viscosity on molecular weight.

On the other hand, in the course of studies of anionic polymerization of siloxanes it was found, rather accidentally, that in the living potassium siloxanolate system upon aging a precipitate appeared, which was soluble only in strongly polar solvents. When derivatized with trimethylchlorosilane the material was identified on GC as a mixture of short linear oligomers - disiloxanolates with the number of siloxane units from 3 to 5. More detailed studies of the equilibrium of polydimethylsiloxane followed in the systems having exclusively silanolate end groups or in which these groups eventually constitute an appreciable portion of the total amount of chain ends. In such systems we found that the equilibrium of the reaction shows remarkable deviation from that predicted by the theory of random reorganization. This apparently iconoclastic opinion is well-supported by the experimental results [22].

Polymerization of hexamethylcyclotrisiloxane initiated with potassium trimethylsilanolate eventually leads to an equilibrium mixture which contains, besides a series of cyclic siloxanes, three series of linear polymeric species differing with respect to the structure of their end-groups.

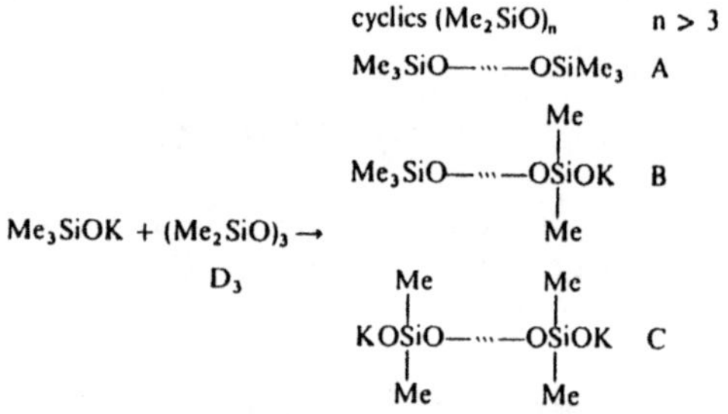

In this particular system, the number of silanolate groups equals the number of trimethylsilyl groups; thus the polymerization mixture should show the molar ratio A:B:C for any polymerization degree equal to 1:2:1. GC analysis of the system functionalized by capping showed that the total amount of oligomers A, B and C exceeds by more than one order of magnitude the amount of linear oligomers anticipated from the most probable distribution. It was also found that the amount of short C oligomers exceeds by at least one order of magnitude the concentration of A oligomers, while B are hardly detectable. Thus most of the short chains are of type C, while most long chains are of type A.

When -dipotassium siloxanolate was used to initiate the polymerization of cyclic siloxanes (D_3 or D_4), apart from cyclics, only linear species, C, having silanolate groups on both ends were present. In the system it was found that about 50% of the total amount of potassium end-groups appear in oligomers having the number of siloxane units within the range 1-8. The phenomenon is only slightly dependent on the initial concentration of siloxanolate within the range of MW of the linear fraction studied (2,000-60,000). A distinct maximum on the oligomer size distribution curves is always observed for potassium counterion at n=4 or 5. Sodium counterion shifts the maximum to n=4, while cesium to n=5. Neither temperature nor medium has any considerable effect on the proportion of short linear oligomers.

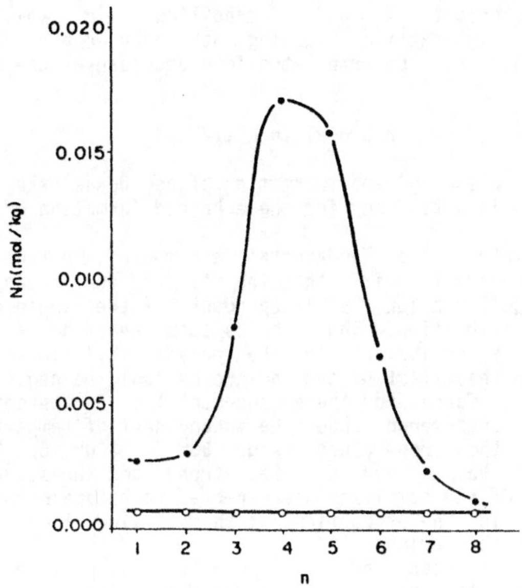

Figure 6. Concentration of living oligomers $KO[SiMe_2O]_nK$ in equilibrated PDMS living polymers having exclusively potassium siloxanolate end-groups observed; anticipated from the most probable distribution, [22].

The general explanation of the phenomenon [22] is based on the assumption that silanolate groups in the polymerization system are associated, with only a small fraction of free silanolate groups able to undergo propagation, depropagation or chain transfer. The aggregates of silanolate might form as a result of interaction between two end groups belonging to two different chains or as a result of the interaction of the two ends of the same chain, which adopts a ring conformation. One should also consider the formation of ring structures containing two or more aggregates.

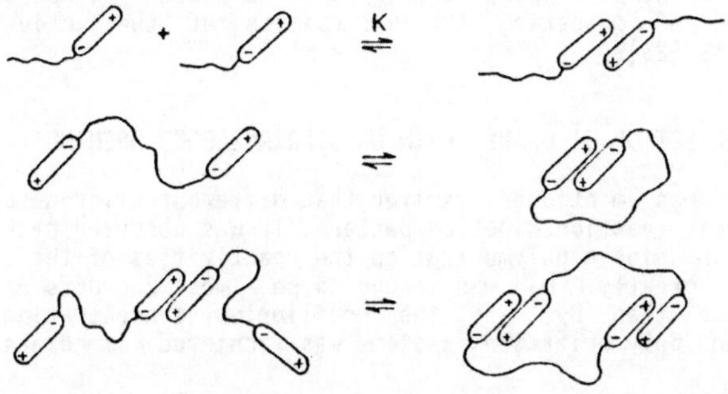

The formation of these cyclic structures is strongly preferred for short chains because the end-groups of these chains always operate at a short mean distance from each other. Their formation accounts for the increase of the effective equilibrium constant K_{eff} over the equilibrium constant K under conditions in which only the intermolecular interaction leading to open chain structures is considered. This can be translated into decrease of the thermodynamic potential

$$-\Delta G = RT \ln K_{eff}/K$$

When constant breaking and formation of new bonds takes place in the system, there is a tendency for the enhanced formation of living short chain species.

This violates the fundamental assumption in the Flory-Scott statistical treatment, viz. that the standard free energy change in a process of equilibration is independent of the length of the chains undergoing equilibration. Thus the system cannot be treated as being thermodynamically equivalent to the polydimethylsiloxane equilibrium mixture wherein the effect of the end-groups could be neglected.

The driving force for the enhancement lies in the entropy factor; therefore the phenomenon should be independent of temperature, as was confirmed for the temperature range 50° - 100° C. Complexes of silanolates of Na, K and Cs are strong and the structures of the bonding parts of the complexes are expected to be analogous. Thus, only minor changes in the proportions of short living oligomers were found for these three counterions.

Interaction between end-groups may strongly perturb the molecular distribution of the polymer in a dynamic living system. Since M_n of the polymer is not affected, the oligomer enhancement phenomenon leads to an increase of M_v, M_w and M_z of the polymer. Therefore all properties related to these parameters are affected. If there are also neutral end-groups in the system, there is a tendency for these groups to occupy the position in the longer chains. Some kinetic effects of the polymerization can also be expected, as living species in the longer chains might be more reactive (less tightly aggregated) than ones in the shorter chains.

In view of the results proving enrichment of living short oligomers at equilibrium, it is evident that the system is not suitable for studying aggregation phenomena by the methods based on viscometry.

It was found recently that in polymeric systems with lithium silanolate end-groups, where equilibration with formation of short living chains is strongly retarded, no perturbation of the thermodynamics was observed. Direct confirmation of aggregates formation could be obtained for this system by capping the living polymers with capping agents with different numbers of functional groups and comparing the viscosities of the active and terminated polymers [23].

III. EFFECT OF SILOXANE CHAIN ON SILOXANE BOND OPENING

It was mentioned earlier that different siloxane oligomers varied in their reaction kinetics pattern. It was observed by Laita et al.[24] that in anionic polymerization the reactivities of the cyclic siloxanes differ greatly. D_7 was found to be almost 2 orders of magnitude more reactive than D_4, and the equilibrium concentration of D_7 and D_8 in the polymerization system was achieved immediately. The authors provided a rather sophisticated interpretation of the data based on the differences in the entropy of activation in the polymerization of various cyclic dimethylsiloxanes. Equipped with the kinetic tools for analysis of the polymerization systems involving aggregation, we decided to re-evaluate the experimental data concerning the reactivities of siloxanes [25]. Laita et al. obtained their data by studying the kinetics of propagation. We chose to study the initiation reaction, since we proved by mathematical analysis that the initial rates of the reaction, as found from the first order kinetic plots, can be considered as the quantitative measures of the reactivity of the specific siloxane species, as at low conversions of the initiator the following equation is valid

$$d(a-x)/dt = k_{app}[M]\ (a-x)/a^{(n-1)/n}$$

where $k_{app} = k_{abs}/(n \times K_0)^{1/n}$. Since K_0, [M], a and n are constant within the series studied, k_{app} can be considered as a measure of k_{abs}, at least for the comparative studies of reactivities.

The experiments confirmed very large differences in reactivities between different cyclic siloxanes. D_3 reacts much faster than D_4. Reactivity increases with increase in the size of the molecule to D_7 - D_9. Surprisingly, a very similar trend was found for the series of linear permethylsiloxanes. A series of oligomers from dimer to dodecamer was studied. The data shows (Table 1) that reactivity dramatically increases on going from MM to MD_4M and continues to grow through $MD_{10}M$. Reactivities of larger cyclic and linear siloxanes (calculated for siloxane unit) are very similar. This was an indication that the peculiar behavior has the same origin in both series. If so, Laita's interpretation was unacceptable. Being at most conceivable for cyclics, it could not account for the same dramatic changes in the reactivities of linear molecules. Studies of the activation parameters (see: Table 1) proved that differences in reactivity are due to the decrease in the activation enthalpy value. Entropy of activation shows large negative values suggesting tight structures of the transition state, especially since the entropy of the system in the ground state must be low due to the aggregation of silanolate groups.

In the interpretation of this phenomenon it was assumed that some kind of specific interaction between siloxane and silanolate exists which facilitates the cleavage of the siloxane bond. The interaction of the cation-multipoint solvation type was proposed, which accords well with the fact that the intensive growth in reactivity with increasing siloxane size is observed only up to 5 or 7 siloxane units in the linear and cyclic siloxanes, which number makes the size of the molecules sufficient to encircle the counter ion of the silanolate group.

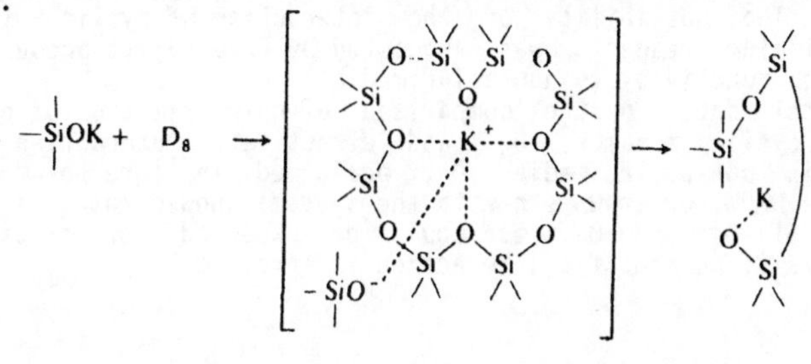

Table 1. The observed first order rate constants (30°C) and the activation parameters for the reaction of $PhMe_2SiOK$ with series of linear and cyclic siloxanes; heptane/dioxane (95:5 v/v); $[I]_0$ 9.5 10^{-3} mol/dm^3; $[(Me_2SiO)]$ 0.45 mol/dm^3. (Adapted from Ref. [25], by courtesy of the publishers, Hüthig and Wepf Verlag)

Siloxane	$10^3 \cdot k_{obs.}$ / s^{-1}	$\Delta H^{\ddagger}$ / kcal mol^{-1}	$\Delta S^{\ddagger}$ / e. u.	$\Delta G^{\ddagger}$ / kcal mol^{-1}	Temp. range in °C
MM	0,0015 (70°C)				
MDM	0,0018				
MD_2M	0,023	19,6 ± 1,6	− 14,5 ± 2,4	24,0 ± 1,6	30,0–70,0
MD_3M	0,17	10,7 ± 0,5	− 40,5 ± 0,8	23,0 ± 0,5	30,0–50,0
MD_4M	1,2				
MD_5M	1,3				
MD_6M	2,0				
MD_7M	2,1	7,0 ± 0,4	− 47,0 ± 0,6	21,2 ± 0,4	9,9–40,0
MD_8M	2,4				
MD_9M	3,1				
$MD_{10}M$	4,0				
D_3	2,9				
D_4	0,033	15,6 ± 0,6	− 27,6 ± 1,0	24,0 ± 0,6	30,1–69,0
D_5	0,035				
D_6	0,39	13,0 ± 1,3	− 31,0 ± 3,6	22,4 ± 1,3	20,0–41,0
D_7	5,3	12,5 ± 0,8	− 27,9 ± 1,4	21,0 ± 0,8	10,0–31,0
D_8	6,0				
D_9	4,1				

One can expect in this case energetically favorable conditions for the reaction on both sides: the ion pair of the silanolate group would be loosened and the electron density on the silicon atoms would be decreased, making easier the nucleophilic attack of the silanolate anion. Such a mechanism would certainly account for large negative values of the entropy of activation, larger for the linear molecules than for the cyclics.

Mazurek and Chojnowski [25] argued that siloxanes, being far less basic than ethers, were not able to interact with counter-ions as typical polydentate ligands, but still are able to assist the siloxane bond cleavage by multipoint interaction with the silanolate group counter-ion in the transition state of the reaction. Later work by Olliff et al. [26,27] on ion binding using diorganosiloxanes, and by Ricard [28] on sodium complexing with siloxanes, studied by NMR technique, showed that some higher molecular weight siloxane oligomers, and also cyclic trimers, can in fact effectively interact with counter-ions. The possibility of the interaction of cyclic siloxane with organolithium reagents was suggested by Frye [8] to account for their very high reactivity in non-polar media.

Experimental data on the comparison of polymerization of poorly interacting cyclic monomer, D_4, and effectively interacting with counter-ion in non-polar media, D_7, performed in more polar media (addition of HMPT or crown ether to the system) showed rather similar reactivities of D_4 and D_7, as could be expected for strainless cyclic monomers if no specific interaction is involved.

The idea of polymer chain interactions with the active center of the chain propagation is being used in the anionic polymerization of oxiranes [29], where it was demonstrated that reactivity in the polymerization system is a function of the chain length of the growing molecule.

IV. KINETICALLY CONTROLLED POLYMERIZATION

Most of the complications discussed above can, however, be effectively overcome by the proper choice of the polymerization system. D_3, the only member of the homologous series of cyclic dimethylsiloxanes having a considerably strained ring, can be polymerized in the presence of specific catalysts to obtain a very high yield of polydimethylsiloxane polymer with a negligible amount of cyclic byproducts. Initiators with lithium counter-ion (BuLi, R_3SiOLi) can polymerize D_3 in non-polar media, while the reaction with siloxane bonds in strainless systems is very slow. To some extent, in polymerization with complex counter-ions (tetramethylammonium silanolate, tetrabutylphosphonium silanolate [30]), advantage can also be taken of the differences in the reactivity of D_3 and linear or cyclic strainless siloxane molecules, since these bulky counter-ions are not able to be effectively solvated by polydimethylsiloxane molecules. These catalytic systems, lithium silanolates in particular, have very important practical applications, as they permit the synthesis of polydimethylsiloxane polymers and block copolymers, e.g. with polystyrene, with narrow molecular weight distribution. This possibility has been recently explored in the preparation of siloxane macromers [31]. The very large difference in reactivity between lithium trimethylsilanolate and oligomeric counterparts (siloxanolates), allowed Yamashita et al. to synthesize with high yield the functional siloxane oligomers, namely tetramers, in the reaction of D_3 with lithium silanolates.

V. EFFECT OF STRUCTURE OF SILANOLATE GROUP ON COPOLYMERIZATION

Cyclic monomers with a heterogenous composition of reactive groupings constitute an interesting class of monomers since they permit the synthesis of copolymers of a specific chain structure in homopolymerization processes. These monomers have more than one site of ring opening which are not regio- or chemoequivalent, and their polymerization involves two or more active centers of different structures. Siloxanes offer a variety of such monomers, e.g., dimethyldiphenyl, phenylmethyldimethyl, methyltrifluoropropyldimethyl, and many others.

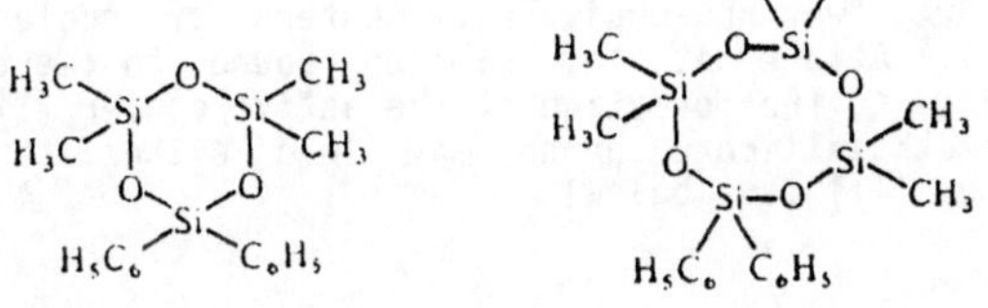

1 2

It was found in the polymerization of 2,2,4,4-tetramethyl-6,6-diphenylcyclo - trisiloxane (1) with two different catalytic systems - potassium silanolate and lithium silanolate with HMPT that the polymerization pattern is rather different in the two cases [32]. When the lithium catalyst was employed, polymer of pretty regular structure was obtained, as indicated by C^{13} NMR. The use of potassium silanolate, however, led to a complicated kinetic pattern and, consequently, to a less regular polymer structure. In the latter case, the transient formation of large quantities of the cyclic eight-membered oligomer (2) was observed, in competition with the formation of the polymer, during the monomer conversion. The amount of the oligomer reaches a maximum when nearly all monomer is converted. At that moment it amounts to as much as 45 wt-% of all siloxanes in the system, and its concentration then is an order of magnitude higher than its equilibrium concentration (Figure 7a). Similar behavior was observed in the copolymerization of the monomer with D_3 (Figure 7b). This was an interesting observation on its own. There are not very many systems known in which the concentration of the intermediate goes through a maximum to achieve equilibrium in the subsequent repolymerization.

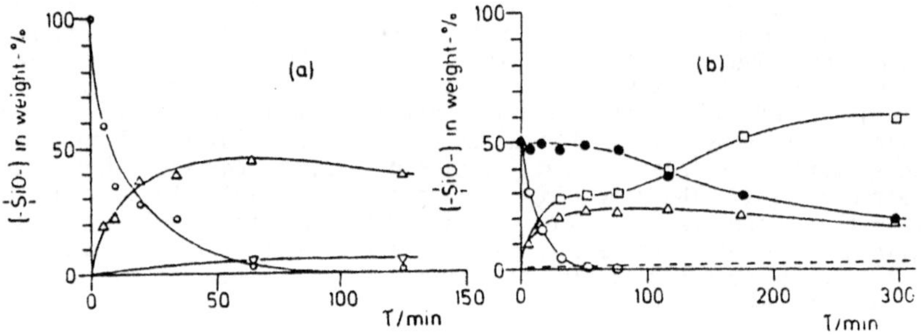

Figure 7. (a) - Polymerization of (1) on potassium siloxanolate in bulk at 70°C. K/Si = $1.2 \cdot 10^{-4}$ mol/dm^{-3}; (○) - (1); (△) - (2); (□) - polymer; (b) - Copolymerization of (1) with D_3 (●) under the same conditions. (Ref. [32], by courtesy of the publishers, Hüthig & Wepf Verlag)

These data, together with the experimental results on the comparison of the reactivity of D_3 and the monomer (1) suggested that the following generalizations concerning the mechanism of the anionic polymerization and copolymerization of (1) can be drawn.

There are two non-equivalent centers of nucleophilic attack in monomer (1). Attack at the silicon bound to the two phenyl groups always leads to the formation of the active center (I), whereas attack at the dimethylsilicone group may lead either to the formation of center (I) or (II) (see below).

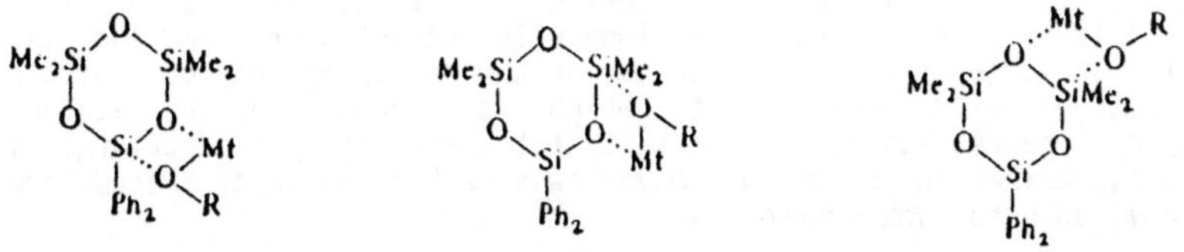

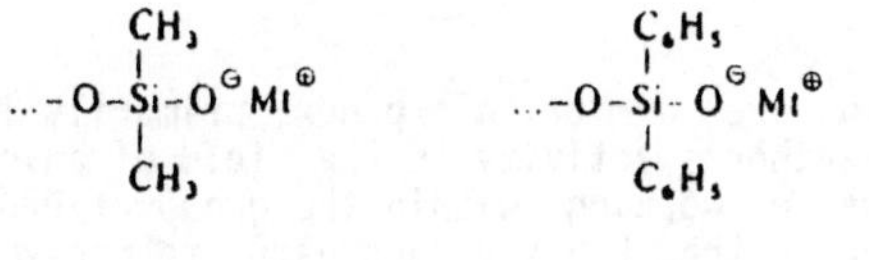

(I) (II)

In the polymerization with lithium catalyst, the ring opening step determines the structure of the living end, since no secondary processes leading to the transformation of one living end into another are involved. In the system with potassium catalyst, however, the active centers can undergo transformation from one into another by means of back-biting depolymerization and chain transfer reactions. Since the active center (II) is more stable then (I) there is a driving force to the formation of the former one. All the experimental findings fit the following mechanistic scheme:

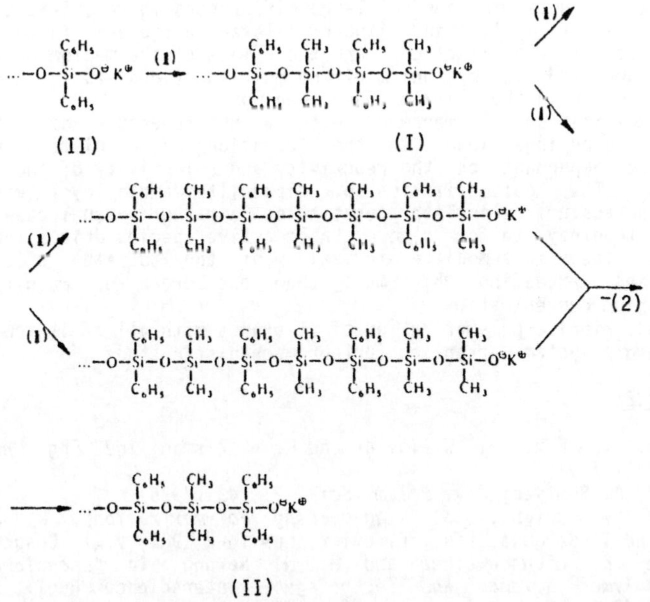

(II)

The monomer adds to the more stable living center mostly by attack on the dimethylsilicone group (steric reasons). A chain end fragment with active center (I) is formed which again reacts mainly toward propagation by attack on either the dimethyl or diphenyl group, leading in both cases to structures capable of undergoing depolymerization by back biting, with the formation of the more stable diphenyl living group and tetramer as an intermediate. In each case, it requires at least two acts of monomer ring opening for the depolymerization step. Thus, polymer growth is observed along with the tetramer formation. The results indicate that both propagation and depropagation occur with a high degree of regioselectivity which is dependent on the catalyst employed. The results also indicate that the usually applied treatments of copolymerization based on Mayo-Lewis or Finneman-Rose methods are not applicable to such systems.

VI. CONCLUSIONS

It has been the aim of this paper, primarily intended as a brief summary of the author's activity in the field of anionic polymerization of siloxanes while working within the group of Prof.J.Chojnowski, to demonstrate that the kinetic approach to polymerization systems complicated by aggregation phenomena can be successfully applied in mechanistic studies. Mathematical analysis of the kinetic schemes provides the adequate equations to describe the reaction pattern, when cross- aggregation is involved. The proper choice of the model systems for studying the component reactions is critical.

Aggregation phenomena lead to the thermodynamic enhancement of the formation of short difunctional oligomers in the polymerization system, which accounts for the perturbation of the molecular weight distribution, as well as for the unusual viscosity behaviour of "living" polymers.

Analysis of the general kinetic schemes of reactions involving cross-aggregation of the active species provided the basis for a quantitative determination of large differences in reactivities within the series of cyclic and linear siloxane oligomers. The activation parameters of the reactions and the effect of the medium on reaction rates supported the mechanism of the multifunctional assistance of the siloxane chain in the siloxane bond opening.

Polymerization of monomers with a heterogenous composition of reactive groupings leads to the formation of copolymers having a structure dependent on the reactivity and selectivity of the reactive species. The polymerization of diphenyltetramethylcyclotrisiloxane using potassium silanolate constitutes a rather unusual case where a strong tendency to form a more stable active species drives the system through the intermediate formation of the tetramer siloxane in quantities exceeding by more than one order of magnitude its equilibrium concentration.

The anionic polymerization of siloxanes with all of its complexity is indeed a system warranting further mechanistic study.

REFERENCES

1. H. W. Fox, P. W. Taylor and W. A. Zisman, *Ind. Eng. Chem.*, 39, *1401 (1947)*
2. J. A. Semlyen, *Adv. Polym. Sci.*, 21, *43 (1976)*
3. P. V. Wright, in *"Ring-Opening Polymerization"*, K. J. Ivin and T. Saegusa, Eds., Elsevier, New York 1984, v.II, Chapter 14
4. H. K. Lichtenwalner and M. N. Sprung, in *"Encyclopedia of Polymer Science and Technology"*, Interscience Publ., New York 1970, v.12, p.503
5. W. T. Grubb and R. C. Osthoff, *J. Amer. Chem. Soc.*, 77, *1405 (1955)*

6. M. Szwarc, *"Carbanions, Living Polymers and Electron-Transfer Processes"*, Interscience Publ., New York 1968
7. G. V. Schultz, *Adv. Chem. Ser.*, 128, 1 (1973)
8. C. L. Frye, R. M. Salinger, F. W. G. Fearon, J. M. Klosowski and T. DeYoung, *J. Org. Chem.*, 35, 1308 (1970)
9. Z. Laita, unpublished results (1970)
10. Z. Laita and M. Szwarc, *Macromolecules*, 2, 412 (1969)
11. Yu. A. Yuzhelevskii, E. Kogan and N. N. Fedoseeva, *Dokl. Akad. Nauk SSSR*, 190, 647 (1970)
12. J. Chojnowski, M. Mazurek, *Makromol. Chem.*, 176, 2999 (1975)
13. T. G. Selin, *3rd International Symposium on Organometallic Chemistry, Munich, 1967, Abstracts, p. 67*
14. W. A. Fessler and P. C. Juliano, *Ind. Eng. Chem., Prod. Res. Develop.*, 11, 407 (1972)
15. J. Chojnowski, S. Rubinsztajn, W. Stańczyk and M. Ścibiorek, *Makromol. Chem., Rapid Commun.*, 4, 703 (1983)
16. M. Morton, *"Anionic Polymerization: Principles and Practice"*, Academic Press, 1983, Chapter 5
17. J. E. Figueruelo and D. J. Worsfold, *Europ. Polym. J.*, 4, 439 (1968)
18. P. Sigwalt and N. Spassky, in *"Ring-Opening Polymerization"*, K. J. Ivin and T. Saegusa, Eds., Elsevier, New York 1984, v.II, Chapter 9
19. M. Mazurek and J. Chojnowski, *Macromolecules*, 11, 347 (1978)
20. S. Boileau, in *"Ring-Opening Polymerization. Kinetics, Mechanisms, and Synthesis"*, J. E. McGrath, Ed., ACS Symposium Series 286, Washington 1985 Chapter 2,
21. J. Chojnowski and M. Mazurek, unpublished results
22. M. Mazurek, M. Ścibiorek, J. Chojnowski, B. G. Zavin and A. A. Zhdanov, *Europ. Polym. J.*, 16, 57 (1980)
23. J. P. Kennedy and L. Wilczek, private communication (paper in preparation)
24. Z. Laita and M. Jelinek, *J. Polym. Sci., Part B*, 4, 739 (1966)
25. M. Mazurek and J. Chojnowski, *Makromol.Chem.*, 178, 1005 (1977)
26. C. J. Oliff, G. R. Pickering and K. J. Rutt, *J. Inorg. Nucl. Chem.*, 42, 288 (1980)
27. C. J. Oliff, G. R. Pickering and K. J. Rutt, *J. Inorg. Nucl. Chem.*, 42, 1201 (1980)
28. A. Ricard, *Europ. Polym. J.*, 15, 1 (1979)
29. T. Tsuruta, *J. Polym. Sci., Part D*, 6, 179 (1972)
30. A. R. Gilbert and S. W. Kantor, *J. Polym. Sci.*, 40, 35 (1959)
31. Y. Kawakami and Y. Yamashita, in *"Ring-Opening Polymerization. Kinetics, Mechanisms, and Synthesis"*, J. E. McGrath, Ed., ACS Symposium Series 286, Washington 1985, Chapter 19
32. M. Mazurek, J. Ziętera, W. Sadowska and J. Chojnowski, *Makromol. Chem.*, 181, 777 (1980)

SYNTHESIS

SYNTHESIS AND APPLICATIONS OF MACROMONOMERS

PAUL REMPP and EMILE FRANTA
Institut Charles Sadron (CRM/EAHP), CNRS,
6, rue Boussingault, 67083 Strasbourg Cedex France

ABSTRACT

The methods of synthesis of macromonomers are reviewed with emphasis on those based on anionic polymerization, and capable of yielding samples of well defined molecular weights and low polydispersity. Homopolymerization of macromonomers - either free radical or anionic - yields highly branched and compact macromolecules. The chief interest of macromonomers is to provide an easy access to graft copolymers upon free radical copolymerization with an acrylic or vinylic comonomer.

INTRODUCTION

The field of macromolecular monomers - generally referred to as macromonomers or as macromers™ - was opened ten years ago by the pioneering work of the late Ralph Milkovich [1,2]. Ever since, the interest devoted to these polymer species fitted at chain end with a polymerizable entity - such as those shown below - increased and developed later in universities and in industries.

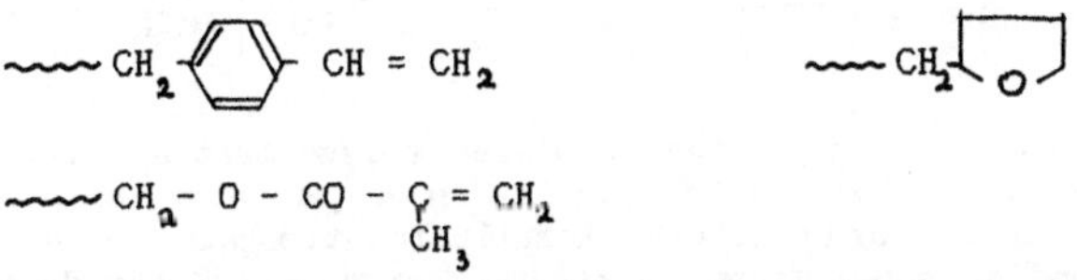

Macromonomers are in effect useful intermediates in polymer synthesis , especially for the preparation of graft copolymers. It is appropriate to include macromonomers into the scope of a symposium devoted to anionic polymerizations, as the synthesis of macromonomers owes much to "living" anionic polymerization techniques. Several review articles [3,4] have already been published in the domain concerned. The purpose of this paper is rather to select some trends and to discuss the chief orientations of research on macromonomers, with some emphasis on those making use of anionic polymerization.

SYNTHESIS OF MACROMONOMERS

Macromonomers have been obtained in a large number of ways. However the methods based on ionic "living" polymerizations are best suited for functionalizations at chain end , because of the absence of spontaneous transfer and termination reactions . The number of growing polymer molecules is constant and - when initiation is fast - it is equal to the number of initiator molecules introduced. This allows for a control of the molecular weight of the species formed, and for the sharpness of their molecular weight distribution. Since high functionalization yields can be attained, these methods are suited for the synthesis of well defined macromonomers.

The nature of the polymerizable entity deserves a few comments. Though a vast majority of macromonomers carry end-standing styrene (or α-methyl-

Recent Advances in Anionic Polymerization
T.E. Hogen-Esch and J. Smid, Editors

styrene) or methacrylic ester functions, that can participate in free radical polymerizations, there are also macromonomers exhibiting heterocyclic end groups , which can undergo ring opening polymerization. A third category of macromonomers is fitted with two alcohol (or two acid) endgroups at the *same* chain end , and can thus get involved in polycondensation processes.

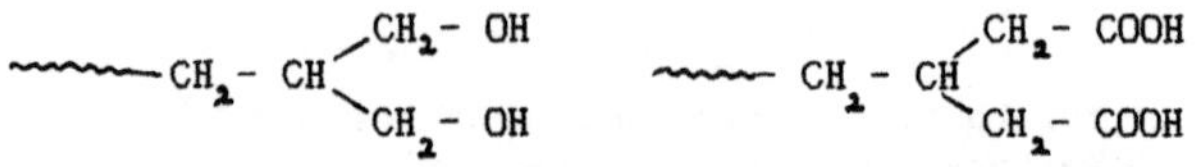

In principle , two methods can be used to introduce a polymerizable unsaturation at chain end, using ionic "living" polymerization techniques : upon initiation or upon deactivation of the living sites [5].

1- Use of an unsaturated initiator - provided initiation occurs by addition onto the monomer - is the most obvious procedure , but it is restricted to a small number of cases , since it implies that no side reaction involves the unsaturation. This method proved to be efficient in some cases:

* The cationic ring opening polymerization of oxolane (THF) or of N-substituted aziridines can be initiated by methacryloyl oxocarbenium salts [6,7]. The methacrylic ester unsaturation is insensitive to cationic sites, and polyoxolanes (poly-THF) macromonomers are obtained in good yields.

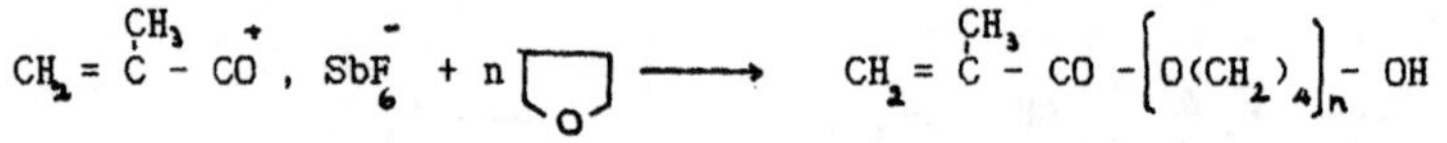

The same is true for the oxolane polymerization initiated by p.vinylbenzylium salts [5], which can be made in situ by reaction of p.vinylbenzyl bromide and silver hexafluoroantimonate. The initiator efficiency is good as shown from the kinetic curves. Once the desired chain length has been reached, the growing oxonium sites are deactivated and the macromonomer is recovered.

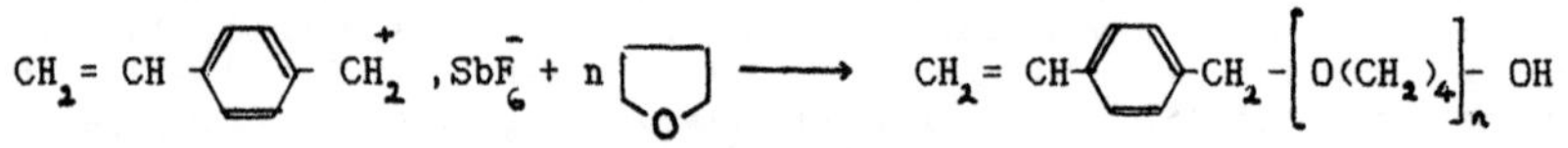

* The use of an unsaturated anionic initiator - such as potassium p.vinylbenzoxide - is possible for the ring opening polymerization of oxirane [8]. Although initiation is generally heterogeneous, the polymers exhibit the molecular weight expected and a low polydispersity. In this case the styrene type unsaturation at chain end cannot get involved in the process, as the propagating sites are oxanions.

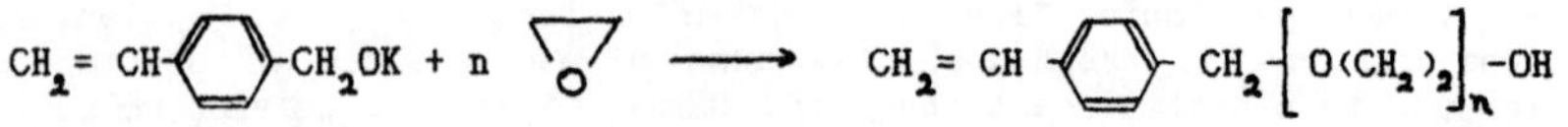

Similarly, ω-oxazolinyl poly(ethylene oxide) samples have been made by Saegusa [9] by anionic initiation ; the oxazoline cycle is insensitive to anionic attacks, but it polymerizes cationically quite effectively.

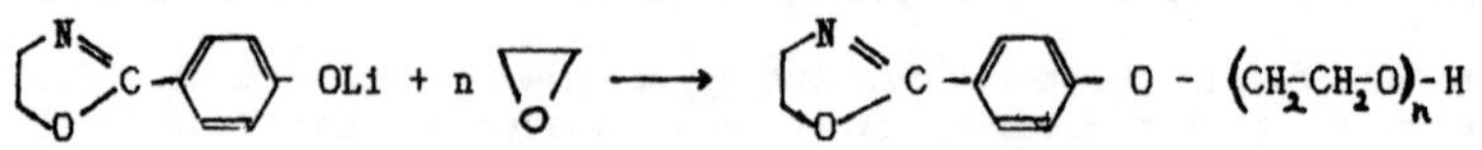

More recently the same procedure was used to synthesize polyester macromonomers by ring opening polymerization of caprolactone initiated with an unsaturated alkoxide. In that case, however, back-biting reactions do occur [10] yielding cyclic oligomers, whereby both the polymer yield and the molecular weight of the polyester macromonomer are lowered as compared to those of a regular living system. Conditions are being searched to minimize these side reactions [11].

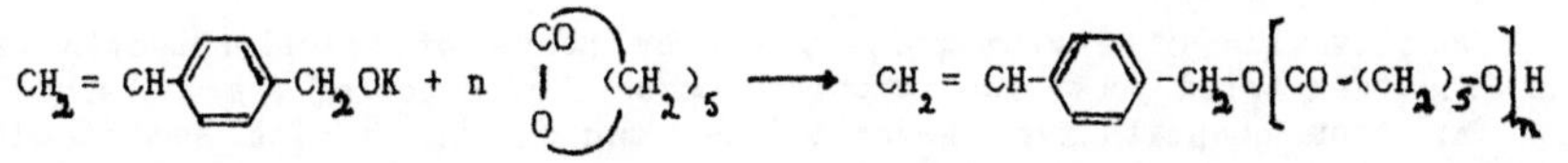

Along the same line, experiments of Yamashita [12] and of Fock [13] aimed at the synthesis of poly(ethylene oxide) chains fitted at one end with two hydroxyl functions. During the oxirane polymerization the two OH functions are protected by acetalization, and they are released afterwards by acid hydrolysis. Such macromonomers can take part in polycondensations with diacids, or diisocyanates, or dichlorosilanes, thus giving access to graft polycondensates.

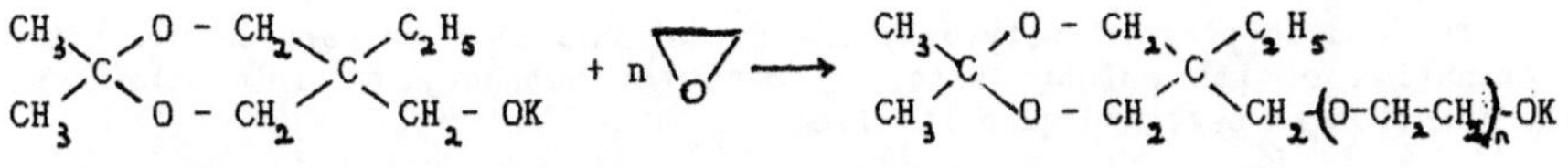

2 - Use of an unsaturated deactivator to end-cap the chains of a living polymer is a method far more general than the previous one [1,2,5]. In that case the ionic polymerization - efficiently initiated - proceeds to completion (or to the conversion searched for) and then the unsaturated reagent is added : a nucleophile if the polymerization is cationic, an electrophile in the case of anionic polymerization. Again, unwanted side reactions involving the unsaturation are to be prevented, which can be achieved more easily than in the previous case. An intermediate addition of a monomer - such as 1,1 diphenyl ethylene or oxirane - may however be necessary to get less reactive sites at chain end [2,14].

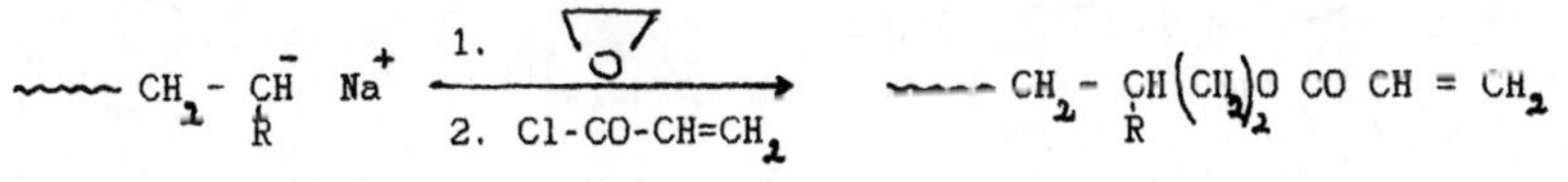

Numerous examples of such procedures are found in the literature. It is the method generally used to make macromonomers from anionically polymerizable vinylic or acrylic monomers [2,5,15,16]. It can also be applied to a number of monomers able to undergo ring opening polymerizations, such as oxirane [8] ,some lactones [11], and hexamethyl cyclotrisiloxane [17,39].

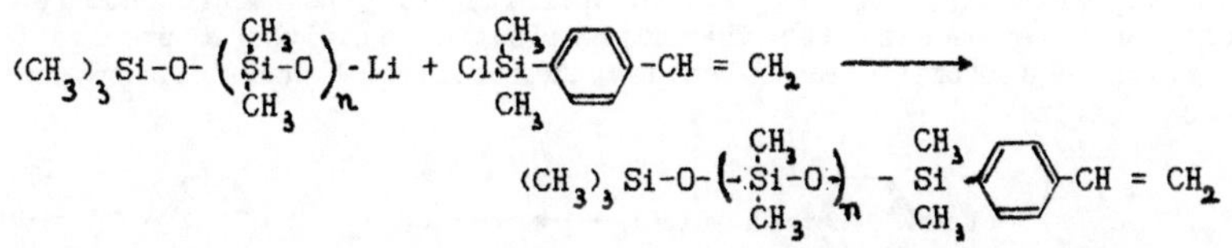

The deactivators most commonly used to end-cap the anionic sites are methacryloyl chloride and vinylbenzyl halides. The latter proved efficient even with carbanions of low nucleophilicity (such as those derived from methacrylic esters [18]), or with oxanions [19]. Deactivators involving SiCl bonds have also been attempted, e.g. dimethylchlorosilyl propyl methacrylate [20], which is obtained in good yields from allyl methacrylate and chlorodimethyl silane, by hydrosilylation.

$$CH_2=C(CH_3)-COO-CH_2-CH=CH_2 + H-Si(CH_3)_2-Cl \longrightarrow CH_2=C(CH_3)-COO-CH_2-CH_2-CH_2-Si(CH_3)_2-Cl$$

Deactivation of living polystyrene by means of epichlorohydrin [21] has also been tried, but side reactions occur , as expected , and the yields are far from quantitative. Nevertheless, macromonomers with end standing epoxy functions were characterized as such.

Acetylene can react with the alcoholate functions of a living poly(ethylene oxide), whereby a vinyl ether end group is formed [22]. This unsaturation can undergo alternating free radical copolymerization with monomers such as maleic anhydride.

$$CH_3-(O-CH_2-CH_2)_n-OK + CH{\equiv}CH \xrightarrow{(+H^+)} CH_3-(O-CH_2-CH_2)_n-O-CH=CH_2$$

The end-capping of poly(ethylene oxide) can also be carried out after protonation of the anionic site, by means of methacryloyl imidazole [23]. The reaction is neat and quantitative.

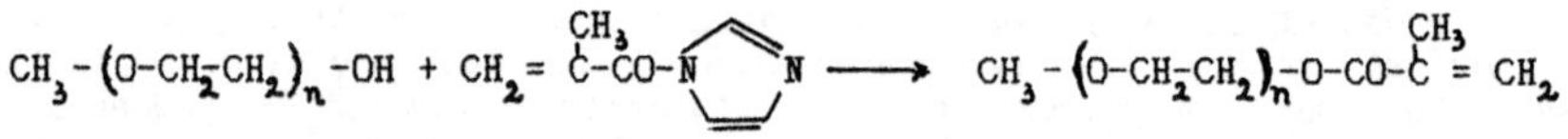

Polyamide macromonomers have been obtained in a similar manner using p.vinylbenzyl amine to end-cap the growing sites [24]. The reaction mechanism is somewhat different here, as the active sites are acyl-lactam functions. Advantage is taken of the higher nucleophilicity of the imide function derived from benzyl amine with respect to that of the lactam anion.

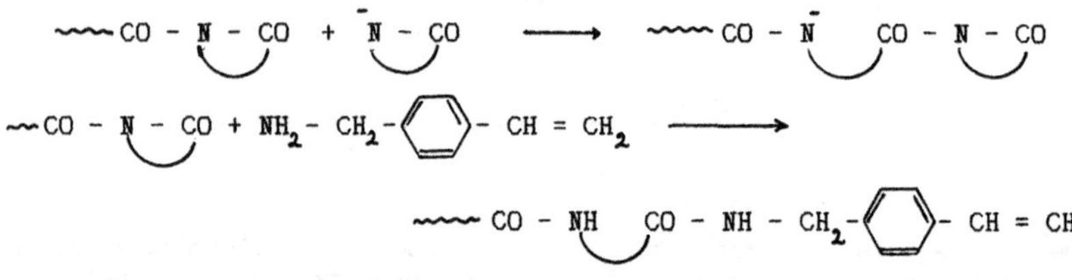

Cationic living polymers (poly-THF, poly N-t.butylaziridine) can also be end-capped with unsaturated nucleophiles such as alcoholates [25,26], or amines, or even salts derived from methacrylic acid. The reaction is fast and quantitative; it should be carried out to rather small conversion of the oxolane polymerization if low polydispersity of the macromonomers is required. This is because the THF polymerization involves a propagation-depropagation equilibrium, and because reshuffling becomes important at higher conversions.

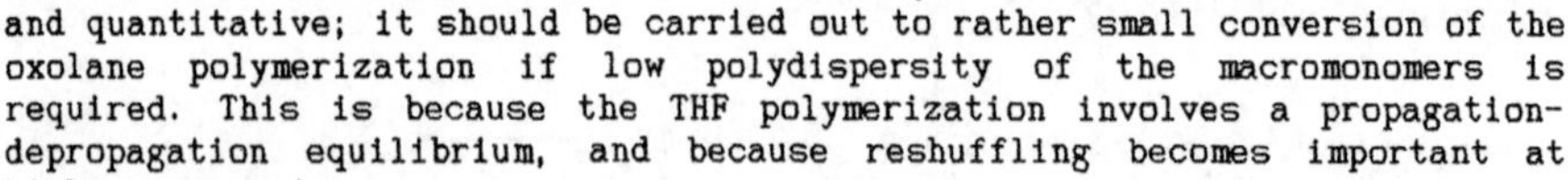

3 - Polycondensations can also serve for the synthesis of macromonomers, as they yield functional macromolecules. In that connection an interesting two step method was designed by Tsuruta [27] : p.divinyl

benzene, when reacted in a basic medium with a secondary diamine, yields a monoadduct in near to quantitative yields. This adduct can undergo self condensation to yield polymer molecules bearing one styrene end-group and one secondary amine chain end. This self-condensation can also be considered as an anionic polymerization involving proton transfer after each growth step.

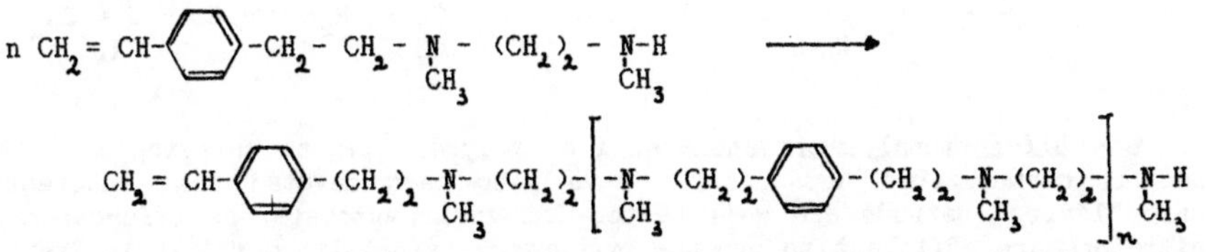

4 - Cationic polymerization of vinyl monomers generally involves transfer to the monomer , which makes it impossible to carry out direct functionalizations. However the synthesis of macromonomers has been made possible by the so called "inifer" method, developed by Kennedy [28]. The initiating system involves BCl_3 and an alkyl halide that can give rise to transfer reactions. Since both the transfer process with RCl and the ion-splitting of the BCl_4^- anion yield the same species, and since both these reactions are more likely to occur than transfer to the monomer, polymers (especially polyisobutene) with an end-standing C-Cl linkage are obtained in high yields.

$$R\,Cl + B\,Cl_3 \longrightarrow R^+, B\,Cl_4^-$$

$$R^+, B\,Cl_4^- + M \longrightarrow R\,M^+, B\,Cl_4^-$$

$$R\,M_n^+, B\,Cl_4^- \longrightarrow R\,M_nCl + B\,Cl_3$$

$$R\,M_n^+, B\,Cl_4^- + R\,Cl \longrightarrow R\,M_nCl + R^+, B\,Cl_4^-$$

A three-step procedure enables then to fit the chain end with an acrylic ester function, thus yielding polyisobutylene macromonomers.

5 - Free radical polymerization has also been used for the synthesis of macromonomers. Owing to the short lifetime of the free radical sites, functionalizations cannot be carried out directly. Advantage has to be taken of transfer reactions with functional transfer agents to get well defined functions at chain end. It is even better to have the process started with a functional initiator, in order to have all macromolecules formed (those arising from transfer, and those arising from primary radicals) carry the same function at chain end [29]. High functionalization yields have been attained. In addition some control of the molecular weight is provided by a proper choice of the [transfer agent]/[monomer] mole ratio. The polymers formed bear acid or alcoholic end groups.

$$\sim\sim CH_2-\underset{R}{C}H + H\text{-}S\text{-}CH_2-COOH \longrightarrow \sim\sim CH_2-\underset{R}{C}H + {}^{\bullet}S\text{-}CH_2-COOH$$

$$HOOC-CH_2\text{-}S^{\bullet} + CH_2=\underset{R}{C}H \longrightarrow HOOC-CH_2\text{-}S-CH_2-\underset{R}{C}H^{\bullet} \xrightarrow{+M}$$

To get macromonomers these functions have to be reacted with acrylic or methacrylic derivatives under conditions such as to get high yields.

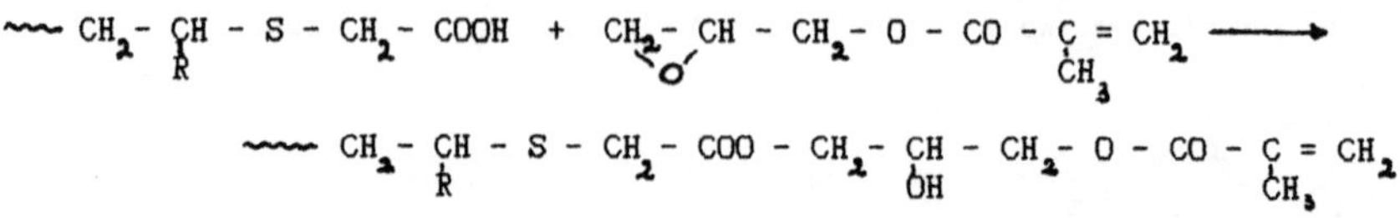

6 - <u>Bifunctional macromonomers</u>, i.e. polymer species carrying at both ends a polymerizable unsaturation have also been devoted some interest. Ionic "living" methods are well suited for such synthesis, as bifunctional initiators are efficient to produce precursor chains with carbanionic, resp. oxonium sites at both chain ends. The end-capping procedure by means of an unsaturated deactivator is the same as that described above. Difunctional anionic initiators include Na naphthenide, α-methylstyrene "tetramer", or the diadduct of BuLi onto diisopropenyl benzene [30]. The oxocarbenium salts derived from adipic acid or terephthalic acid are efficient bifunctional cationic initiators [31].

POLYMERIZATION OF MACROMONOMERS

Homopolymerization of macromonomers should result in branched macromolecules exhibiting a high segment density, as each unit of the main chain carries a graft. However, the ability of macromonomers to undergo polymerization can be questioned for two reasons :

1 - The molar concentration of a macromonomer (i.e. that of the polymerizable species) is low, even in the bulk, owing to its relatively large molecular weight (generally between 1000 and 20000).

2 - The lower accessibility of the growing sites to the double bonds should impede the polymerization, especially at elevated concentrations, at which the reaction medium is very viscous.

Nevertheless, polymerization does occur - sometimes even unwanted, when macromonomers are stored for longer times -. Only few attempts to homopolymerize macromonomers are described in the literature [2,10,25, 32,33]. They chiefly refer to polystyrene, poly(ethylene oxide) and polyisobutene macromonomers.

Free radical polymerization usually generates small yields of polymacromonomer of low polymerization degree. This can be accounted for by the rather short life-time of free radicals, which allows each of them only a small number of growth steps.

In some cases, transfer reactions involving the macromonomer chain cannot be excluded, especially in the case of poly(ethylene oxide). It sometimes happens that insoluble (yet swellable) material is formed [35].

Cationic homopolymerization of ω-oxazolinyl poly(ethylene oxide) was investigated by Saegusa [9]. Here again, the number of units linked together stays low, and no gel formation is observed.

Anionic polymerization of macromonomers is expected to be more efficient, provided their chains do not contain any function that could deactivate the carbanionic sites. Polymerization of polystyrene macromonomers was carried out to high yields and led to polymers with up to 50 macromononomer units.

These highly branched macromolecules exhibit, as expected, very low hydrodynamic volumes in solution, as compared to those of linear homologues of same molecular weight. This is evidenced by their smaller radii of gyration - as determined by small angle neutron scattering -, by their longer SEC retention times, and by their lower limiting viscosity numbers [35]. Further characterization work is in progress.

COPOLYMERIZATION OF MACROMONOMERS WITH VINYLIC COMONOMERS

The interest for macromonomers arises from the fact that they provide an easy access to well defined graft copolymers, upon free radical copolymerization with a vinylic or acrylic monomer. Account has to be taken of the low values of the mole ratio of macromonomer to comonomer, due to the difference in molecular weight between the two species. The total weight of the grafts may amount to 50 weight %, while the mole ratio of grafts can stay below 5 %.

The kinetics of this free radical copolymerization was investigated by a number of authors [2,5,35-39], who attempted to determine the radical reactivity ratios of the system. Two procedures were used :

- Copolymerizations were carried out to low conversions, and the compositions of the graft copolymers were thoroughly characterized, by analytical methods.
- Size exclusion chromatography was used to determine quantitatively the amount of macromonomer converted, and the amount of copolymer formed, as a function of time. The data were processed according to Fineman-Ross , or to Kelen-Tüdös.

The reactivity ratios r_A of the comonomer were generally obtained with a relatively good accuracy, but the values of r_M for the macromonomer are very inaccurate, which is easy to understand. The classical copolymer composition equation can be written as :

$$\frac{d[A]}{d[M]} = \frac{[A]}{[M]} \cdot \frac{r_A[A]/[M] + 1}{r_M + [A]/[M]}$$

Since the mole ratio [A]/[M] is very large, it can be assumed that the following inequalities hold :

$$r_A[A]/[M] >> 1 \qquad ; \qquad [A]/[M] >> r_M$$

Then the equation reduces to

$$\frac{d[A]}{d[M]} = r_A \frac{[A]}{[M]}$$

This would imply that such a copolymerization be governed by a single parameter (as in an "ideal" copolymerization), within the range of the experimentally accessible [A]/[M] mole ratios. However, this procedure has been questioned recently by Cameron [39], since the values of r_A for a given macromonomer/comonomer system are not found independent of the [A]/[M] mole ratio, as it should be if the process were governed by a single parameter. Further experimental work is needed to settle this point.

The variation of the reactivity ratio of the comonomer, as a function of the molecular weight of the macromonomer, could indicate whether the difficult access of the growing sites to the macromonomer unsaturation impedes the process. Other explanations could be considered, as well. It was found that r_A increases with the molecular weight of the macromonomer [39], though it had been claimed earlier [37] that the value of r_A remained practically unchanged as the length of the macromonomer chain varies. Since the free radical homopropagation rate constant of a monomer is independent of the other species present, the increase of r_A is a consequence of a decrease of k_{AM}, the rate constant of the reaction between $A^{\bullet}$ and M. This result means that the higher the molecular weight of the macromonomer, the more difficult the access of the radicals to the unsaturation of the macromonomer. Other authors have suggested that the participation of macromonomers to the process is diffusion limited [37], because it implies a reaction between two macromolecular entities : growing chain and macromonomer. This would imply that the activation energy be low, as it is the case for diffusion controlled processes. More experiments have to be performed, at various temperatures, and with macromonomers of widely different molecular weights, to settle this point and to reach definite conclusions.

CONCLUSION

A large amount of work has already been performed on the synthesis and characterization of macromonomers and on their uses as intermediates for the synthesis of well defined graft copolymers. Much remains to be done in order to understand the special features of macromonomer copolymerization, and to explore other potential applications of macromonomers in polymer synthesis.

Nevertheless, the numerous applications of graft copolymers have been and are still the chief motivation for research in the field of macromonomers. These applications arise from the incompatibility between homopolymer chains of different chemical nature. The phase separation occuring intramolecularly between grafts and backbone leads to the so-called surface accumulation, a phenomenon which is frequently used in coatings or paints to have a surface modified, as for example : to render it hydrophilic or hydrophobic, to lower the surface tension, to make it antistatic or to provide for adhesion. Other applications of graft copolymers as compatibilizers in polymer blends, or as emulsifying agents or as dispersion stabilizers are also of great interest.

The authors wish to express their deep appreciation to Drs. Lutz, Gnanou and Masson for their participation in research projects on macromonomers. They also want to thank Drs. Beinert and Chaumont, Mrs Graff and Mr. Lamps.

REFERENCES

1 . R. Milkovich, US Patent 3 786 116 (1974)
2 . G.O.Schulz, R.Milkovich, J.Appl.Polym.Sci. 27, 4773 (1982)
J Polym. Sci. Chem. Ed 22 3795 (1984); 22 1633 (1984)
3 . P. Rempp, E. Franta, Adv. Polym. Sci. 58 1 (1984)
4 . Y. Yamashita, J. Appl. Polym. Sci. 36 193 (1981)
5 . P. Rempp, P. Lutz, P. Masson, E. Franta,
Makromol. Chem. Suppl. 8 3 (1984)
6 . J. Sierra-Vargas, J.G. Zilliox, P. Rempp, E. Franta
Polym. Bull. 3 83 (1980)
M. Takaki, R. Asami, T. Kuwabara, Polym. Bull. 7 521 (1982)
7 . E.J. Goethals, M. Vlegels, Polym. Bull. 4 521 (1981)

8 . P. Masson, G. Beinert, E. Franta, P. Rempp, Polym. Bull. 7 17 (1982)
9 . S. Kobayashi, M. Kaku, T. Mizutani, T. Saegusa,
Polym. Bull. 9 169 (1983)
10. K. Ito, H. Hashizuka, Y.Yamashita, Macromolecules 10 821 (1977)
11. Y. Gnanou, P. Rempp, to be published
12. Y. Chujo, T. Tadsuda, Y. Yamashita,
Polym. Bull. 5 361 (1981) ; 8 239 (1982)
13. J. Fock, E. Esselborn, G.E. Hahn, and G. Riess, to be published
14. P. Masson, E. Franta, P. Rempp,
Makromol. Chem. Rap. Comm. 3 499 (1982)
15. R. Asami, M. Takaki, H. Hanahata, Macromolecules 16 628 (1983)
16. P.R. Rao, P. Masson, P. Lutz, G. Beinert, P. Rempp
Polym. Bull. 11 115 (1984)
17. Y. Kawakami, Y. Yamashita, Polym. J. 14 915 (1982) ;
ACS Polym. Preprints 25 245 (1984)
18. B.C. Anderson, W.H. Sharkey, et al., Macromolecules 14 1599 (1981)
19. W.W. Schulz, J Kaladas, D.N. Schulz,
J. Polym. Sci. Chem. Ed. 22 3795 (1984)
20. Y. Kawakami, Y. Miki, T. Tsuda, R. Murthy, Y. Yamashita,
Polym. J. 14 913 (1982)
21. Hongquan Xie, Wenbo Son, ACS Polym. Preprints 25 67 (1985)
22. T. Suzuki, T. Tomono, J.Polym. Sci. Chem. Ed. 22 2829 (1984)
23. H.G. Batz, G. Franzman, H. Ringsdorf, Makromol. Chem. 172 27 (1973)
24. K. Hashimoto, H. Sumitomo, M. Kawasumi,
J. Macromol. Sci. Chem. A21 1049 (1984)
Polym. J. 17 1045 (1985)
25. J. Sierra-Vargas, P. Masson, G. Beinert, P. Rempp,E. Franta,
Polym. Bull. 7 277 (1982)
26. R. Asami, R. Takaki, K. Kita, E. Asakura,
Polym. Bull. 2 713 (1980) ; Pol. J. 15 139 (1983)
27. T. Nishimura, M. Maeda, Y. Nitadori, T. Tsuruta,
Makromol. Chem. 183 29 (1981)
28. J.P. Kennedy, M. Hiza, J. Polym. Sci. Chem. Ed.21 1033 (1983)
29. F.A. Waite, J. Oil Coll. Chem. Assoc. 54 342 (1971)
30. P. Lutz, E. Franta, P. Rempp, Polymer 23 1953 (1982)
31. E. Franta, L. Reibel, J. Lehmann, J. Penczek,
J. Polym. Sci. Symp. 56 139 (1976)
32. R. Asami, M. Takaki, K. Kyuda,
Polym. J. 13 141 (1981) ; 15 261 (1983)
33. K. Ito, N. Usami, Y. Yamashita, Macromolecules 13 216 (1980)
34. Ph. Gramain, Y. Frère, Polymer Comm. 27 16 (1986)
35. P. Rempp, P. Lutz, P. Masson, P. Chaumont, E. Franta,
Makromol. Chem. Suppl. 13 47 (1985)
36. T. Hamaide, A. Revillon, A. Guyot, Polym. Bull. 6 235 (1982) ;
Europ. Polym. J. 20 855 (1984)
37. J.P. Kennedy, C.Y. Lo, Polym. Bull. 8 63 (1982)
38. K. Ito, Y. Masuda, T. Shintani, T. Kitano, Y. Yamashita,
Polym. J. 15 443 (1983)
39. G.G. Cameron, M.S. Chisholm, Polymer 26 437 (1985)

INVESTIGATIONS OF HYDROCARBON SOLUBLE DIFUNCTIONAL ORGANOLITHIUM INITIATORS BASED UPON 1,3-BIS(PHENYL ETHENYL) BENZENE

A. D. BROSKE, T. L. HUANG*, R. D. ALLEN**, J. M. HOOVER, AND J. E. McGRATH***
Department of Chemistry and the Polymer Materials and Interfaces Laboratory
Virginia Polytechnic Institute and State University
Blacksburg, VA 24061

INTRODUCTION

A difunctional anionic initiator that will polymerize vinyl and diene monomers has been an academic as well as a commercial goal in elastomer technology since pioneering work in anionic polymerization was reported 30 years ago [1-9]. The desire to prepare difunctional reactive oligomers as well as to synthesize tri- or multiblock copolymers has been an incentive to conduct this work. One advantage of using a difunctional anionic initiator is that triblock copolymers can be synthesized in two, or possibly only one monomer addition step. Thus, if the center block is prepared first, followed by the subsequent addition of a second monomer, one may prepare the two end blocks concurrently. A second feature relates to functionalization reactions, such as the reaction of the living chain ends with ethylene oxide. With proper conditions, perfectly difunctional linear (f=2.0) hydrocarbon based polyols can be produced. These polydiene oligomers can be used as "soft" segments in chain extended high molecular weight copolymers. In addition, hydrogenation can be conducted which will enhance thermal and oxidative stability. Difunctional initiators also are important where the second monomer is incapable of re-initiating the first monomer (e.g. diene or styrenic-methacrylic systems). Preparation of diene polymers displaying a high 1,4 microstructure has also emphasized the need to avoid, or at least minimize, polar additives, which might otherwise solubilize and enhance reactivity in these systems.

One important method of synthesizing difunctional organolithium reagents has been by addition of a monofunctional organolithium reagent (such as sec-butyllithium) to a compound that contains two double bonds [10-12]. The divinyl compound should have reactive double bonds which are not conjugated with each other in order to allow the sites to react essentially independently. Equal reactivity is obviously also desirable. In addition, steric and/or electronic structural features should be designed into the system that would restrict possible oligomerization of the diene which could result in either a gain or loss of perfect difunctionality.

The use of 1,3 divinylbenzene, **1**, as a possible precursor was first patented by Morrison and Kamienski [13] and further studied by Lutz [14] and Popov [15]. Unfortunately, this compound can easily oligomerize, which leads to ill-defined functionalities that are <u>greater</u> than two.

<u>Current Address</u>:

*3M Center, Traffic Control Dept., St. Paul, MN 55144

**IBM Almaden Research Laboratory, 650 Harry Road, San Jose, CA 95120-6099

***To whom correspondence should be addressed

Recent Advances in Anionic Polymerization
T.E. Hogen-Esch and J. Smid, Editors

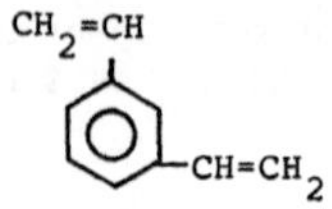

1

Sigwalt and coworkers [16,17] describe the use of the dilithium initiators formed from the reaction of butyllithium with bis (p-isopropenyl-phenyl) alkanes or diphenyl-alkadienes which were prepared in hydrocarbon solvents without polar additives. Unfortunately, one major drawback with this system is their insolubility in hydrocarbon solvents. They also reported using an excess of butyllithium which would clearly result in diblock and/or homopolymer contamination. Moreover, when a stoichiometric amount of the simple organolithium was employed, the reaction was reported to not go to completion even after one week at room temperature. Nevertheless, these initiators were then used to prepare triblock copolymers of styrene-isoprene-styrene and styrene-butadiene-styrene. Mechanical properties of these copolymers were reported to be comparable to those for commercial Kraton standards.

Another precursor, m-diisopropenyl benzene (DIB), 2, has been reacted with alkyllithiums to form a soluble difunctional initiator. Foss and coworkers [18] report that the product of the reaction of DIB with sec-butyllithium in the presence of between 0.1 and 0.2 equivalents of

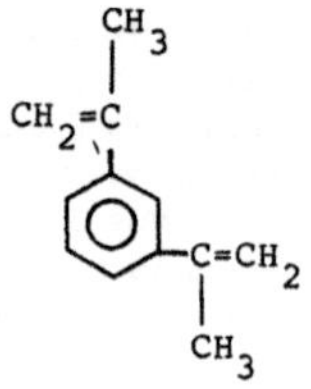

2

triethylamine, followed by reaction with isoprene improves solubility in essentially hydrocarbon solvents. The reaction rate is also increased due to higher solvent polarity. However, for the best case, the reported polydispersity of the resulting polyisoprene is broader (Mw/Mn = 1.3) than for typical anionic polymerizations and in the worst case, a bimodal distribution was obtained. Apparently, the presence of higher concentrations of the amine promotes the formation of species with functionalities greater than 2, leading to uncontrolled molecular weight and increased vinyl content in polydienes. Cameron, et al. [19], have also investigated the reaction of DIB with butyllithium in benzene or cyclohexane. Even at elevated temperatures, purely difunctional species did not result, but rather a mixture of di-, and polyfunctional products was observed. Even in the presence of tertiary amines, the products were similar. They concluded that utilization of this system as an efficient difunctional anionic initiator should be considered highly suspect.

In contrast, Lutz, et al. [20], reached a completely different conclusion for this system. They state that in benzene solution, a difunctional initiator was formed even without added polar modifiers. Also, polymerization of styrene and isoprene was achieved. Unfortunately, their own data appear to indicate that a mixture of mono-, di-, and trifunctional species were observed from GPC results. Also, an excess of sec-butyllithium was added to increase yields of difunctional species, which must result in residual, unreacted sec-butyllithium and precipitation.

Tung and coworkers [10-12] reported the use of difunctional initiators based on soluble difunctional 1,1-diphenylethylenes where the reactive double bonds are separated by a variety of moieties such as p-phenyl, biphenyl, p,p-biphenylether, and m-phenyl. These divinyl compounds when reacted with sec-butyllithum produce a dilithium initiator which can be used for block copolymer synthesis. Hocker and Lattermann [21-22] studied the use of 1,3- and 1,4 bis(phenylethenyl) benzene in polycombination electron transfer reactions to form multifunctional lithium species. These compounds do not homopolymerize under normal anionic conditions. More recently, they also reported [23-34] that the meta isomer will quantitatively react with sec-butyllithium in toluene to form a hydrocarbon soluble dilithium initiator which is suitable for anionic polymerizations.

We believe these latter meta linked soluble difunctional systems have high potential as difunctional initiators and have already independently reported our preliminary results at several meetings of the Polymer Division of the American Chemical Society [26-28]. This paper represents a more formal contribution. Our work has largely been conducted in cyclohexane, which though kinetically slower, eliminates the possible metalation side reaction of the benzylic methyl group of toluene [6] which becomes significant at elevated temperatures.

EXPERIMENTAL

Kinetic and mechanistic studies of anionic polymerization require the rigorous purification of all materials used. In fact, results and conclusions may be highly dependent on experimental techniques. Any reactive impurities, primarily water and air, must be removed from monomers, solvent, initiators, and glassware by appropriate purification procedures.

In general, organolithium initiators were used at concentrations of between 10^{-5} to 10^{-4} M, so the level of any impurities which could react with the organolithium initiator or living chain end must be kept below 10^{-6} M. An all glass high vacuum system (10^{-5} torr) employing a series of stopcocks similar to the system described by Morton and Fetters [24] and our own studies [7] was used for most of the kinetic studies in order to maintain these reaction conditions. Premature termination of the active species was prevented as judged by the generation of predictable molecular weights and narrow molecular weight distributions.

Cyclohexane is known to contain traces of olefinic impurities which must be removed before use. Cyclohexane (Phillips, 99%) was first stirred over concentrated sulfuric acid (about a 10% solution) for at least a week at room temperature. This removes the olefinic impurities by converting them to sulfonates which then become soluble in the acid layer. The cyclohexane layer was decanted into a round bottom flask containing crushed calcium hydride. It was then distilled into a round bottom flask containing a sodium dispersion and stirred overnight before being placed on the vacuum line. Following the degassing procedure, the cyclohexane was distilled into a flask containing low molecular weight polystyryl-lithium which has a characteristic red color and is used simply as an indicator which insures high purity. Cyclohexane was then vacuum distilled from this flask, as needed, into various reactors.

A variety of polar compounds were investigated as additives to study the spectra of the completely reacted difunctional anionic initiator. In general, the molar ratio of organolithium to polar additive was 0.2 or less. Purification of tetrahydrofuran (THF) involved distillation from calcium hydride into a flask containing the purple colored complex formed from the addition of benzophenone to sodium dispersion and was then distilled as needed.

The reduced volatility and increased reactivity of some of the other polar additives such as triethylamine (TEA) and tetramethylethylene diamine (TMEDA) required utilization of alternate procedures. These amines were first stirred over crushed calcium hydride overnight in a round bottom flask equipped with a drying tube and then distilled under high vacuum at room temperature. The purified distillate was then split down into ampules. Dipiperidinomethane and 1,2-dipiperidinoethane (Di-PIP) (Aldrich, 98%) were purified using an analogous procedure.

The purification of sec-butyl lithium (Lithium Corp., 1.4N in cyclohexane) was necessitated by the fact that the commercially available material contains varying amounts of alkoxide impurities, which may alter the reaction kinetics, microstructure and copolymerization parameters. Purification was achieved by distillation (after removal of solvent) under high vacuum ($<10^{-4}$ torr) at 65°C. The purified sec-butyl lithium was diluted with cyclohexane and also split down into ampules.

The 1,3-bis(phenylethenyl) benzene (DDPE) **4** was kindly provided by Dow Chemical Co. as a yellow liquid which upon standing will slowly crystallize. The resulting yellow crystals are quite oxygen sensitive and were further purified before use. An effective purification method was to distill from n-butyl lithium under high vacuum conditions. Distillation at about 140°C yields a clear, colorless liquid. Transferring the desired amount to a second flask followed by dilution with a measured amount of cyclohexane under vacuum and splitting down into ampules yields the final solution of approximate concentration. The purity of the white crystallized solid was also verified by differential scanning calorimetry (DSC). A sharp melting point at 39-40°C was observed.

An all-glass reactor was used for studying the kinetics of the reaction between sec-butyllithium and DDPE under high vacuum conditions. The reactor illustrated in Figure 1 was divided into two sections, one being the main reactor and the other the purge section as described earlier [7,24]. The former consisted typically of a 500 ml round bottom flask to which two side arms have been attached. These side arms allow various ampules or dilatometers to be attached. Two break seals were provided which allowed for addition of additional solvent or attachment of more ampules which may contain the various polar additives. Two quartz UV/Visible cells of 10 mm and 1 mm path lengths were attached at the top of the reactor, which allows a spectrum to be obtained.

The purge system consists typically of a 200 ml round bottom flask which contained a solution of n-butyllithium in cyclohexane in a side arm. This solution was used to condition the main reactor and to thus remove any reactive impurities from the surface of the flamed glassware. The reactor was assembled by heat sealing on the reagent ampules containing the DDPE, sec-butyllithium, the wash n-butyllithium and if desired, any polar additives. Magnetic stirring bars were used as the break nails. After several rinses with cyclohexane, complete removal of the excess butyllithium from the glassware was achieved. Then, the desired amount of cyclohexane was distilled from the purge system to the main reactor.

Subsequently, the solutions were cooled and the purge section was heat sealed from the main reactor.

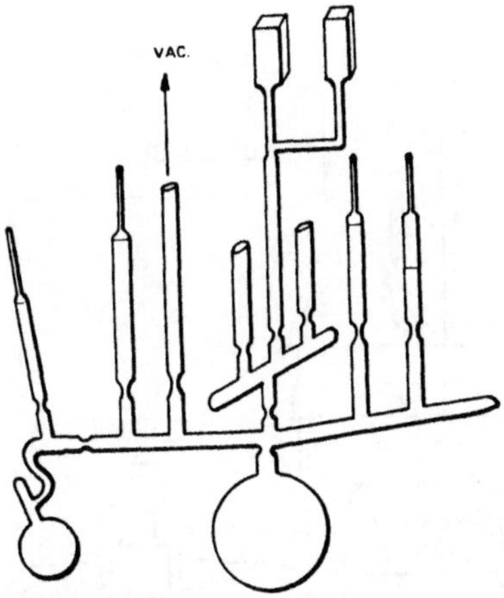

FIG 1. Anionic Polymerization Reactor for High Vacuum Studies

The main reactor was placed in a circulating water bath which had its temperature controlled to ±0.1°C, until the cyclohexane has reached the desired temperature. When this was achieved, the two ampules containing the two reagents were broken via the break seal and were allowed to mix in the reactor. Thorough mixing of the two reagents was achieved by shaking the reactor and rinsing the ampules with this solution to ensure complete mixing of all components.

A series of UV/Visible spectra were taken at specified times during the reaction by inverting the reactor and placing the now filled quartz cells in the spectrophotometer. A Perkin-Elmer (Model 552) microprocessor controlled UV/Visible spectrophotometer equipped with a Model C 550-0555 thermoelectrically controlled cell was used for this analysis. Spectra were obtained by scanning from 650 to 190 nm at 120 nm/min.

Though quite sensitive, the UV/Visible method does not distinguish between the mono- and dianion reaction products. Therefore, complementary GC, FT-IR and ^{1}H-NMR methods were utilized employing a high pressure nitrogen polymerization reactor [25]. A schematic diagram of the high pressure nitrogen reactor is shown in Figure 2. In this case, the desired amounts of DDPE and sec-butyllithium solutions were added to the reactor containing cyclohexane via syringe through the initiator inlet. Samples were removed from the reactor at specific times during the reaction. The samples were terminated with N_2 purged methanol and then analyzed via gas chromatography (GLC), FT-IR (Nicolet MX-1) and proton NMR (IBM 270 MHz).

Initially a Gow-Mac series 550 GLC equipped with a thermal conductivity detector was utilized. The detector and injector temperatures were both set at 250°C and the column was maintained at 220°C. Helium was used as the carrier gas with a flow rate of 80 ml/min through the column. The column was a 4'x ¼" aluminum column packed with 3% OV-101 on 100/120 mesh Anachrom/ABS. Subsequently, studies with a more sensitive Varian Model 6000 equipped with a capillary column were conducted.

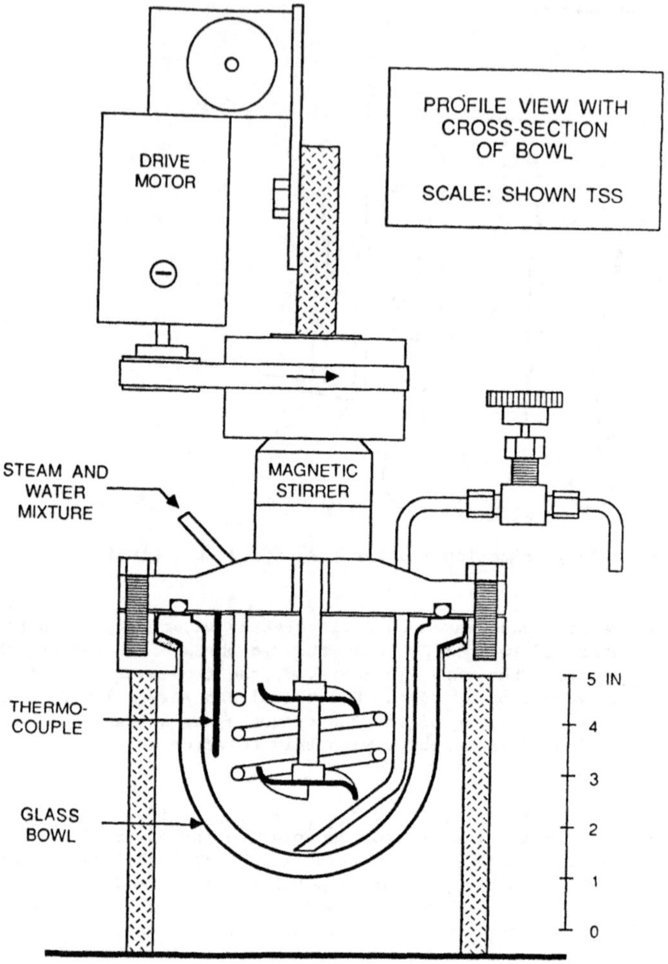

FIG. 2. Glass Bowl Pressurized Reactor

RESULTS AND DISCUSSION

Several aspects of anionic polymerization were of interest in the current investigations. Firstly, the kinetics of the reaction between sec-butyl lithium and DDPE was studied as outlined in Scheme 1. Secondly, the effect of polar additives on the nature of the ion pair of the dilithium species was investigated by observing the wavelength shift of the anion as a function of concentration of the additive. Thirdly, this initiator 5 was used to polymerize butadiene and isoprene for purposes of synthesizing hydrocarbon polyols and triblock copolymers based on hydrocarbon monomers, and/or alkyl methacrylates.

Scheme 1

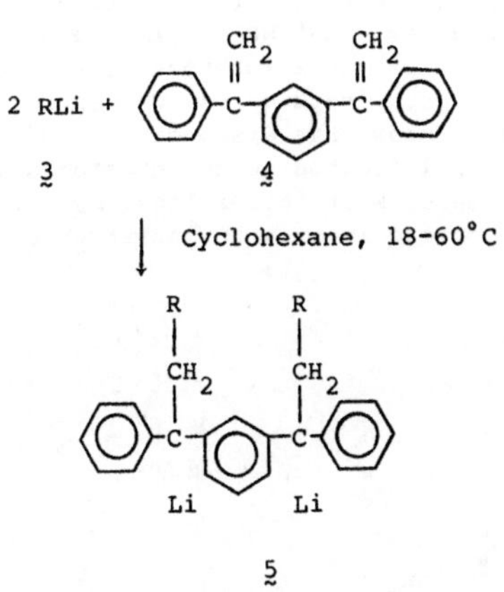

Kinetic experiments were first conducted under high vacuum conditions utilizing the glass apparatus previously described in the Experimental Section. UV spectroscopic analysis of the reaction mixture shows two peaks of interest. One is centered at about 227 nm and assigned to the unreacted non-conjugated DDPE 4 which decreases with time. The other at about 430 nm is assigned to the combined mono- and dianion 5 species. The latter increases with time and/or reaction temperature. Figure 3 shows a series

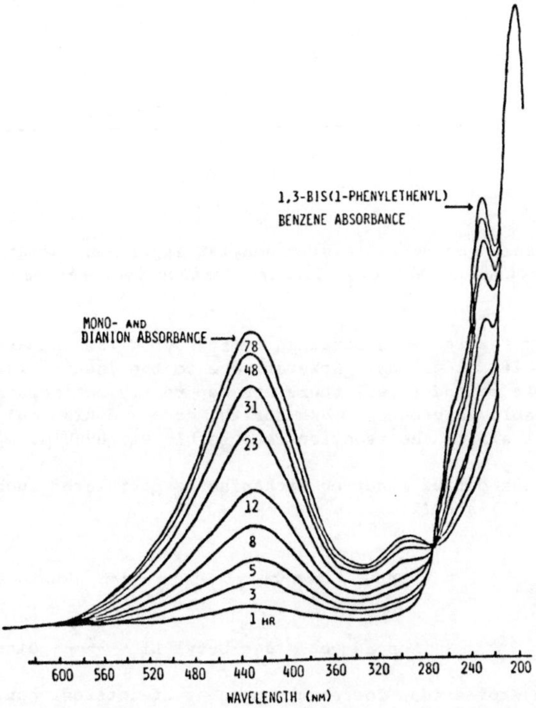

FIG. 3. Spectrophotometric Study of the Reaction of 1,3-Bis(1-Phenylethenyl) Benzene and Sec-Butyl Lithium in Cyclohexane at 18°C

of UV/Visible spectra overlayed which were measured with the reaction temperature maintained at 18°C. The reaction is quite slow at this temperature and requires about 80 hrs for complete reaction, as evidenced by the time to obtain aximum absorbance at 430 nm. Reactions were also conducted at several temperatures using identical concentrations of reactants. Figure 4 shows the disappearance of DDPE as a function of reaction time at the indicated reaction temperatures. As expected, the reaction rate increases with increasing temperature and one may note that the reaction is complete in cyclohexane after only two hours at 60°C.

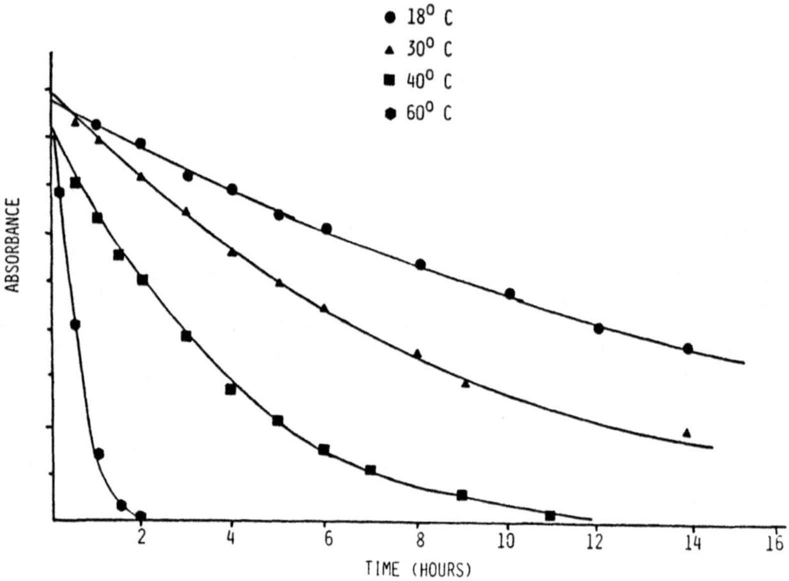

FIG. 4. Absorbance of 1,3 Bis(1-Phenylethenyl) Benzene at 227.5 nm as a Function of Reaction Time at Various Temperatures

Similar results were obtained by following the appearance of the anionic species at 430 nm. However, due to the insensitivity of this spectroscopic method (i.e., there are apparently not separately distinguishable absorbance peaks for the mono and dianion) only an overall kinetic analysis of the reaction is possible via UV-VIS.

If a consecutive reaction mechanism is postulated such as illustrated in equation (1) and (2),

$$\text{DDPE} + \text{sec-butyl Li} \xrightarrow{k_1} \text{Monoadduct} \qquad (1)$$

$$\text{Monoadduct} + \text{sec-butyl Li} \xrightarrow{k_2} \text{Diadduct} \qquad (2)$$

the kinetic expressions for each component of interest could be given in general form as:

$$-\frac{d(DDPE)}{dt} = k_1[DDPE]^{\alpha}[RLi]^{\beta} \qquad (3)$$

$$\frac{-d[Mono]}{dt} = k_2[Mono]^{\gamma}[RLi]^{\delta} - k_1[DDPE]^{\alpha}[RLi]^{\beta} \qquad (4)$$

$$\frac{d[Di]}{dt} = k_2[Mono]^{\gamma}[RLi]^{\delta} \qquad (5)$$

where [Mono] and [Di] refer to the concentrations of the mono- and diadduct, respectively, and [RLi] refers to the sec-butyl lithium concentration.

Because the mono- and dianion absorbances are not separated from one another and the absorbance for unreacted sec-butyl lithium is obscured by the solvent, equations (4) and (5) cannot easily be directly evaluated. However, one notes from Figure 3 that an absorbance peak attributed to DDPE is well separated from the anions, which allows for kinetic evaluation of at least the first step of the reaction.

Using equations (1)-(3), some assumptions must be made regarding the order of the reaction. Realistic test orders in DDPE concentration as one [19] and one quarter [20] in sec-butyl lithium concentration were assumed. Secondly, since sec-butyllithium cannot be observed by UV-Vis spectroscopy, its concentration term was equated with the DDPE concentration as indicated in equation (6), which implies that the monoanion concentration remains low.

$$[\text{sec-butyl Li}] = 2[DDPE] \qquad (6)$$

Making these substitutions into equation (3) and simplifying affords:

$$-\frac{d[DDPE]}{dt} = 2^{1/4}\, k_1[DDPE]^{1.25} \qquad (7)$$

which when integrated over the appropriate limits yields

$$\frac{4}{2^{1/4}\,[DDPE]^{0.25}} + \text{constant} = k_1\, t \qquad (8)$$

This equation was used to first confirm the validity of the initial assumptions and secondly to evaluate the apparent global rate constant.

Figure 5 shows the influence of temperature on the disappearance of DDPE obtained using the UV/Visible method. It is consistent with first order reaction kinetics, in terms of DDPE concentration. Straight line behavior is observed at all reaction temperatures. This plot supports one of our first assumptions, that the reaction should be first order in DDPE concentration. Figure 6 illustrates a plot of equation (8) where the initiator concentration, expressed as $[DDPE]^{0.25}$, is plotted versus reaction time. Here again, at all temperatures straight line behavior is observed and a correlation coefficient >0.99 was calculated. This demonstrates that our assumption of the one-quarter reaction order in sec-butyl lithium concentration was reasonable. Other orders were tested and

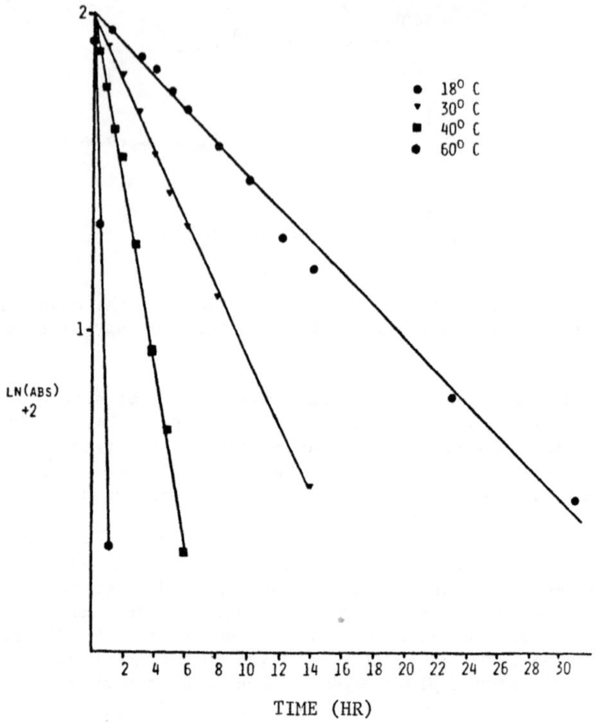

FIG. 5. Influence of Temperature on the Disappearance of 1,3 Bis(1-Phenylethenyl) Benzene

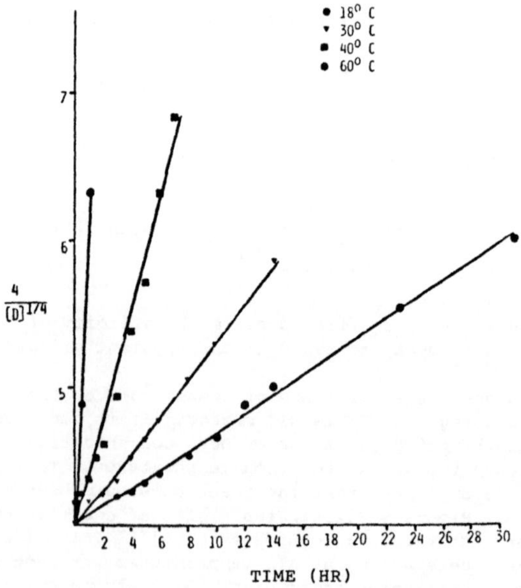

FIG. 6. Relationship Between Initiator Concentration and Reaction Time

did not yield as good a fit. The slope of these straight lines then is a direct measure of the apparent rate constant of this reaction at each temperature.

To further evaluate the kinetic analysis, the rate constants at each temperature were plotted versus reciprocal temperature as shown in Figure 7. A straight line was again obtained as evidenced by the high correlation coefficient (>0.99). The slope of this line is proportional to the energy of activation which was calculated to be about 78.5 (kJ/mole) or 19 kcal/mole in cyclohexane.

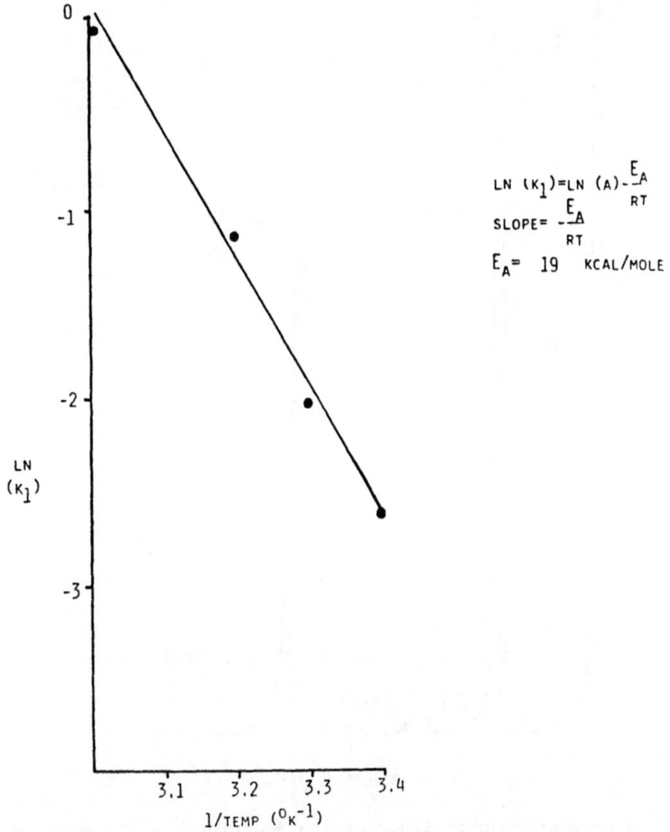

FIG. 7. Arrhenius Plot for the Sec-Butyl Lithium Reaction with 1,3-Bis(Phenyl Ethenyl) Benzene in Cyclohexane

Additional kinetic studies utilized reactions conducted in the pressure reactor in conjunction with analytical data obtained by gas liquid chromatography GLC, FT-IR and ^{1}H-NMR. Both FT-IR and ^{1}H-NMR at first glance appear to be effective methods for studying this reaction by either following the disappearance of carbon-carbon double bond stretching in the IR or the disappearance of vinyl protons in the NMR. However, distinguishing the double bonds of the monoadduct and starting material is impossible. On the other hand, gas chromatography affords one the ability to separate all components of interest since the boiling points of the three components of interest are different from one another.

Figure 8 shows three overlayed chromatograms taken at different times during the reaction. The first unattenuated peak decreases with reaction time and is due to unreacted DDPE. The last eluting peak is assigned to the terminated dianion species which increases with reaction time. The intermediate peak is attributed to the terminated monoadduct. The peak assignments were verified by primarily injecting standards of known structure. The unreacted divinyl material was easily identified by injecting a solution of the purified DDPE in cyclohexane.

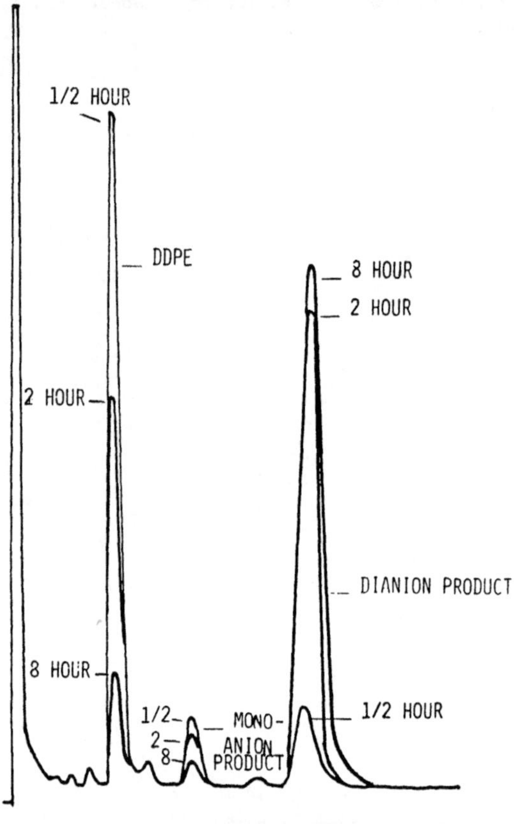

FIG. 8. Gas Chromatographic Study of the Reaction Mixture with Time at 60°C

The diadduct standard was prepared by adding a one molar excess of sec-butyl lithium to DDPE in cyclohexane which maximized the formation of the diadduct. The reaction product was washed with water, dried, dissolved in cyclohexane and injected on the gas chromatograph.

The monoadduct is more difficult to prepare. Statistically, addition of 1 mole sec-butyl lithium to 1 mole of DDPE should produce primarily the monoadduct. However, if the reaction is conducted in cyclohexane, predominantly the diadduct and unreacted DDPE are observed as major products in the reaction. Remarkably, conducting the same reaction in THF at -78°C produces the monoadduct in about 90-95% purity as evidenced by GC analysis. This solution was then used to identify the monoadduct peak. To confirm the identification of all the components of interest, small amounts of the standards were spiked into samples taken during a kinetic run.

Figure 9 shows the concentration of all species of interest plotted versus reaction time. This is clear evidence for a consecutive reaction mechanism. It is of interest to note that the concentration of the monoadduct remains relatively low throughout the reaction.

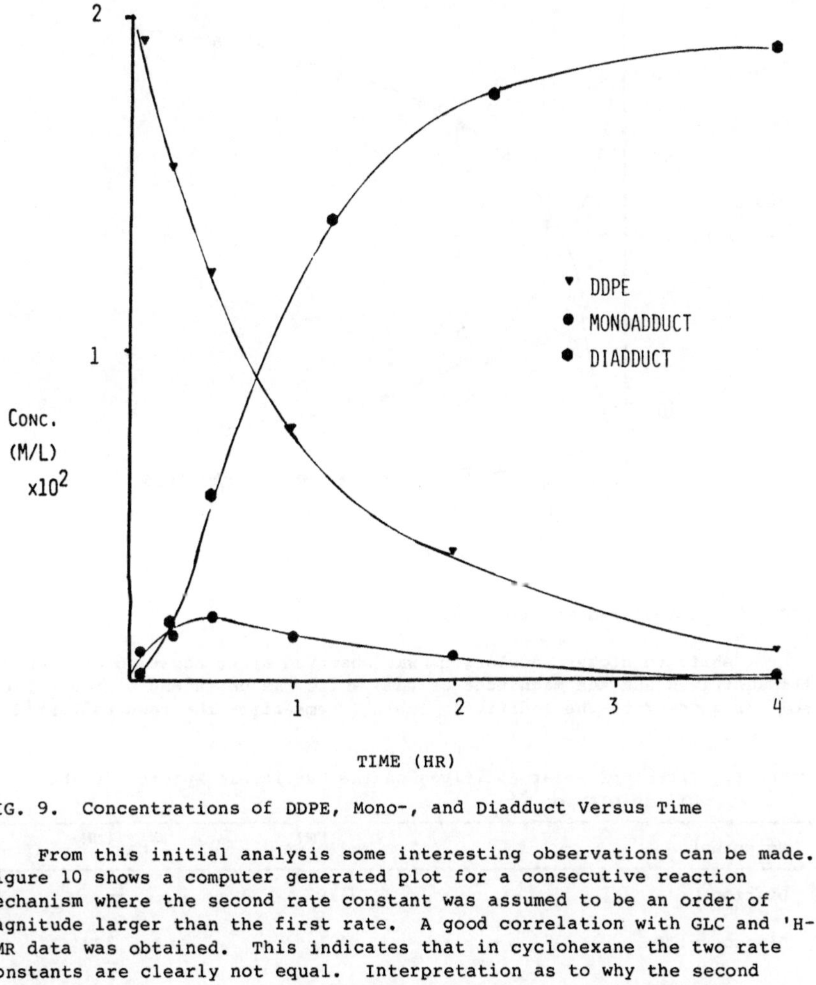

FIG. 9. Concentrations of DDPE, Mono-, and Diadduct Versus Time

From this initial analysis some interesting observations can be made. Figure 10 shows a computer generated plot for a consecutive reaction mechanism where the second rate constant was assumed to be an order of magnitude larger than the first rate. A good correlation with GLC and 'H-NMR data was obtained. This indicates that in cyclohexane the two rate constants are clearly not equal. Interpretation as to why the second adduct is formed at such greatly enhanced rates has not been firmly established. However, we suggest that chain end association [3,6,7] is involved. For example, preliminary reaction measurements in partially polar solvents [29] are quite different.

A second area of interest is in fact to study the effect of polar additives on the UV/Visible spectra of the dianionic species. This was easily accomplished by attaching an ampule which contains the additive to the glass reactor assembly which was then added to a completed reaction from the kinetic study.

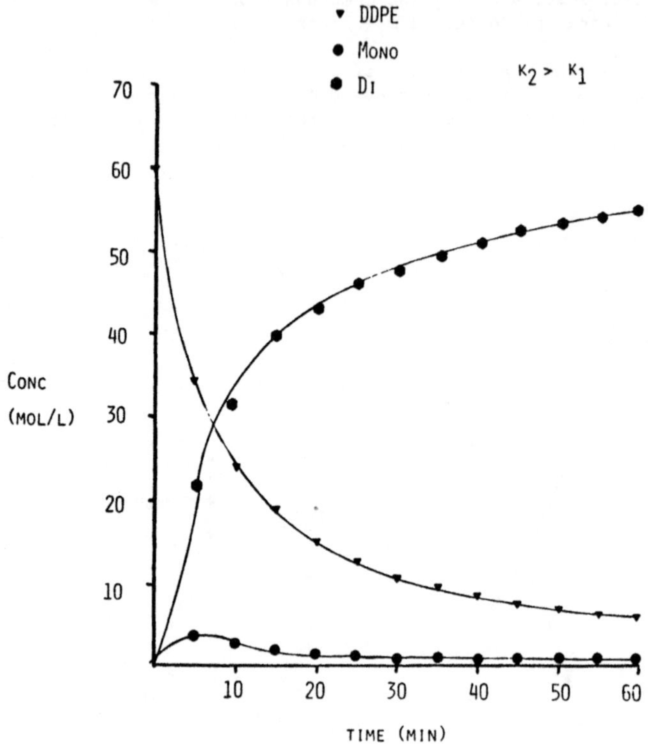

FIG. 10. Calculated Kinetic Data

A shift to higher wavelengths was observed after addition of each of the additives and the magnitude of this shift was dependent on the relative solvating power of the additive. Table I summarizes the results tested.

TABLE I. Effect of Polar Additives on the UV-Visible Spectra of the Dilithio Anion(1)

Component	λ_{Max}(NM)$_\Delta$	λ_{Max}(NM)
Dianion	430	---
Dianion + TEA	462	32
Dianion + THF	465	35
Dianion + Dipip(ethane)	468	38
Dianion + Dipip(methane)(2)	---	--
Dianion + TMEDA(2)	---	--

(1) Molar ratio of additive to chain end was ≤0.2.
(2) Premature termination was observed for the variety of additives

One may notice that the more powerful solvating liquids [30] show a shift to higher wavelengths. These results are interpretable in terms of the structure of the ion pair which is emplified by the typical Winstein spectrum where a series of equilibria probably exists between tight ion pairs, loose ion pairs, and perhaps free ions. A major factor influencing this series of equilibria is the solvating ability of the solvent for the cation primarily due to the steric bulk of the anion. Thus, these solvents raise the concentration of the various ion pairs and free ions and a shift in the wavelength maximum is observed. This behavior can be explained in terms of the relationships between frequency and energy. In nonpolar solvents, tight ion pairs are favored. Since these species are relatively energetic species (i.e., the energy difference between the ground and excited states is large) higher energy radiation is required to observe a transition and the observed transition occurs at a lower wavelength. As conditions are modified to favor the lower energy differences between these states, the observed transition shifts to higher wavelengths.

Another major objective of our work was to use this difunctional initiator to polymerize diene and styrene based monomers. Initial experiments were directed towards polymerizing butadiene to high molecular weight in the nitrogen pressure nitrogen reactor. This was most convenient since samples could be removed from the reactor during the reaction as a function of time, etc. Figure 11 shows two size exclusion chromatograms of

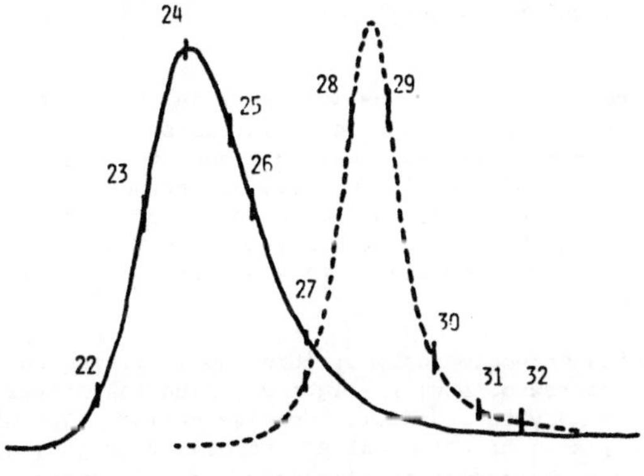

FIG. 11. Size Exclusion Chromatograms of Narrow Distribution Polybutadiene and Broadened Species Obtained after Aging at 60°C

samples taken from the reactor during and after the polymerization was complete. The narrow peak at high elution volumes (lower molecular weight) corresponds to a sample taken during the polymerization. Notice the narrow molecular weight distribution which is typical for anionic polymerizations. However, after the polymerization reaches quantitative conversions the molecular weight distribution apparently broadens considerably as evidenced by the second peak at low elution volumes. It is suggested that at 60°C, the dianion chain end is less stable after all of the monomer is consumed. Partial elimination of lithium hydride could generate a diene-like species capable of reacting with residual living-ends, to generate the broader, possibly branched structure [31,32]. Additional results in this area will be reported later [29]. Table II summarizes the SEC results on several polybutadiene samples taken as a function of reaction time. Again, narrow molecular weight distributions are maintained initially, while later, broadening of the distribution occurs. Most likely the branching probably occurs when little or no monomer is present.

TABLE II. Influence of Reaction Time, Conversion and Polybutadiene Dianion Aging Time at 60°C on $\bar{M}n$ and Polydispersity

Reaction Time	$\bar{M}n$	Poly-Dispersity
10 Min	26,700	1.16
20 Min	54,200	1.09
30 Min	70,300	1.12
60 Min	86,400	1.28
90 Min	86,700	1.64
180 Min	115,600	1.58

* 10% solids (weight to weight)

Triblock copolymers of styrene and isoprene (S-I-S) or butadiene (SBS) were synthesized using this difunctional initiator. A one step synthesis was utilized wherein both monomers were charged at the same time. This method takes advantage of reactivity ratio differences between dienes and styrene [2-7]. In cyclohexane, the center block of diene is substantially polymerized first followed by the two styrene end blocks. Because both monomers are added simultaneously, a tapered triblock copolymer structure is expected [8,10].

Size exclusion chromatography of this copolymer showed a single monomodal peak and a narrow molecular weight distribution, clearly indicating the absence of homopolymer and diblock contamination. Dynamic mechanical thermal analysis (DMTA) of a typical SIS copolymer (Figure 12) shows behavior typical of thermoplastic elastomers; namely separate Tg's for both the isoprene and styrene phase and the presence of an extended rubbery plateau which indicated a phase separated system. Further work on the synthesis, structure and property behavior of these elastomers is continuing. In particular, hydrogenation of the polydiene blocks is of interest.

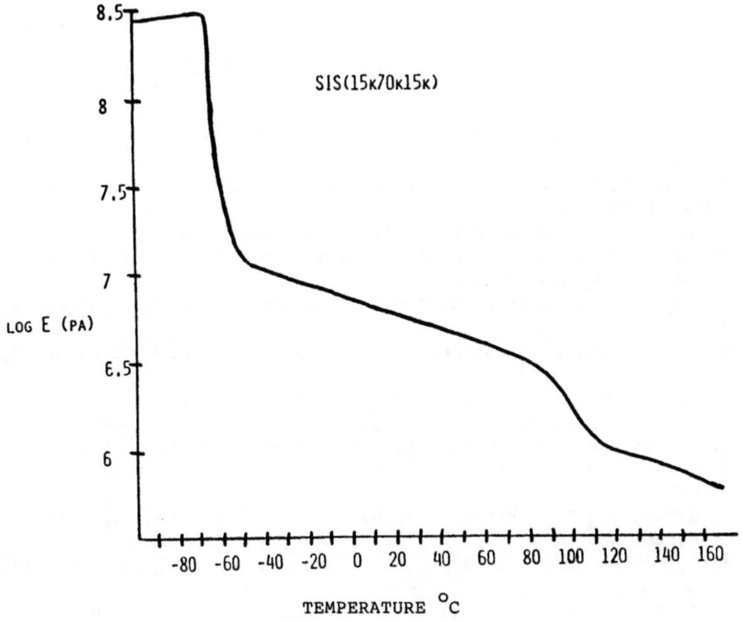

FIG. 12. Dynamic Mechanical Behavior of an S-I-S Triblock Copolymer

ACKNOWLEDGMENT

The authors would like to thank both the Dow Chemical Co. and TACOM for support of this research. Helpful discussions with Prof. David Hewitt, Drs. L. H. Tung and A. L. Gatzke are also acknowledged with appreciation.

REFERENCES

1. M. Szwarc, M. Levy, and R. M. Milkovich, J. Am. Chem. Soc. 78, 2656 (1956).
2. M. Szwarc, "Carbanions, Living Polymers, and Electron-Transfer Processes", Wiley-Interscience, New York, 1968.
3. M. Morton, "Anionic Polymerization, Principles and Practice", Academic Press, New York, 1983.
4. M. Morton and L. J. Fetters, Macromolecules, 2, 453 (1969).
5. M. Szwarc, Adv. in Poly. Sci. 49, 1 (1983).
6. R. N. Young, R. Quirk and L. J. Fetters, Adv. Poly. Sci. 56, (1984).
7. J. E. McGrath, Editor, "Anionic Polymerization: Kinetic, Mechanisms and Synthesis, ACS Symposium Series No. 166, 1981.
8. A. Noshay and J. E. McGrath, "Block Copolymers: Overview and Critical Survey" Academic (1977).
9. G. Reiss, G. Hurtrez, and P. Bahadur, Encyclo. Poly. Sci. and Eng., 2nd Edition, Vol. 2, John Wiley & Sons, New York, 1985.
10. L. H. Tung, G. Y. S. Lo, and D. E. Beyer, Macromolecules 11, 616 (1978).

11. L. H. Tung, G. Y. S. Lo, U.S. Patents 4,182,818 (1980); 4,196,153 (to Dow Chemical) (1980).
12. L. H. Tung, G. Y. S. Lo, J. W. Rakshys, and D. E. Beyer, U.S. Patents, 4,200,718 (1979); 4,201,729 (1980); 4,205,016 (to Dow Chemical) (1980).
13. R. C. Morrison and C. W. Kamineski, U.S. Patent (to Lithium Corp.) 3,725,368 (1973).
14. P. Lutz, E. Franta, and P. Rempp, C. R. Seances Acad. Sci., Ser. C. 283, 123 (1970).
15. G. Popov and G. Schwarchula, Plast. U. Kautch. 26, 263 (1979).
16. P. Guyot, J. C. Favier, H. Uytterhoeven, M. Fontanille, and P. Sigwalt, Polymer, 22, 1724 (1981).
17. P. Guyot, J. C. Favier, M. Fontanille, and P. Sigwalt, Polymer 23, 73 (1982).
18. R. P. Foss, H. W. Jacobson, and W. H. Sharkey, Macromolecules 10, 287 (1977).
19. G. G. Cameron, G. M. Buchan, Polymer 20, 1129 (1979).
20. P. Lutz, E. Franta, and P. Rempp, Polymer 23, 1953 (1982).
21. H. Hocker and G. Latterman, Makromol. Chem. 158, 191 (1972).
22. H. Hocker and G. Latterman, J. Polym. Sci., Polym. Symp. 54, 361 (1976).
23. H. Hocker and G. Latterman, Angew Chem. Int. Ed. Eng. 19, 219 (1980).
24. M. Morton and L. J. Fetters, Rubber Rev. 48, 359 (1975).
25. J. M. Hoover and J. E. McGrath, Polymer Preprints 27(2), 150 (1986); J. Poly. Sci., in press, 1986.
26. A. D. Broske, T. L. Huang, J. M. Hoover, R. D. Allen, and J. E. McGrath, Polymer Preprint 25(2), 85 (1984).
27. A. D. Broske and J. E. McGrath, Polym. Preprint 26(1), 241 (1985).
28. J. E. McGrath, R. D. Allen, J. M. Hoover, T. E. Long, A. D. Broske, S. D. Smith, and D. K. Mohanty, Polymer Preprints 27(1), 183 (1986).
29. A. D. Broske and J. E. McGrath, J. Poly. Sci., in preparation; A. D. Broske, Ph.D. Thesis, Virginia Polytechnic Institute and State University, 1987.
30. P. Bres, M. Viguier, J. Sledz, F. Schue, P. E. Black, D. J. Worsfold, and S. Bywater, Macromolecules 19, 1325 (1986).
31. J. N. Anderson, W. J. Kern, T. W. Bethea, and H. E. Adams, J. Appl. Polym. Sci. 16, 3133 (1972).
32. W. J. Kern, J. N. Anderson, H. E. Adams, T. C. Bouton, and T. W. Bethea, J. Appl. Polym. Sci. 16, 3123 (1972).
34. E. Leitz and H. Hocker, Makromol. Chem. 184, 1893 (1983).

SYNTHESIS OF POLY(METHYL METHACRYLATE) MACROMONOMER BY GROUP TRANSFER POLYMERIZATION, AND POLYMERIZATION OF THE MACROMONOMER

R. Asami, Y. Kondo, and M. Takaki
Department of Applied Chemistry Nagoya Institute of Technology
Gokiso-cho, Showa-ku, Nagoya 466, Japan

INTRODUCTION

Macromonomers which can be defined as polymers endowed with a polymerizable functional group at their chain end have received considerable interest during the last ten years. Both their usefulness in the preparation of graft or comb polymers and their reactivity in homo and copolymerizations have been studied [1].

The syntheses of macromonomers by means of living polymers are unequivocally preferred to other methods since they yield well-defined macromonomers with the desired molecular weight and a narrow molecular weight distribution. By using this living polymer technique including both anionic and cationic polymerization polystyrene- [2], polyisoprene- [3], polyvinylpyridine- [4], poly(ethylene oxide)- [5], and poly(tetrahydrofuran)- [6] macromonomers have already been synthesized.

The preparation of the living polymer of polyMMA by anionic polymerization is not easy in contrast to that of polystyrene living polymer. Brief papers on the preparation and functionalization of living polyMMA have been reported by Anderson et al. [7] and Rempp et al. [8] but no details were given. Recently, the group transfer polymerization (GTP) of MMA was discovered by researchers at du Pont [9]. This polyMMA has been shown to be a stable living polymer with a silyl ketene acetal group as an active living end. The preparation of these acrylic living polymers was conducted to obtain telechelic or end functionalized polymers [10]. Since the discovery of group transfer polymerization, the authors have attempted to prepare a well-defined polyMMA macromonomer by using living polymers made by group transfer polymerization of MMA.

To synthesize macromonomers by means of anionic polymerization, two pathways can be proposed. The first one, termed termination method, is an end capping technique. It is carried out by introducing a polymerizable group such as a vinyl group, at the end of the polymer chain by terminating the living polymer after the completion of polymerization. In the second pathway, classified as initiation method, unsaturated compounds are used as initiators for the living polymer. If the initiation proceeds by addition to the monomer, and the double bond of the initiator is retained during the polymerization, the polymer obtained after quenching by a proton should be a macromonomer.

The synthesis of polyMMA macromonomer using living polyMMA formed by group transfer polymerization with silyl ketene acetal as initiator was carried out by both the termination and initiation method. The results obtained in the preparation of the well-defined macromonomers, and their polymerization behavior will be presented.

SYNTHESIS AND CHARACTERIZATION OF POLYMMA MACROMONOMER

Synthesis of Vinylbenzyl PolyMMA Macromonomer by Termination of Living PolyMMA formed by Group Transfer Polymerization

Webster et al. [10], the discoverers of group transfer polymerization, have conducted the coupling reaction of living polyMMA with *p*-xylylene dibromide to prepare telechelic functional polyMMA. As an application of this reaction we tried to prepare the polyMMA macromonomer by the coupling

Recent Advances in Anionic Polymerization
T.E. Hogen-Esch and J. Smid, Editors

reaction of the living end with electrophiles such as vinylbenzyl halide or tosylate (VBTs). The reactions are shown in Scheme I.

A preliminary experiment was carried out to prepare living polyMMA by polymerizing MMA with dimethylketene methyl trimethylsilyl acetal (DMKTSA) as initiator and tris(dimethylamino)sulfonium difluoro trimethylsilicate ($TASF_2SiMe_3$) as catalyst in THF, following procedures outlined by the du Pont group. The data shown in Table I confirm the easy control of the molecular weight and a low polydispersity of polyMMA. Monodisperse living polyMMA is obtained under the conditions of r.t. and at [monomer] to [initiator] ratios of less than 50. The macromonomer was then synthesized by end-capping the silanolate end of the living polyMMA with vinylbenzyl bromide (VBB) or VBTs under various conditions. The functionality (f) of the styryl polyMMA macromonomer is a measure of its purity, and indicates the extent of the coupling reaction. It was measured by UV, the same technique used in the preparation of styryl PTHF and styryl polystyrene macromonomers [2, 6].

The results of the monodispersity, M_w/M_n, and the purity of the polyMMA macromonomer, f, are shown in Tables II and III. The macromonomers are highly monodisperse, $M_w/M_n = 1.1$, with a molecular weight of 4400 by GPC analysis. However, it can be seen that the functionality depends on the coupling conditions and at least an equivalent amount of catalyst to the silanolate end is necessary to obtain quantitative coupling. In the case of VBB, Table II, the highest f value found was 0.83. VBTs is considered to be more electrophilic than VBB, and is expected to give a high f value. However, this was not the case, the highest f value being 0.73 (Table III). The reason why quantitative coupling (f = 1) can not be attained under these conditions is not yet clear.

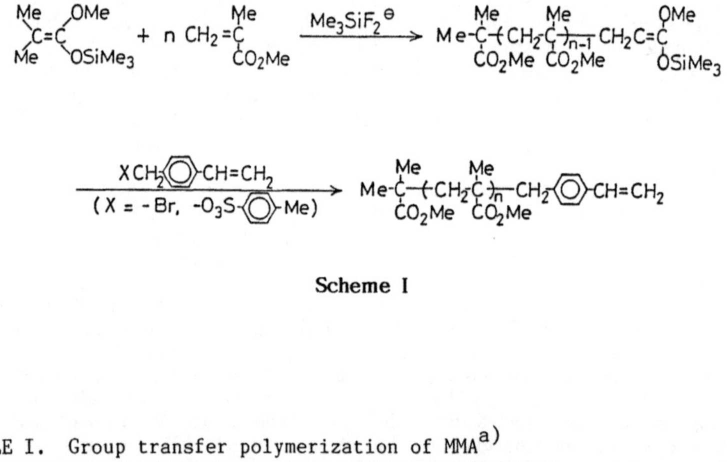

Scheme I

TABLE I. Group transfer polymerization of MMA[a)]

RUN no.	MMA (mol/L)	KTSA[b)] (mmol/1)	[M]/[I]	Yield (%)	M_n x 10^{-3}	M_w/M_n
1	1.17	27.3	42.7	100	4.4	1.0_6
2	1.11	32.5	34.2	100	4.2	1.0_8
3	1.52	47.4	32.0	100	3.3	1.0_7
4	0.839	8.44	99.4	92	8.8	1.2_4

a) In THF at r.t. for 1 h; catalyst, $TASF_2SiMe_3$. b) Initiator, dimethylketene methyl trimethylsilyl acetal.

TABLE II. Preparation of PMMA macromonomer by termination of living PMMA with p-vinylbenzyl bromide[a)]

	Living PMMA				Coupling reaction			Product	
Run no.	(mmol/L)	vol (mL)	M_n x 10^{-3}	M_w/M_n	[VBB[b)]]/ [LE]	[cat[c)]]/ [LE]	time (h)	yield (%)	f[d)]
1	27.3	1.59	4.4	1.0_6	10.6	0	3	100	0
2	27.3	1.70	4.4	1.0_6	9.92	0.482	3	100	0.25
3	27.3	1.71	4.4	1.0_6	10.4	0.456	41	79	0.50
4	27.3	1.08	4.4	1.0_6	5.77	1.69	3	90	0.75
5	47.4	1.44	3.3	1.0_7	2.79	1.52	24	68	0.83
6	27.3	1.83	4.4	1.0_6	9.98	1.36	72	98	0.70

a) In THF at r.t. b) p-Vinylbenzyl bromide. c) Catalyst, $TASF_2SiMe_3$.
d) Functionality determined by UV.

TABLE III. Preparation of PMMA macromonomer by termination of living PMMA with p-vinylbenzyl tosylate[a)]

	Living PMMA				Coupling reaction			Product	
Run no.	(mmol/L)	vol (mL)	M_n x 10^{-3}	M_w/M_n	[VBT[b)]]/ [LE]	[cat]/ [LE]	time (h)	yield (%)	f[c)]
1	27.3	1.01	4.4	1.0_6	7.08	2.58	3	95	0.69
2	27.3	1.67	4.4	1.0_6	2.85	2.04	24	100	0.71
3	27.3	1.00	4.4	1.0_6	7.11	1.42	48	72	0.73

a) In THF at r.t. b) p-Vinylbenzyl tosylate. c) Functionality determined by UV.

It may be due to side effects in the reaction of living polyMMA with an equimolar amount of the unique catalyst, $TASF_2SiMe_3$, rather than to a side reaction occurring in the coupling reaction with the vinylbenzyl derivatives which has been shown to take place in the reaction of polystyryl anion with benzyl derivatives [2].

Preparation of Vinylbenzyl PolyMMA by Initiation Method of Group Transfer Polymerization [11]

The termination method described in the preceding section is a useful procedure to produce monodisperse polyMMA macromonomer. However, it requires two steps, the preparation of the living polyMMA with the regular initiators, followed by the termination of the living polymer with an unsaturated electrophile. The initiation method involves only the polymerization of MMA with an unsaturated initiator which must first be synthesized. Hence, in the study of the initiation method much work has been done for the syntheses of novel unsaturated initiators. We succeeded in synthesizing initiators with a vinyl group, such as vinylphenylketene methyl trimethylsilyl acetal and trimethylsilylated vinylbenzyl cyanide. These silyl compounds were shown to polymerize MMA in the presence of the catalyst. The syntheses of these initiators and the group transfer polymerization of MMA with them are presented below.

Preparation of Vinylbenzyl PolyMMA Macromonomer with Vinylphenylketene Methyl Trimethylsilyl Acetal (VPKTSA)

a) Synthesis of VPKTSA VPKTSA was synthesized according to Scheme II, using *p*-vinylbenzyl chloride (VBC) or *m*, *p*-VBC which is commercially available as a starting reagent. The new initiator was identified by elemental analysis, IR spectrum and ^{1}H NMR spectrum which is shown in Figure 1. VPKTSA was distilled at b.p. 110°C (3 mmHg) and the overall yield from VBC was more than 40%. An alternative method shown in Scheme III gives the same initiator but at a lower yield than in the former route.

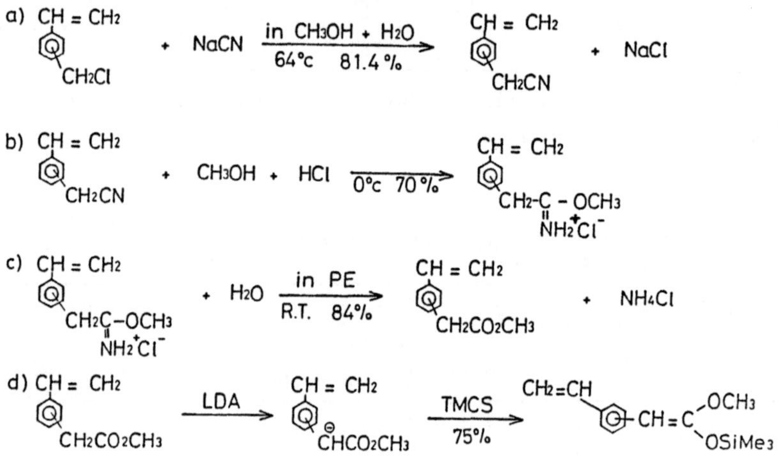

Scheme II

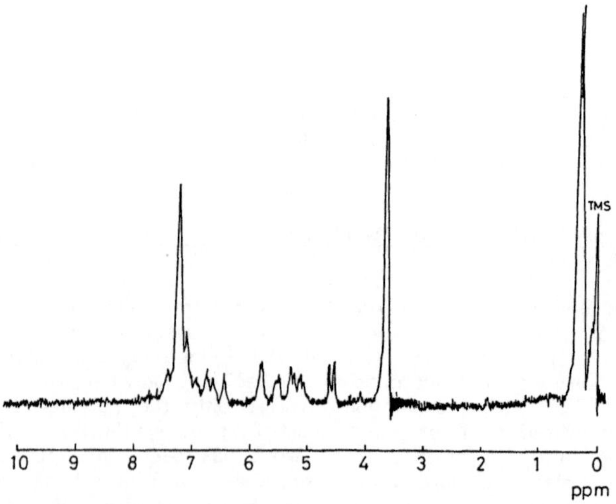

FIG. 1. ^{1}H NMR of p-vinylphenylketene methyl trimethylsilyl acetal (in $CDCl_3$, 60 MHz).

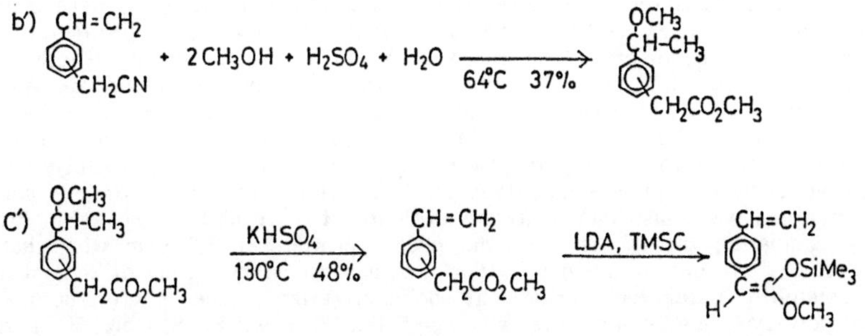

Scheme III

b) Group Transfer Polymerization of MMA with VPKTSA The group transfer polymerization of MMA was carried out in THF solution using VPKTSA and $TASF_2SiMe_3$ as initiator and catalyst, respectively. The reactions are shown in Scheme IV. The results obtained in this polymerization under various conditions are summarized in Tables IV and V.

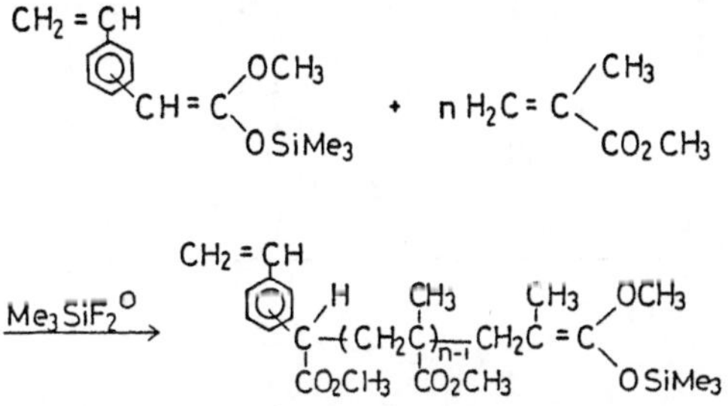

Scheme IV

TABLE IV. Preparation of PMMA macromonomer by initiation method[a)]

Run no.	MMA [M] (mol/L)	VPKTSA[b)] [I] (mmol/L)	[M]/[I]	Time (h)	Yield (%)	M_n x 10^{-3}	M_w/M_n
1	0.538	17.9	30.1	1	89	3.5	1.0_8
2	0.338	11.3	29.9	3	90	3.5	1.0_9
3	0.290	9.61	30.2	6	90	3.6	1.0_9
4	0.367	12.2	30.2	12	90	3.7	1.0_8
5	0.422	14.0	30.1	24	91	3.6	1.0_9

a) In THF at r.t.; catalyst, $TASF_2SiMe_3$. b) Initiator, vinylphenylketene methyl trimethylsilyl acetal.

The polyMMAs recovered after the termination with methanol were characterized by GPC and ^{1}H NMR spectra. Figure 2 shows GPC profiles of the polyMMA macromonomers. The sharp peaks in the GPC charts, whether obtained by RI or UV detection, indicate that the polyMMA macromonomers have quite high monodispersities, and the molecular weights are close to those calculated theoretically from the [monomer]/[initiator] ratios. In the range of ratios from 10 to 10^2 the polymers are found to have unequivocally a vinyl group at each end of the polyMMA by UV. The vinyl groups of the macromonomer end were also quantitatively evaluated by ^{1}H NMR as shown in Figure 3, by comparing the intensity of the phenyl protons at 7.2 ppm with that of the methoxy protons at 3.6 ppm. It shows that the styryl group of the initiator has remained unreacted during the polymerization. The microstructure of these polyMMA macromonomers was I = 5.4%, H = 36.0%, S = 58.6% of triad

TABLE V. Preparation of macromonomer by initiation method at various temperatures and [M]/[I] values

Run no.	MMA [M] (mol/L)	VPKTSA [I] (mmol/L)	[M]/[I]	Time (h)	Temp. (°C)	Yield (%)	M_n x 10^{-3}	M_w/M_n
1	0.422	14.0	30.1	24	r.t.	91	3.6	1.0_9
2	0.377	7.54	50.0	24	r.t.	95	5.5	1.1_7
3	0.883	12.7	69.6	24	r.t.	68	4.5	1.3_2
4	0.613	6.05	101	24	r.t.	46	4.4	1.3_3
5	0.434	4.37	99.3	24	0	70	7.0	1.0_8
6	0.499	4.95	101	48	0	100	10.7	1.1_0
7	0.544	5.29	103	72	0	100	11.4	1.1_0

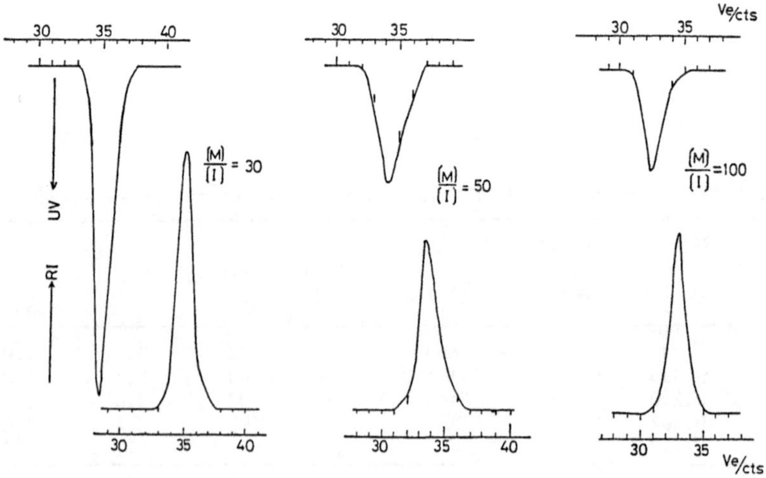

FIG. 2. GPC of PMMA macromonomer by initiation method.

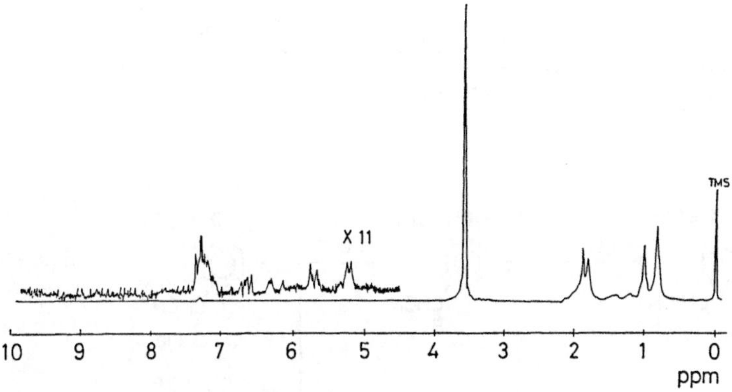

FIG. 3. ^{1}H NMR of PMMA macromonomer by initiation method with VPKTSA (in $CDCl_3$, 200MHz).

by ^{1}H NMR, a result similar to that of the du Pont group. Table V indicates that it is better enough to keep polymerization temperature at 0 °C to obtain a high yield and a highly monodisperse macromonomer.

In conclusion, the novel initiator, VPKTSA, is a useful initiator of high efficiency in the group transfer polymerization of MMA, producing the living styryl polyMMA macromonomer which is monodisperse and has the desired molecular weight up to 10^4. Therefore, this initiation method is recommended as a better pathway than the termination method for the preparation of a polyMMA macromonomer by group transfer polymerization, in terms of simplicity of procedure, the purity of macromonomer and the applicability of the "living" macromonomer such as ω-end functionalization of the macromonomer.

Preparation of Vinylbenzyl PolyMMA Macromonomer by Using Trimethylsilylated Vinylbenzyl Cyanide

a) Synthesis of N-Trimethylsilyl Vinylphenyl Keteneimine (TMSVPKI) and α-Trimethylsilyl Vinylbenzyl Cyanide (TMSVBCN) N-silyl keteneimines are known to be as useful as silyl ketene acetals in the organic syntheses. As already mentioned we found that in group transfer polymerization silyl vinylphenylketene acetal, VPKTSA, is an excellent initiator of high efficiency, producing a high quality, polyMMA macromonomer. By the analogy of ester and alkylnitrile, the synthesis of N-silyl vinylphenyl keteneimine from vinylbenzyl cyanide was carried out according to Scheme V. The trimethylsilylate compound isolated by vacuum distillation was identified by IR and ^{1}H NMR spectra as the mixture of N-silylated and C-silylated benzylcyanide, i.e., N-trimetylsilyl vinylphenyl keteneimine and α-trimethylsilyl benzyl cyanide. ^{1}H NMR analysis showed the product to be composed of a 2 : 1 molar ratio of the C-silyl and N-silyl derivatives.

b) Group Transfer Polymerization of MMA with TMSVBCN The polymerization of MMA was performed in THF solution using the mixture of TMSV-PKI and TMSVBCN as initiator and $TASF_2SiMe_3$ as catalyst to prepare the styryl polyMMA macromonomer under various conditions. The results of this polymerization, summarized in Table VI, are similar to those of Tables IV and V. The polyMMA macromonomers have a high monodispersity and the molecular weights are slightly larger than those calculated from the ratios of [monomer] to [initiator]. This fact suggests that in the presence of the catalyst, both the N-silyl and C-silyl compound in the initiator mixture have initiated the polymerization of MMA with a high efficiency. It may be assumed

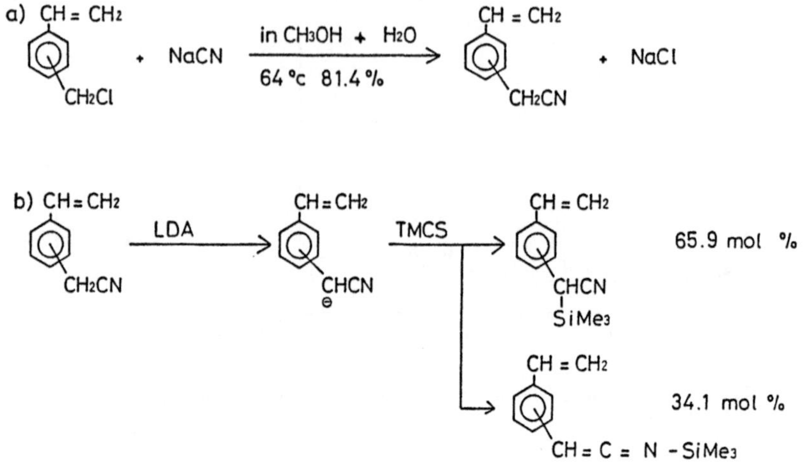

Scheme V

TABLE VI. Preparation of PMMA macromonomer by initiation method with TMSVBCN[a)]

Run no.	TMSVBCN[b)] [I] (mmol/L)	MMA [M] (mol/L)	[M]/[I]	Time (h)	Temp. (°C)	Yield (%)	M_n x 10^{-3}	M_w/M_n
1	10.9	0.329	30.2	24	0	trace		
2	16.5	0.500	30.3	24	0	70	4.9	1.1_9
3	24.7	0.740	30.0	24	0	100	4.6	1.2_2
4	34.8	1.03	29.6	24	0	100	3.9	1.1_6
5	29.2	0.876	30.0	24	-20	99	3.7	1.1_5
6	30.0	0.905	30.2	24	-40	85	3.4	1.1_4
7	31.3	0.933	29.8	24	-78	trace		
8	20.4	1.02	50.1	24	0	100	8.5	1.2_1
9	23.8	1.60	67.2	24	0	100	9.0	1.2_0
10	12.3	1.28	104	24	0	100	13.0	1.2_4

a) In THF; catalyst, $TASF_2SiMe_3$. b) Initiator, mixture of C-trimethylsilyl- and N-trimethylsilyl-vinylbenzyl cyanides.

that an equilibrium exists between the two compounds, and that the N-silyl keteneimine functions as the actual initiator of the group transfer polymerization. According to this assumption the proposed reaction scheme is shown in Scheme VI. A monomer concentration effect was observed, i.e., the polymerization of MMA did not take place in a MMA concentration lower than 0.3 mol/L. However, the details were not studied. In conclusion, the trimethylsilylated vinylbenzyl cyanide can also be used as an effective initiator for the group transfer polymerization of MMA, producing polyMMA macromonomer of high purity and monodispersity.

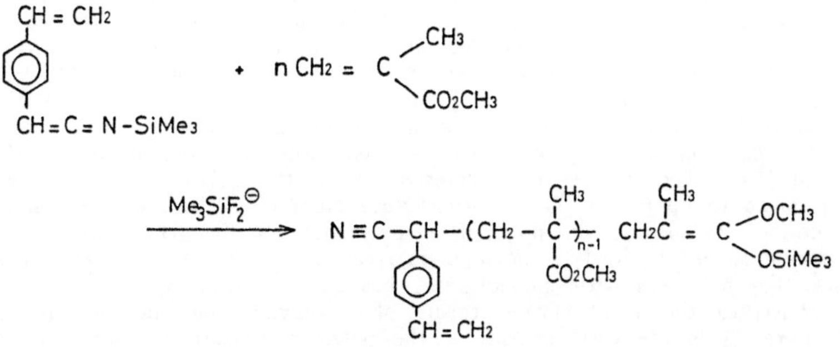

Scheme VI

RADICAL POLYMERIZATION OF VINYLBENZYL POLYMMA MACROMONOMER

The styryl polyMMA macromonomer was earlier reported by Anderson et al. [7] and Rempp et al. [8]. However, the polymerization of this macromonomer has not been reported yet. We have studied the polymerizability of the macromonomers and published the results [12]. In this study we carried out the radical polymerization of the styryl polyMMA macromonomers which had been prepared by either the termination or initiation method using the living group transfer polymerization of MMA.

A polyMMA macromonomer prepared by the termination method and characterized as $f = 0.83$, $M_n = 3300$, and $M_w/M_n = 1.0_7$ (Table II, run no. 5), was copolymerized with styrene or MMA as comonomer using AIBN at 60°C. Figure 4 shows the GPC of the copolymers recovered after polymerization for 40 hrs. The GPC profiles demonstrate that the macromonomer copolymerized with both styrene and MMA. Moreover, there exists a small peak corresponding to the 17% polyMMA with no polymerizable end group and which contaminated the poly(PMMA macromonomer).

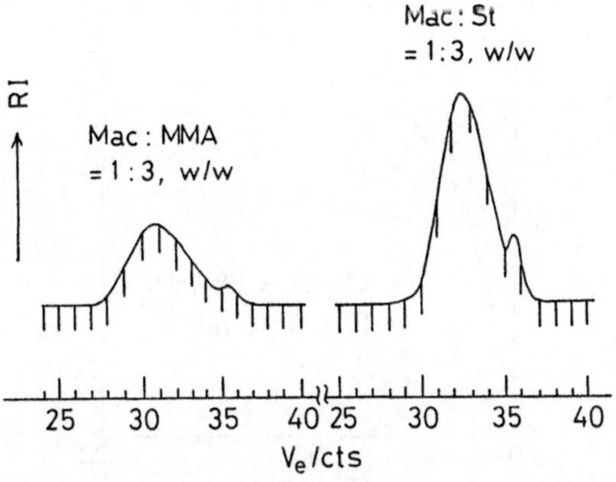

FIG. 4. Radical copolymerization of PMMA macromonomer with MMA and styrene in benzene (f = 0.83, AIBN, 60°C, 40 h).

The radical homopolymerizations of styryl polyMMA macromonomer were performed chiefly with macromonomers obtained by the initiation method whose functionality is nearly stoichiometric. A macromonomer obtained with VPKTSA (M_n = 3600, M_w/M_n = 1.1) was homopolymerized in benzene with AIBN. The results are shown in Table VII and Figure 5. The high conversion of the macromonomer (up to 94%) verifies its high functionality, and its homopolymerizability is consistent with that of polystyrene and polyTHF macromonomers already reported [10]. The degree of polymerization of the polyMMA macromonomers is estimated to be 6-7 by GPC data, values similar to those of the homopolymerized polystyrene [2] and polyTHF macromonomers [6]. Therefore, the poly(macromonomer)s obtained by homopolymerization can be regarded as comb-like or star-like polymers with monodisperse branches. The GPC curves of Figure 6 demonstrate the fractionation result of a polymerized macromonomer. In this figure, A is the GPC profile of the polymer mixture recovered after the polymerization, B is that of the fully polymerized macromonomer and C is that of the original, unconverted macromonomer. The fractionation was performed by extraction with methanol which dissolves only the original macromonomer.

TABLE VII. Radical homopolymerization of PMMA macromonomer in benzene[a)]

Run no.	VB-PMMA[b)] (mg)	AIBN (mg)	Benzene (mg)	Time (h)	Conv. (%)	D.P.
1	36.6	0.4	91.5	6	29	8
2	30.7	0.3	78.4	12	53	7
3	28.9	0.3	75.9	24	73	6
4	28.9	0.3	73.8	50	91	7
5	30.8	0.3	77.0	100	94	6

[a)]Polymerization temp., 60°C. [b)]M_n, 3.6 x 10^3; M_w/M_n, 1.1.

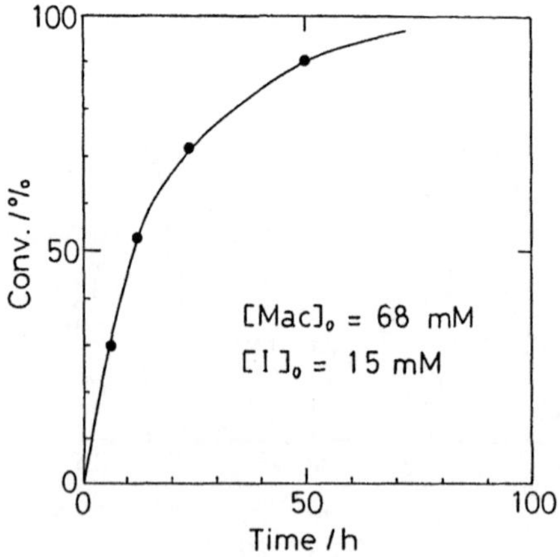

FIG. 5. Rate of radical polymerization of PMMA macromonomer in benzene.

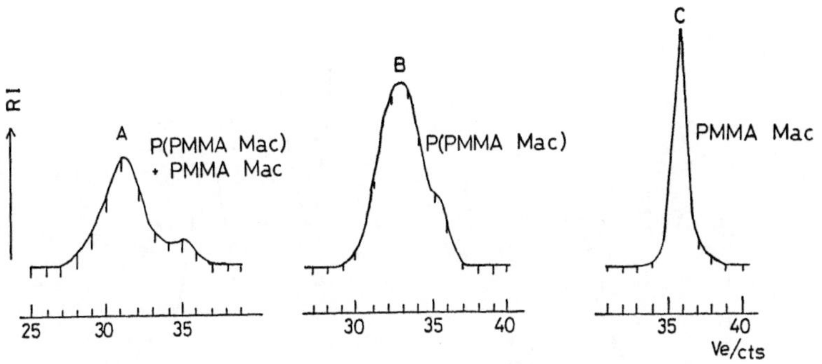

FIG. 6. Fractionation of polymerized PMMA macromonomer.

The copolymerization of styryl polyMMA macromonomer with styrene or MMA as comonomer was conducted in benzene with AIBN. The results shown in Table VIII and Figure 7 suggest a most interesting behavior of the macromonomer in the radical copolymerization with small monomers. The styryl macromonomer and MMA were copolymerized to completion after 24 hrs which is similar to the bulk polymerization of MMA. The same macromonomer copolymerized with styrene did not go to completion even after 48 hrs, although the macromonomer and styrene have the same vinyl group. While the styrene converted to the copolymer quantitatively, 20-30% of the styryl polyMMA macromonomer did not polymerize. This fact suggests that the primary factors which determine the reactivity of the macromonomer in copolymerization are not only the chemical properties of the functional group but also properties related to the conformations of the macromonomer and the graft copolymer formed, their compatibility and the diffusion phenomena in the viscous solutions. More studies on the copolymerization of the macromonomers are planned.

TABLE VIII. Radical copolymerization of PMMA macromonomer with vinyl monomers in benzene[a]

Run no.	VB-PMMA[b] (M_1) (mg)	Comonomer (M_2) (mg)		AIBN (mg)	Benzene (mg)	Time (h)	Conv., (%) M_1	M_2
1	49.9	St	110	1.60	160	46	72	100
2	103	St	124	2.26	226	46	83	100
3	101	St	52.0	1.54	154	46	87	100
4	51.0	MMA	102	1.55	155	24	100	100
5	105	MMA	107	2.12	212	24	100	100
6	106	MMA	80.4	2.39	239	24	100	100

[a]Polymerization Temp., 60°C. [b]M_n, 3.7 x 10^3; M_w/M_n, 1.0_8.

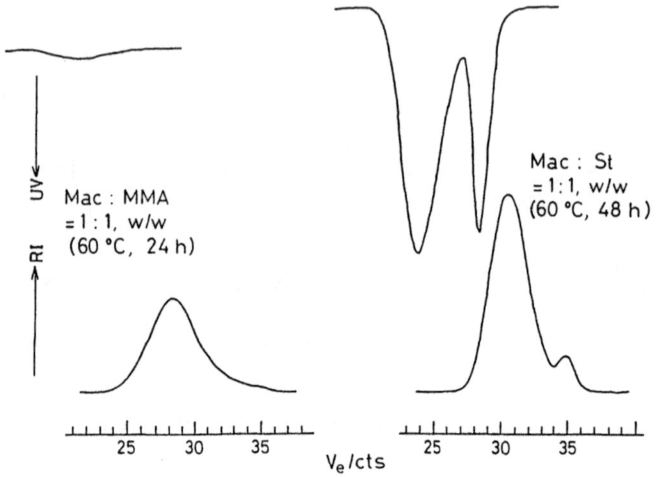

FIG. 7. GPC of copolymerized PMMA macromonomer.

Acknowledgment. We would like to thank Mr. T. Hirano and Mr. K. Kojima for their participation in this study.

REFERENCES

1. P. Rempp and E. Franta, Advances in Polymer Science **58**, 1-53 (1984).
2. R. Asami, M. Takaki, and H. Hanahata, Macromolecules **16**, 628-631 (1983).
3. R. Asami and M. Takaki, Makromol. Chem., Suppl. **12**, 163-173 (1985).
4. a) P. R. Rao, P. Masson, P. Lutz, G. Beinert, and P. Rempp, Polym. Bull. **11**, 115-120 (1984).
   b) M. Takaki, R. Asami, S. Tanaka, H. Hayashi, and T. E. Hogen-Esch, Macromolecules **19** (12) in press.
5. P. Masson, G. Beinert, E. Franta, and P. Rempp, Polym. Bull. **7**, 17-22 (1982).
6. R. Asami, M. Takaki, K. Kita, and E. Asakura, Polym. Bull. **2**, 713-718 (1980).
7. B. C. Anderson, G. D. Andrews, P. Jr. Arthur, H. W. Jacobson, L. R. Melby, A. J. Playtis, and W. H. Sharkey, Macromolecules **14**, 1599-1601 (1981).
8. P. Lutz, P. Masson, G. Beinert, and P. Rempp., Polym. Bull. **12**, 79-85 (1984).
9. O. W. Webster, W. R. Hertler, D. Y. Sogah, W. B. Farnham, and T. V. RajanBabu, J. Am. Chem. Soc. **105**, (1983).
10. O. W. Webster, W. R. Hertler, D. Y. Sogah, W. B. Farnham, and T. V. RajanBabu, J. Makromol. Sci.-Chem. **A21**, 943-960 (1984).
11. Recently styryl polyMMA macromonomer was synthesized by an anionic polymerization using vinylbenzyl Grignard reagents as initiator: K. Hatada, H. Nakanishi, K. Ute, and T. Kitayama, Polym. J. **18**, 581-591 (1986).
12. R. Asami, M. Takaki, K. Kyuda, and N. Sukenaga, Polym. J. **15**, 261-266 (1983).

LIVING COUPLING AGENTS. RATIONAL SYNTHESIS OF HETEROARM STAR-BRANCHED POLYMERS

Roderic P. Quirk and Frederick Ignatz-Hoover
Institute of Polymer Science
University of Akron
Akron, Ohio 44325

INTRODUCTION

One of the goals of synthetic polymer chemistry is to prepare polymers with control over the major variables affecting polymer properties. These variables include molecular weight, molecular weight distribution, copolymer composition and microstructure, tacticity and diene microstructure, chain end functionality, branching and morphology. Anionic polymerization methods have approached this goal of synthesizing polymers with well-defined structures in monomer/initiator/solvent systems which proceed in the absence of chain termination and chain transfer reactions.[1-4] For example, under certain carefully controlled conditions, the anionic polymerizations of styrene, diene, methacrylate, epoxide, and lactone monomers proceed without significant contributions from chain termination and chain transfer reactions to produce polymers which retain their reactive anionic chain-end functionality when all of the monomer has been consumed. In these system, the major variables affecting polymer properties can indeed be controlled. The scope and limitations of these anionic polymerization methods are described in several recent books and reviews.[1-5]

One consequence of the non-terminating nature of many anionic polymerizations is the generation of polymer chains with stable carbanionic chain ends which can be reacted with linking reagents to generate star-branched polymers and copolymers as illustrated with silicon tetrachloride as the linking agent in eq 1.[6-8] The resulting homogeneous star-branched polymers can be formed with arms characterized by well-defined, controlled

$$4PLi + SiCl_4 \longrightarrow P_4Si + 4LiCl \qquad (1)$$

molecular weights and narrow molecular weight distributions.[6,7] Unfortunately, these methods are not readily adaptable to the preparation of star-branched polymers in which the arms differ in either molecular weight or composition. The results reported herein describe a novel approach to the synthesis of star-branched copolymers with compositionally heterogeneous arms using the concept of living coupling agents.

EXPERIMENTAL

Styrene, isoprene, benzene and tetrahydrofuran were purified as described previously.[9] Solutions of sec-butyllithium (Lithium Corporation of America, 12.0 wt. % in cyclohexane) and polymeric organolithiums were analyzed using the double-titration procedure with 1,2-dibromoethane[10] or the sec-butyl alcohol titration with 1,10-phenanthroline as indicator[11], respectively. The concentration of poly(styryl)lithium in benzene was also determined by ultraviolet absorption spectrophotometric analysis using the 335 nm band which is reported to exhibit an extinction coefficient of 13,000.[12]

1,3-Bis(1-phenylvinyl)benzene (MDDPE) was synthesized using procedures analogous to those reported by Schulz and Hocker.[13]

sec-Butyllithium-initiated polymerizations were carried out at 30°C in all glass, sealed reactors using break-seals and standard high vacuum

Recent Advances in Anionic Polymerization
T.E. Hogen-Esch and J. Smid, Editors

techniques.[14] Coupling reactions with MDDPE and subsequent chain extension reactions were performed in an argon atmosphere glove box. The argon atmosphere was purified by continuous circulation through towers of molecular sieves and manganous oxide as described by Brown et al.[15].

Size exclusion chromatographic analyses (Waters 150-C) were performed in THF using four μ-Styragel columns (Two-100, 500 and 1000A) after calibration with standard polystyrene samples. Ultraviolet absorption spectra were obtained by using a Varian DMS-90 UV-vis spectrophotometer. HPLC analyses (Varian 5000 liquid chromatograph) were performed by using a Micro-Pak CH-10 column (30x4 mm) in 70/30 CH_3CN/H_2O at 120 atm with a Varian UV-50 variable-wavelength detector.

RESULTS AND DISCUSSION

A living carbanionic coupling reagent is a species which can react with polymers which have carbanionic chain ends to generate a coupled polymer product which retains the active center stoichiometry, i.e., the coupled product is dicarbanion.[16] A generalized scheme illustrating the concept of living coupling agents is shown below in eq 2, where C is a living coupling agent, P_1Li is a living carbanionic polymer chain and M_2 is

$$2P_1Li + C \longrightarrow (P_1)_2C(Li)_2 \xrightarrow{nM_2} (P_1)_2C(P_2)_2 \qquad (2)$$

a second monomer which can be anionically polymerized by the active centers in the living coupled product to generate polymer chains, P_2, after quenching. It is obvious from eq 2 that the use of living coupling agents provides a method for the controlled synthesis of star-branched copolymers where two arms differ in molecular weight and/or composition from the other two arms, i.e., heteroarm star-branched polymer and copolymers.

Several criteria must be satisfied for a species to be useful as a living coupling agent:

(1) The living coupling agent must react quantitatively with living carbanionic chain ends.

(2) The coupled product must retain the carbanionic active centers stoichiometrically.

(3) The living coupled product must be capable of reinitiating polymer chain growth rapidly (relative to propagation) and stoichiometrically.

Unfortunately, although there are several reports describing the use of various divinyl species to couple living carbanionic chains and maintain active center stoichiometry, we are unaware of any successful application of the above principles for the controlled synthesis of heteroarm star-branched copolymers. Eschwey and Burchard[17] utilized the coupling of poly(styryl)-lithium chains with p-divinylbenzene to generate coupled products which retain their active center stoichiometry. This was demonstrated by chain extension reactions of the living coupled products with styrene and isoprene. However, this method suffers from the lack of precise stoichiometric control of the functionality of the living coupled products. In another example of the application of this concept, Yamagishi, Szwarc, Tung and Lo[18] examined the kinetics of the addition of poly(styryl)lithium to double diphenylethylenes, 1-3, in benzene solution. The monoadducts and diadducts were characterized by ultraviolet and visible spectroscopy. For example, the addition of poly(styryl)lithium to 1 proceeds stepwise to generate the monoadduct, 4, and then the diadduct, 5, (eq 3). They reported that the addition of the second mole of poly(styryl)lithium was substantially slower than that of the first one. This work is an example of a living coupling reaction since the products of the coupling reaction retain their active

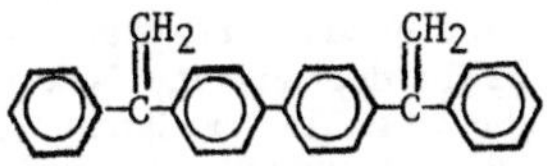

1

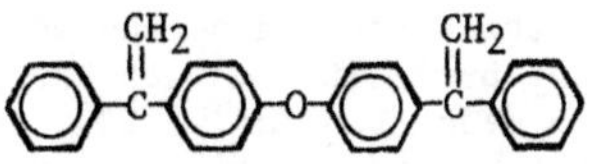

2

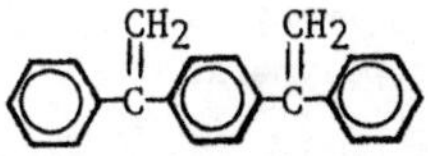

3

center stoichiometry. Unfortunately, the efficiency of the coupling reaction of poly(styryl)lithium with the double diphenylethylenes, 1-3, was not examined for stoichiometric quantities of the reactants, so that the efficiency of the coupling reaction cannot be deduced [Criterion (1) above].

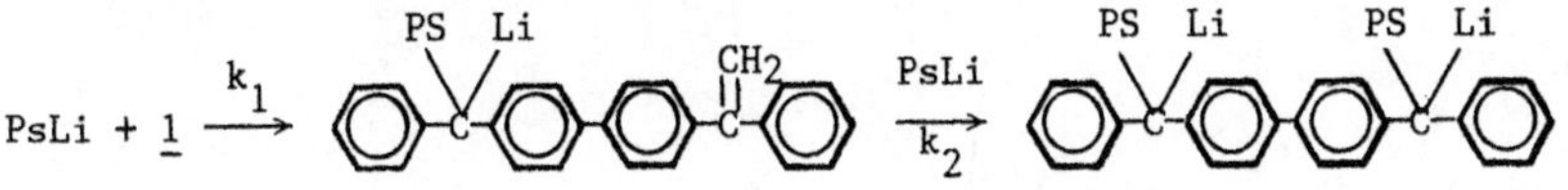

4 5 (3)

These workers also examined the kinetics of the addition of "a large excess of styrene" to the diadduct, 5, (eq 4).[18] The absorption spectrum recorded after completion of the polymerization indicated the presence of only the half crossed-over adduct, 6, and unreacted diadduct, 5. It was

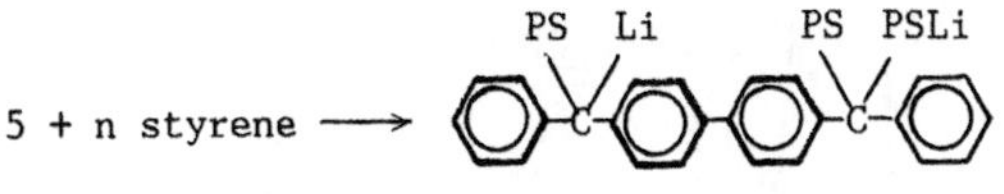

(4)

6

concluded that the addition of styrene takes place mainly on only one of the ends of the diadduct, 5, since only 50% of the active centers in 5 were estimated to have crossed over to styryl anions. Obviously, these results indicate that the diadduct, 5, is not synthetically useful for the preparation of star-branched copolymers since the living coupled product is not capable of stoichiometrically reinitiating the polymerization of added monomer [Criterion (3) above].

There is evidence in the literature that the meta-substituted double diphenylethylenes behave quite differently from the para-substituted analogs. For example, Leitz and Hocker[19] have reported that for the addition of sec-butyllithium to the para-double diphenylethylene, 3, the rate constant for the addition of the second mole of sec-butyllithium is 13.6 times slower than the addition of the first mole; however, for the addition of sec-butyllithium to the meta-double diphenylethylene, MDDPE, 7, the rate

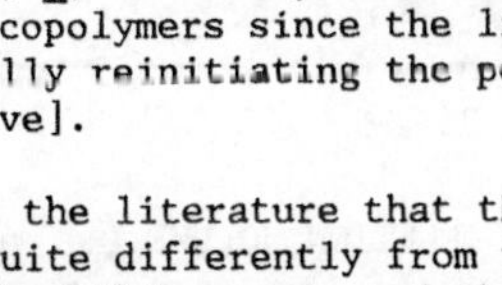

7

constants for the first and second additions are equal. Furthermore, the diadduct of sec-butyllithium with the para-double diphenylethylene, 3, is insoluble in hydrocarbon media[20]. In contrast Schulz and Hocker[21] have reported that the diadduct of sec-butyllithium with MDDPE is soluble in hydrocarbon media for weeks even at low temperatures. Therefore, it was of interest to examine the use of MDDPE as a living coupling agent for the synthesis of star-branched copolymers.

The use of MDDPE as a coupling agent for polymeric organolithium compounds was examined by a combination of ultraviolet spectroscopy and size exclusion chromatography. An aliquot of a solution of poly(styryl)lithium with $\bar{M}_n$=1,000 in benzene solution was reacted with a stoichiometric amount of MDDPE. Size exclusion chromatograms of the base polymer and the product resulting from methanol quenching of the product of this coupling reaction are shown in Figure 1. It is estimated by curve resolution that the

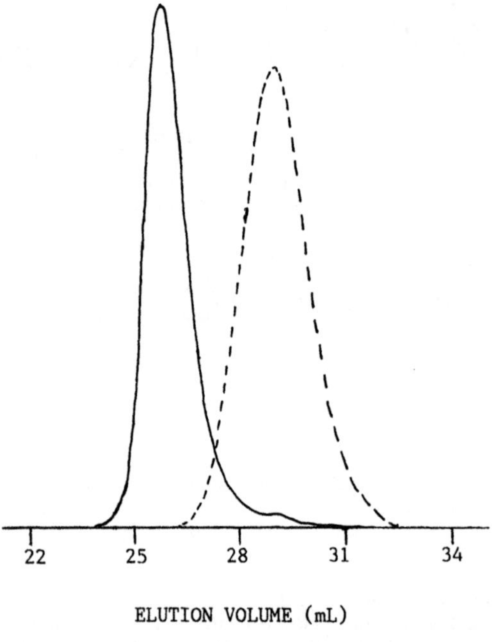

Figure 1. Size exclusion chromatogram for the polystyrene base polymer with $\bar{M}_n$=1,000 (---) and the coupled polymer product with $\bar{M}_n$=2,000 (——).

coupling efficiency for this reaction is greater than 99%. In order to obtain further evidence regarding not only the efficiency of this coupling procedure but also information regarding the coupling rates, the ultraviolet spectrum of the coupled product was determined one hour after mixing another poly(styryl)lithium ($\bar{M}_n$=5,000, $\bar{M}_w/\bar{M}_n$=1.07) with a stoichiometric amount of MDDPE as shown in Figure 2; the ultraviolet spectrum of the base poly(styryl)-lithium is also shown in Figure 2 for comparison. The ultraviolet absorption for the MDDPE living coupled adduct exhibits a λ_{max} at 438 nm and ε=21,000. The results from both size exclusion chromatography and ultraviolet absorption spectroscopy are consistent with quantitative coupling of the poly-(styryl)lithium chains by MDDPE and retention of active center stoichiometry

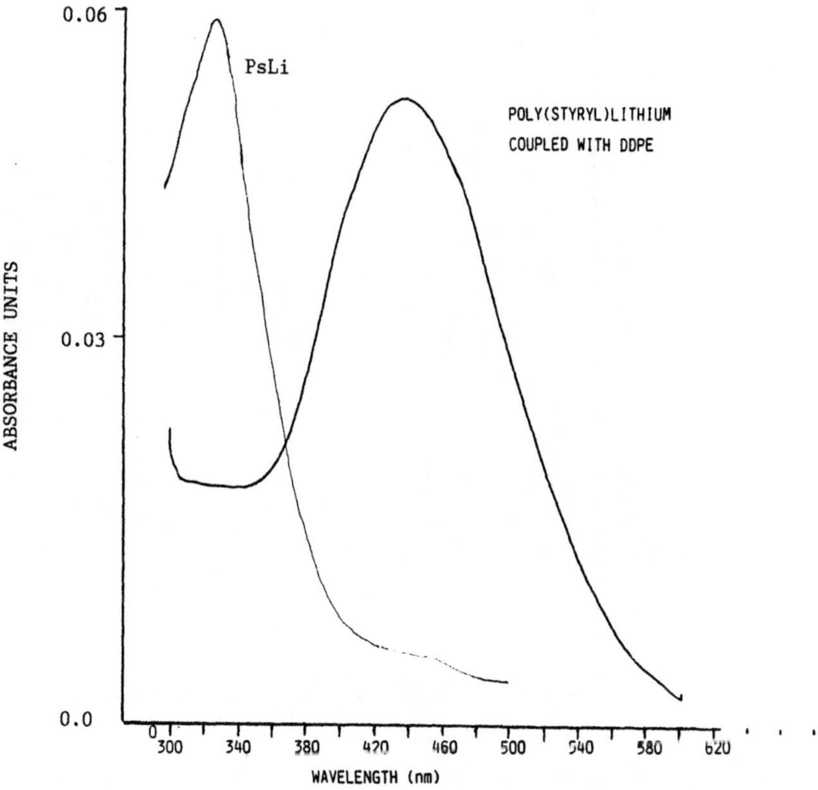

Figure 2. Ultraviolet absorption spectra in benzene for poly(styryl)-lithium ($\overline{M}_n$=5,000) and the product of coupling of MDDPE with this poly(styryl)lithium.

[Criteria (1) and (2) above]. However, these results do not show that the living coupled adduct is capable of quantitatively and rapidly reinitiating anionic polymerization to generate two new active polymer arms [Criterion (3) above].

The crossover reaction of the adduct of MDDPE and poly(styryl)lithium ($\overline{M}_n$=5,000) was examined by a combination of ultraviolet absorption spectroscopy and size exclusion chromatography. An aliquot of this adduct was mixed with sufficient styrene to generate two new polystyrene branches, each with $\overline{M}_n$=10,000. The ultraviolet absorption spectrum of the product of this chain-extension reaction one hour after mixing the reactants is shown in Figure 3. The absorption at 438 nm for the living coupled adduct has been replaced by the absorption at 335 nm for poly(styryl)lithium. It is estimated that this result corresponds to a minimum of 95% crossover to styrene based on analysis of the absorption at 438 nm. The size exclusion chromatogram of the product of the polymer produced from this crossover reaction is shown in Figure 4. The product exhibits a monomodal distribution with $\overline{M}_w/\overline{M}_n$ of 1.25. The molecular weight distribution is somewhat broader than the original base polymer ($\overline{M}_w/\overline{M}_n$=1.07). This indicates that the crossover

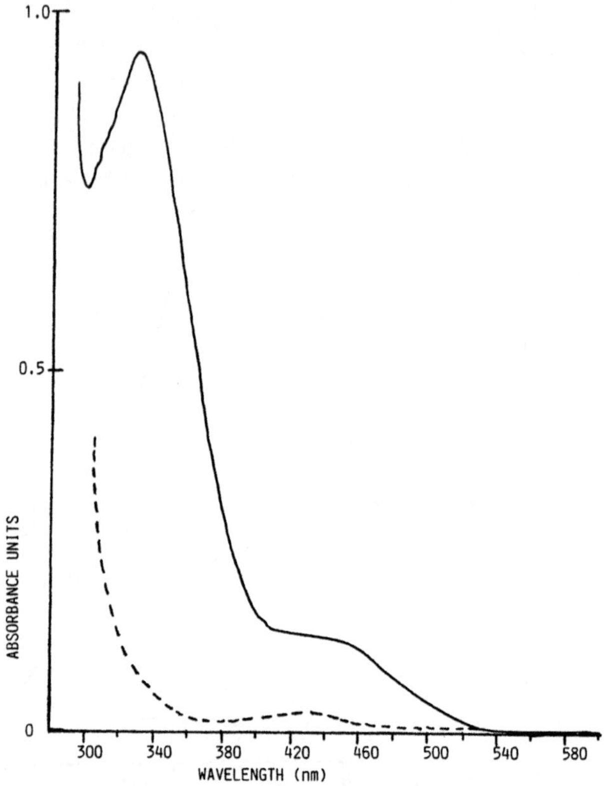

Figure 3. Ultraviolet absorption spectra of the products of reaction of the coupling product, generated from MDDPE with poly(styryl)-lithium, with styrene monomer (—) and isoprene monomer (---).

reaction to styrene monomer from the living coupled adduct is not faster than propagation, but this distribution is still reasonably narrow. A second aliquot of the base polymer adduct with MDDPE was reacted with sufficient styrene to generate two new polystyrene branches, each with $\overline{M}_n^{calc}$=24,000. The size exclusion chromatogram of the polymer product from this chain extension reaction is shown in Figure 4. This polymer exhibits a monomodal molecular weight distribution with $\overline{M}_w/\overline{M}_n$=1.32. Again the broadening of the molecular weight distribution indicates that the crossover reaction to styrene is not necessarily rapid relative to propagation. These ultraviolet absorption spectroscopic and size exclusion chromatographic experiments are consistent with the formation of heteroarm polystyrene star-branched copolymers. It will be necessary to carry out further experiments to carefully determine the composition of these polymers and the uniformity of the two branches formed by the chain extension reactions.

Preliminary evidence is also available concerning the crossover from the MDDPE adduct of poly(styryl)lithium with isoprene. The crossover reaction was also examined by a combination of ultraviolet absorption spectroscopy and size exclusion chromatography. The ultraviolet absorption spectrum of the crossover reaction of the adduct one hour after mixing with isoprene is shown in Figure 3. This spectrum is consistent with efficient crossover to poly(isoprenyl)lithium chains.

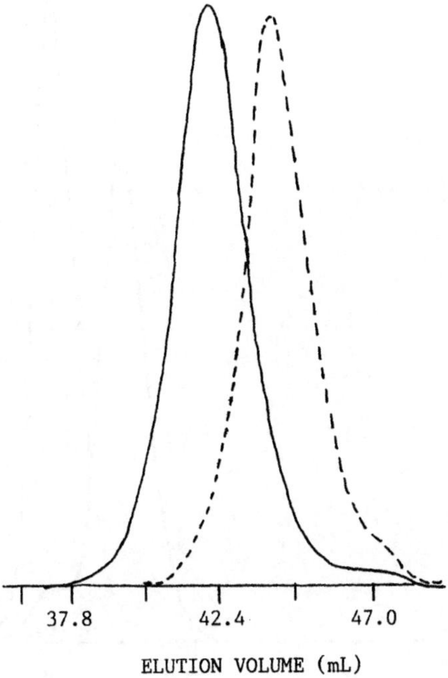

Figure 4. Size exclusion chromatogram for the products with $\bar{M}_n$=30,000 (---) and $\bar{M}_n$=58,000 (—) from the reaction of styrene monomer with the coupling product generated from MDDPE and poly(styryl)lithium ($\bar{M}_n$=5,000).

The rational synthesis of a star block copolymer with compositionally heterogeneous arms is shown in the scheme below. The results of this type of experiment are shown in Figure 5. In Figure 5, the size exclusion

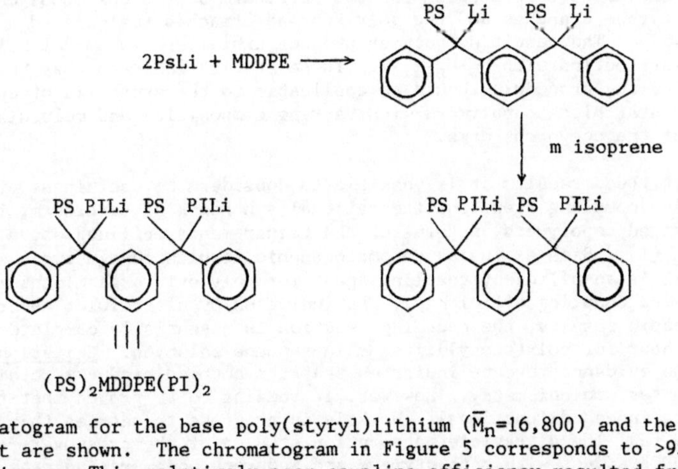

chromatogram for the base poly(styryl)lithium ($\bar{M}_n$=16,800) and the MDDPE adduct are shown. The chromatogram in Figure 5 corresponds to >95% coupling efficiency. This relatively poor coupling efficiency resulted from imprecision in the determination of the concentration of active poly(styryl)-lithium chain ends by titration methods; ultraviolet absorption analysis is

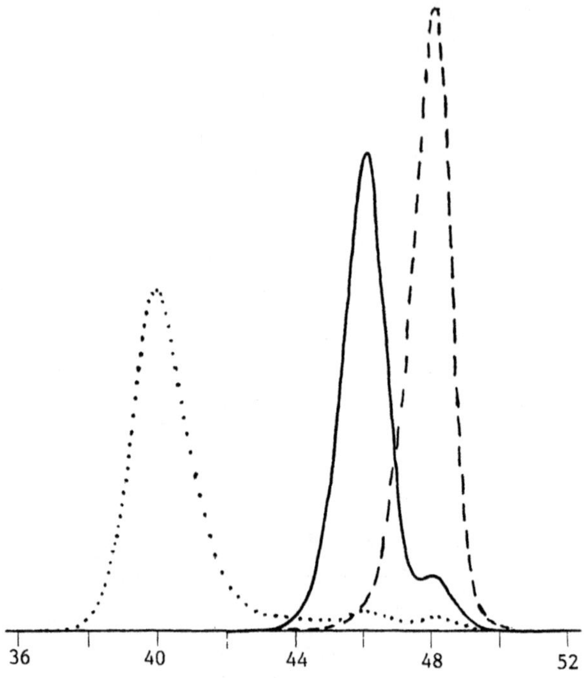

Figure 5. Size exclusion chromatograms for the polystyrene base polymer, $\bar{M}_n$=16,800 (--); coupling product of poly(styryl)lithium with MDDPE, $\bar{M}_n$=33,600 (—); and the product of reaction of isoprene monomer with the coupling product, $\bar{M}_n{}^{calc}$=267,200 (....).

now our preferred method to obtain >99% coupling efficiencies. In Figure 5, the size exclusion chromatogram for the heteroarm star-block copolymer with two polystyrene branches and two polyisoprene branches (calculated $\bar{M}_n$=267,200). The resulting polymer product exhibits a monomodal molecular weight distribution with $\bar{M}_w/\bar{M}_n$=1.17. These results indicate that the MDDPE living coupling procedure should be applicable to the synthesis of analogous heteroarm star-block copolymers with varying composition and molecular weights of the component arms.

With these results it is possible to consider the usefulness of MDDPE as a living coupling agent for the rational synthesis of heteroarm, four-arm, star-branched copolymers in terms of the requirements outlined above. Criterion (1). Size exclusion chromatographic results conclusively show that MDDPE is an efficient coupling agent for polymeric organolithium compounds (>99% coupling efficiency). As indicated by ultraviolet absorption spectroscopic results, the coupling reaction is essentially complete in less than one hour for poly(styryl)lithium in benzene solution. Criterion (2). All of the evidence to date indicates that the coupled products retain their active center stoichiometry. However, it remains to be proven that the chains are indeed difunctional. Work is in progress to address this question. Criterion (3). The ultraviolet absorption studies of the crossover reactions of the MDDPE adduct with both styrene and isoprene monomers indicate that the crossover reactions are efficient and relatively rapid. Size exclusion chromatographic analyses indicates that there is some broadening of the molecular weight distribution compared to the base polymers. However, it is important to note that monomodal and relatively narrow molecular weight distributions are obtained. This is consistent with complete and uniform crossover and chain growth with the added monomers. It will be necessary to provide further experimental evidence regarding the uniformity of the arms which grow from the MDDPE adduct.

In conclusion, the results reported herein are consistent with the hypothesis that MDDPE is a useful living coupling agent. However, it is obvious that additional careful experimental work will be required to characterize the resulting star-branched polymers and to delineate the scope and limitations of the method for the rational synthesis of heteroarm star-branched copolymers.

ACKNOWLEDGEMENT

We gratefully acknowledge the generous support of this research by the Exxon Education Foundation, Owen-Corning Fiberglas Corporation, a Material Science Fellowship Grant from the 3M Foundation, and a fellowship (F.I.-H.) from the Goodyear Tire and Rubber Company.

REFERENCES

1. S. Bywater, "Anionic Polymerization," in "Encyclopedia of Polymer Science and Engineering," 2nd ed., Wiley-Interscience, New York, 1985, Vol. 2, p 1.
2. M. Morton, "Anionic Polymerization: Principles and Practice," Academic Press, New York, 1983.
3. R.N. Young, R. P. Quirk and L.J. Fetters, Adv. Polym. Sci., 56, 1 (1984).
4. M. Szwarc, Adv. Polym. Sci., 49, 1 (1983).
5. "Anionic Polymerization: Kinetics, Mechanism and Synthesis," J.E. McGrath, Ed., ACS Symposium Series 166, American Chemical Society, Washington, D.C., 1981.
6. S. Bywater, Adv. Polym. Sci., 30, 89 (1979).
7. B.J. Bauer and L.J. Fetters, Rubber Chem. Tech., 51, 406 (1978).
8. H.L. Hsieh, Rubber Chem. Tech., 49, 1305 (1976).
9. R.P. Quirk and W.-C. Chen, Makromol. Chem., 183, 2071 (1982).
10. H. Gilman and F.K. Cartledge, J. Organomet. Chem., 2, 447 (1964).
11. S.C. Watson and J.F. Eastham, J. Organomet. Chem., 9, 165 (1967).
12. D.J. Worsfold and S. Bywater, Can. J. Chem., 38, 189 (1960).
13. G.G. H. Schulz and H. Hocker, Makromol. Chem., 178, 2589 (1978).
14. M. Morton and L.J Fetters, Rubber Chem. Tech., 48, 359 (1975).
15. T.L. Brown, D.W. Dickerhoff, D.A. Bafus and G.L. Morgan, Rev. Sci. Instrum., 33, 491 (1962).
16. This is a much more general concept which applies to linking reactions (i.e., joining of 2 or more arms) of any polymer chain which has reactive chain-end activity (i.e., is living).
17. H. Eschwey and W. Burchard, Polymer, 16, 180 (1975).
18. A. Yamagishi, M. Szwarc, L. Tung and G.Y.-S. Lo, Macromolecules, 11 607 (1978).
19. E. Leitz and H. Hocker, Makromol. Chem., 184, 1893 (1983).
20. L.H. Tung, G.Y.-S. Lo and D.E. Beyer, Macromolecules, 11, 616 (1978).
21. G. Schulz and H. Hocker, Angew, Chem. Int. Ed. Engl., 19, 219 (1980).

SYNTHESIS AND SOLUTION PROPERTIES OF MACROCYCLIC POLYMERS

Pierre LUTZ, Claude STRAZIELLE and Paul REMPP
Institut Charles Sadron (CRM/EAHP) CNRS
6, rue Boussingault 67083 Strasbourg-Cedex, France.

ABSTRACT

Cyclic macromolecules can be synthesized in two different ways : either taking advantage of ring-chain equilibria, or by end-to-end cyclization. The latter procedure is based upon coupling of anionically obtained αω-dicarbanionic polystyrene, by means of a difunctional electrophile. The solution behavior of these cyclic polymers is compared with theoretical expectations issued many years ago.

INTRODUCTION

Cyclic macromolecules have been subject to interest for more than three decades. The effect of cyclization on the size of a macromolecule in dilute solution has been treated theoretically in several papers [1-3]. The absence of end groups in a cyclic macromolecule is also expected to have consequences on such properties as the glass transition temperature [4], which is known to vary linearly with the reciprocal of the molecular weight. Finally the concept of reptation of a macromolecule through a curvilinear tube, recently brought about by de Gennes [5] to account for the self-diffusion of a polymer in the bulk would imply that the diffusion of a cyclic macromolecule be highly impeded, which may be seen as an unexpected consequence of the absence of chain ends.

Until recently, the theoretical predictions on cyclic macromolecules could not be tested adequately, because no well defined macrocyclic samples covering a sufficiently broad range of molecular weights had been made available. During the past decade much experimental work has been performed along that line, in a number of laboratories, to provide for well defined cyclic polymer samples, and to allow an accurate check of the theoretical expectations. It is the purpose of this paper to review the methods aiming at the synthesis of macrocycles, with special emphasis on the end-to-end cyclisations based on ionic polymerization processes. The special features of the solution properties of these species will be briefly reported in the second part of this article.

SYNTHESIS OF MACROCYCLIC POLYMERS.

Two very different methods have been used to get access to ring-shaped macromolecules : Ring-chain equilibria and end-to-end cyclization.

1 - Ring - chain equilibria have been studied intensively for many years [6-8]. Prerequisites for the occurence of such an equilibrium are the presence of functions -C- along the chain, and the reversibility of the condensation reaction leading to these functions.

$$\sim\sim A \quad + \quad B \sim\sim \quad \rightleftharpoons \quad \sim\sim C \sim\sim$$

Recent Advances in Anionic Polymerization
T.E. Hogen-Esch and J. Smid, Editors

The end-standing functions A or B of the polycondensate should be able to react with any functional link C, provided the experimental conditions allow these exchanges to occur . Such reactions can take place intermolecularly - they are referred to as scrambling or reshuffling - whereby the total number of molecules remains unchanged, and the number average molecular weight as well. They can also occur intramolecularly - they are called backbiting -, and cycles are formed, whereby the number of molecules is increased and their average molecular weight is lowered.

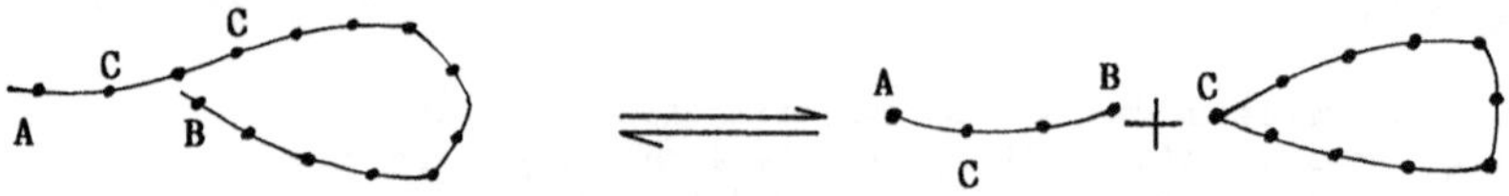

The probability of backbiting obviously increases as the overall polymer concentration decreases. If the backbiting reaction is allowed to proceed, the ring-chain equilibrium, characteristic for the temperature and concentration chosen, is eventually reached.

These equilibria have been studied theoretically by Jacobson and Stockmayer [6]. The molecular weight distribution of both the linear and the cyclic constituents at equilibrium are rather broad, because the process is random. The probability for a cycle with y monomer units to be formed is related to the probability for a chain with y units to have its chain ends overlap. The equilibrium constant K_y can be expressed as :

$$K_y = \left(\frac{3}{2\pi \langle r^2\rangle}\right)^{3/2} \cdot \frac{1}{2\, N_A\, y}$$

where $\langle r^2\rangle$ is the mean square end-to-end distance and N_A is Avogadro's number. The smaller cycles are thus more likely to be formed.

The development of efficient fractionation techniques [9] was a prerequisite to the syntheses of well defined cyclic polymer samples. Preparative size exclusion chromatography also contributed to the success of the separations required [10].

Polymers that can undergo ring-chain equilibration include a number of polycondensates, such as polyesters [6, 7], and several polymers arising from ring-opening polymerization : polydioxolane, polycaprolactone, and poly(dimethyl siloxane), among others.

Semlyen [8 - 10] has established that upon heating a dilute PDMS solution in the presence of an appropriate basic catalyst, the cyclization equilibrium can be reached eventually. Cycles covering a wide range of molecular weights are formed, the smallest being still the most numerous. Under well chosen conditions (high dilution) the amount of remaining linear constituent is in some cases vanishingly small. Very careful fractionation of the reaction mixture allowed Semlyen to quantitatively separate the linear constituents from the cycles, and to fractionate the latter according to their molecular weights. Well defined cyclic PDMS fractions of low polydispersity were thus obtained, covering a molecular weight range extending up to 30,000 approximately. Semlyen used these samples for a comparative investigation of the solution properties of PDMS cycles and of their linear homologues. The limited range of molecular weights accessible does not harm too much, as the PDMS chains are very flexible. Semlyen was able to bring convincing experimental evidence to support the theoretical expectations [9 - 11].

2 - Induced ring closure reactions (or end-to-end cyclisations) are based upon an entirely different principle [12-16] : A polymer molecule carrying reactive functions at both chain ends is reacted stoichiometrically with a bifunctional coupling agent, whereby links are formed between chain ends. The process has to be fast, quantitative and free of any side reaction. Coupling can occur intramolecularly, yielding cyclization, or

intermolecularly to give chain extension. The probability of intramolecular reaction depends obviously on both the concentration and the molecular weight of the precursor.

To be more specific, the probabilities of both types of couplings can be estimated, as it will be shown, and one can define a concentration c_{eq} at which both are equal. The behavior of the precursor chain being supposedly gaussian, the probability of cyclization is determined by the probability for the two chain ends to be in immediate vicinity of each other (within a volume element τ) ,c being the concentration, M the molecular weight, $\langle r^2 \rangle$ the mean square end-to-end distance of the precursor polymer and V the total volume of the solution :

$$p_{intra} = \left(\frac{3}{2\pi \langle r^2 \rangle} \right)^{3/2} \cdot \frac{c\, V\, N_A\, \tau}{M}$$

The probability of intermolecular reaction is related to the probability for two precursor molecules to be in the same volume element τ, account being taken of the fact that only four of the six possible coupling reactions involve sites belonging to different precursor molecules.

$$p_{inter} = \frac{2\, n^2\, \tau^2}{3 \cdot 2V^2} = \frac{c^2\, N_A^2\, V\, \tau}{3\, M^2}$$

The concentration at which both types of reactions occur with the same probability, c_{eq} is then given by :

$$c_{eq} = 3 \left(\frac{3}{2\pi \langle r^2 \rangle} \right)^{3/2} \cdot \frac{M}{N_A}$$

Under θ conditions, $\langle r^2 \rangle$ is proportional to the molecular weight M, which means that the probability of cyclization is smaller for high molecular weight precursors. In a good solvent, owing to chain expansion, the average end-to-end distance is increased, and the probability of intramolecular coupling is lowered.

The induced cyclization method was applied to anionic "living" polystyrenes [12-16], fitted at both chain ends with carbanionic sites. Potential electrophilic coupling agents include compounds such as phosgene, bis-(1.4-bromomethyl)benzene and dichloro-dimethyl-silane. The reactions are fast, quantitative and free of side reactions. In the case of phosgene, however, the cycles would contain a carbonyl function, which is sensitive to oxidative degradation. The two other coupling agents were used successfully. An advantage of the method is that the stoichiometry can be controlled visually, even at low concentration, as the carbanionic sites of the polystyrene are colored.

The experimental procedure involves first the synthesis of a bifunctional "living" polystyrene of known molecular weight, in the usual concentration range (5-10%), in tetrahydrofurane solution, using potassium naphthenide as the initiator. The living polymer is stepwise sampled out, and transferred into another reactor where it is diluted to 1/100 of its initial concentration. A dilute benzene solution of the coupling agent is then added dropwise until the coloration of the solution just vanishes. Next, another volume of the living polymer solution is transferred, and so on. All these operations are carried out under inert atmosphere (or under high vacuum), using highly purified reagents and solvents. Efficient stirring is essential to prevent local excess of reagents.

The raw reaction product contains both the cyclic macromolecules and the "polycondensate" resulting from the chain extension process (Fig 1). Advantage is taken of the large difference in molecular weight between the two species, to have them separated efficiently by fractionate precipitation. Each fraction is characterized by SEC and by light scattering, to check the accuracy of the fractionation process. This is an important point, as contamination of the cyclic samples with some linear constituents would badly affect the comparisons to be made between linear and cyclic homologues. The limit of accuracy of these checks is better than 5%.

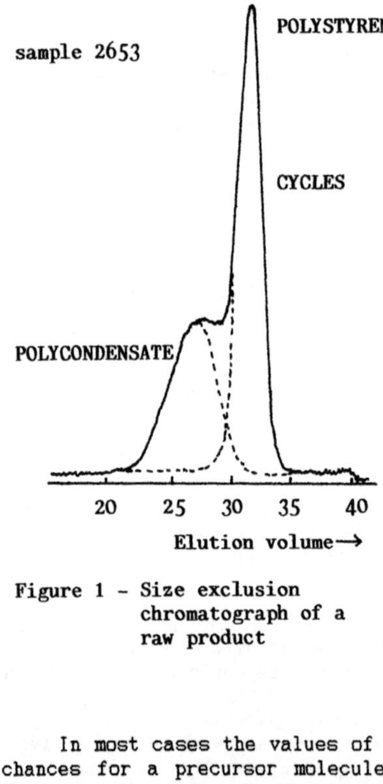

Figure 1 - Size exclusion chromatograph of a raw product

The cyclization yields generally range from 25 to 55%, and are somewhat lower than expected. This may be attributed to the fact that if a given precursor molecule is end-capped at both chain ends with two different coupling agent molecules, it is lost for the cyclization : it can only be incorporated into a "polycondensate". It is therefore very important to have the reactor fitted with efficient stirring, to prevent local excess of coupling agent during the reaction.

It may also happen that accidental deactivation occurs, especially when high molecular weight cycles are searched for. The overall probability of reaction p (whether intramolecular or intermolecular) can be evaluated from the cyclization yield x (measured by SEC on the raw product) and the degree of chain extension DP_n attained in the polycondensate (resulting from molecular weight determinations on the linear fractions). The expression used to calculate p is given below [17] :

$$p = x + (1 - x) (1 - 1/DP_n)^{1/2}$$

In most cases the values of p are close to 90%, which implies that the chances for a precursor molecule to be protonated at both chain ends are less than 1%, except in the upper range of molecular weights.

The possibility of formation of cyclic macromolecules having twice the molecular weight of the precursor has been questioned. Never has any measurable amount of such species been isolated, or even detected. This result is not very surprising, as the formation of larger cyclic molecules would imply inter- and intramolecular reactions occurring successively. The same explanation can be taken into consideration to account for the absence of interpenetrating cycles.

The molecular weights experimentally accessible for cyclic polystyrene range from 10,000 to approximately 300,000. When the molecular weight of the precursor reaches very high values the molar concentration of the active sites becomes very low (less than 10^{-5} mole/l), and the chances for accidental deactivation can no more be neglected. Moreover, as the molecular weight of the precursor increases, the magnitude of chain extension gets smaller. Consequently the separation of the cycles from the raw product - by fractionate precipitation - becomes more difficult. Roovers [15] has attempted to overcome this difficulty by a method of his own, but the limit of accessible molecular weights is not shifted to much higher values.

SOLUTION BEHAVIOR OF MACROCYCLIC POLYMERS

An interesting feature of the end-to-end cyclization method is that the linear homologue of a cyclic polymer is nothing else but the corresponding precursor polymer, after it has been deactivated protonically. Accurate comparisons of the solution behavior between cyclic and linear polymers of same molecular weight are thus facilitated. The main goal of the characterization measurements was to get access to the hydrodynamic dimensions of the polymers, in θ solvents as well as in good solvents, and to compare the results with the theoretical expectations issued many years ago.

The cyclic polystyrene fractions have been submitted to measurements of their radii of gyration (by small angle neutron scattering), of their limiting viscosity numbers, and of their SEC retention volumes [18-20]. Some results are presented in tables I and II.

TABLE I . Molecular weights and limiting viscosity numbers of cyclic (C) and linear (L) polystyrenes

Samples		M_w (LS)	M_w/M_η (SEC)	$[\eta]$ ml/g cyclohexane 35°C	$[\eta]$ ml/g THF 25°C	$[\eta]_c / [\eta]_L$ cyclohexane 35°C
2653	C	11,100	1.09	5.55	7.0	0.68
	L	10,500	1.15	8.20	10.3	
2857	C	19,400	1.16	7.55	10.1	0.68
	L	19,400	-	11.1	14.1	
2648	C	27,000	-	8.7	-	0.66
	L	37,000	1.10	13.2	19.5	
2654	C	37,800	1.13	9.9	-	0.64
	L	37,800	-	15.6	22.4	
2774/7	C	79,400	1.12	15.9	-	0.70
	L	79,500	1.10	22.6	38.6	
2774/8	C	73,000	-	-	24.2	
2829	C	106,000	-	17.8	35.5	0.68
	L	107,000	1.16	26.0	46.4	
2780	C	181,500	1.10	23.5	53.0	0.69
	L	182,000	1.08	34.1	67.3	
2881/9	C	211,000	1.15	-	-	-
	L	217,000	1.20	-	-	

<u>Radii of gyration</u> (Fig 2). Calculations of Cassasa [3] have predicted that under θ conditions (in the absence of excluded volume), the mean square radius of gyration of a macrocycle should be exactely half that of its linear counter-part. This result is generally supported by the experimental data obtained [18-20]. However, account has to be taken of the fact that the θ temperature is lowered by several degrees for cyclic polymers [18-19], an experimental fact that is not well accounted for at the present time.

TABLE II. Radii of gyration of cyclic polystyrenes

Samples	M_w (LS)	M_w / M_n (SEC)	R_L[Å] linear (a)	R_c[Å] cyclic(b)	R_c^2 / R_L^2
2857	19,400	1.16	40.4	30.3	0.56
2654	37,800	1.13	56.4	39.0	0.48
2774/8	73,000	-	78.3	56.0	0.51
2829	106,000	-	94.4	66.0	0.49
2780	181,500	1.10	123.5	85.0	0.47
2881/9	211,000	1.15	133.2	91.0	0.47

(a) interpolated from the plot of R versus M for linear polystyrene at the actual θ-conditions [23].
(b) measured in perdeutero cyclohexane at the actual θ temperature.

Limiting viscosity numbers. (Fig 3). Theoretical work has predicted that under θ conditions the [η] value of a cyclic polymer should be approximately 2/3 of that of the corresponding linear polymer. The measurements carried out on linear and cyclic polystyrene samples in cyclohexane at 35° are in good agreement with this expectation. It should be observed, however, that at this temperature the cycles are already slightly expanded. This is shown by the slope of the diagram representing [η] versus M in logarithmic scale. There are still some controversies about this point [18-19].

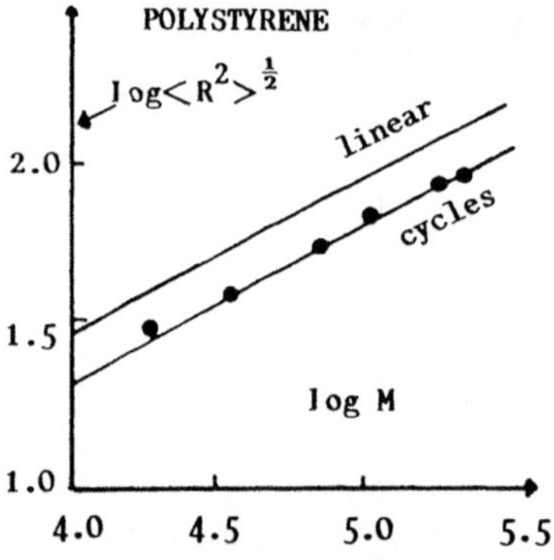

Figure 2 - Radii of gyration (cyclohexane)

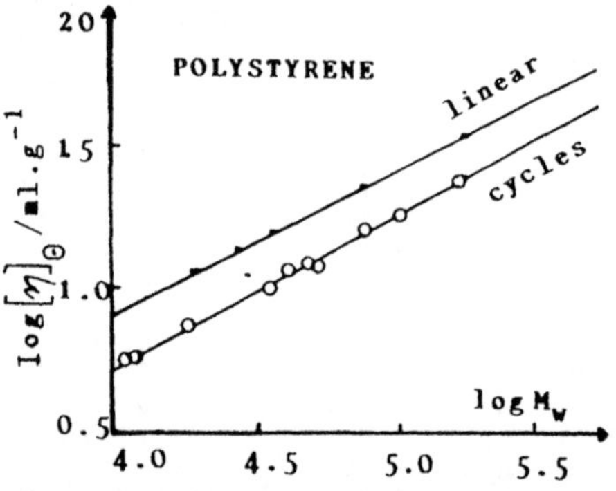

Figure 3 - Limiting viscosity numbers - cyclohexane -34.5°C

The Stockmayer-Fixman-Burchard plot [21], that is known to give access to the unperturbed dimensions from measurements carried out in good solvents, was also applied and led to values quite compatible with those quoted above (Fig 4).

It is noteworthy that the effect of cyclization does not increase much upon switching from a θ solvent to a good solvent. The ratio of the limiting viscosity numbers for cyclic and linear polymers of same molecular weight is found only slightly higher than the value at θ-conditions (0.7 instead of 0.66).

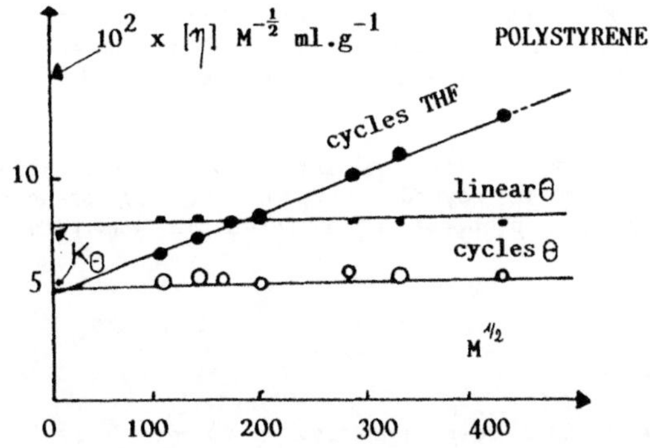

Figure 4 - Stockmayer-Fixman-Burchard plots for linear and cyclic polystyrenes in THF and in cyclohexane

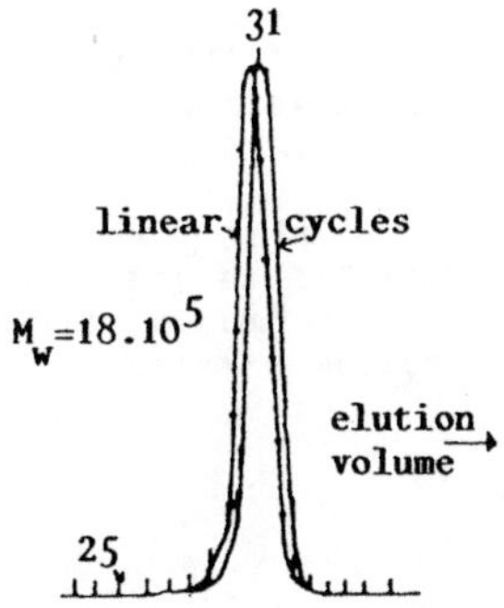

Figure 5 - SEC curves

Size exclusion chromatography. The SEC chromatograms of ring and linear polymers of same molecular weight do not exactly overlap (Fig 5), even when the fractionation has been accurate. This is again a consequence of the smaller hydrodynamic volume of the cycles [18-19]. The apparent molecular weight M_{ap} obtained by SEC from the calibration curve set up with linear polystyrene is found too small by approximately 22%.

To account for this difference, it should be remembered that the factor that determines chromatographic retention is the hydrodynamic volume of the solute molecules, which can be characterized by the product [η]*M.

It follows that the ratio of the apparent molecular weight to its true value is given by :

$$\frac{M_{ap}}{M} = \left(\frac{[\eta_c]}{[\eta_L]}\right)^{\frac{1}{1+a}}$$

Knowing the exponent a of the viscosity - molecular weight relationship of the polymer in the elution solvent (THF), and the ratio of the viscosities for cycles and their linear homologues (0.66), one can calculate the ratio of the apparent molecular weight to its true value. The results fit well with the experimental data.

CONCLUSION

Anionic "living" polymerization is quite helpful to achieve end-to-end cyclization reactions on αω-dicarbanionic polystyrene. Adequate fractionation techniques are required to quantitatively separate the cyclic constituents from the "polycondensate" formed simultaneously. Very accurate characterization of the fractions obtained has demonstrated that the cycles are generally free of any linear contaminants. Until now, all the efforts were focused on the synthesis of macrocyclic polystyrenes. The extension of the method to other polymers is presently under investigation.

Well defined samples of cyclic polymer made it possible to check the theoretical predictions issued 30 years ago, concerning the hydrodynamic dimensions of the species in θ solvents as well as in good solvents. The results obtained are generally in satisfactory agreement with the expectations.

The bulk properties of these cyclic polymers [22] are also investigated in great detail. This is beyond the scope of the present article, primarily devoted to the problems encountered in end-to-end cyclization processes.

The authors wish to express their deep appreciation to Mrs Graff, to Mrs Mottin, to Mr Lamps and Mr Roth for their technical assistance, and to Drs A.J. Kovacs, G.B. McKenna and G. Hadziioannou for stimulating discussions.

REFERENCES

1 - B.H. Zimm, W.H. Stockmayer, J. Chem. Phys. 17 1301 (1949)
2 - V. Bloomfield, B.H. Zimm, J. Chem. Phys. 44 315 (1966)
3 - E. Cassasa, J. Polym. Sci. A 3 605 (1965)
4 - C.G. Berry, T.G. Fox, Adv. Polym. Sci. 5 261 (1968)
5 - P.G. De Gennes, J. Chem. Phys. 55 572 (1971)
6 - H. Jacobson, W.H. Stockmayer, J. Chem. Phys. 18 1600, 1607 (1950)
7 - F.R. Jones, L.E. Scales, J.A. Semlyen, Polymer 15 738 (1974)
8 - J.A. Semlyen, Adv. Polym. Sci. 21 43 (1976)
9 - K. Dogson, J.A. Semlyen, Polymer 18 1265 (1977)
10 - K. Dogson, D. Sympson, J.A. Semlyen, Polymer 19 1285 (1978)
11 - J.S. Higgins, K. Dogson, J.A. Semlyen, Polymer 20 553 (1980)
12 - G. Hild, A. Kohler, P. Rempp, Europ. Polym. J. 16 525 (1980)
13 - G. Hild, C. Strazielle, P. Rempp, Europ. Polym. J. 19 721 (1983)
14 - D. Geiser, H. Höcker, Macromolecules 13 653 (1980)
15 - J. Roovers, P.M. Toporowski, Macromolecules 16 843 (1983)
16 - B. Vollmert, J.X. Huang,
Makromol. Chem. Rap. Comm. 1 333 (1980) ; 2 467 (1981)
17 - Y. Gnanou, Ph. Chaumont, P. Lutz, G. Hild, P. Rempp
Makromol. Chem. 185 2647 (1984)
18 - P. Lutz, G.B. McKenna, P. Rempp, C. Strazielle
Makromol. Chem. Rap. Comm. in press
19 - J. Roovers, J. Polym. Sci. (Phys. Ed). 23 1117 (1985)
20 - M. Ragnetti, D. Geiser, H. Höcker, R. C. Oberthür,
Makromol. Chem. 186 1701 (1985)
21 - W.H. Stockmayer, M. Fixman, J. Polym. Sci. C 1 137 (1963)
W. Burchard, Makromol. Chem. 50 20 (1961)
22 - G. B. McKenna, G. Hadziioannou et al.
Macromolecules, in press
23 - J.P. Cotton, J. Physique 41 231 (1980)

ANIONIC POLYMERIZATIONS OF 2-ISOPROPENYLNAPHTHALENE

H. L. Hsieh and I. W. Wang
Research and Development
Phillips Petroleum Company
Bartlesville, Oklahoma 74004

ABSTRACT

The anionic polymerizations of 2-isopropenylnaphthalene (IPN) with butyllithium in toluene are described. Because of its low ceiling temperature at 23.9°C, high monomer concentration, low polymerization temperature at 5°C, and the addition of THF or dimethoxyethane as solvating agents, were utilized in order to increase the polymer yield. The copolymerization of IPN and styrene can lead to higher conversion, partially because the styrene comonomer can act as "in-situ" capping agent to prevent the depolymerization of IPN sequence in the backbone of S/IPN copolymer. The Tg of the poly(S/IPN) (50/50 wt ratio) segment was found to be 130°C which is much lower than the Tg of 221°C reported for poly(IPN) at infinite, molecular weight. The Tg of IPN segment in the IPN-B diblock copolymer of this study was found to be about 200°C. Some mechanical properties were also measured for these diblock co- and terpolymers.

INTRODUCTION

The polymerization of α-substituted styrenes generally exhibits a low ceiling temperature, *i.e.*, they undergo equilibrium polymerization except at very low temperatures. The anionic equilibrium polymerization, for example, of α-methylstyrene has been studied extensively,[1-4] and its absolute ceiling temperature has been found to be 54°C.[5] Among the analogs, 2-isopropenylnaphthalene (IPN)

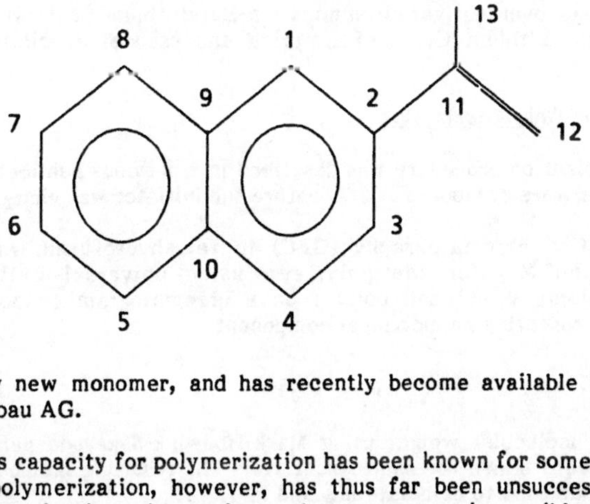

is a relatively new monomer, and has recently become available in high purity from the Bergbau AG.

The IPN's capacity for polymerization has been known for some years.[3,6,7] Its radical homopolymerization, however, has thus far been unsuccessful, although radical copolymerization with various other monomers is possible.[6,8,9] Cationic polymerization leads only to oligomers,[10,11] or product with a low degree of polymerization.[12]

Recent Advances in Anionic Polymerization
T.E. Hogen-Esch and J. Smid, Editors

We have studied in some detail the homo- and co-polymerization of IPN by anionic means, because the monomer in pure form was made available to us and the potential of the high Tg of this polymer was attractive. We also briefly examined the properties of the final products.

EXPERIMENTAL

Materials

The monomer, 2-isopropenylnaphthalene, was made available to us by Bergbau-Forschung GmbH, Essen-Kray. The elemental analysis for IPN: C, 92.8%; H, 7.2%; found: C, 92.61%; H, 7.03%, N, 0.00%; S, 0.49%. The ^1H NMR spectrum shows following signals: δ 7.0-8.0 (m. 7H, naphthyl); δ 5.55 (s. 1H, CH_2); δ 5.2 (1H, CH_2); δ 2.25 ppm (s. 3H, CH_3) and the ^{13}C NMR spectrum gives:

δ	143.0	C-11	δ	126.1	C-6
δ	138.4	C-2	δ	125.8	C-7
δ	133.4	C-9	δ	124.3	C-1
δ	132.8	C-10	δ	123.9	C-3
δ	128.2	C-4	δ	113.0	C-12
δ	127.7	C-8	δ	21.8	C-13
δ	127.5	C-5			

There was no indication of any impurity from these NMR results.

Since the quantity of monomer IPN was limited, it was used as received. The effective amount of n-BuLi initiator was the total charged subtracting the predetermined scavenger level. The scavenger level is the amount of n-BuLi destroyed by impurities including water in 100 g of IPN in a 1 liter solution in the polymerization bottle. Styrene, polymerization grade, was distilled twice over CaH_2 under reduced pressure. Phillips special purity butadiene was flashed first from its dimer or inhibitor before passing through the column of activated alumina.

The solvent employed, toluene and Phillips' polymerization grade cyclohexane, was dried by countercurrent scrubbing with prepurified nitrogen, followed by passage over activated alumina. n-Butyllithium (n-BuLi) in heptane was obtained from Lithium Corp. of America and used after diluting in dry cyclohexane.

Polymerization and Characterization

The polymerization procedure was described in a previous publication.[13] The monomer solutions were precooled at 5°C before the initiator was charged.

Gel permeation chromatography (GPC) in tetrahydrofuran was used to determine M_n and M_w for the polymers, using universal calibration.[14] Hydrodynamic volume V at each point i on a chromatogram or portion of a chromatogram representing an individual component

$$V_i = KM_i^{\alpha+1}$$

was converted to molecular weight using Mark-Houwink-Sakarada parameters K and α previously determined for each structure.[14] The resulting molecular weight distribution was then used to calculate M_w and M_n.

RESULTS AND DISCUSSION

A number of IPN-containing polymers were prepared by the method described in the experimental section. We found little characterization data of the poly(IPN) in the literature except for a few German reports mentioned earlier.[3,6,7,9-11] Hardy[9] cites 250-280°C for the melting range. The ceiling temperature has been determined by Hopff and Lüssi[3] as 23.9°C in THF. Additional data of the intrinsic viscosity/molecular weight relationship, isotactic triads content and decomposition temperature at 300°C have been reported by Engel and Schulz.[7]

Homopolymerization of IPN

Anionic homopolymerization of 2-isopropenylnaphthalene (IPN) was carried out in a hydrocarbon medium, using n-butyllithium as the initiator. Because of its low ceiling temperature[3] the polymerization temperature was thus set at 5°C. Toluene was the solvent used. Cyclohexane is a nonsolvent for poly(IPN) at ambient temperature. The presence of THF as cosolvent in trace amount which can be as low as 0.15 M in toluene is necessary, since otherwise the polymerization will not proceed.

The importance of THF as well as other factors such as monomer concentration and polymerization temperature with respect to the polymer conversion are shown in Table I.

TABLE I. Effects of Monomer Concentration, THF, and Temperature on the Anionic Homopolymerization of 2-Isopropenylnaphthalene in Toluene.

[IPN] M	[THF] M	[n-BuLi][a] M x 10^{-3}	Temp. °C	Time hr.	Conv.[b] %	$M_w \times 10^{-3}$ [c]	$M_n \times 10^{-3}$ [c]	M_w/M_n
2.76	0.32	9.9	5	17	87	47	41	1.13
2.76	0.32	6.1	25	17	21	19	16	1.19
5.16	0.60	2.0	5	17	91	-	-	-
2.83	0	1.0	5	1	0	-	-	-
2.80	0.16	7.8	5	1	81	56	49	1.13
2.76	0.32	7.2	5	1	85	64	55	1.18
2.69	0.63	3.1	5	1	86	167	127	1.32
2.56	1.21	8.2	5	1	88	55	46	1.20
2.23	2.63	5.8	5	1	91	69	59	1.17

a)Effective initiator.
b)Precipitation in isopropanol.
c)[η]/GPC universal calibration curve measurement.

It was found later that at 5°C and with THF of 0.32 M in toluene, the equilibrium poly(IPN) conversion in the range of 80-85% can be reached in less than one hour (bottom curve of Fig. 1). 1,2-Dimethoxyethane was also very effective (Table II).

FIG. 1 -CONVERSION VS. POLYMERIZATION TIME FOR (1) HOMOPOLYMERIZATION OF 2-ISOPROPENYLNAPHTHALENE (□) AND (2) COPOLYMERIZATION WITH EQUAL WEIGHT OF STYRENE (○), 5°C IN TOLUENE.

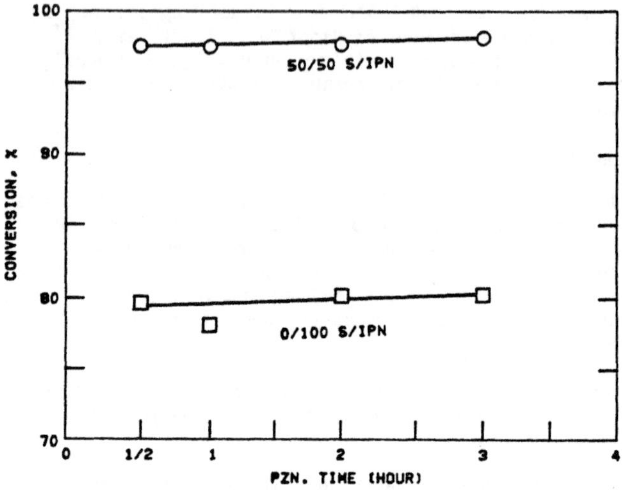

TABLE II. Effect of 1,2-Dimethoxyethane on the Anionic Polymerization of 2-Isopropenylnaphthalene in Toluene.[a]

[IPN] M	[DME] M	[n-BuLi][b] M x 10^{-3}	Conv.[c] %	M_w x 10^{-3} [d]	M_n x 10^{-3} [d]	M_w/M_n
2.80	0.13	7.7	83	60	51	1.18
2.76	0.25	8.6	83	55	45	1.23
2.68	0.50	7.5	83	76	50	1.52
2.55	0.95	8.3	89	68	46	1.48

a) At 5°C for 1 hour. b) Effective initiator. c) Coagulation in Cyclohexane
d) [η]/GPC universal calibration curve measurement.

In most cases, a molecular weight distribution <1.2 can be claimed (Fig. 2). It is, of course, the consequence of termination-free "living" polymerization and extremely rapid initiation. The living poly(IPN) anion shows its dark-green color immediately upon the addition of n-BuLI into the monomer solution.

Copolymerization of 2-Isopropenylnaphthalene with Styrene and/or Butadiene

Styrene in copolymerization with IPN can serve two purposes: (1) as a capping agent, which minimizes the depolymerization of IPN sequence in the backbone; and (2) as a comonomer to produce random copolymers with lower Tgs than the IPN homopolymer. The copolymer conversions, as seen in Table III and Figure 1, can be near quantitative. We were unable to verify each monomer content in the IPN/S copolymer by UV because of interference.

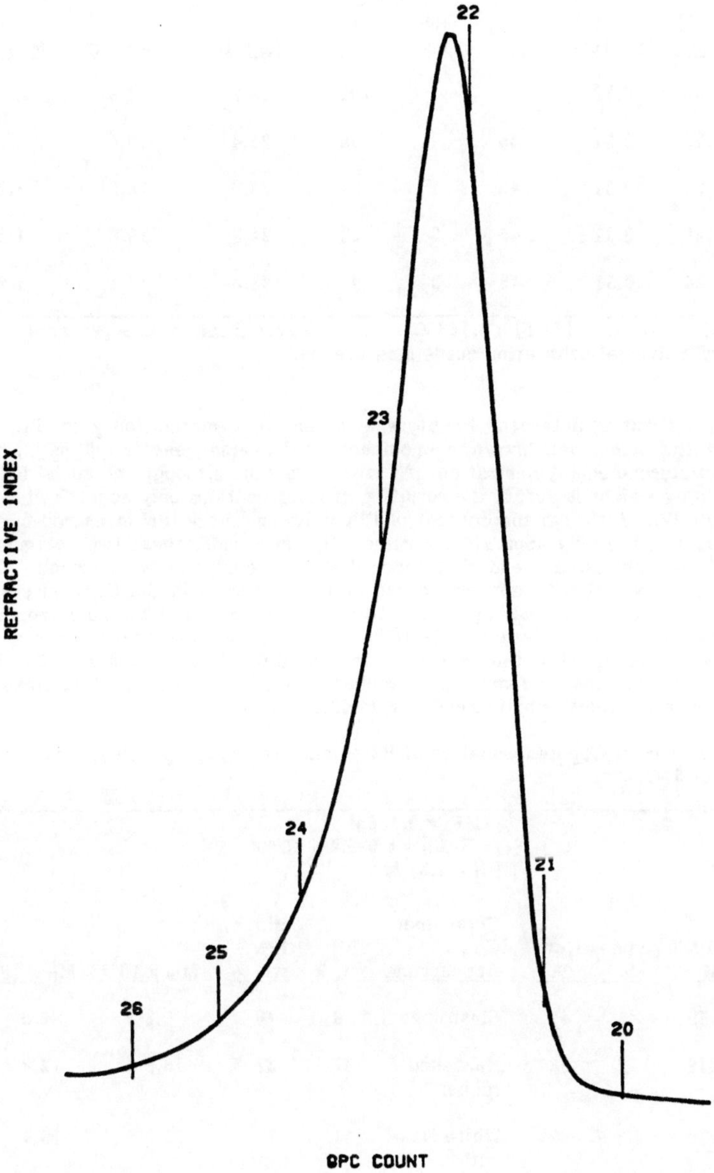

FIG. 2 -GPC CURVE OF POLY(IPN)

TABLE III. Direct Copolymerization of Styrene and 2-Isopropenylnaphthalene.

[IPN] M	[S] M	[n-BuLi][a] $M \times 10^{-3}$	Temp. °C	Time hr.	Copolymer Conv.[b] %	$M_w \times 10^{-3}$ [c]	$M_n \times 10^{-3}$ [c]	M_w/M_n
0	4.44	0.32	26	1	100	18.2	17.6	1.03
2.64	4.26	0.31	45	½	98	24.4	19.4	1.36
2.64	4.26	0.31	45	1	97	23.2	18.2	1.27
2.64	4.26	0.31	45	2	98	24.3	19.2	1.26
2.64	4.24	0.31	45	3	97	23.3	18.4	1.26

a) Effective initiator. [THF] = 0.156M. b) Precipitation in isopropanol.
c) [η]/GPC universal calibration curve measurement.

In substituting butadiene for styrene in the copolymerization with IPN, we observed the same well-known phenomenon of "inverse reactivity" as in the butadiene/styrene copolymerization. At low conversion, although the initial B/IPN feed ratio by weight is 50/50, the resulting product contains only about 9 wt % of IPN (Table IV). Although the content of IPN units and block IPN measured by the UV absorption, are only approximate values, the data under these two columns in Table IV do manifest a trend that tapered B/IPN copolymer was formed. The increase of poly IPN with conversion can also be observed in the GPC curves of Figure 3. The total polymer appears with the RI detector and the poly aromatic IPN appears with the UV detector. Note how the relative area under the cures (UV area/RI area) increased as the conversion increased (49-86%) in runs A to D. The resulting products change from a pure rubbery material to opaque white flake as copolymerization conversion increases from 49% to 86%.

TABLE IV. Direct Copolymerization of Butadiene and 2-Isopropenylnaphthalene in Toluene at 5°C.

[THF] = 0.156M
[n-BuLi] = 5.0-5.4 x 10^{-4}M
[B] = 4.32 M

Run	[IPN][a] M	Time hr.	Conv. %	Copolymer Conv. Appearance	IPN wt. %	Block IPN wt. %	$M_w \times 10^{-3}$	$M_n \times 10^{-3}$
A	1.39	1	49	Elastomer	9	6	9.3	9.0
B	1.39	2	68	Hardened rubber	37	27	13.8	12.8
C	1.39	3	82	White flake with elasticity	44	32	16.8	15.4
D	1.39	4	84	"	49	42	18.3	16.7
E	1.39	24	86	"	50	41	17.1	15.8

a) Weight ratio IPN/B = 1/1.

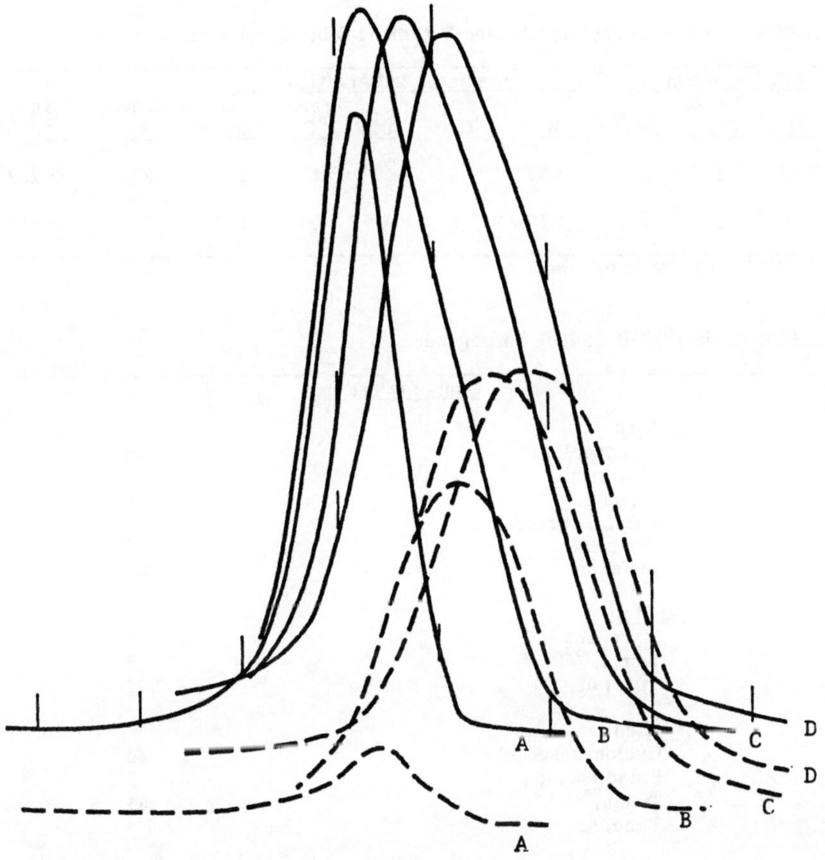

FIG.3-GPC CURVES FOR RUN A - D OF TABLE IV.
(————, RI), (- - - - - -, UV)

Attempts were made to prepare B-IPN diblock copolymer and (S/IPN)-B type diblock terpolymer (Tables V and VI). The GPC trace in Fig. 4 demonstrates that block polymerization can be carried out. They are not perfect due to the impurities in IPN. Transition temperatures measured via Rheovibron at 11 Hz on cured samples were 190-200°C for a poly(IPN) segment, and 130°C for a poly(S/IPN)(50/50 wt ratio) segment. The tensile properties of the (S/IPN)-B terpolymer (Dicup cured) are typical of diblock polymer with high tensile strength and elongation.

TABLE V. 2-Isopropenylnaphthalene-Butadiene Diblock Polymer.[a]

IPN, 1st Addition			Butadiene, 2nd Addition						
M	Temp. °C	hr.	M	Temp °C	hr.	Temp °C	hr.	Conv. %	Tg °C
1.39	5	1	4.32	5	1	50	1	83	-4; 202
1.39	5	1	4.32	5	1	50	1	77	-2; 190

a) [THF] = 0.156M in toluene.

TABLE VI. (S/IPN)-B Diblock Terpolymer.

Charge Order	
1st Step	
Toluene,[a] ml	10
IPN, g	5
Styrene, g	4
n-BuLi, mmole	0.3
Temp, °C	5
Time, hr.	0.5
2nd Step	
Styrene, g	1
Temp., °C	5
Time, hr	0.17
3rd Step	
Cyclohexane, ml	60
Butadiene, g	9
Temp.,°C	50
Time, hr.	1.0
Mechanical Properties[b]	
Tg, °C	
lower	-50
upper	130
E' x 10^9 @ 20°C, dyn/cm^2	7.5
Yield, MPa	12.9
Tensile, MPa	18.1
Elongation, %	42

a)[THF] = 0.156M b)Rheovibron measurement on cured sample

FIG. 4 - GPC OF 2-IPN-BD DIBLOCK COPOLYMER

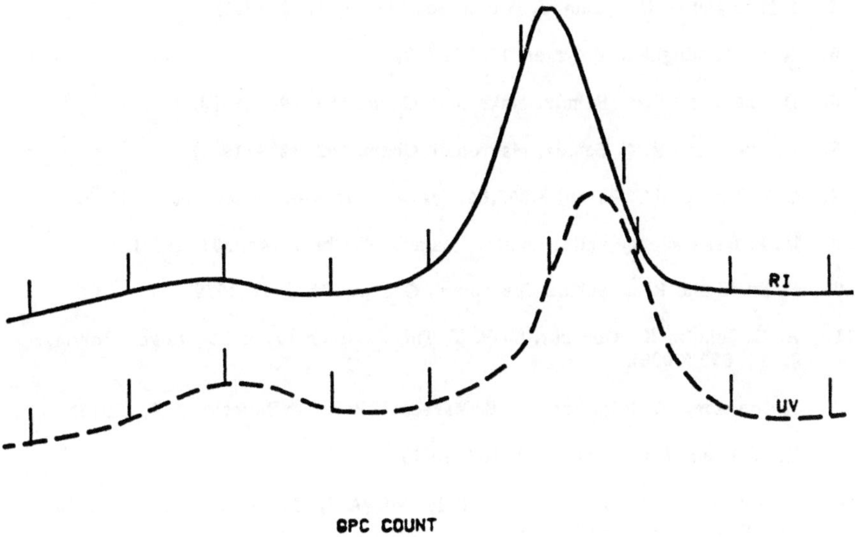

CONCLUSIONS

2-Isopropenyl naphthalene-containing polymers can be prepared anionically using n-BuLi as initiator in toluene. High IPN concentration, low polymerization temperature, and the presence of polar cosolvent are necessary to increase the polymer yield. The copolymerization of IPN and styrene can give quantitative conversion, while copolymerization of IPN and butadiene shows an "inverse reactivity" phenomenon and tapered copolymer was produced. The Tg's of the poly(IPN) segment in IPN-B diblock and of the poly(IPN/S) segment in the S/IPN)-B terpolymer were confirmed to be ~200°C and 130°C, respectively. The mechanical properties of the cured (S/IPN)-B diblock sample were as expected for diblock polymers.

Acknowledgements

We express our thanks to G. Kölling (Bergbau-Forschung GmbH, Essen-Kray) for providing the IPN monomers, and to Mr. K. W. Rollmann for the measurement of dynamic mechanical properties.

REFERENCES

1. D. J. Worsfold and S. J. Bywater, J. Polym. Sci., 26, 299 (1957).
2. A. Vrancken, J. Smid, and M. Szwarc, Trans. Faraday Soc., 58, 2036 (1962).
3. V. H. Hopff and H. Lüssi, Makromol Chem. 62, 31 (1963).
4. I. Mita and H. Okuyama, J. Polym. Sci., A-1, 9, 1983 (1971).
5. R. E. Cunningham, Polymer, 19, 729 (1978).
6. D. Engel and R. C. Schulz, Makromol. Chem. 180, 2987 (1978).
7. D. Engel and R. C. Schulz, Makromol. Chem. 182, 3279 (1981).
8. G. F. Hardy, U.S. Patent 3,202,642 (assigned to Allied Chem. corp.) 1965.
9. M. L. Hallensleben and I. Lumme, Makromol. Chem. 144, 201 (1971).
10. D. Engel and R. C. Schulz, Makromol. Chem., 180, 2991 (1979).
11. R. C. Schulz, K. Albrecht, Q. V. T. Thi, J. Nienburg, and D. Engel, Polymer J., 12, 639 (1980).
12. J. Coudane, M. Grigodiot, and E. Marechal, Polymer Bulletin 1, 465 (1979).
13. H. L. Hsieh, J. Poly. Sci. A-3, 163 (1965).
14. G. Kraus and C. J. Stacy, J. Poly Sci. A-2, 657 (1972); J. Poly. Sci. Symposium No. 43, 329 (1973).

ANIONIC POLYMERIZATION OF α-(TRIFLUOROMETHYL)ACRYLATE

Hiroshi Ito and Reinhold Schwalm
IBM Almaden Research Center
650 Harry Road
San Jose, California 95120

ABSTRACT

Methyl α-(trifluoromethyl)acrylate (MTFMA) readily undergoes anionic polymerization by initiation with organic and inorganic salts in the presence of 18-crown-6 to provide high molecular weight polymers in good yields. These weak anionic initiators are effective because of the low electron density (e=2.5) on the β-carbon of MTFMA. Amines are also effective initiators. Pyridine forms a very stable charge transfer complex with MTFMA. While wet 2,6-di-t-butylpyridine can initiate the polymerization, the hindered amine is not effective when dry, suggesting that the initiating species is a hydroxide in this case. However, dry pyridine gives rise to a vigorous polymerization with a color development, which suggests that pyridine itself is capable of initiating anionic polymerization of MTFMA. n-Butyllithium readily adds to the β-carbon of MTFMA but the adduct species further reacts with the initiator to introduce additional n-butyl group into the carbon bearing the fluorine atoms. Phenylmagnesium bromide and α -methylstyrene tetramer dianion fail to induce the anionic polymerization presumably due to the attack of the initiator onto the CF_3 group. The thermal degradation behavior of the polymers depends very much on the initiator employed, suggesting the occurrence of depolymerization from the initiator side of the chain.

INTRODUCTION

The polymerization behavior of methyl α -(trifluoromethyl)acrylate (MTFMA) has been studied by Ito et al. [1] and it has been reported on the basis of radical homopolymerization experiments and copolymerization analysis that this fluoromonomer is reluctant to undergo radical homopolymerization. The studies on the radical copolymerization with methyl methacrylate (MMA, M_2) have yielded the monomer reactivity ratios (r_1=0 and r_2=2.36) [1]. The relative reactivity of MTFMA toward a cyclohexyl radical has been determined by the "mercury method" [2], which has indicated that MTFMA is 53 times more reactive than MMA toward the electron-rich radical [3]. The combination of the copolymerization analysis and the mercury method has allowed the determination of Alfrey-Price Q-e values for MTFMA (Q=0.74 and e=2.5) [3]. Shortly after Ito's report, Iwatsuki et al. published their results of radical copolymerizations of MTFMA with three different comonomers including MMA [4]. The monomer reactivity ratios in the bulk copolymerization of MTFMA with MMA reported by Iwatsuki et al. (r_1=0.1 ± 0.1 and r_2=2.3 ± 0.2) agree very well with the values obtained by Ito et al. Iwatsuki et al. have estimated the Alfrey-Price Q-e values of MTFMA on the basis of Lussi's method [5] to be Q=0.8 and e=2.9. The Q and e values determined by the two different methods are in good agreement.

Owing to its very high e value, MTFMA is expected to undergo a facile anionic polymerization. This is evidenced by successful polymerization by pyridine initiation. However, typical anionic initiators for MMA were not as

Recent Advances in Anionic Polymerization
T.E. Hogen-Esch and J. Smid, Editors

effective as pyridine [1]. In order to understand the unique behavior of MTFMA in anionic polymerization, we have continued our studies on this monomer. This paper presents anionic polymerization of MTFMA with organic and inorganic salts as well as with amines and also discusses the complication observed in the n-butyllithium (nBuLi) initiation. Narita et al. have recently reported their preliminary results on anionic polymerization of ethyl α-(trifluoromethyl)acrylate [6].

EXPERIMENTAL

Materials

MTFMA was custom-synthesized by Parish Chemical according to the method previously described [1], fractionally distilled, and dried over molecular sieves 3A, phosphorus pentoxide (P_2O_5), or calcium hydride (CaH_2). Drying with CaH_2 occasionally resulted in polymerization especially upon long standing. Potassium t-butoxide was prepared by reaction of t-butanol and a potassium mirror in THF. t-Butanol used for the reaction was dried on sodium and distilled three times with the center-cut taken each time under high vacuum with use of a break-seal technique. Potassium benzoate was prepared by reaction of benzoic acid with potassium in toluene, and the precipitated product was filtered and washed with toluene. Inorganic salts were commercially obtained and used without further purification. Pyridines were distilled and dried over either molecular sieves 3A, KOH pellets, or CaH_2. In one experiment, both MTFMA and pyridine were titrated separately with nBuLi after drying over CaH_2 to establish a scrupulously dry condition. 18-Crown-6 (Aldrich) was purified by recrystallization from n-hexane.

Polymerization

Polymerizations were carried out under high vacuum with use of a break-seal technique and terminated by adding methanol. Polymers were purified by precipitation in methanol except for low molecular weight materials, which were precipitated in hexanes, and dried in vacuo at about 40°C.

Measurements

NMR spectra were recorded in deuteriochloroform ($CDCl_3$) unless otherwise indicated on Varian EM-390 (proton), IBM NR-80 (F-19), and Varian CFT-20 (proton and C-13) spectrometers. IR spectra were recorded on a Perkin-Elmer 983 infrared spectrometer. UV measurements were performed on a Hewlett-Packard 8450A UV/Vis spectrophotometer. Thermal analysis was performed on a Du Pont 1090 thermal analyzer at a heating rate of 5°C/min. Molecular weight determination was made by gel permeation chromatography with a Waters Model 150 chromatograph equipped with six μStyragel columns at 30°C in THF and the reported values are polystyrene-equivalent. Gas chromatographic analysis was performed on an OV-1 column using a Hewlett-Packard 5830A GC equipped with a 18850A data terminal. GC/MS analysis was carried out using Hewlett-Packard 5995A Gas Chromatograph/Mass Spectrometer.

RESULTS AND DISCUSSION

Organic and Inorganic Salt Initiators

Because organometallic initiators typically employed for anionic polymerization of MMA fail to provide high molecular weight polymers of MTFMA in a good yield [1], we tested several other weaker initiators as shown in Table I. All the polymerizations were carried out in bulk at room temperature unless otherwise indicated. MTFMA was mixed with an initiator and stirred for 3 hours without showing any evidence of polymerization. However, upon addition of an equimolar amount of 18-crown-6, a highly exothermic reaction immediately took place and the system solidified in 30 minutes in all the bulk polymerizations carried out with potassium salts as initiators. The solution polymerization with potassium t-butoxide/18-crown-6 in THF (THF/MTFMA=5/1 v/v) at -78°C provided the highest molecular weight (M_n=67,200 and M_w=147,600, relative to polystyrene as determined by GPC) in a very good yield of 92%. Although the polymerization was carried out for 25 hours for convenience, the system completely solidified in 15 minutes. As Table I indicates organic and inorganic potassium salts are excellent initiators for the anionic polymerization of MTFMA when used in conjunction with 18-crown-6. It is noteworthy that potassium carbonate containing 1.5 moles of water and potassium fluoride containing 2 moles of water produced high molecular weight polymers in good yields. The reaction was very fast in both cases. The polymerization system initiated with $KF•2H_2O$ solidified in 5 minutes.

Table I Bulk Anionic Polymerization of MTFMA with Organic and Inorganic Salts in the Presence of 18-Crown-6 at Room Temperature

Initiator (mole %)	Time (hr)	Yield (%)	Mn	Mw
KOH (2.0)	24	64	8,200	11,400
KCN (2.0)	24	82	11,600	21,300
$K_2CO_3·1½ H_2O$ (2.0)	24	84	13,300	20,200
K O_2C Ph (2.0)	24	70	11,700	16,200
K O Bu^t (2.0)	15	76	20,300	41,600
$KF·2H_2O$ (2.0)	17	50	19,200	52,400
K O Bu^t (2.0)/THF (-78°C)	25	92	67,200	147,600
Na O_2C CH_3 (2.0)	72	44	18,700	36,900
Na SCN (2.0)	113	43	14,700	21,600
LiF (2.0)	24	0		
Et_4NBr (2.0)[a]	8 weeks	81	45,200	120,400

a: without 18-crown-6.

Sodium acetate and sodium thiocyanate also initiate anionic polymerization of MTFMA in the presence of 18-crown-6 but are not as effective as potassium salts. The sodium acetate system took about 3 days to solidify and the NaSCN system started becoming viscous after 4 days but did not solidify even after 5 days.

The polymerization with tetraethylammonium bromide was very slow due to the low solubility of the salt in MTFMA. The mixture was, however, completely solidified upon long standing at room temperature and provided a high molecular weight polymer (M_n=45,200 and M_w=120,400) with a broad molecular weight distribution. MTFMA by itself gives only a trace of oligomeric or polymeric product when allowed to stand at room temperature for as long as one year.

Narita et al. have recently carried out anionic polymerization of ethyl α-(trifluoromethyl)acrylate with lithium and potassium t-butoxides as the initiator in toluene as well as in THF [6]. After running the reactions for 7 days, they recovered products by drying the organic phase after washing processes. They obtained a solid polymer in a quantitative yield by carrying out the polymerization with lithium t-butoxide as the initiator in toluene at -20°C. When potassium t-butoxide was used, the best result (60% yield) was obtained in THF at -20°C. Crown ethers were not used in their polymerizations.

As the C-13 NMR spectrum of MTFMA (Fig. 1) indicates, the β -carbon resonates at much lower field than that of MMA (133.2 ppm vs. 125.0 ppm), resulting from the low electron density on the β -carbon of MTFMA reflected by the high e value (2.5). Proton NMR also reflects the lower electron density of the MTFMA vinylic bond (βCH_2: 6.40 and 6.67 ppm vs. 5.49 and 6.04 ppm of MMA). Owing to this low electron density on the β-carbon, MTFMA readily undergoes addition reaction with weak nucleophiles shown in Table I. These weak initiators tend to minimize side reactions involving the ester functionality and the CF_3 group. Thus, in spite of the fact that the polymerizations were carried out at room temperature in bulk, the molecular weight distribution is in general fairly narrow, with the dispersity ranging typically from 1.3 to 2.0.

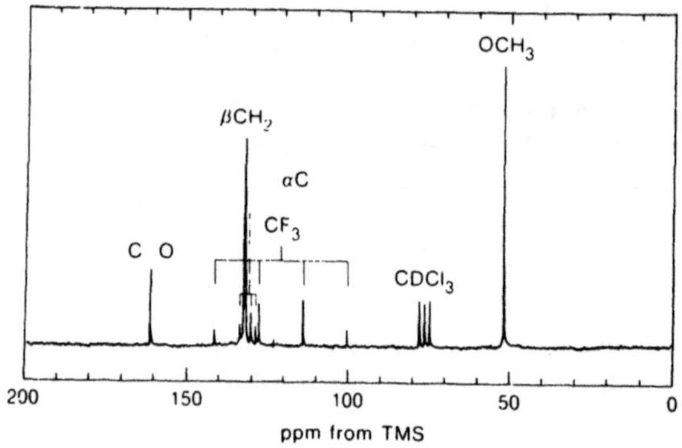

Fig. 1 20 MHz C-13 NMR Spectrum of MTFMA in $CDCl_3$.

Proton NMR studies of a mixture of MTFMA and potassium acetate (1:1 molar ratio) in CD_3OD indicated that after complete consumption of the monomer 80 % of potassium acetate had remained unreacted, which suggests that the propagation is much faster than the initiation in this system. In consequence, some oligomers were produced in the reaction. The slower initiation appears to be responsible for the molecular weight distribution greater than 1.3.

nBuLi Initiation

Ito et al. have reported that initiation by organometallic reagents such as nBuLi and phenylmagnesium bromide (PhMgBr) results in a very sluggish polymerization [1]. Narita et al. have recently observed a similar behavior in the anionic polymerization of ethyl α-(trifluoromethyl)acrylate [6]. In order to understand why nBuLi is not a good initiator, we analyzed products formed at room temperature in the reaction of MTFMA with nBuLi (0.2 mole% to MTFMA) by GC/MS. Because of the ester group participation, the reaction is quite complex. However, the GC/MS studies show that the n-butyl group adds to the β-carbon of MTFMA readily and that the adduct further reacts with nBuLi to introduce additional n-butyl group into the carbon bearing the fluorine atoms. We have synthesized a model compound of the adduct by reacting MTFMA with n-butylmercuric chloride [7] and treated the model compound with nBuLi. After metallization, the model compound undergoes the same sequence of reactions. When PMTFMA synthesized by pyridine initiation was treated with nBuLi, the F-19 NMR spectrum exhibited a slight change in the CF_3 region due to the reaction on the ester group but CF_2 and/or CF absorptions were not detected, indicating that the direct substitution reaction does not take place. The addition-elimination sequence in the reaction between organolithium compounds and fluoro-olefins has been known. For example, Feiring has reported that 1-(trifluoromethyl)ethenyl phenyl sulfide reacted at -70°C with nBuLi to give the difluoromethylene compound in 79% yield, which reacted at higher temperature with additional nBuLi, giving

5-fluoro-6-(phenylsulfenyl)undec-5-ene [8]. The lithium enolate, however, does not seem to eliminate LiF spontaneously, as Narita et al. [6] polymerized ethyl α-(trifluoromethyl)acrylate with lithium t-butoxide. It is speculated that nBuLi undergoes the substitution reaction onto the CF_3 group of the lithium enolate to eliminate LiF.

Pyridine Initiation

As previously reported [1], pyridine is a very good initiator for MTFMA. The polymerization carried out at room temperature with the use of 1.3 mole% of pyridine produced a high molecular weight polymer (M_n=21,300 and M_w=30,600) in 94 % yield (Table II). The polymerization system solidified in 15 minutes. The increase in the pyridine concentration to 12 mole% resulted in only a small reduction in the molecular weight (M_n=13,400 and M_w=25,700). Even when a drop of pyridine was added to MTFMA in an ampoule open to the atmosphere, a rapid polymerization took place, resulting in solidification in 15 minutes and production of a polymer with M_n=7,900 and M_w=12,100. As shown in Table II, a copolymer of 4-vinylpyridine (4VP) and styrene can nicely initiate the polymerization and triethylamine is also an effective initiator.

Table II Bulk Anionic Polymerization of MTFMA with Amine at Room Temperature

Initiator (mole %)	Time	Yield (%)	Mn	Mw
Pyridine (1.3) wet	24 hrs	94	21,300	30,600
P(4VP-co-ST) (0.14g)/THF		82	9,300	18,200
Crosslinked P4VP (0.1g)	3 weeks	85	10,100	28,000
Et_3N (2.0)		76	16,900	24,800
Pyridine (2.0) dry	4 hrs	70	60,300	109,900
2,6-Di-tButylpyridine (1.25) dry	3 weeks	0		
(1.25) wet	2 weeks	35	36,300	63,000

The polymer obtained by initiation with 1.3 mole% of pyridine did not show any proton NMR absorption due to the pyridine residue. In order to unequivocally know whether pyridine is attached to the polymer end, we carried out a polymerization of MTFMA with crosslinked poly(4-vinylpyridine) (P4VP) as an initiator. The reaction was slow because swelling of the crosslinked initiator in the monomer is required for effective initiation. The product was extracted with dichloromethane and precipitated in n-hexane. The soluble product amounted to 85 % of the MTFMA feed and the insoluble fraction isolated after the extraction did not show any significant weight increase (0.1 g → 0.101 g). This experiment clearly indicates that the majority of the PMTFMA chains thus formed is not attached to the crosslinked P4VP. Furthermore, when MTFMA dried over P_2O_5 and 2,6-di-t-butylpyridine dried over CaH_2 were mixed together, no polymerization occurred (3 weeks), whereas the hindered pyridine which had not been treated with a drying agent was capable of initiating the polymerization of MTFMA dried over molecular sieves 3A. Because of the bulky t-butyl groups located at the 2 and 6 positions, this amine can not add to or complex with electrophiles but can accept a proton to generate hydroxides, which are responsible for the initiation of the MTFMA polymerization in this case. The polymerization with wet 2,6-di-t-butylpyridine carried out for two weeks in an open vial at room temperature produced a high molecular weight polymer (M_n=36,300 and M_w=63,000) in 35 % yield. Since Saegusa et al. have shown that pyridine adds to the β -carbon of acrylic acid to form a zwitterionic betaine [9], it was expected that the nitrogen lone-pair of pyridine would directly add to the more electron-deficient β-carbon of MTFMA. However, the above experimental results seemed to indicate that the initiating species in the polymerization with pyridine was also pyridinium hydroxides instead that the polymerization was via a zwitterionic mechanism.

The addition of pyridine to MTFMA results in the development of a yellow color. In Fig. 2 is shown a UV spectrum of a 1:1 mixture of MTFMA and pyridine. The intense absorption extending to 500 nm may indicate a formation of a charge transfer complex. A mixture of MMA and pyridine remains colorless under the same conditions. GC/MS studies also suggest that MTFMA and pyridine forms a 1:1 adduct (or complex, MW=233) that is stable even above 180°C (abundance of mass 233.15=23-24 %). However, our attempts to isolate the adduct have not been successful yet. Proton NMR spectrum of MTFMA measured in pyridine shows that the vinylic CH_2 peaks become smaller upon standing as MTFMA is converted to oligomers, indicating that the propagation is faster than the initiation. Thus, the charge transfer complex has not been detected by NMR presumably due to the low concentration. 2,6-Di-t-butylpyridine does not induce the color development when mixed with MTFMA because the sterically hindered amine can not form such a complex with MTFMA.

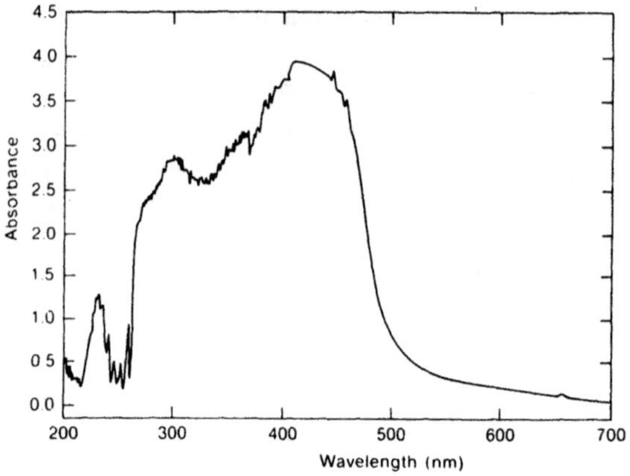

Fig. 2 UV Spectrum of MTFMA/Pyridine Mixture.

In one polymerization experiment, both MTFMA and pyridine were titrated separately with nBuLi after drying over CaH_2, and then distilled into the polymerization ampoule. It was assumed that nBuLi should react with residual water faster than with MTFMA and pyridine. Within 5 minutes after mixing in a tube sealed under high vacuum, a highly exothermic polymerization took place with development of a yellow color and increase in viscosity. The system solidified in 10 minutes. This apparently indicates that pyridine is capable of initiating anionic polymerization of MTFMA under "anhydrous" conditions. It has been reported that the anionic ring-opening polymerization of β-propiolactone with tertiary amines is via a zwitterionic mechanism under anhydrous conditions and is initiated by ammonium hydroxide in the presence of a trace of water [10]. Our experimental results appeared to be consistent with this report. However, the polymerization with "wet" pyridine is as fast as that with "anhydrous" pyridine, but much faster than that with "wet" 2,6-di-t-butylpyridine, which seems to suggest that pyridine itself is a more effective initiator than pyridinium hydroxide and that pyridine itself may be responsible for initiating polymerization of MTFMA even under "wet" conditions. Since PMTFMA is not attached to the crosslinked P4VP used as the initiator and PMTFMA obtained by "wet" pyridine does not show proton NMR signals due to the pyridinium residue, the zwitterionic mechanism is not likely to be the major pathway in these polymerizations. The most likely mechanism would involve one-electron transfer from pyridine to MTFMA through the charge transfer to form a pyridine radical cation (Scheme I). Detection of pyridine radical cations by spin-trapping has been reported by Ledwith and Russell in the oxidation of pyridine with potassium peroxydisulphate [11].

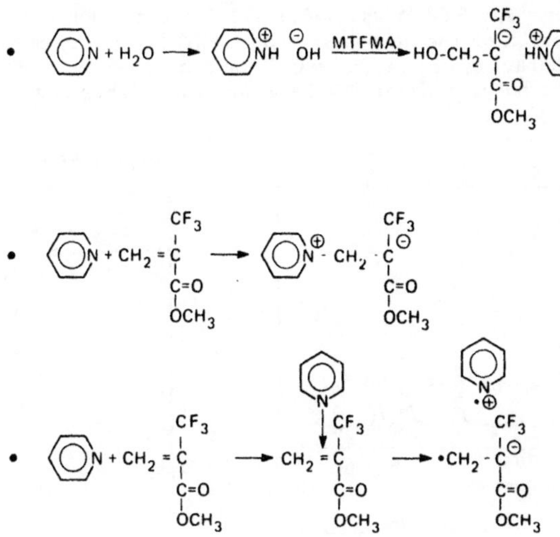

Scheme I Initiation Mechanism

Because MTFMA undergoes a rapid polymerization even in an open ampoule, we investigated the effect of water (Table III). Addition of water to the MTFMA/pyridine system results in reduction of molecular weight. However, the effect is surprisingly small and the molecular weight distribution remains quite narrow, ranging from 1.3 to 1.6. The polymer yield is not very much affected by the addition of water. These observations seem to indicate that water does not inhibit the polymerization.

Table III Bulk Anionic Polymerization of MTFMA with Pyridine in the Presence of H_2O at Room Temperature

Pyridine mole %/H_2O mole %	Time	Yield (%)	Mn	Mw
12.0/0	24 hr	91[a]	13,423	25,689
5/5	12 hr	76[b]	1,840	2,610
10/10	12 hr	56[b]	1,230	1,650
20/20	12 hr	41[b]	1,020	1,320
5/1	40 min	66[a]	6,370	10,230
5/2	45 min	28[a]	5,950	8,370
5/3	60 min	76[b]	2,733	4,300
5/4	70 min	64[b]	2,460	3,900

a: precipitated in MeOH
b: precipitated in Hexane

Polymer Characterization

C-13 NMR spectrum of PMTFMA obtained by pyridine initiation is provided as Fig. 3. The 20 MHz spectrum measured in $CDCl_3$ at room temperature did not exhibit splittings due to tacticity. One interesting feature in Fig. 3 is that the carbonyl group of PMTFMA resonates at about 10 ppm higher field than that of PMMA (166.1 ppm for PMTFMA vs. 176.7-178.1 ppm for PMMA). This indicates that the polarization of the carbonyl bond in PMTFMA is very much suppressed by the electron withdrawing CF_3 group, which is confirmed by IR studies as shown in Fig. 4. The carbonyl band of PMTFMA absorbs at higher wavenumber than that of PMMA (1751 vs. 1725 cm^{-1}).

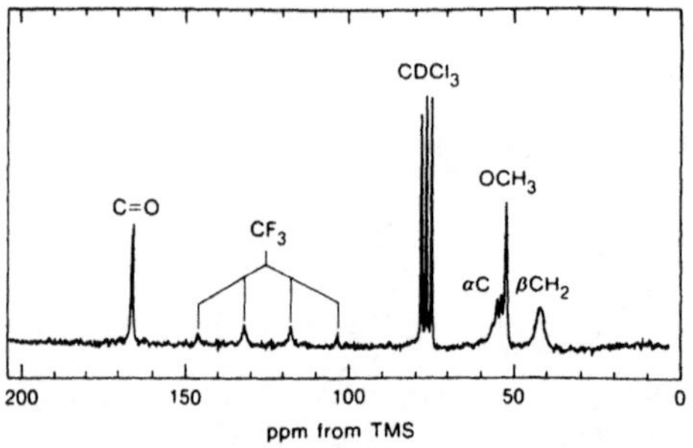

Fig. 3 20 MHz- C-13 NMR Spectrum of PMTFMA Obtained by Pyridine Initiation (in $CDCl_3$).

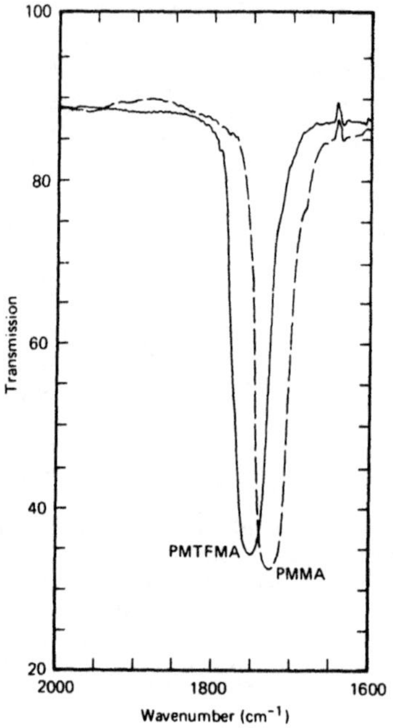

Fig. 4 IR Carbonyl Bands of PMTFMA and PMMA Films.

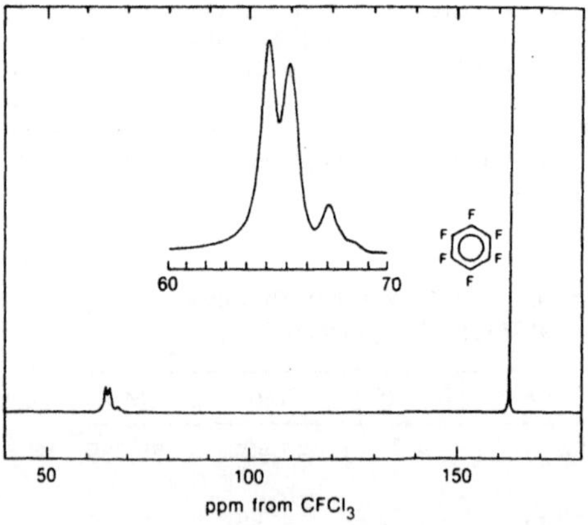

Fig. 5 73 MHz F-19 NMR Spectrum of PMTFMA Obtained by Pyridine Initiation (in hexafluorobenzene).

As demonstrated in Fig. 5, the F-19 NMR of PMTFMA is indicative of tacticity. The CF_3 resonance shows well-resolved peaks probably assignable to triad tacticities (pentad sequences may have to be considered as a small fourth peak is observed at 68.5 ppm). The polymer made by pyridine initiation in bulk at room temperature seems to be rich in syndiotactic and heterotactic contents and poor in isotactic triad. We are now engaged in more systematic tacticity studies using high field C-13 and proton NMR as well as F-19 NMR and hope to make the tentative assignment more unequivocal.

The glass transition temperature of PMTFMA obtained by "wet" and "dry" pyridine and by "wet" 2,6-di-t-butylpyridine is about 95°C as determined by DSC. As shown in Fig. 6, the thermal decomposition behavior significantly differs from one sample to another. The dependence of the PMTFMA degradation on the mode of preparation seems to indicate that some of the end groups derived from initiators are detached from the PMTFMA chain at about 140°C, which then results in depolymerization from the initiator side of the chain. A random main chain scission occurs at about 300°C. The polymers obtained with potassium cyanide and 2,6-di-t-butylpyridinium hydroxide show a similar thermal stability and do not lose much weight at 280°C. In contrast, the polymer prepared by K_2CO_3 initiation loses 40 % of its weight by the time the temperature reaches 280°C, presumably because the decarboxylation of the thermally unstable carbonate group triggers the depolymerization. PMTFMA obtained with Et_3N as the initiator rapidly degrades, losing 50 % of its weight below 220°C. The degradation behavior of the polymers obtained with "dry" and "wet" pyridine does not differ much from each other, but exhibits the two-stage weight loss, though to much lesser extents than in the case of the Et_3N initiation. It has been known that the stability of polymers are dependent on the method of preparation, particularly on the initiator used [11]. PMMA prepared by free radical polymerization decomposes in two stages; the first stage is due to depolymerization initiated at the unsaturated chain ends and the second stage due to depolymerization initiated by random chain scission [12]. Although it may be premature to draw a conclusion from the TGA studies of PMTFMA, the thermal degradation behavior of the polymer seems to provide some information with regard to the mechanism of initiation by amines. It would be safe to assume that the polymerization with "wet" 2,6-di-t-butylpyridine is initiated by hydroxide anions as discussed earlier. We believe that the initiation by pyridine is primarily by the one-electron transfer. However, a small amount of polymer may be produced via the zwitterionic mechanism, which may be responsible for the weight loss below 240°C. In the polymerization initiated by Et_3N, the primary structure of the growing chain may be zwitterionic. The quarternary ammonium ends would be cleaved off the polymer chain upon heating to ca. 140°C, triggering depolymerization.

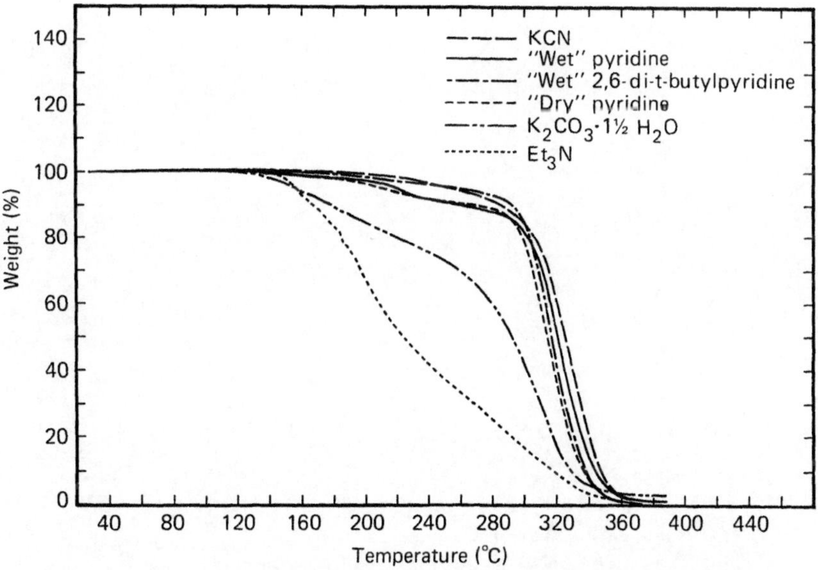

Fig. 6 TGA of PMTFMA Obtained with Various Initiators.

ACKNOWLEDGMENT

The authors gratefully thank W. Fleming and C. Hrusa for NMR measurements, C. Cole for GPC work, R. Siemens and E. Hadziioannou for thermal analysis, and R. Herbold for GC/MS analysis.

REFERENCES

1. H. Ito, D. C. Miller, and C. G. Willson, Macromolecules, 15, 915 (1982).
2. B. Giese and J. Meister, Angew. Chem. Int. Ed. Engl. 16, 178 (1977).
3. H. Ito, B. Giese, and R. Engelbrecht, Macromolecules, 17, 2204 (1984).
4. S. Iwatsuki, A. Kondo, and H. Harashina, Macromolecules, 17, 2473 (1984).
5. H. Lussi, Makromol. Chem. 103, 68 (1967).
6. T. Narita, T. Hagiwara, H. Hamana, and T. Nara, Makromol. Chem., Rapid Commun. 6, 301 (1985).
7. n-Butylmercuric chloride was synthesized by reaction of n-butylmagnesium bromide with mercuric chloride and then reacted with MTFMA in the presence of sodium borohydride (ref. 2).
8. A. E. Feiring, J. Org. Chem. 45, 1962 (1980).
9. T. Saegusa, S. Kobayashi, and Y. Kimura, Macromolecules, 7, 139 (1974).
10. V. Jaacks and N. Mathes, Makromol. Chem. 131, 295 (1970).
11. A. Ledwith and P. J. Russell, J. Chem. Soc., Perkin II, 582 (1974).
12. for example, G. G. Cameron, J. M. Meyer, and I. T. McWalter, Macromolecules, 11, 696 (1978).
13. H. H. Jelinek, "Aspects of Degradation and Stabilization of Polymers," Elesevier, Amsterdam, 1978.

SYNTHESIS AND CHARACTERIZATION OF POLYSULPHIDE POLYMERS

Jean Marie CATALA, Jean Marc PUJOL and Jean BROSSAS
Institut Charles Sadron (CRM-EAHP)(CNRS-ULP)
6,rue Boussingault, 67083 Strasbourg Cedex, France

INTRODUCTION

The dihydroxy- or dicarboxy telechelic polyethers or polyesters are very well-known reactive polymers. Another type concerns the polyalkyl-polysulphides or polyalkyl-polyether polysulphide with thiol end groups. These thiol functions can be used in condensation reaction with isocyanate, or in addition reaction on double bonds in the presence of a radical initiator; they can be coupled also with oxydizers to give chain extension.

It was interesting for us to combine the advantage of thiol functions and the presence of sulphur atoms in the main chain with those of dienic or vinylic blocks. With this aim in mind, we have prepared new dithiol telechelic polysulphide polymers. We have previously studied in a homogeneous phase the reaction of carbanions with elemental sulphur S_8 |1|. The reaction proceeds by a nucleophilic attack of the cyclic S_8 to give polysulphide polymer |2|. However, in the homogeneous phase,in order to obtain high sulphur content polymers, it is necessary to prepare low molecular weight dicarbanionic species before deactivation on sulphur. This synthesis implies the use of a large amount of initiator such as biphenyl sodium which is then difficult to remove from the resulting polymers. For this reason, we studied a one step process in which initiation of the monomer by an alkali metal, propagation and deactivation on S_8 take place simultaneously |3|.

SYNTHESIS DESCRIPTION

The synthesis is carried out with an alkali metal in tetrahydrofuran (THF) in the presence of elemental sulphur S_8. The 1.3-butadiene monomer (or another one) is added dropwise into the initial mixture. After 4 hours under stirring, the solution is filtered and poured into an acidic solution. The polymer is obtained after standard treatments.

MECHANISM

The mechanism proposed in the heterogeneous phase is the following:

- initiation: the alkali metal transfers its electron to the monomer to produce a radical anion which dimerizes. The dicarbanionic dimer can propagate or attack the elemental sulphur.
- propagation:

$$^{-}MM^{-} + M \xrightarrow{kp} {}^{-}MMM^{-}$$

- polycondensation on sulphur:

$$p\ {}^{-}MM^{-} + p\ S_8 \xrightarrow{kd} [MM - S_x]_p + p\ S_{8-x}^{-2}$$

The repeating unit is the following:

$$[M_n - S_x]_p$$

Recent Advances in Anionic Polymerization
T.E. Hogen-Esch and J. Smid, Editors

POLYSULPHIDE-POLYBUTADIENE

Characterization of the polymer

Determination of 1.2 and 1.4 structures in the repeating unit

The amount of 1.2 structure is determined by H.NMR, using the ratio of the integration at 1.2 ppm (a) corresponding to the protons on saturated carbons to the total integration (T):

$$(\%\ 1.2) = 3\ a/T \qquad (\%\ 1.4) = 1 - 3\ a/T$$

Determination of the sulphur rank : x

Some model molecules were prepared and the chemical shifts of protons on the carbon linked to the sulphur atoms were determined. The chemical shifts of these protons are an increasing function of the number of sulphur atoms in the polysulphide bridge:

Sulphur rank x	ppm: chemical shifts
1	3.05
2	3.28
3	3.47
4	3.56
5	3.65

From the integration of the different peaks of these protons it is possible to determine:

. the fraction of each polysulphide type bridge:

a_1 : fraction of monosulphide, a_2 : fraction of disulphide, etc...

. the average number of the sulphur rank:

$$x = a_1 + 2\ a_2 + 3\ a_3 + 4\ a_4$$

Determination of n and p : $[M_n - S_x]_p$

The sulphur content in the polymer allows us to determine the average organic chain length: n

$$(\%\ S) = \frac{x\ .\ 32}{x\ .\ 32 + n\ .\ 54} \quad \text{and} \quad n = \frac{32\ .\ x\ .\ [1 - (\%\ S)]}{54\ (\%\ S)}$$

From x, n and $\bar{M}_n$ values the degree of condensation index can be calculated:

$$p = \frac{\bar{M}_n}{n\ .\ 54 + x\ .\ 32}$$

RESULTS

1 - Functionality of the polysulphide polymer

The thiol content of several polymers was determined and compared with theoretical values of the corresponding linear polymers.

Table 1 - Percentage of linear and cyclic chains

Polymer	$\bar{M}_n$	% $SH_{theor.}$	% $SH_{exp.}$	% linear	% cyclic
1	5 100	1.3	0	0	100
2	2 000	3.3	0.4	12	88
3	4 000	1.7	0.3	6	94
4	12 500	0.5	0.1	20	80
5	12 000	0.6	0	0	100
6	5 500	1.2	0	0	100
7	2 000	3.3	0.5	15	85
11	8 000	0.8	0.1	13	87
15	9 000	0.7	0	0	100

The results (Table 1) show that the experimental percentage of linear polymer is very small and tends to zero. Two experiments confirm these cyclic structures. The use of a classical oxidizing agent to form disulphide has not led to chain extension. In the same order, the use of diisocyanate has not given chain extension.

2 - Influence of different parameters on the characteristics of the polysulphide-polybutadiene

2.1 - Influence of the reaction time

An evolution of the sulphur rank is observed if the polymerization is carried out during 4 hours, while the monomer has been added during the first 2 hours (Table 2). By H.NMR, the variation of the peaks corresponding to the protons on the carbon linked to sulphur atoms was followed.

Table 2 - Evolution of the sulphur rank in relation to the time

time (hours)	% (S_4 + S_3)	% S_2	% S_1
1	39	41	20
2	20	51	29
3	7	54	39
4	0	31	69

$[M] = 0.4.mol.l^{-1}$ $T = -40°C$

At the beginning (t = 1 h) the carbanions react with the cyclic sulphur S_8, and nevertheless monosulphide bridges are found even in the presence of a large excess of sulphur. The alkali polysulphides formed are soluble in THF. After 2 hours there is no elemental sulphur remaining in the mixture and the carbanions attack the sulphur bridge in the middle where the electronic density is lower. The proportion of bridges with 4 and 3 sulphur atoms decreases sharply.

$$-MM^{-} + -MM\text{-}S\text{-}S\text{-}S\text{-}MM- \longrightarrow -M\text{-}M\text{-}S\text{-}S\text{-}MM- + {}^{-}S\text{-}MM-$$

After 3 hours, the bridges with 3 and 4 sulphur atoms disappear. At the end of the reaction bridges with 2 sulphur atoms begin to be attacked. Simultaneously, the peak corresponding to the monosulphide increases.

2.2 - Influence of the ratio $K = [Na]/[S_8]$ on the sulphur rank

Table 3 - Influence of K on the sulphur rank

Run	[Na] $mol.l^{-1}$	$[S_8]$ $mol.l^{-1}$	K	% S	$\bar{M}_n$	n	x	K'
1	0.4	0.20	2.0	30	5 000	4.0	2.9	1
2	0.4	0.16	2.4	26	2 000	4.4	2.6	1.1
3	0.4	0.10	4.0	20	4 000	4.0	1.7	2
4	0.4	0.06	6.6	16	12 500	4.3	1.4	3.1

$[M] = 0.4\ mol.l^{-1}$ $\quad$ T = - 40°C

When the ratio K increases from K = 2 to K = 6.6 the sulphur rank decreases from 2.9 to 1.4 and the sulphur content decreases from 30% to 16%. Under certain experimental conditions alkali initiator remains in the mixture after the reaction and it is necessary to use another ratio K' to characterize the evolution of the sulphur rank. K' is calculated from the amount of carbanions $|C^-|$ actually formed during the entire reaction.

$$n = 2\,[M]/[C^-] \qquad K' = [C^-]/[S_8] = 2\,[M]/n\,.\,[S_8]$$

From the plot $x = f(K')$ (Figure 1) it is observed that x tends to 1 for K' = 8. This result is similar to that obtained in a homogeneous phase and corresponds to the attack on polysulphide bridges when the elemental sulphur concentration decreases in the solution.

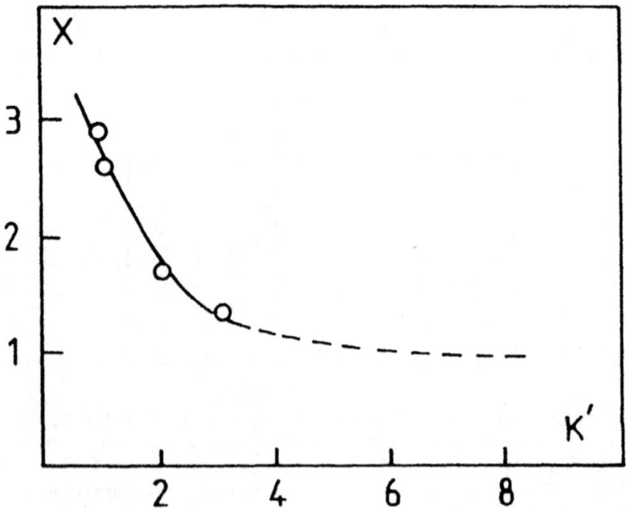

Figure 1 - Effect of K' on the sulphur rank

2.3 - Influence of the deactivating agent

. on the organic chain length:

We observe in Table 3 that the n values are similar for different K values (n = 4). This observation implies that the deactivation rate of the carbanions by cyclic sulphur at the beginning of the reaction, is similar to that of the carbanions by the polysulphide bridge or by the alkali polysulphide.

. on the microstructure of the butadiene unit in the α position to the sulphur bridge:

There are two possible structures for the butadiene units linked to the sulphur bridges:

. 1.4 unit: $-CH_2-CH=CH-CH_2-S_x-$ with 2 H atoms in one half repeating unit

The proportion of these protons is given by the ratio:

$$2/(6n/2) = 2/3\ n$$

. 1.2 unit: $-CH-CH-S_x-$ with 1 H atom per half repeating unit

```
-CH-CH-Sx-
 |
 CH
 ||
 CH2
```

The proportion of these protons is: 2/6 n.

If b is the content of 1.4 units linked to sulphur atoms and c the total proportion of protons on the carbon bearing the sulphur atoms, the relation giving b is the following:

$$c = b\ .\ (2/3\ n) + (1-b)(1/3\ n) \qquad b = 3\ n\ c\ -1$$

Then , the nature of the anionic species which preferentially have reacted with the elemental sulphur S_8 was determined.

Table 4 - Percentage of 1.4 units in the position vicinal to the sulphur links

Run	$\bar{M}_n$	T (°C)	n	total % 1.4	b
1	12 000	- 72	5.6	27	100
2	5 100	- 40	4	50	100
3	8 000	- 40	10.1	33	99
4	12 500	- 40	4.3	57	100
5	5 500	- 10	3.3	66	100
6	1 000	20	2.4	84	97
7	3 000	20	2.3	74	95

The results of Table 4 show that the microstructure of the units deactivated by the cyclic sulphur or by sulphur bridges is essentially of the 1.4 type. Even though the propagation occurs from the secondary carbanions to give 1.2 units (run 3) the deactivation on sulphur takes place from the primary carbanion.

2.4 - Influence of the global concentration

Table 5 shows the possibility to use different K values to obtain polymers with different sulphur rank from x = 1.4 to x = 3 in dilute solutions.

Table 5 - Characteristics of obtained polymers in dilute solutions

Run	K	Yield %	% S	$\bar{M}_n$	n	x	% 1.4 total	p
1	2.0	99	30	5 100	4.0	2.9	50	16.5
2	2.4	80	26	2 000	4.4	2.6	50	6
3	4.0	60	20	4 000	4.0	1.7	50	15
4	6.7	75	16	12 500	4.3	1.4	57	45

[M] = 0.4 $mol.l^{-1}$ [Na] = 0.4 $mol.l^{-1}$ T = - 40°C

It is possible to obtain long organic chains between the sulphur bridges. If the reaction is carried out at high concentration at - 40°C, and a high K value, two reactions are favored: the initiation step and the anionic propagation step. Moreover, at these high concentrations, the solution is viscous and the deactivation of the carbanionic chain ends on the polysulphide bridges becomes difficult compared to the easy diffusion of the monomer towards the anionic sites. With K = 6.7 (run 13 - Table 5) the molecular weight reaches 20,000 and n = 11.7

Table 6 - Characteristics of polymers obtained in concentrated solutions

N°	[M]	[Na]	$[S_8]$	K	Yield %	% S	$\bar{M}_n$	n	x	% 1.4	p
11	0.8	0.56	0.06	8.7	99	11	8 000	10.1	2.1	33	13
12	0.8	0.80	0.26	3.1	92	30	8 000	3.6	2.6	57	29
13	1.6	1.60	0.21	6.7	99	12	20 000	11.7	2.7	33	28
14	1.6	1.60	0.40	4.0	99	21	8 000	5.4	2.4	49	22
15	1.6	1.60	0.52	3.1	99	22	9 000	4.6	2.2	52	28

T = - 40°C

CHARACTERIZATION OF OTHER POLYSULPHIDE POLYMERS

1 - Polysulphide polyisoprene

In this case, after the introduction of elemental sulphur and sodium in the THF, the monomer is added rapidly. The solution becomes orange colored. The yield is quantitative and the characteristics of the polymer are given in the Table 7.

The chemical shifts of the protons on the carbon linked to the sulphur atoms range from δ = 2.8 ppm to δ = 3.68 ppm, very close to the chemical shifts observed for the polysulphide-polybutadiene.

When the low K value is used, the polymer obtained is a mixture of linear and cyclic macromolecules. For high K values, the carbanions attack the polysulphide bridges and the polymers obtained are linear, the composition of which are, for example:

. run 24 - Table 7 : $HS-(C_5H_8)_4-SH$

. run 25 - table 7 : 66 % $HS-(C_5H_8)_4-SH$ 34% $HS-(C_5H_8)_4-S-(C_5H_8)_4-SH$

Table 7 - Characterisation of polysulphide-polyisoprene

Run	[M] $mol.l^{-1}$	[Na] $mol.l^{-1}$	[S_8] $mol.l^{-1}$	K	% S	% SH	$\bar{M}_n$	linear chain
16	0.21	0.38	0.13	2.9	35.3	0.45	930	6
17	0.37	0.36	0.12	3	33.1	0.70	900	9.5
18	0.74	0.37	0.12	3	32.6	0.61	760	7
19	1.48	0.34	0.11	3	23.2	0.54	1 230	10
20	1.99	0.22	0.10	2.2	20.4	0.24	1 450	5
21	0.32	0.72	0.12	6	36.4	0.19	720	2
22	1.30	0.68	0.11	6.1	21.1	2.61	970	38
23	1.33	0.68	0.11	6.1	25.3	0.80	1 140	14
24	0.74	0.52	0.06	8.6	21.0	17.80	360	97
25	0.78	0.55	0.06	9.1	20.4	14.66	440	98

T = - 40°C

2 - Polysulphide-polystyrene

The polysulphide-polystyrene is prepared under the same experimental conditions as the polybutadiene one. The propagation rate for polystyrene is higher than that for polybutadiene and the molecular weight of the resulting polymers is also higher.

Table 8 - Characterization of polysulphide-polystyrene

Run	[M] $mol.l^{-1}$	[Na] $mol.l^{-1}$	[S_8] $mol.l^{-1}$	K	T°C	Yield %	n	% S	x	$\bar{M}_n$
39	0.40	Na 0.4	0.13	3	- 10	85	2	22	2	20 000
40	0.80	Li 0.5	0.06	8	- 40	99	3	9	1	32 000
41	0.34	Na 0.35	0.17	2	20	27	-	29	-	1 800
42	0.05	Li 0.1	0.05	2	20	76	3.7	22	1.8	2 000

In the experiment N° 39 or 42, the K value is low and the sulphur rank corresponds to polysulphide-polystyrene. When K increases to 8, the polymer otbained is a monosulphide (run 40). In this case, the organic chain length corresponds on the average to a styrene trimer.

3 - Polysulphide-polymethylmethacrylate

The GPC curves of the polymer obtained vary with the temperature. At room temperature, the deactivation rate of the carbanion is very high compared to the propagation rate and deactivation takes place on the same S_8 molecule. The mass spectroscopy of the mixture shows cyclic molecules containing the methylmethacrylate dimer unit cyclized: with one sulphur atom (M = 232), with two sulphur atoms (M = 264), with three sulphur atoms (M = 296) and, finally with four sulphur atoms (M = 328).At low temperature the propagation rate is higher than the deactivation rate, and the organic chain becomes longer.

Table 9 - Characterisation of polysulphide-polymethylmethacrylate

Run	[M] $mol.l^{-1}$	[Na] $mol.l^{-1}$	$[S_8]$ $mol.l^{-1}$	T°C	Yield %	% S	$\bar{M}_n$
28	0.4	0.4	0.13	- 80	1	17	1 700
29	0.4	0.4	0.13	- 30	42	35	600
30	0.4	0.4	0.13	- 30	45	30	530
31	0.4	0.4	0.13	20	17	47	450
32	0.8	0.8	0.26	- 30	68	28	1 250
33	0.8	0.8	0.13	- 30	96	11	1 500

4 - Polysulphide-polyacrylate

With this monomer the experimental results are very similar as those obtained with methylmethacrylate.

Table 10 - Characterisation of polysulphide-polyacrylate

Run	T°C	Yield %	% S	$\bar{M}_n$	% SH
36	- 80	1	32	600	0
37	- 30	44	16	350	0
38	20	26	32	245	0

$[M] = 0.4\ mol.l^{-1}$ $[Na] = 0.4\ mol.l^{-1}$ $[S_8] = 0.13\ mol.l^{-1}$ $t = 4\ h$

At room temperature, cyclic structures containing acrylic dimer with one or two sulphur atoms are obtained. At low temperature cyclic structures of higher molecular weight are produced. With this monomer, the deactivation reaction of the carbanion by S_8 is theoretically in competition with the deactivation of the carbanions by the labile hydrogen of the acrylate. However the first reaction is faster than the second one, because no protonated oligomers have been identified.

The NMR and the mass spectroscopy of the oligomers are in good agreement with the polysulphide-polyacrylate structures.

REDUCTION OF THE POLYSULPHIDE POLYMERS

In order to functionalize the cyclic polymers a reduction was carried out by Clemmensen's reaction. All the polysulphide bridges are reduced into thiol functions except the monosulphide which remains in the final oligomer the structure of which is the following:

$$HS - [M_n - S]_{p'} - H$$

This dithiol oligomer has been used to prepare new materials (4).

Table 11 - Characterization of reduced polysulphide polymers

Run	Before reduction			After reduction				
	Polymer	$\bar{M}_n$	% S	$\bar{M}_n$	% S	% SH	p'	n
10	P Bu	3 000	30	550	22	12.2	2.7	2.8
11	P Bu	8 000	11	800	9	8.3	1.3	10.5
14	P Bu	8 000	21	600	15	10.9	1.8	5.2
15	P Bu	9 000	22	580	16	11.3	2.1	3.7
27	PI	1 700	19	770	13	8.6	2.1	3.7
39	PSty	2 000	22	430	17	15.3	1.3	3.7
51	PMMA	600	35	400	17	16.6	1.1	2.1

CONCLUSION

The single step synthesis enables us to prepare polysulphide polymers or linear dimercaptotelechelic monosulphide polymer. Moreover it is possible to control the sulphur rank of the polymer obtained by varying the ratio K. An increase of the K ratios leads to a decrease of the sulphur rank from $x = 3$ to $x = 1$. The organic chain length can be adjusted by changing the monomer concentration. Finally, it is possible also with this process to use monomers containing labile hydrogen atoms if the deactivation reaction on sulphur is faster than the deactivation reaction on the labile hydrogen.

REFERENCES

1. J.F. Boscato, J.M. Catala, J. Brossas, Makromol.Chem. 180, 1571 (1979)
   J.M. Catala, J.F. Boscato, E. Franta, J. Brossas, ACS Symp.Series 166, Anionic polymerization, J.E. McGrath 483 (1981)
   J.M. Catala, J.F. Boscato, J. Brossas, J.Organometal.Chem. 276, 155 (1984)
2. J.F. Boscato, Thesis (October 1980) - Strasbourg
   J. Brossas, J.M. Catala, J.Oil and Color Chem. Ass. 66(9), 259 (1983)
3. J.M. Pujol, Thesis (October 1985) - Strasbourg
4. J.M. Catala, J.M. Pujol, J. Brossas, French Pat. N° 85.02742 (1985)
   St. Gobain Vitrage

* * * *

SYNTHESIS OF CELLULOSE-g-POLYSTYRENE BY ANIONIC POLYMERIZATION

Ramani Narayan and Margaret Shay
Laboratory of Renewable Resources Engineering
A. A. Potter Engineering Center
Purdue University
West Lafayette, IN 47907

INTRODUCTION

The forced union by block and graft polymerization of synthetic polymers to natural polymers such as cellulose can lead to materials with unusual and useful properties for differing applications. Graft copolymerization is an attractive means for modifying base polymers because grafting frequently results in the superposition of properties relating to backbone and side chains [1]. Grafting is especially useful when the backbone and side chain homopolymers are incompatible. Special properties of graft copolymers that lead to potential or actual applications are changes in permeability, solubility, compatibility, and the mosaic structure of the solid state [2].

The properties of cellulose grafts depend on the location of the grafted side chains [3]. Properties such as abrasion resistance, enhanced adhesion and wetting only need surface modification of the base polymer. Flame retardancy and water sorbency need bulk grafting. Other properties, including, ion exchange ability, dimensional stability and electrical properties can also be imparted to cellulose. The presence of cellulose has also been shown to impart biodegradability to cellulose--synthetic polymer grafted materials.

Many applications have been proposed for cellulose--synthetic polymer grafts. For example, cellulose acetate--polystyrene grafts have been suggested as replacements for cellulose acetate desalination membranes [4]. These modified membranes would be more mechanically stable under the high pressures used in desalination. Cotton fabric grafted with polystyrene to give improved crease resistance and soil release properties has also been investigated [5,6].

Much work has been done studying the grafting of synthetic polymers to cellulosic backbones and several extensive reviews written on this subject [7-11]. However, almost all of the grafting methods involve radical polymerization approaches. This approach has several major disadvantages: 1) The molecular weight and molecular weight distribution of the grafts is difficult to control or change. 2) The molecular weight is often very high and the molecular weight distribution wide. 3) Grafting of only a few high molecular weight grafts occurs. 4) Considerable amounts of homopolymer are sometimes formed some of which may get embedded in the cellulose matrix and are difficult to remove. 5) No knowledge exists of the nature of the linkage between the cellulose backbone and the graft, nor the ability to control it.

In this paper we present a method for the synthesis of cellulose--polystyrene graft copolymers involving anionic polymerization techniques that would overcome these difficulties. Anionic polymerization methods provide an extensive and unprecedented control over polymerization processes. This includes polymer composition, microstructure, *molecular weight* and *molecular weight distribution*, and monomer

Recent Advances in Anionic Polymerization
T.E. Hogen-Esch and J. Smid, Editors

sequence distribution. The method involves i) preparation of "living" polystyrene of the desired molecular weight by anionic polymerization, ii) introduction of tosyl groups, onto the cellulose backbone by chemical modification, and iii) reaction of the "living" polystyrene with the modified cellulose under homogenous reaction conditions, in a nucleophilic displacement reaction.

BACKGROUND

Cellulose

Cellulose is a naturally occurring polymer consisting of D-anhydroglucose units joined by β-1,4-glycosidic bonds (Figure 1). The viscosity--average molecular weight for natural cellulose can range as high as 800,000 [7], with a degree of polymerization of 5000.

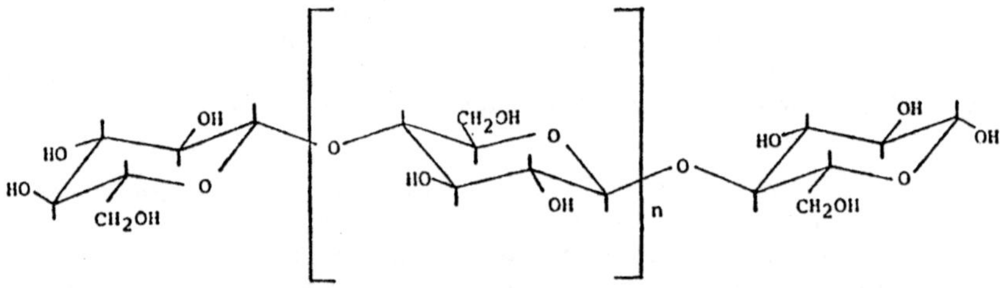

Figure 1. Structure of Cellulose

Cellulose exhibits a high degree of hydrogen bonding which results in a highly rigid, crystalline structure. It is polymorphic, and crystallizes naturally in two predominant forms, cellulose I and cellulose II. The most common form of cellulose I is cotton, of which 70-80% is in crystalline regions. Wood fiber is a common form of cellulose II, about 60% of which is in crystalline regions [7].

The strong intermolecular forces present in cellulose cause it to decompose before it softens or melts [7]. These forces also make cellulose insoluble in simple solvents. It is soluble in aqueous solutions of metal complex compounds such a cuprammonium hydroxide, which are able to overcome the hydrogen bonding.

Cellulose reacts as trihydric alcohol with one primary and two secondary hydroxyl groups per anhydroglucose unit. Because of the hydrogen bonding, as well as general dispersion forces determined by neighboring atoms, the C-6 in many hydroxyl groups reacts much faster than the secondary hydroxyl groups, and the C-2 group reacts faster than the C-3 group.

The crystalline regions of cellulose react slowly. For most processes, the crystalline structure must be disturbed before the reaction can proceed at a reasonable rate. This disruption can be accomplished by treating cellulose with a swelling agent such as water or dimethylsulfoxide. It can also be accomplished by chemically modifying the cellulose. Many modifications have been reported, but cellulose acetate is most frequently used.

Cellulose is an abundant, low cost material. It has high strength, high thermal stability, good mechanical properties, and the ability to absorb moisture. Its drawbacks include poor solubility in common solvents, which makes spinning processes almost impossible, lack of thermoplasticity, which makes extrusion processes impossible, and poor dimensional stability. Cellulose use is very widespread. Its uses range from textiles to structural materials.

Polystyrene

Polystyrene is a versatile thermoplastic resin available in a wide range of formulations, from general-purpose crystal and compact grades to highly specialized resins for foam molding and extrusion. As one of the main general-purpose plastics, polystyrene is surpassed in usage only by PVC and polyolefins [12].

When prepared by radical polymerization in bulk, solution, emulsion, suspension, or anionic polymerization, polystyrene is completely amorphous. This is because the bulky benzene rings do not permit the close approach of the polymer chains, so a random arrangement of chains results. Crystalline polystyrene can be prepared by using stereospecific catalysts which cause the phenyl groups to assume a regular arrangement along the carbon backbone. Each molecule has a helical distribution conformation and the chains form an orderly arrangement, producing a crystalline structure.

Polystyrene is a colorless, transparent thermoplastic. It is hard, has a fairly high tensile strength and a high refractive index. It has excellent electrical properties, including a very low power factor and high dielectric strength. It is resistant to water and strong acids and bases, and is dimensionally stable. Microorganisms have little effect on its properties.

Polystyrene is intrinsically rather brittle, so toughened grades have been developed, in which rubbery materials are incorporated into the polystyrene. Prolonged exposure to sunlight causes yellowing of clear polystyrene, often with crazing and impairment of mechanical properties. Ultra-violet stabilizers give some protection.

Polystyrene can be easily extruded or injection molded. It is thermally stable at processing temperature and has relatively low shrinkage on cooling. It can also be foam molded by the use of physical or chemical blowing agents. Applications for polystyrene range from simple household items to sophisticated applications in the electronics industry. Expandable polystyrene is used as building insulation, and it is also used as a composite material with cement for light weight concrete [9].

GRAFTING OF POLYSTYRENE ONTO DERIVATIZED CELLULOSE

Acetone-soluble cellulose acetate (DS = 2.47, DP = 110) was modified by tosylation of the unesterified hydroxyl groups using literature procedures (Scheme 1 (1)) [14]. Introduction of the tosyl groups provided: (1) a more reactive group than acetate in nucleophilic displacement, and (2) capping of the free hydroxyl groups on the cellulose backbone, thus eliminating the possibility of homopolymer formation due to quenching of the polystyryl carbanions by the hydroxyl protons. Since 40-60% of the free hydroxyls in the cellulose acetate are primary hydroxyls ($-CH_2OH$) on the C-6 carbon [15], a significant number of the tosyl groups would be present on the C-6 carbon. Thus, tosylation not only introduced a better leaving group, but also put the tosyl groups on the least sterically hindered carbon position

(C-6), which made it easier to introduce the bulky polystyryl chain. "Living" polystyrene of desired molecular weight was prepared by anionic polymerization in tetrahydrofuran (THF) using n-butyllithium as the initiator (Scheme 1 (2)).

However, when the tosylated cellulose acetate (TCA) was reacted with the polystyryl carbanions, no grafting occurred. There was significant homopolymer formation with concurrent degradation of the cellulose chain (Scheme 1 (3)). It appears that the very basic polystyryl carbanion is cleaving the glycosidic bond.

To circumvent this problem, the basicity of the polystyryl carbanion was reduced by capping with 1,1-diphenylethylene (Scheme 2). Reaction of this modified polystyryl carbanion with tosylated cellulose acetate proceeded smoothly and gave excellent yields of the graft polymer (Scheme 2). The graft polymer was extracted with toluene to remove any unreacted polystyrene. The amount of polystyrene grafted onto the cellulose backbone was determined by UV spectroscopy [16]. Table 1 summarizes the amount of polystyrene grafted and the number of anhydroglucose units per grafted side chain (or number of PS chains per cellulosic chain) for three different molecular weight polystyrenes.

Table 1. Results of Grafting Polystyrene Onto Tosylated Cellulose Acetate

Sample No.	Molecular Weight of PS	% PS (by wt)	Anhydroglucose Units Per Grafted Side Chain	PS Chains Per *Cellulosic Chain
1	3,400	39	16	6.7
2	10,000	51.8	28	3.9
3	32,200	42.2	135	0.81

*DP = 110; DS of acetyl =2.47; DS of tosyl = 0.42

Proof of Grafting

Thin Layer Chromatography (TLC). TLC established that there was no unreacted polystyrene occluded with the cellulose graft polymer. Thus, comparison of a graft polymer sample with a graft polymer-polystyrene mixture by TLC using silica gel plates with cyclohexane-benzene-acetone (12:4:1) as the developing solvent system showed no spot corresponding to polystyrene in the graft polymer sample. Polystyrene had a normalized migration rate value, R_f, of about 0.75 while the graft polymer did not move.

Solubility. Solubility tests showed that the TCA-g-PS was insoluble in toluene and glacial acetic acid, while PS and TCA are soluble in toluene and and glacial acetic acid respectively. However, the graft polymer was soluble in THF, acetone and pyridine, which are solvents for both PS and TCA. This solubility behavior supported the formation of a true graft polymer.

Infrared Spectra (IR). Figure 2 shows the IR spectra of TCA and the TCA-g-PS. The IR spectrum of the TCA-g-PS showed the aromatic ring vibration bands at 1500-1600 cm^{-1} and the aromatic -CH stretching vibrations above 3000 cm^{-1} corresponding to the polystyrene component and the characteristic carbonyl band at 1720 cm^{-1} due to the acetate group of the TCA.

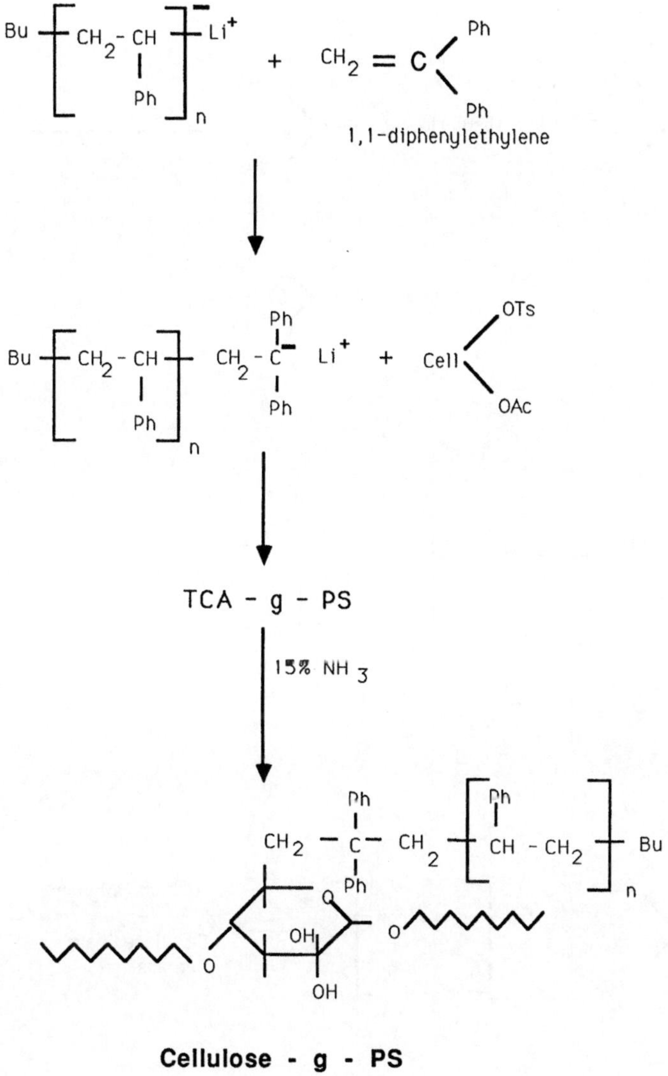
Bu
CH2 - CH
Ph
Li+
n
+
CH2 = C
Ph
Ph
1,1-diphenylethylene
Bu
CH2 - CH
Ph
n
CH2 - C
Ph
Ph
Li+
+
Cell
OTs
OAc
TCA - g - PS
15% NH 3
CH2 - C - CH2
Ph
Ph
CH - CH2
n
Bu
O
OH
OH
O
O
Cellulose - g - PS

SCHEME 2

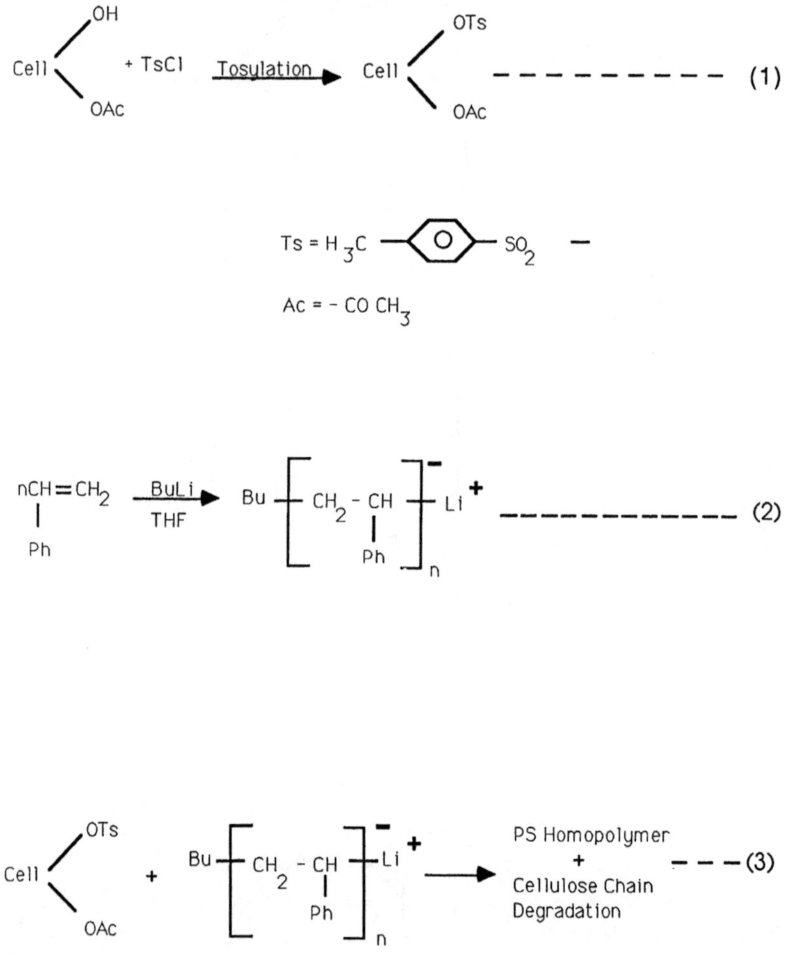
OH
Cell
OAc
+ TsCl
Tosylation
OTs
Cell
OAc
(1)
Ts = H3C
SO2
Ac = - CO CH3
nCH=CH2
Ph
BuLi
THF
Bu
CH2 - CH
Ph
Li+
n
(2)
OTs
Cell
OAc
+
Bu
CH2 - CH
Ph
Li+
n
PS Homopolymer
+
Cellulose Chain
Degradation
(3)

SCHEME 1

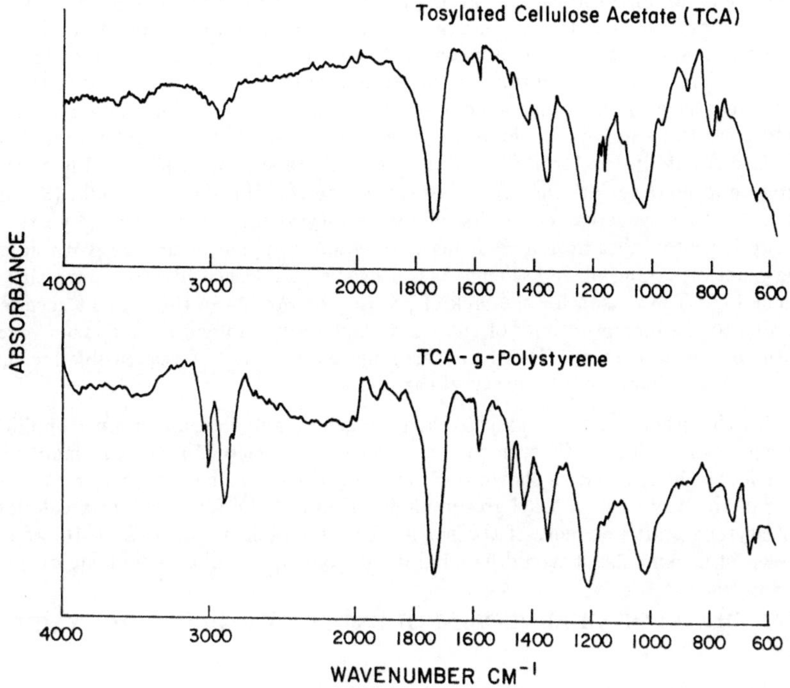

Figure 2: IR Spectra

Elemental Analysis. The elemental analysis of the tosylated cellulose acetate (TCA) and the TCA-g-PS is shown below.

Table 2. Elemental Analysis

Element	TCA	TCA-g-PS Sample No. 1
C%	49.51	73.56
H%	5.48	7.22
S%	4.10	2.30
O%	40.91	16.92

TLC and solubility has already established that there is no free unreacted polystyrene contaminating the graft polymer. Thus, the substantial increase in the carbon and hydrogen content coupled with a marked decrease in the oxygen and sulfur content of the graft polymer as compared to the starting TCA supported the formation of a true TCA-g-PS copolymer.

Reaction Mechanism

The polystyryl carbanion can react with the tosyl group on the cellulose backbone by a nucleophilic displacement reaction. This results in the formation of a carbon-carbon bond with the C-6 carbon of the cellulose molecule. The polystyryl carbanion can also attack the carbonyl carbon of the acetate group. Attack on the carbonyl carbon by a carbanion is a reversible step and a subsequent equilibrium displacing step to product is necessary. For example, a dehydration step is necessary to displace the equilibrium in the Aldol and Perkin reactions, and in the Claisen ester condensation the equilibrium displacing step is the conversion of the keto ester to its anion. In the present case, attack on the carbonyl carbon of the acetate results in the formation of an intermediate with no other good leaving group and no other driving force to displace the equilibrium to the right (Scheme 3b). Another conceivable reaction of the polystyryl carbanion with the acetate group is proton abstraction resulting in only homopolymer formation with no grafting reaction taking place (Scheme 3a). However, as described earlier, grafting is occurring. The reason for the lack of proton abstraction in the present case may be due to the increased bulk of the 1,1-diphenyl alkyl anion and the rapid reaction with the tosylate group. However, this reaction may well be responsible for the protonation of unreacted polystyryl carbanions.

On the other hand, the tosyl groups are good leaving groups in nucleophilic displacement reactions. Thus, in the present case, a nucleophilic displacement reaction should be favored as shown in Scheme 4. If this mechanism is correct hydrolysis of the acetate and tosyl groups should result in the formation of a cellulose-polystyrene graft polymer. If the polystyrene is attached via the acetate group as shown in Scheme 3a, it would be eliminated from the cellulose backbone during hydrolysis.

SYNTHESIS OF CELLULOSE-POLYSTYRENE GRAFT POLYMER (Cell-g-PS)

The TCA-g-PS was subjected to mild hydrolysis with 15% NH_3 solution to remove tosyl and acetyl groups remaining on the cellulose backbone (Scheme 2). In the IR spectrum of the hydrolysis product (Figure 2) the disappearance of the carbonyl band at 1720 cm^{-1} and the appearance of the -OH stretching vibrations at 3500 cm^{-1} confirmed the hydrolysis of acetate and tosyl groups. More importantly, the product spectrum still contained the aromatic -CH stretching vibrations above 3000 cm^{-1} and aromatic ring vibrations in 1500-1600 cm^{-1} region. This confirmed that the polystyrene graft chain was still attached to the cellulose backbone, and the product was cellulose-g-PS. The elemental analysis showed C = 79.34%; H = 7.21%; S = 0.77%; and O = 12.68% which is in agreement with a cellulose-polystyrene graft polymer. The product was insoluble in all the above solvents, a behavior typical of cellulose and unlike that of the TCA-g-PS copolymer.

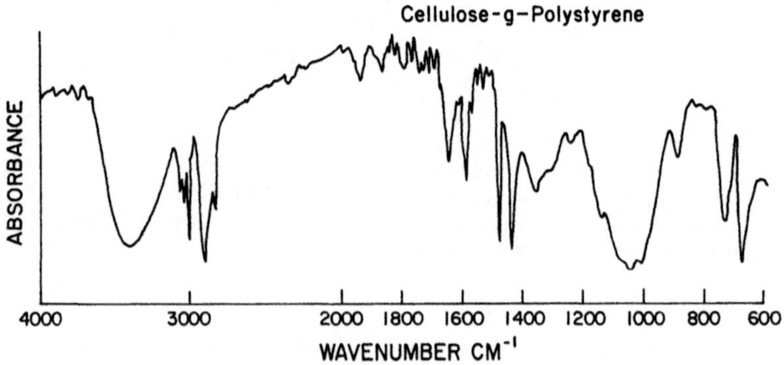

Figure 3: IR Spectra

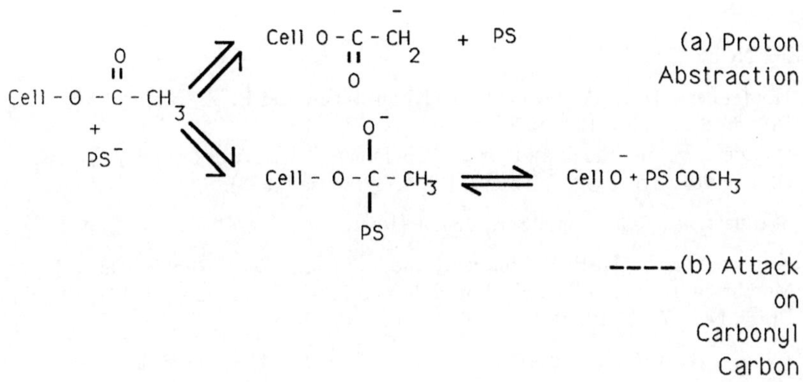

SCHEME 3

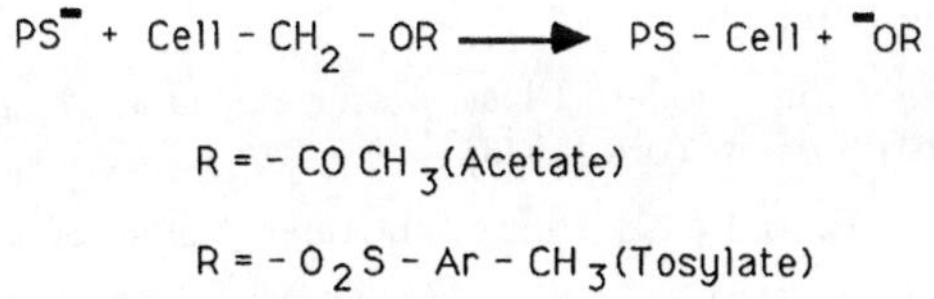

SCHEME 4

ACKNOWLEDGMENTS

This material is based upon work supported by the National Science Foundation under Grant No. (CBT-8502498).

REFERENCES

1. Hopfenberg, H. B., V. Stannett, F. Kimura-Yeh and P. T. Rigney, Appl. Polym. Symp., *13*, 139 (1970).
2. Stannett, V., in "Block and Graft Copolymers," J. J. Burke and V. Weist, eds., p. 281, Syracuse University Press, Syracuse (1973).
3. Stannett, V., ACS Symp. Ser., *187*, 1 (1982).
4. Kimura-Yeh, F., H. B. Hopfenberg, and V. Stannett, in "Reverse Osmosis Membrane Research," H. K. Lonsdale and H. E. Podall, eds., p. 177, Plenum Press, New York (1972).
5. Avny, Y. and R. F. Schwenker, Jr., Text. Res. J., *37*, 817 (1967).
6. Sakurada, I., T. Okada and K. Kaji, J. Polym. Sci., Part C, *37*, 1 (1972).
7. Arthur, J. C., Jr., Appl. Polym. Symp., *36*, 201 (1981).
8. Mansour, O. Y. and A. Nagaty, Prog. Polym. Sci., *11*, 91 (1985).
9. Bhattacharyya, S. N. and D. Maldas, Prog. Polym. Sci., *10*, 171 (1984).
10. Morin, B. R., I. P. Breusova and Z. A. Rogovin, Adv. Polym. Sci., *42*, 139 (1982).
11. Hebish, A. and J. T. Guthrie, "The Chemistry and Technology of Cellulosic Copolymers," Springer-Verlag, New York (1981).
12. Roff, Scott and Pacutti, "Handbook of Common Polymers," p. 45, CRC Press, Cleveland (1971).
13. Swett, R. M., in "Modern Plastics Encyclopedia," J. Agranoff, ed., p. 72, McGraw-Hill, New York (1983).
14. Cramer, F. B. and C. B. Purves, J. Amer. Chem. Soc., *61*, 3458 (1939).
15. Tanghe, L. J., L. B. Genung and J. W. Mench, "Methods in Carbohydrate Chemistry," Vol. III, R. L. Whistler, ed., Academic Press, New York (1963).
16. Mansson, P. and L. Westfelt, J. Polym. Sci., Polym. Chem. Ed., *19*, 1509 (1981).

ION-CONTAINING STYRENE-METHOXYPOLYETHYLENEGLYCOL METHACRYLATE ABA TRIBLOCK COPOLYMERS

ISHRAT M. KHAN, DARYLE FISH and JOHANNES SMID
Polymer Research Institute, College of Environmental Science and Forestry, State University of New York, NY 13210, USA

INTRODUCTION

The discovery of anionic living polymerization by Szwarc and coworkers (1) has led to the synthesis of polymers with well defined structures. The sequential addition of monomers to anionic initiators produces block copolymers which, by virtue of their molecular design, exhibit many interesting properties (2). Anionic polymerization is by far the most versatile technique to synthesize model heterophase block copolymers. For example, McGrath et.al. (3) obtained polystyrene-polycarboxylate block ionomers from the partial hydrolysis of a styrene-alkylmethacrylate block copolymer synthesized by anionic polymerization techniques. They reported a large enhancement in the rubbery plateau of the thermomechanical analysis curve for a block copolymer consisting of a polystyrene block attached to an isobutylmethacrylate-potassium carboxylate copolymer, relative to that of the non-hydrolyzed polymer.

An interesting type of ion-containing polymer can be generated by dissolving salts of low lattice energy into polymers with heteroatoms like O, N, or S (4). Our recent interest has been in ionomers obtained by solubilizing lithium perchlorate or other alkali or alkaline earth salts into poly(methacrylate)s or poly(siloxane)s carrying pendant oligo-oxyethylene chains as cation complexing moieties (5-6). Such solid polymer electrolytes exhibit interesting ion-conducting properties and have potential application in high energy density storage devices (7). The salt free homopolymers with relatively short oligo-oxyethylene side chains (e.g., eight ethylene oxide units) are viscous liquids. Incorporation of salt raises the glass transition temperature of these materials, and amorphous or crystalline solids are formed depending on the nature of the salt and length of the oligo-oxyethylene chain (5,6,8). It should be of interest to incorporate non salt complexing blocks such as polystyrene into polymer electrolytes in order to form materials with separate conducting and non-conducting microdomains. If the salt free block has a low T_g such as polyisoprene or polysiloxane, the block ionomers may function as a thermoreversible elastomer. Also, a high T_g block like polystyrene would impart mechanical strength to polymers with pendant oligo-oxyethylene chains. In random copolymers with monomers like styrene or methacrylate, the side chains tend to act as plasticizers. This paper reports on the synthesis, characterization and some properties of block copolymers of styrene and -methacryloyl- -methoxypoly(oxyethylene).

EXPERIMENTAL

Synthesis. The synthetic pathway for the block copolymerization is outlined in Scheme 1. All polymerizations were carried out using high vacuum line breakseal techniques. The macromonomer, MG8 (abbreviation for methoxypolyethyleneglycol methacrylate, $CH_2{=}C(CH_3)COO(CH_2CH_2O)_nCH_3$, with average n=8), was dried by several azeotropic distillations with benzene.

Recent Advances in Anionic Polymerization
T.E. Hogen-Esch and J. Smid, Editors

SCHEME I

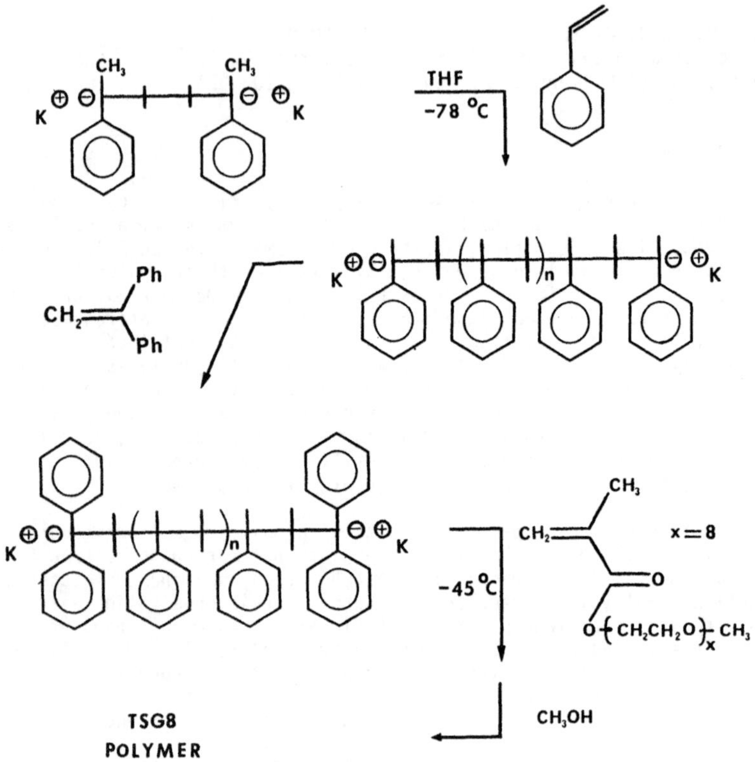

The two-ended living polystyrene was capped with 1,1-diphenylethylene to minimize nucleophilic attack of the carbanion on the carbonyl group of MG8(9). The initial concentration of the macromonomer during the polymerization was 0.02M or lower. The triblock copolymers, abbreviated as TSG8, were purified by precipitation with a 90/10(v/v)hexane-THF mixture in which MG8 is slightly soluble. Block copolymers with high styrene content (>55% by weight) were further purified by precipitation from methanol. Finally, any polystyrene present could be removed by precipitating the product from diethylether, which, according to the GPC analysis, left the block copolymer free of polystyrene (10). Random copolymers of styrene with MG8, abbreviated as RSG8, were prepared in benzene at 70°C using AIBN as the initiator. The polymers were precipitated from methanol.

Measurements: The number average molecular weights (M_n) were determined by membrane osmometry in toluene using a Wescan recording osmometer. Gel permeation chromatograms were obtained in THF using a Waters GPCIA system. Experimental details for obtaining DSC thermograms and conductances have been described elsewhere (5). Salt complexes were prepared by slow evaporation of homogeneous polymer-salt mixtures in THF. Routine ^1H NMR spectra were taken on a Varian XL-100 spectrometer at ambient temperature.

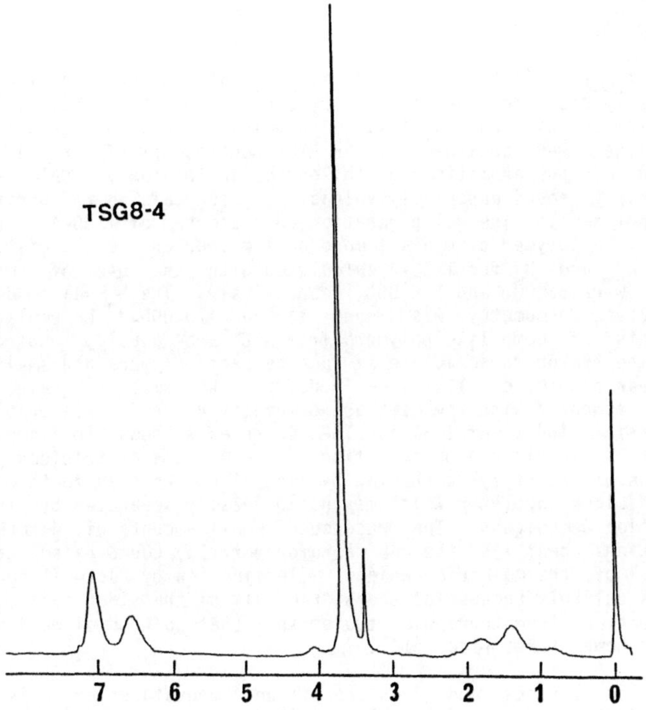

Figure 1. 100 MHz ^1H NMR of TSG8-4 in $CDCl_3$ at 25°C.

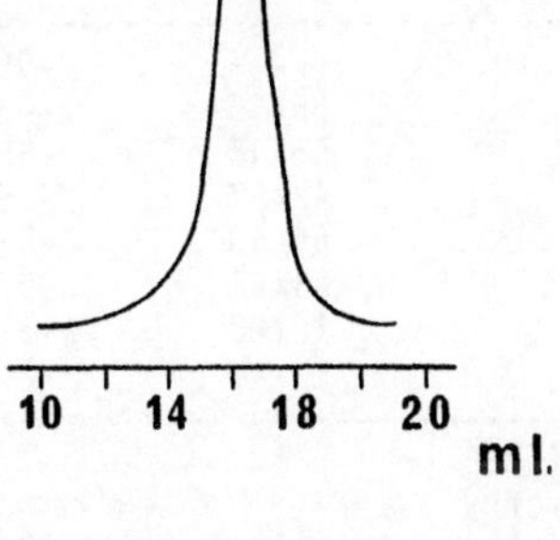

Figure 2. Gel permeation chromatogram of the TSG8-4 triblock copolymer in THF.

RESULTS AND DISCUSSION

Styrene-Methoxypoly(ethyleneglycol) Methacrylate Block Copolymers. The block copolymers were characterized by GPC, membrane osmometry, NMR and differential scanning calorimetry. The weight percent of MG8 and styrene in the polymers were obtained from ^{1}H NMR spectroscopy (figure 1), by using the ratio of the aromatic and ethylene oxide protons. Gel permeation chromatography proved especially valuable in ascertaining the purity of the block copolymers. The gel permeation chromatogram of TSG8-4 is shown in figure 2. The polymer displays a monomodal molecular weight distribution. The GPC M_n and M_w for TSG8-4 calculated with the use of polystyrene standards were 96,000 and 126,000, respectively. The M_n value determined from membrane osmometry was found to be 170,000. Molecular weight measurements of comb-like polymers from GPC are usually underestimated because the hydrodynamic volume of such compact polymers are smaller than their linear homologues (11,12). Therefore, the smaller M_n value obtained from GPC compared with the membrane osmometry value is reasonable. The polydispersity index of 1.31 for TSG8-4 reveals possible complications during the synthesis and purification of these types of triblock polymers. Termination of the styryl carbanion by impurities can lead to the formation of some diblock copolymer which may not be easily separated by fractional precipitation techniques. The presence of small amounts of dimethacrylate cross-linking agent in the MG8 macromonomer (13) could also cause some broadening of the molecular weight. Termination by back-biting in this system is unlikely because of the steric bulk of the side chains. Also, gel permeation chromatographic studies show that no homopolymerization of MG8 macromonomer takes place (14).

The compositions and glass transition temperatures of a series of triblock (TSG8) and random copolymers (RSG8) are compared in Tables 1 and 2. The block copolymers contain a soft and a hard phase as indicated by two T_g's, while only one glass transition temperature is observed for the random copolymers. However, there is considerable phase mixing in the TSG8 block copolymers as indicated by the depression in the T_g of the polystyrene phase. Similar decreases in the Tg have been noted for styrene-isoprene diblock and triblock copolymers and have been attributed

Table 1. Glass Transition Temperatures and Conductivities of Styrene-MG8 Block Copolymers.

Polymer	$M_n^{b} \times 10^{-3}$	MG8/Styrene (weight %)	$T_g(1)$ C	$T_g(2)$ C	σ^{d} $ohm^{-1}cm^{-1}x10^6$
PMG8[a]	--	--	-58	--	11.0
TSG8-1	30	65/35	--	--	1.2
TSG8-2	64	55/45	-55	72	0.9
TSG8-3	72	43/57	-55	72	0.5
TSG8-4	170	60/40	-57	71	0.7
TSG8-5	60	33/67	-57	71	--
TSG8-6	--	18/82	-57	93	--
Polystyrene	--	--	--	110	--

a) Homopolymer of MG8
b) Membrane osmometry
c) Homogeneous mixtures of polymer and $LiClO_4$, molar ratio of ethylene oxide to $LiClO_4$ = 17
d) $25^{o}C$

Table 2. Glass Transition Temperatures of Styrene-MG8 Random Copolymers.

Polymer	MG8/Styrene (wt%)	Styrene/MG8 Molar ratio	T_g
RSG8-1	38/62	8	12
RSG8-2	25/75	15	39
RSG8-3	20/80	19	49

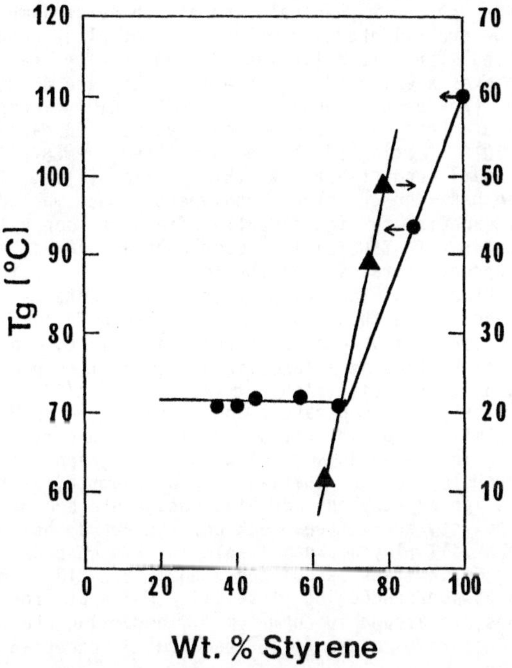

Figure 3. Glass transition temperature of the random copolymers (▲) and of the polystyrene domain of the triblock copolymers (●) as a function of styrene content.

to strong dynamical interactions at the interface between the chains of the two microphases (15,16). In our system the short oligo-oxyethylene side chains could possibly penetrate the polystyrene microdomains and act as a plasticizer. Such a plasticizing action has been reported for copolymers of styrene and methacrylate terminated polydimethylsiloxane macromers, the effect being larger with shorter polydimethylsiloxane side chains (17). As seen in figure 3, the decrease in the Tg of the polystyrene phase with increasing weight percent of MG8 levels off around 71°C. It is quite plausible that once the styrene microphase is saturated with ethylene oxide units, increasing the weight percent of MG8 has no further effect.

Interestingly, regardless of the block copolymer composition, the Tg of the soft phase is essentially unaffected. Hence the internal plasticization in these types of block copolymer is an interesting phenomenon.

The presence of the hard polystyrene block in the TSG8 polymers gives them superior physical properties relative to the PMG8 homopolymers. Also, the RSG8 random copolymers, up to a ratio of ten styrene to one MG8 unit, are sticky melts, while in the TSG8 block copolymers a 2.5:1 styrene to MG8 ratio results in a white powdery material with good film-forming properties.

Conducting Properties of Solid TSG8-$LiClO_4$ Complexes: We earlier reported on the conducting properties of solid electrolyte complexes with polymethacrylates and polysiloxanes containing pendant oligo-oxyethylene chains (5,6) which are potentially useful in high energy density storage devices (7). For proper mechanical strength these polymer electrolyte complexes must be cross-linked; however, use of block copolymers could eliminate the need for cross-linking. Conductivity values of solid TSG8/$LiClO_4$ complexes are listed in Table 1. They are about an order of magnitude below the value found for PMG8/$LiClO_4$ complexes. The conductivity measurements on the block copolymers were carried out at an ethylene oxide/$LiClO_4$ ratio of 17 because the PMG8/$LiClO_4$ complex displayed a maximum conductance at this ratio (5). This may not necessarily be the case for the block copolymers, but small differences in the ratio are not expected to significantly affect the conductivities. The lower conductivities of TSG8/$LiClO_4$ complexes are a result of the non-conducting polystyrene domains and the lower overall polarity of the system. Longer blocks of the two polymer constituents may give better conductivity values since this would decrease the effect of mixing the conducting and non-conducting domains at their interface. In figure 4 the conductivities are plotted as a function of temperature according to the Vogel-Tamman-Fulcher (VTF) relationship $\sigma = AT^{-1/2}\exp[-K/(T-T_o)]$ where T_o is the ideal glass transition temperature and A_o and K are constants (7). The straight line relationship indicates that ion transport occurs in an amorphous matrix. In the absence of electron micrographs we are unable to comment on the morphology of the complexes, for example, as to whether the salt-containing oligo-oxyethylene domains constitute a continuous phase. The fraction of polystyrene in the block copolymer and the length of the hydrophobic blocks will play a crucial role in this respect. In spite of the morphological complexities associated with TSG8/$LiClO_4$ complexes, they are of interest as ion-conducting plastics. While at room temperature conductivity values are around $10^{-6}\,ohm^{-1}cm^{-1}$, increasing the temperature to 80^oC yields conductivity values ($\sim 10^{-5}\,ohm^{-1}cm^{-1}$) acceptable for solid polymer electrolytes.

CONCLUSIONS

The synthesis of the block copolymers TSG8 and their subsequent conversion into polymer electrolyte complexes has made available heterophase ion-containing polymeric systems with interesting properties. It would be especially pertinent to compare the dynamical mechanical behavior of these salt-containing materials with the ion-containing block copolymers reported by McGrath (3). The properties of our polymers can be readily adjusted by solubilizing different salts in varying quantities into the oligo-oxyethylene containing phase.

Acknowledgement: The authors gratefully acknowledge the financial support of the National Science Foundations Polymers Program under grant no. DMR 8504999.

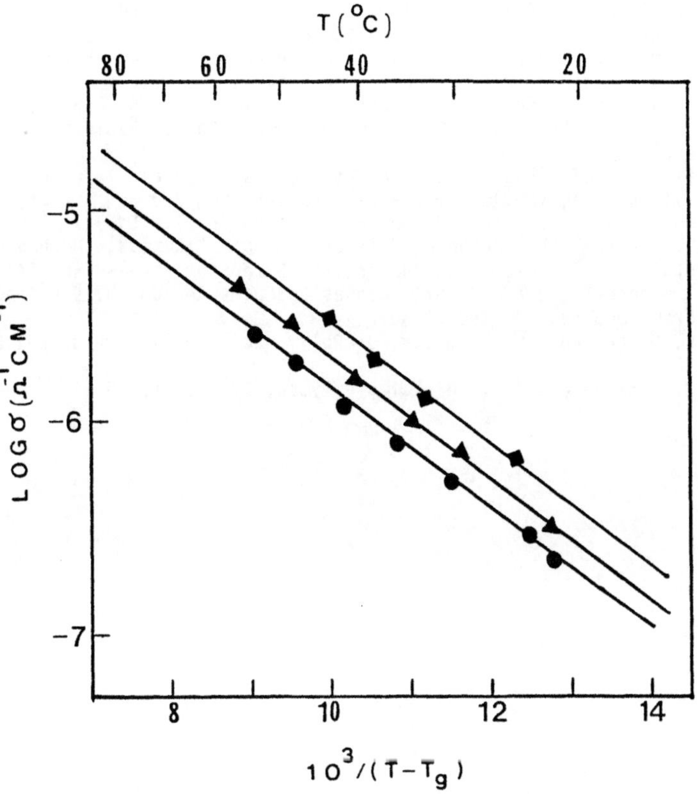

Figure 4. Conductivity (VTF) plots for the triblock copolymers. TSG8-2 (◆), TSG8-3 (●), and TSG8-4 (▲). Ratio EO/Li = 17.

REFERENCES

1. M. Szwarc, M. Levy and R. Milkovich, J. Am. Chem. Soc. 78, 265 (1956).
2. A. Noshay and J.E. McGrath, "Block Copolymers: Overview and Critical Survey", Academic Press, New York, 1977.
3. R.D. Allen, T.L. Huang, D.K. Mohanty, S.S. Huang, H.D. Qin and J.E. McGrath, Polym. Prepr., Am. Chem. Soc., Div. Polym. Chem. 24(2), 41 (1983).
4. See for example a) R. Minne and E.G. Rochow, J. Am. Chem. Soc. 82, 5628 (1960); b) M. Watanabe, S. Ohashi, K. Sanui, N. Ogata, T. Kobayashi and Z. Ohtaki, Macromolecules, 18, 1945 (1985); c) A. Eisenberg, K. Ovans and H.N. Yoon in "Ions in Polymers" A. Eisenberg, editor, Advances in Chemistry Series, (1980) p.267.
5. D. Fish, D.W. Xia and J. Smid, Makromol. Chem., Rapid Commun. 6, 761 (1985).
6. D. Fish, I.M. Khan and J. Smid, Makromol. Chem., Rapid Commun. 7, 115 (1986).
7. M.B. Armand, J.M. Chabagno and M.J. Duclot in "Fast Ion Transport in Solids", edited by P. Vashista, J.N. Mundy and G.K. Shenoy, North Holland, New York, 1979, p.131.

8. D.W. Xia and J. Smid, J. Polym. Sci., Polym. Lett. Edit. 22, 617 (1984).
9. D. Freyss, P. Rempp and H. Benoit, Polym. Lett. 2, 217 (1964).
10. J.J. O'Malley, R.G. Crystal and P.F. Erhardt in "Block Polymers", A.L. Aggarwal, editor, Plenum Press, New York, 1970, p.163.
11. Z. Grubisic, P. Rempp and H. Benoit., Polym. Lett. 5, 753 (1967).
12. P. Mason, E. Franta and P. Rempp, Makromol. Chem., Rapid Commun. 3, 499 (1982).
13. Analysis of hydroxyl content of a series of commercial poly(ethyleneglycol)monomethyl ethers show [OH]/[OCH_3] as being larger than 1. R.D. Vos and E.J. Goethals, Polym. Bull, 15, 547 (1986).
14. A.H. Muller in "Anionic Polymerization: Kinetics, Mechanism and Synthesis", J.E. McGrath, editor, ACS Symposium Series 166, 441 (1981).
15. B. Morese-Seguela, M. St. Jacques and J.M. Renaud, J. Prud'homme, Macromolecules, 13, 100 (1980).
16. G.C. Meyer and J.M. Widmaier, J. Polym. Sci., Polym. Phys. Ed. 20, 389 (1982).
17. G.G. Cameron and M.S. Chisholm, Polymer, 27, 437 (1986).

AUTHOR INDEX

SUBJECT INDEX

C

D

E

F

G

H

I

N

O

P

Q

R

S

T

U

V

W

X

Z

GPSR Compliance

The European Union's (EU) General Product Safety Regulation (GPSR) is a set of rules that requires consumer products to be safe and our obligations to ensure this.

If you have any concerns about our products, you can contact us on ProductSafety@springernature.com

In case Publisher is established outside the EU, the EU authorized representative is:

Springer Nature Customer Service Center GmbH
Europaplatz 3
69115 Heidelberg, Germany

Batch number: 10823811

Printed by Printforce, the Netherlands